10
Plant Cell Monographs

Series Editor: David G. Robinson

Available online at
SpringerLink.com

Plant Cell Monographs

Recently Published Titles

Plant Growth Signaling
Volume Editors: Bögre, L., Beemster, G.
Vol. 10, 2008

Cell Division Control in Plants
Volume Editors: Verma, D. P. S., Hong, Z.
Vol. 9, 2008

Endosperm
Volume Editor: Olsen, O.-A.
Vol. 8, 2007

Viral Transport in Plants
Volume Editors: Waigmann, E., Heinlein, M.
Vol. 7, 2007

Nitric Oxide in Plant Growth, Development and Stress Physiology
Volume Editors: Lamattina, L., Polacco, J.
Vol. 6, 2007

The Expanding Cell
Volume Editors: Verbelen, J.-P., Vissenberg, K.
Vol. 5, 2007

The Plant Endoplasmic Reticulum
Volume Editor: Robinson, D. G.
Vol. 4, 2006

The Pollen Tube
A Cellular and Molecular Perspective
Volume Editor: Malhó, R.
Vol. 3, 2006

Somatic Embryogenesis
Volume Editors: Mujib, A., Šamaj, J.
Vol. 2, 2006

Plant Endocytosis
Volume Editors:
Šamaj, J., Baluška, F., Menzel, D.
Vol. 1, 2005

Plant Growth Signaling

Volume Editors: László Bögre, Gerrit Beemster

With 1 Table and 38 Figures, 6 in Color

Springer

Volume Editors:

Dr. László Bögre
University of London
School of Biological Sciences
Royal Holloway & Bedford
New College
Egham, Surrey TW20 0EX
UK
L.Bogre@rhul.ac.uk

Dr. Gerrit Beemster
Flanders Interuniversity
Institute for Biotechnology (VIB)
Dept. Plant Syst. Biol.
Technologiepark 927
9052 Zwijmaarde
Belgium
gerrit.beemster@psb.ugent.be

Series Editor:

Professor Dr. David G. Robinson
Ruprecht-Karls-University of Heidelberg
Heidelberger Institute for Plant Sciences (HIP)
Department Cell Biology
Im Neuenheimer Feld 230
D-69120 Heidelberg
Germany

ISBN 978-3-642-09632-7 e-ISBN 978-3-540-77590-4
DOI 10.1007/978-3-540-77590-4

Plant Cell Monographs ISSN 1861-1370

Editor: Dr. Christina Eckey, Heidelberg
Desk Editor: Anette Lindqvist, Heidelberg
Cover design: WMXDesign GmbH, Heidelberg

Printed on acid-free paper

9 8 7 6 5 4 3 2 1 0

springer.com

Editors

László Bögre is a full Professor in Plant Cell and Molecular Biology at the School of Biological Sciences, Royal Holloway, University of London, United Kingdom. He obtained his B.Sc degree in agricultural sciences from Gödöllõ University, Hungary and his Ph.D. in Biology from the Biological Research Centre, Szeged, Hungary. His research pioneered the mitogen-activated protein kinase signaling pathways in plants and the regulation of the cell cycle. His current research is focused on studying growth signaling networks using the model plant, *Arabidopsis thaliana*.

Gerrit T.S. Beemster is a group leader at the department of Plant Systems Biology at the Flemish Institute for Biotechnology (VIB) and the University of Gent, Belgium. He obtained his B.Sc degree in horticultural science from Wageningen University, The Netherlands and his Ph.D. in Plant Science from the Australian National University, Canberra, Australia. His research pioneered the implementation of kinematic analyses for the analysis of cell division and cell expansion in growing organs, including primary roots and leaves of *Arabidopsis thaliana* and several species of monocotyledonous leaves, the use of profiling techniques to study growth process specific expression patterns and the development of dynamic growth models based on experimentally obtained data.

Preface

Plant growth provides the basis for life on earth and is a process that is intimately linked with human civilizations. Continuous development in agricultural practices and in plant breeding allows us to keep plant production in line with demands. Industrialization, based on an enormous input of energy, mostly fossil fuels derived from biomass produced in the past, made it possible to reach our current standard of living. However, the uncontrolled use of plants and fossil fuels is an important contributor to global warming due to the associated production of CO_2. Furthermore, the reserves of this energy source are rapidly being depleted, so alternatives need to be found. Improved plant production is seen at least as a partial solution, as photosynthesis enables the "recycling" of CO_2 and fixing of energy. Thus, in recent years, we have seen a rapidly emerging market for bio-energy as well as the production of a myriad of natural products from plants. These bio-energy and bio-product producing crops, however, compete for the available agricultural land used for the production of food and feed, which is already starting to affect market prices of these commodities. Therefore, there is a renewed pressure on plant scientists to find solutions to increase plant productivity in a sustainable way.

Plant growth is intimately connected to the capacity of source organs to produce assimilates. Light is a key energy source and environmental cue controlling development, predominantly via leaves. It is known that growth-promoting signals are perceived in mature leaves and transmitted via unknown signals to developing leaves to regulate their growth. The nature of this transmissible signal is not known, but assimilates, such as sugars are thought to play a key role. Plant hormones also provide long-distance signaling to interface environmental conditions and organ growth. At the cellular level extracellular signals are sensed, transmitted and integrated by intracellular signaling pathways, which on one hand can directly regulate metabolic enzymes and other cellular functions, while on the other hand they feed into the regulation of gene transcription, protein stability protein modifications to quantitatively fine-tune cellular components or behavior. However, little is known about the intracellular signaling pathways in plants that regulate growth or its various components. Genetic approaches are difficult when genes function in an interconnected complex network, and regulate processes that are quantitative, such as growth. Novel methods, together with systems approaches, are needed for multiplex measurements of the outputs of signaling pathways at various complexity levels.

Growth of new organs requires a combination of cell division in or near meristems, cell growth, differentiation and cell expansion. Both developmental and environmental inputs influence organ growth by altering the pool of proliferating cells. These developmental pathways are composed of individual modules consisting of signal(s), transducers, transcriptional regulator(s) and targets. Viewed this way, plant development is a cascade of events that, by continual external and internal input, direct the orderly activation of the hierarchically arranged modules. How these processes are linked and coordinated is not understood.

To gain a systems-wide understanding of any developmental or physiological process, an increasing number of methodologies to obtain "omics" data at various levels and of computational and network-modeling techniques are available. However, a key, sometimes overlooked issue is the precise experimental approach and is the exact source of the "omics" data. To understand a system, one should be able to produce, as far as possible, a list of its parts, to introduce perturbations in the system and to monitor the behavior of the parts following the perturbation. A further source of critical information is time-resolved data, because it can be assumed that changes in concentration/activity of the regulator will inevitably precede the changes in the regulated component.

First and foremost, the sequencing of the genome of *Arabidopsis thaliana* has launched plant science into the genomics era and provided a gathering platform for plant scientists. This is now rapidly followed by the sequencing of other plant genomes with agricultural importance, including rice, poplar, grapevine, tomato and maize. The impact of having the full list of coding and regulatory sequences for understanding the behavior of plant growth is enormous, as investigators can shift their attention from gene-identification to functional analysis of these genes at the molecular, cellular and whole plant levels. Genomic sequence availability also allowed the development of profiling technologies to monitor gene expression, protein abundance, localization and modifications on a genome-wide scale under a wide range of experimental conditions and in specific cells or tissues. Our ability to simultaneously study the function of virtually all genes encoded by the plant genome, has led to a new more holistic approach to biology named systems biology. Rather than focusing on the function of a few genes in a particular pathway, the emphasis in systems biology is to understand which are the key components regulating specific processes and how such components are connected in "regulatory networks".

As outlined above, plant growth is a particularly intriguing phenomenon as it is under the control of a multitude of interacting regulatory pathways. In this monograph several of the contributing pathways are reviewed, including light signaling (López-Juez and F. Devlin, Chapter 11), the classical hormones auxin (Zago et al., Chapter 8), ethylene (Dugardeyn and Van Der Straeten,

Chapter 10), and brassinosteroids (Clouse, Chapter 9), and signaling pathways including the TOR pathway (Anderson, Chapter 12), Armadillo repeat proteins (Coates, Chapter 15) and the MAPK cascades (Suzuki and Machida, Chapter 13), and protein dephosphorylation mechanisms (Schweigenhofer and Meskiene, Chapter 14). Devoto and Paccanaro (Chapter 17) describe the use of profiling and modeling to analyze signaling pathways on a genome-wide level. Downstream of those signaling pathways,, several key aspects of growth regulation itself are discussed, starting from the unicellular perspective of algae (Bišová, Chapter 18) to the regulation of cell growth, cell division (Doerner, Chapter 1), the switch between division and differentiation (Magyar, Chapter 5), the endoreduplication processes (Yoshizumi et al., Chapter 6) and interactions between cell size and cell numbers (Ferjani et al., Chapter 3) in higher plants. At the whole organ level the role of the epidermal layer in growth control is reviewed (Ingram, Chapter 7) and overall organ size control mechanisms are explored (Anatasiou and Lenhard, Chapter 2). Finally, emerging experimental approaches as proteomics (Schulze, Chapter 16) and kinematic analysis of growth (Walter, Chapter 4) are described.

We think it is timely to bring together this overview of the developments in various areas of plant-growth research in this monograph, firstly to give the reader a comprehensive insight into the current state of knowledge in the field. Reading through, it is possible to see common themes emerging from different fields of research and therefore we hope that this book will also stimulate an integrating perspective for future research aimed to better understand the fascinating process that plant growth represents.

March 2008

László Bögre and
Gerrit T.S. Beemster

Contents

Plant Cell Monogr (10)
L. Bögre and G. Beemster: Plant Growth Signaling
DOI 10.1007/7089_2007_142/Published online: 28 July 2007

Signals and Mechanisms in the Control of Plant Growth

Peter Doerner[1,2]

[1]Institute of Molecular Plant Science, School of Biological Sciences, University of Edinburgh, Daniel Rutherford Building, King's Buildings, Edinburgh EH9 3JH, UK

[2]*Present address:*
ICMB, University of Edinburgh, Daniel Rutherford Building, Edinburgh EH9 3JH, UK
Peter.Doerner@ed.ac.uk

Abstract Plant growth is mediated by three fundamental processes: cell growth, division, and expansion. The mechanistic analysis of their contributions are complicated by the observation that the balance of their contributions to organ growth are not hard-wired. Reduced cell proliferation, irrespective of whether this is caused by decreased cell growth or diminished cell division, can be, at least partially, compensated for by increased cell expansion. It is therefore argued that for a functional understanding of how gene regulatory networks control growth of the plant body, it is essential that all cellular parameters contributing to organ growth are quantified in concert. Plant growth behavior is exquisitely responsive to environmental change. Cell growth, division, and expansion, in aggregate, are promoted by nutrient availability and inhibited by abiotic stress. Recent studies that address how light intensity, CO_2 concentration, water activity, and temperature have complex effects on proliferation, cell expansion, and endoreplication that affect leaf organ growth are reviewed. Root growth rates and patterns are also very sensitive to mineral nutrient concentration and distribution. The mechanistic basis of plant organ growth still remains unknown; but such knowledge is critical for rational approaches to manipulate plant growth. Critical steps towards this goal are discussed.

1 Introduction and Background

Plants adapt exquisitely to their environment: physiology and metabolism change diurnally and in response to many environmental conditions, and reproductive development is generally sensitive to day length, temperature, or other proxies of seasonal change. The most fundamental adaptation to environmental change in plants is altered growth behavior, involving changes to root or shoot growth patterns, rates, or both.

Despite their fundamental importance for our understanding of plant growth, for rational approaches to sustainably enhance yields in agriculture and forestry and, ultimately, for human welfare, we still understand surprisingly little about the mechanisms that govern growth in plants. In this chapter, I will consider the signals and genetic mechanisms involved in controlling growth in aerial and underground organs.

1.1
Distinct Processes Contribute to Plant Growth

At the whole plant level, growth of the plant body proceeds by the linear extension of stems and branches, the production of leaf or floral organs, and the elongation and branching of roots, mediated by apical, axillary, and lateral meristems. Secondary growth, or radial thickening mediated by cambial cells, contributes to body size increase in many plants, but will not be considered further in this chapter. Primary stem or leaf and root organ growth, here defined simply as an increase in volume, proceeds in two stages, which I will call phase I, during which cells multiply in cycles of growth and division; and phase II, during which cells cease dividing but expand until differentiation is completed. High rates of proliferation are observed in meristems, in young leaf and floral primordia, but not in stem cells and the stem cell niche (Fig. 1).

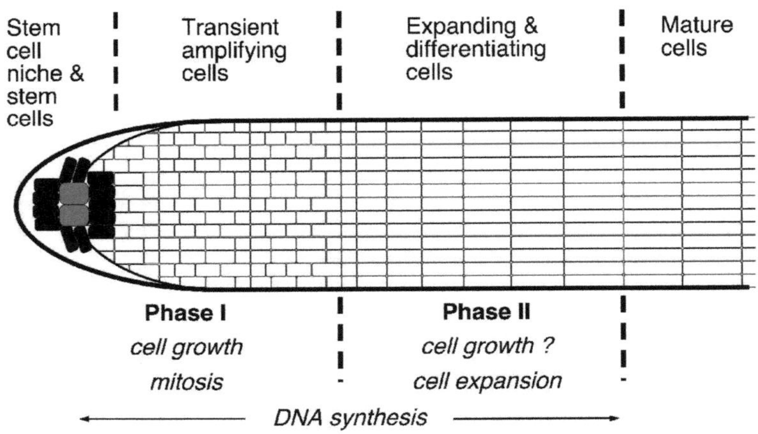

Fig. 1 Schematic representation of the root apical meristem illustrating the different zones of growth and the positional extent of various growth processes

1.2
Cell Growth

In phase I, cell growth alternates with division in mitotic cells. Cell growth is a prerequisite for division in meristems and organ primordia, and is driven by the increase of cell mass by synthesis of macromolecular cell constituents (Jorgensen and Tyers 2004). Ribosomes limit macromolecular synthesis and, therefore, their synthesis and its regulation is at the nexus of growth control. For example, yeast cells commit ∼50% of their total transcription activity and a large fraction of their energy budget towards building ribosomes (Warner 1999) and quantitative studies reveal a strong positive correlation of ribosome synthesis with cell growth (Planta 1997; Warner 1999). There is good evidence that impaired ribosome biosynthesis reduces plant growth (Van Lijsebettens et al. 1994; Weijers et al. 2001; Horvath and Bogre, this volume), but no detailed information is yet available on how well ribosome biosynthesis correlates with growth activity in plants. The expression of many components of the plant ribosome is regulated transcriptionally (McIntosh and Bonham-Smith 2006), but it is still poorly understood how ribosomal RNA and protein synthesis for ribosome production are coordinated mechanistically (for review, see McIntosh and Bonham-Smith 2006).

Cell growth is under control of the target of rapamycin (TOR) pathway, which couples nutritional cues to the regulation of ribosome biosynthesis, the rates of protein synthesis and proliferation. The TOR pathway interacts with the PI-3-kinase pathway, which mediates growth factor cues, and this interaction insures coordinate cellular growth responses (Arsham and Neufeld 2006; Jorgensen and Tyers 2004). The TOR pathway has been well characterized in animal and yeast systems, but much detail remains to be uncovered in plants: Orthologs of the TOR kinase, and of some additional components of the TOR signaling pathway have been identified in plants (Bogre et al. 2003; Menand et al. 2002; Wang et al. 2003), but their functional significance for plant cell growth control, specifically for coupling environmental change to growth responses, are only beginning to be examined in detail (Mahfouz et al. 2006). Likewise, plant homologs of PI-3-kinases and their effectors, the AGC kinases have been identified (Wang et al. 2003). At least one AGC kinase has been shown to be responsive to auxin and cytokinin growth regulator inputs (Anthony et al. 2004), and IRE (an AGC kinase) positively regulates root hair tip growth (Oyama et al. 2002). However, many gaps need to be filled until we understand the mechanisms of how growth regulator and nutrient inputs converge on cell growth control in plants.

1.3 Cell Division

In contrast, the mechanisms controlling cell division are much better understood than those regulating cell growth in plants (see Inze and De Veylder (2006) for an excellent recent review). Components of the plant cell cycle machinery (cyclins, cyclin-dependent kinases), orthologs of the retinoblastoma (Rb) gene, and E2F/DP-type transcription factors were identified based on their sequence homology (Vandepoele et al. 2002). Largely based on gain-of-function studies with transgenic plants over- or ectopically expressing cell cycle regulators and expression analysis, the following view is emerging: In association with CDKA (A-type cyclin-dependent kinase), D-type cyclins are involved in controlling the entry into the cell cycle (Menges et al. 2006, Riou-Khamlichi et al. 1999), whereas A- and B-type cyclins, in association with CDKA and CDKB play a major role in S-phase and entry into M-phase, respectively (Doerner et al. 1996; Weingartner et al. 2003). As in animal systems, the E2F/DP and related genes, promote S-phase and DNA synthesis, but are also involved in controlling the switch between mitotic cell cycles and the endoreplication cycle. Likewise, CDK inhibitors function in post-translational control of cyclin–CDK complex activity. Anaphase promoting complex (APC) proteolytic activity at the metaphase-to-anaphase transition insures the irreversible directionality of cell cycle progression, as in other model systems.

Cell growth is coupled to cell division progression by mechanisms that monitor cell size. For example, in yeast, coupling of growth to cell cycle entry converges on the regulation of G1-type CLN3 cyclin abundance (Morgan 2007), although this view may be too simplified (Jorgensen and Tyers 2004). CLN3 abundance is regulated at the transcriptional, translational, and post-translational level (MacKay et al. 2001; Morgan 2007; Polymenis and Schmidt 1997). In aggregate, these mechanisms result in a steep stimulus–response coupling (ultrasensitive response) of CLN3 protein levels, and hence of CLN3–CDK complex activity, to the rate of mRNA translation by ribosomes, which reflects the activity of the TOR and other growth regulating pathways. Cell cycle entry in plants requires D-type cyclins. In Arabidopsis, cyclin D3;1 mediates the stimulatory effect of cytokinins on proliferation, while cyclin D2 abundance is responsive to sucrose levels (Riou-Khamlichi et al. 1999, 2000). Cyclin D3;1 is a labile protein (Planchais et al. 2004), as would be expected of a limiting regulator responsive to potentially rapidly changing environments. Moreover, cyclin D3;1 promotes the G1/S transition (Menges et al. 2006). Based on this limited information, it is therefore reasonable to predict that key aspects of the mechanisms that couple cell growth to cell division are conserved in all eukaryotes.

1.4 Cell Expansion

After cells pass through the domain with high rates of cell growth and division, they cease dividing and cell size rapidly becomes larger. This transition from phase I to II is visually distinct in root meristems, whereas in leaf organs this transition is morphologically less conspicuous. Cell expansion in phase II is not driven by macromolecular synthesis but is the result of turgor-driven water uptake and concomitant cell wall loosening. The generation of increased osmotic pressure requires the activities of three major proteins or protein complexes in the tonoplast membrane: The V-type H^+ATPase, H^+pyrophosphatase and aquaporins (see Maeshima 2001 for review). This is balanced by cell wall loosening that permits the cell to expand mostly in one direction, and which involves several activities including expansins, xyloglucan endotransglycolase/hydrolase (XET), endo-(1,4)-β-D-glucanase, and hydroxyl radicals (see Cosgrove 2005 for review). In quantitative terms, cell expansion contributes most to organ growth: during cell expansion, volume increases from 20- to 1000-fold. Thus, the extent of cell growth and division during phase I define the potential for organ growth by producing the cellular building blocks; during phase II, this latent ability is fulfilled during cell expansion.

The phase I/II boundary marks a transition of the cellular mechanism that mediates organ growth: from growth by cell production to organ growth by cell expansion. However, not all processes associated with organ and plant growth change at this transition. DNA synthesis persists during this transition, but in the absence of division, it leads to endoreplication. Therefore, DNA replication can be considered as the process that frames the entire organ growth process. In Arabidopsis, endoreplication can result in ploidy levels of up to 64C (with 1C being a haploid genome equivalent), indicating that cells undertake up to five additional rounds of DNA synthesis without dividing. In Arabidopsis leaves, cellular DNA content is positively correlated with mature, fully expanded cell size (Melaragno et al. 1993), however, in roots no such correlation was found (Beemster et al. 2002). DNA synthesis, and with it endoreplication and cell expansion, is thought to cease when cells become fully differentiated and primary organ growth is completed.

Although expanding cells increase their size by a different mechanism than cells growing in the proliferative zone, they continue entering the DNA replication cycle as long as they undertake endoreplication cycles. The bulk of the volume increase in expanding cells is mediated by inflation of the vacuole, but it is likely that the cytoplasm must also increase in mass to insure that the necessary concentration of reactants is thermodynamically favorable. This raises the interesting, and as yet unresolved, question whether the onset of S-phase in endoreplicating cells is also coupled to proxies of cell growth such as the rate of mRNA translation.

2 Regulation of Growth

Much progress has been made in identifying and functionally characterizing components of the plant cell division apparatus (Inze and Veylder 2006), and the mechanisms involved in cell expansion are also beginning to be quite well understood (Carol and Dolan 2006; Cosgrove 2005; Tsukaya 2006). In contrast, cell growth control is mechanistically still less well understood. Based on the preceding analysis of plant growth processes, I propose the existence of two major growth control points in plants likely to be sensitive to developmental or environmental inputs. The first is suggested to co-regulate cell growth and the onset of the cell cycle; the second is the switch of growth mechanisms at the phase I/II boundary to suppress mitosis and activate cell expansion. The identification of components involved in these control points, the mechanisms by which they operate and how they are coupled to cues will be major milestones to improve our understanding of plant growth control.

Recently, a few candidates for such components were identified. They are considered in detail below, because each one is a possible target or component of growth control pathways responsive to environmental or developmental cues. However, it will require more extensive analysis to unambiguously establish their specific function in growth control networks.

2.1 Coupling of Cell Growth and Division

There is increasing evidence that cell growth and division are co-regulated: rapidly dividing cells in young leaf primordia and in roots are remarkably uniform in size and recently, possible effector pathways for co-regulation of cell growth and division were identified. The best mechanistic evidence for co-regulation of cell division and cell growth currently comes from the analysis of Arabidopsis TCP20. TCP20, which belongs to a plant-specific class of transcription factors and is thought to promote gene expression, binds in vivo to the promoters of ribosomal protein genes as well as to the promoter of the mitotic cyclin B1;1 (Li et al. 2005). Elevated expression of cyclin B1;1 has been shown to promote organ growth (Doerner et al. 1996). However, the biological function of class I TCP genes in control of organ growth has not been reported yet.

EBP1 genes, identified in potato and Arabidopsis, are a further type of effector gene that affect phase I growth (Horvath et al. 2006). Putative orthologs have been identified in other eukaryotes, where they are thought to regulate ribosome biogenesis (Squatrito et al. 2004), modulate translational activity (Squatrito et al. 2006), as well as DNA replication by binding to the Rb protein (Zhang et al. 2003). This wide range of activities raises the interesting possibility that plant EBP1 genes are involved in promoting phase I growth (by promoting cell growth), as well as phase II growth (by regulating E2F activity). Over-expression of plant EBP1 leads to larger leaves with more cells, while reduced expression results in the opposite (Horvath et al. 2006). In this work, cell size at birth and ploidy were not analyzed and so the direct effects of EBP1 on cell growth and the phase I/II switch are not yet known.

Altered expression of many additional genes has been reported to enhance organ growth, including: ARGOS (Hu et al. 2003), AINTEGUMENTA (Mizukami and Fischer 2000), PEAPOD (White 2006), and BIG BROTHER (Disch et al. 2006). All these genes have opposing effects on organ size when either over- or under-expressed. Elevated expression (ARGOS, AINTEGUMENTA) or reduced expression (PEAPOD, BIG BROTHER) leads to extended phase I growth, with little or no effect on final cell size. However, cell size at birth in these plants (i.e., during phase I growth) was not reported, and therefore it is presently not clear whether these genes specifically control the timing of the phase I/II transition, or also affect the rate of cell growth.

Enhanced expression of some activating cyclin subunits of the CDK complexes that are rate-limiting regulators of cell cycle progression has led to accelerated organ growth without affecting the final size of the plant (Cockcroft et al. 2000; Doerner et al. 1996; Li et al. 2005). These observations raise several intriguing possibilities: It is possible (although there is no experimental evidence yet) that CDK activity feeds back on cell growth control. This could be a parsimonious regulatory mechanism, in which for example, developmental pathways could directly regulate cell cycle activity. This would then suffice to entrain appropriate levels of cell growth activity. Alternatively, it is possible that cell division onset in meristems and organ primordia only occurs significantly later than the attainment of a minimal cell size in plants. In this scenario, CDK activity limits organ growth and the plant can cope with increased proliferation because cell mass is sufficient to sustain division at an earlier time. A third possibility is that a specific CDK activity could be required for mitosis and therefore become limiting at the phase I/II boundary. A delay of the phase I/II transition would enhance the growth capacity of the affected organ or meristem by increasing the size of the dividing cell population. In this scenario, CDK mitotic activity limits organ growth by controlling the switch in cellular growth mechanisms.

There is good evidence that cell division activity positively correlates with organ growth rates: High levels of CDK activity are associated with high proliferation (Granier et al. 2000). Enhanced expression of activating cyclin subunits of the cyclin-dependent kinase (CDK) complexes that are rate-limiting regulators of cell cycle progression has led to accelerated organ growth without affecting the final size of the plant (Cockcroft et al. 2000; Doerner et al. 1996; Li et al. 2005). Further, careful quantitative analysis of CDK kinase activity in relation to root organ growth rates support the notion that the level of CDK activity is a good predictor for the magnitude of organ growth rate (Beemster et al. 2002). Therefore, it appears possible that regulatory networks directly regulate CDK activity as a mechanism for plant growth control.

2.2 The Switch from Mitosis to Endoreplication

The switch from phase I to phase II growth mode involves two known mechanisms: (i) the suppression of mitosis and (ii) the stimulation of cell expansion, during which repeated rounds of DNA synthesis persist until cells are fully expanded. CDK–cyclin complexes control the commitment to S phase, but the execution of S-phase is enabled by a CDK-controlled hierarchy of enforcers that include the plant homolog of the retinoblastoma gene (Rb), and a family of related transcription factors that include E2F, DP, and DEL genes (Gutierrez et al. 2002; Inze and Veylder 2006). E2F and DP gene products heterodimerize to bind their canonical target sites, while DEL proteins can bind these as monomers and lack conspicuous activation domains. Rb keeps E2F proteins in check by binding them through a so-called A/B pocket, but upon hyperphosphorylation by CDKs, releases these so they can directly activate expression of S phase genes. Over-expression of Arabidopsis E2F3 (also known as E2Fa) stimulates expression of S phase genes, and enhances proliferation and endoreplication. This phenotype is exacerbated when a DP gene is co-expressed (De Veylder et al. 2002). However, not all E2F genes promote S phase: E2F2 (also known as E2Fc) lacks an apparent activation domain and hence can suppress S-phase associated gene expression (e.g., CDC6) (del Pozo et al. 2002). Reduced E2F2/c expression results in increased expression of S-phase gene markers, enhanced cell production and plants with more, but smaller cells with reduced levels of endoreplication (del Pozo et al. 2006). Likewise, enhanced E2Fd (also known as DEL1) expression reduces ploidy levels, while reduced E2Fd/DEL1 activity increases endoreplication (Vlieghe et al. 2005). In contrast, altered levels of E2Ff/DEL3 had a modest impact on endoreplication, but elevated expression promoted precocious differentiation in roots, possibly because several expansins and xyloglucan endotransglycolase/hydrolase involved in cell wall extension during phase II growth are E2Ff/DEL3 targets (Ramirez-Parra et al. 2004).

Taken together, these results suggest that E2F/DEL genes are involved in regulation of S-phase-specific gene expression as well as promoting phase I/II transition. Therefore, they are likely targets of regulatory pathways that control the suppression of mitosis and the stimulation of cell expansion. However, how the activity of different E2F-like factors, in some cases possibly on common target genes, is regulated is still not well understood, but at least one of them, E2F2c, is unstable (del Pozo et al. 2002). Further evidence for a possibly pivotal role for regulated protein degradation in controlling the phase I/II transition comes from the observation that CCS52, a regulatory component of the anaphase promoting complex (APC) orthologous to CDH1 and fizzy-related (which functions as an inhibitor of mitosis), is required for endoreplication (Cebolla et al. 1999).

Although the experimental evidence clearly points to a complex involvement of Rb/E2F/DP/DEL proteins, as well as regulated proteolysis in controlling the phase I/II transition, the regulators that orchestrate the deployment of these enforcers, specifically their order of action, have not yet been identified.

3 Plant Growth Responses to Environmental Change

Plant growth patterns and rates adaptively respond to changes in the environment. Such adaptive changes confer competitive advantages and allow the plant to survive adverse conditions. Here, I will focus on adaptation to changes in nutrient availability.

Altered nutrient availability can impact plant growth at the cellular, organ, and whole plant level: Local (in the order of 50–100 μm) differences in soil phosphate availability suffice to alter root hair growth, which in low phosphate, is stimulated in a cell-autonomous manner (Bates and Lynch 1996). Likewise, cell expansion in leaves subjected to low water activity is reduced, but can recover when water is available again (Granier and Tardieu 1999). When mature leaves are exposed to high light or CO_2, leaf growth is stimulated and stomatal density increases. Developing leaf primordia exposed to low light or low CO_2 will develop with the characteristics of the mature leaves, indicating that at least some aspects of leaf growth are controlled by systemic signals (Lake et al. 2001; Yano and Terashima 2001; Ferjani et al., this volume). Unfortunately, the precise mechanisms by which any nutritional cue elicits one or more signals controlling cell growth, division, or expansion are not yet known.

3.1 Shoot Growth

Shoot meristems and developing leaf organs adapt exquisitely to the abundance of light, CO_2, and water by modulating leaf production rate, leaf size and shape, anatomy, and physiology to confer competitive advantages in an environment where competition for light is fierce. Many genes and growth factor signaling pathways have been identified that contribute to specifying final leaf size and shape, but it is not yet known whether or how these mediate specific environmental cues as well.

Most leaves are determinate organs, but monocot and dicot leaves grow differently. In monocots, meristematic cells across the leaf base produce cells until the blade has reached its full longitudinal extent, and lateral growth of the blade does not occur. By contrast, growth is more complex in dicot leaves: all cells initially grow and divide, but quiescence sets in in a basipetal direction from the leaf tip to the base, and cell divisions cease early in leaf development. However, proliferation persists at a low rate in vascular tissues and in isolated cells (e.g., cells of the stomatal lineage), and endoreplication continues. Analysis of dicot leaf organ growth in mutants and transgenic plants has also revealed a compensatory mechanism: reduced proliferation can be mitigated by enhanced cell expansion, thereby maintaining a similar leaf area (Hemerly et al. 1995; Horiguchi et al. 2006). Such compensatory control of final leaf organ area has been suggested to result from an organ size control mechanism (Hemerly et al. 1995), but components of such a regulatory mechanism have remained elusive. Since the extent of cell expansion in leaf epidermal cells positively correlates with ploidy level (Melaragno et al. 1993), it has been proposed that the observed compensatory increase in cell expansion could depend on modulation of ploidy. In a recent study, expansion at the cellular and leaf organ level was analyzed in response to low light and water deficit (Cookson et al. 2006). Plants growing in low light produced smaller leaves comprised of fewer, but larger cells; while those growing in water deficit conditions produced smaller leaves comprised of smaller cells. However, the mean number of endoreplication cycles was reduced under both experimental conditions (Cookson et al. 2006). Taken together, these observations are not consistent with a role for endoreplication in governing final cell size in response to a nutrient and environmental cue such as light.

3.2 Light and CO_2

Light quality and intensity has a profound effect on plant growth. Changes to the red–far red ratio trigger the photoreceptor-dependent shade avoidance response, which involves increased cell expansion and, in extreme cases, an acceleration of plant development (Franklin and Whitelam 2005). Here, I will focus on the effects of light quantity on plant growth, specifically leaf organ growth, and the emphasis will be on Arabidopsis. It should be noted that most Arabidopsis experiments are performed in laboratory growth chamber conditions, where "high light" corresponds to 150–250 $\mu mol\, m^{-2}\, s^{-1}$, and "low light" corresponds to 15–75 $\mu mol\, m^{-2}\, s^{-1}$. However, in nature, exposure to sunlight corresponds to 150–2000 $\mu mol\, m^{-2}\, s^{-1}$, and "shade" in nature can span the whole high-light/low-light range examined in the laboratory. Therefore, the relevance of the observations described below remains to be validated for natural conditions.

Leaf organ growth responds to light intensity in several ways: In constant conditions, leaf initiation rate is reduced by low light (Cookson et al. 2005); blade anatomy is altered such that in "sun" leaves, two layers of palisade cells are produced (Kim et al. 2005); and the density of stomata is increased (Lake et al. 2001) in high light. In low light, leaf blade area is decreased, mediated by a reduction in cell number, but it is not yet known whether this is caused by reduced cell division, or whether cell growth (and as a consequence, cell division) is reduced (Cookson et al. 2005; Granier and Tardieu 1999). However, reduced proliferation is compensated for in part by increased cell expansion (Cookson et al. 2005; Cookson and Granier 2006) Moreover, the growth characteristics of the leaf organs are altered in low light so that maximal organ expansion rates are reduced and delayed (Cookson et al. 2005). Interestingly, a strong correlation was observed between leaf initiation rates and leaf epidermal cell number (Cookson et al. 2005). This raises the intriguing possibility that light intensity generates a signal that acts directly on the meristem to control the rate of primordium formation *and* the number of cells committed to a primordium. Such a possibility is consistent with the observation that all early processes in leaf organ development are correlated with each other (Cookson et al. 2005), implying that they are co-regulated.

Non-stressing levels of high light also increase photosynthesis and carbon assimilation and are therefore likely to also affect whole plant growth. Increased root growth (and an associated improved ability for mineral nutrient assimilation), would positively affect leaf growth. Such indirect effects on Arabidopsis leaf growth parameters in different light intensities have not been reported.

Overall growth of most plants, including Arabidopsis, is promoted in elevated CO_2 concentrations (Pritchard et al. 1999; Tocquin et al. 2006). Leaf organ growth in Arabidopsis is stimulated, and this effect is more pronounced when nitrogen is not limiting (Tocquin et al. 2006). Kinematic analysis of Arabidopsis leaf growth under these conditions has not yet been reported. However, in monocot leaves such analysis is more straightforward. In a detailed analysis of leaf growth kinetics in two wheat cultivars, elevated CO_2 concentration led to enhanced cell production and increased meristem size, but no change of cell size at cytokinesis or of final expanded cell size was observed (Masle 2000). Together, this suggests that cell cycle entry is directly stimulated by CO_2 and that control of this parameter mediates CO_2 concentration-dependent organ growth changes. Interestingly, growth in elevated CO_2 concentration leads to significantly increased foliar concentrations of cytokinins, gibberellins, and auxin, while concentrations of growth-inhibitory ABA are reduced in Arabidopsis (Teng et al. 2006). If leaf growth control in dicotyledonous leaves mirrors that in monocots, then Arabidopsis CYCD3;1, which is one of the D3-type cyclins that limits cell cycle entry (Menges et al. 2006) and is also involved in mediating cytokinin-dependent stimulation of cell division (Riou-Khamlichi et al. 1999), may be a direct target of CO_2 concentration-dependent organ growth control.

3.3 Water Activity and Temperature

Water deficits and elevated temperatures are stressful conditions that negatively affect plant and leaf growth. Detailed kinematic analyses of leaf growth at non-stressful temperatures or a range of water activities have not been reported for Arabidopsis. However, in maize, a good correlation between leaf elongation and cell production rates was observed in a range of temperatures and when comparing watered plants with plants experiencing water stress (Granier et al. 2000). Moreover, CDK activity and cell division rate were strongly correlated, but not $p34^{cdk}$ abundance. This observation suggests that post-translational modification of $p34^{cdk}$ or transcriptional control of cyclin expression are potential targets for these signals. In a recent study (Rymen et al. 2007), the effect of cold nights on leaf growth in maize was examined. These conditions did not affect mature cell size, or the size of the meristem, but major changes were observed in the dividing cells of the meristem: cell size of dividing cells was reduced (hence the meristem had more, but smaller cells), and cell cycle duration was extended. This indicated that cell growth was strongly affected, and furthermore, that low temperature resets the size threshold for division. Interestingly, the latter is reminiscent of the response of budding yeast to growth at different levels of nutrient availability: in low nutrients, the size threshold for division is reduced, but when nutrients are abundant, cells divide with a larger mass (Tyson et al. 1979). Ploidy levels were not significantly changed, suggesting that all cell cycle phases were equally affected by the low temperatures, but the expression of some cell cycle regulators that function in the G1/S transition and in S-phase (e.g., cyclin CYCA3;1, CDKA1;1, E2F) was strongly down-regulated.

4
Root Growth

Roots are indeterminate organ systems that grow apically, potentially indefinitely, and that form lateral roots at a distance from the growing apex. The patterns and rates of root system growth are influenced by the distribution and concentration of mineral nutrients in the soil, the availability of water and the degree of soil compaction. The distribution of some mineral nutrients such as phosphate and iron is very heterogeneous, due to their strong ionic interactions with the soil matrix and the strong pH-dependency of their solubility. The abundance of such immobile minerals can vary by an order of magnitude at scales of a 100 μm (Strawn et al. 2002). In contrast, other nutrients such as nitrate and potassium are at least tenfold more mobile in the soil (Marschner 1995), and therefore tend to accumulate as solutes above water-impermeable clay layers. The distinct physicochemical properties of plant mineral macronutrients implies that there should be at least two distinct growth or foraging strategies in response to limitation of soil minerals: for immobile minerals, the most efficient response to enhance uptake is to increase the surface area of the root to directly contact soil particles in previously unexploited domains of the soil. Increased branching, radial thickening, and growth of root hairs, while suppressing primary root growth, best accomplish this objective. In contrast, for mobile elements, the optimal strategy is to enhance root apical growth to reach deeper layers where such solutes accumulate.

Both syndromes are observed: under conditions of phosphate starvation, reduced primary root growth, enhanced lateral root formation, and stimulation of root hair growth is observed (Lopez-Bucio et al. 2002), while during iron starvation, mostly root hair growth is stimulated (Muller and Schmidt 2004). Both iron and phosphate have low mobility in the soil column. In contrast, root apical growth is stimulated in low nitrate (0.1–2.5 mM), when compared to higher concentrations, and this is due to a larger population of dividing cells and a delayed phase I/II transition (Dubrovsky and Doerner, unpublished). At lower concentrations (< 50 μM), primary root growth persists for a while without stimulation of lateral root growth (López-Bucio et al. 2003). Nitrate is relatively mobile in the soil column and accumulates above water-impermeable layers in the ground.

The analysis of root growth parameters is in many ways more straightforward than in shoots: organ growth is essentially anisotropic, and because the different processes contributing to organ growth occur in spatially distinct domains, they can be more readily analyzed. Kinematic analysis is very powerful in this respect, but has surprisingly only been used in a few cases for root growth analysis in Arabidopsis (Beemster and Baskin 1998, 2000; Beemster et al. 2002). In Sect. 4.1, I will focus on growth responses to phosphate starvation as these have been analyzed in greater detail than for other mineral nutrients.

4.1 Phosphate

Arabidopsis responds to phosphate starvation with a complex adaptive growth response. Initially, this involves a rapid inhibition of cell expansion in roots (Lai et al. 2007; Sanchez-Calderon et al. 2005; Williamson et al. 2001) and stimulation of lateral root initiation and emergence (Lopez-Bucio et al. 2002; Williamson et al. 2001). Prolonged starvation involves progressively reduced cell division, quiescence, and differentiation of cells in the apical meristem (Lai et al. 2007; Sanchez-Calderon et al. 2005; Ticconi et al. 2004). While the sequence of these events appears invariant, their kinetics and severity are quite variable between experiments and laboratories, possibly because it is very difficult to completely remove traces of phosphate from the growth media. This sequence of events implies that signaling networks involved in controlling responses to phosphate starvation target more than one of the fundamental mechanisms regulating organ growth.

Recent work indicates that the timing of onset, rate of progression, and severity of growth responses to phosphate depletion depends on the overall growth activity of the plant. Under phosphate starvation conditions, root growth is promoted by sugars and inhibited by nitrate, osmotic stress, or treatments with plant growth regulators (Lai et al. 2007). The emerging concept is that the scale of organ growth activity determines the level of demand for phosphate, which in turn influences the rate at which the plant goes through the series of adaptive growth responses.

The targets of phosphate signaling pathways involved in controlling cell growth, division, or expansion have not yet been identified. However, mutational dissection of adaptive responses to phosphate starvation has resulted in the identification of two interesting classes of mutants: the *pdr* (phosphate deficiency response) and the *lpi* (low phosphate insensitive) mutants. The *pdr2* mutant is hypersensitive to low phosphate availability and shows a short root phenotype under these conditions that is caused by inhibition of cell expansion and division (Ticconi et al. 2004). The onset of quiescence and terminal differentiation observed in wild-type plants only upon extended phosphate starvation (Sanchez-Calderon et al. 2005), occurs earlier and at higher external phosphate levels, and also leads to cell death. This suggests that *PDR2* might be involved in phosphate sensing or coupling perception to root growth responses.

The *lpi* mutants show the opposite phenotype: these mutants are hyposensitive to phosphate starvation. Four complementation groups have been identified, all of which continue root apical growth in the absence of phosphate (Sanchez-Calderon et al. 2006). However, this is not because these plants do not know that they are experiencing phosphate starvation: these mutants activate physiological and gene expression responses to phosphate starvation to a very similar degree as wild-type (Sanchez-Calderon et al. 2006). The *lpi* mutants have constitutively slightly reduced cell expansion, but dramatically increased cell division activity when compared to the wild-type in phosphate-starved conditions. These phenotypes suggest that *LPI* genes are involved in restraining cell division during phosphate limitation. This would serve two complementary purposes: (i) to insure the functional integrity of the root apical meristem for the longest possible time, and (ii) possibly to direct resources to incipient lateral roots to shift the patterns of root growth in favor of increasing root surface area. The cloning of *PDR* and *LPI* genes has not yet been reported, but their identification will facilitate the identification of their targets in the growth control machinery.

5
Integration of Growth Control

Shoots and roots are interdependent for nutrients, with overall shoot growth limited by nutrients assimilated by the root, and root growth limited by fixed carbon (C) translocated from the shoot. Nitrogen (N) limitation and uptake by the root plays a key role in controlling shoot growth and, taken together, this suggests that just as in heterotrophic multicellular organisms, N (amino acid) and C (sugar) availability provide crucial cues in overall plant growth control (Lorberg and Hall 2004). In limiting conditions, nutrients are re-allocated to meristems and developing organs to sustain growth for the longest period possible. Unfortunately, the kinetics of change in nutrient concentrations, transport, and translocation have not yet been examined in whole plants with cellular or high temporal resolution. Therefore, it is presently not clear whether the growth responses observed in response to altered nutrient abundance are due to direct sensing of nutrient levels in meristematic cells, or whether these cells respond to surrogate systemic or mitogenic signals such as plant growth regulators or miRNAs. Novel tools for such measurements are currently being developed (Deuschle et al. 2006; Gu et al. 2006; Lager et al. 2006), and therefore it will be interesting to re-visit some of the experiments relating to plant-mobile nutrients to carefully re-assess plant growth responses when these nutrients are limiting.

A characteristic feature of plant adaptive growth responses is that different shoot or root apices, or leaf organs, grow at different rates. Growth of organs or meristems directly exposed to the nutrient is promoted. The spatially selective allocation of resources to meristems or organs experiencing conditions more conducive to growth than others in effect constitutes foraging behavior, in which the "winners are fed" and which may be cued by the physiology of the affected tissues. For example, if barley root systems are separated into different compartments, and the bulk of the root system is grown in nutrient-limiting conditions, then roots in a compartment that is provided with higher mineral nutrient levels grow faster and branch more, leading to a more effective exploitation of such localized resource "jackpots" (Drew and Saker 1975). Importantly, if the whole root system is uniformly exposed to optimal mineral nutrient levels, stimulated growth is not observed, indicating that the selective growth stimulation observed upon localized nutrient availability is an internally regulated process. Likewise, it was recently reported that the sun leaves, with their higher rates of photosynthesis and transpiration, import almost three times more cytokinins than shade leaves (Boonman et al. 2007). When cytokinins were applied to shaded or water-deficit leaves, these behaved like sun leaves. Taken together, these data are consistent with a model in which the rate of metabolism or physiology cues changes in plant growth regulator concentrations or flux to regulate growth activities.

All classical plant growth regulators: auxins, cytokinins, gibberellins, brassinosteroids, ethylene, and abscisic acid have been shown to be involved directly or indirectly in controlling adaptive growth responses to environmental change. Auxins are required for the initial specification of lateral shoot organs (Reinhardt et al. 2000) and lateral root initiation (Torrey 1950), but it is less clear how it is mechanistically involved in adaptive growth responses to nutrients. Cytokinins are involved in controlling sink-source relationships and the balance of shoot and root growth (Werner et al. 2001, 2003), and at least partially mediate nitrogen nutrient cues (Miyawaki et al. 2004; Rahayu et al. 2005). They may also be involved in controlling root growth rates by affecting the phase I/II transition. Gibberellins (GA) are required for auxin stimulation of root growth (Fu and Harberd 2003), for organ expansion in shoots, and for maintenance of the indeterminate state in axillary meristems, and hence are possibly involved in determining the dividing cell population size in early leaf primordia (Keller et al. 2006). Biosynthesis of GAs is enhanced in low light (Potter et al. 1999), and in high concentrations of CO_2 (Teng et al. 2006), and therefore they likely play a role in stem and leaf organ growth. Brassinosteroids (BR) are required for cell expansion and cell division in leaves and roots (Nakaya et al. 2002). Their biosynthesis is stimulated by light, but since BR concentration is subjected to complex feedback mechanisms (Nomura and Bishop 2006), it is not clear whether BRs mediate light-intensity signaling. Ethylene is involved in many growth responses, particularly involving cell expansion, but is also involved in adaptive changes to leaf blade growth in low light (Vandenbussche et al. 2003). Abscisic acid (ABA), which mediates water deficiency cues, plays a negative role in leaf and root growth.

With the exception of ABA, which has been shown to stimulate expression of CDK inhibitors (KRP genes) (Wang et al. 1998), the mechanisms by which growth processes are controlled by these regulators are not yet clear. However, it is expected that growth regulators that move between different plant organs, i.e., auxin, cytokinin, ABA, as well as novel and still poorly characterized molecules (Booker et al. 2005), will play a major role in integration of growth responses at the whole plant level.

At the whole plant level, it is presently not clear whether cues that appear to promote growth (e.g., mineral nutrients and high, but not stressful, levels of light) and those that inhibit growth (e.g., water deficit or low temperature) act by the same mechanisms to modulate the activity of common targets. In other words, it is unclear whether promoting growth is relieving growth inhibition. Based on first principles, it is simpler, faster, and more economical to arrest growth, because it would suffice to interfere with an essential step, than to promote growth, which would require coordinate regulation of disparate processes. The principles underpinning plant growth regulation will become clearer once the targets of growth signaling pathways are identified and can be subjected to experimental manipulation.

6
Conclusions and Perspectives

Two significant gaps in our understanding of plant growth control remain:

- How environmental, nutritional, and growth factor cues are perceived and processed by sensory networks
- Mechanistic detail on how such networks control and coordinate the activity of cell growth, division, and expansion

Although increasing numbers of genes involved in these mechanisms are uncovered, very little is still known about how these genes interact to form a regulatory network that couples exogenous and endogenous signals to orchestrate growth responses.

Rapid progress in our understanding of environmental (specifically nutrient) control of adaptive growth responses in plants would be very much facilitated if a minimal set of parameters necessary for analyzing how specific cues effect changes in growth processes were determined in future experiments. These include the establishment of size at cell birth, kinematic analysis of the spatio-temporal scale and pattern of growth, ploidy analysis, and final cell size. The analysis of several of these parameters is still very challenging, but novel technical approaches, for example FRET-based sensors (Looger et al. 2005), and approaches that could help determine ploidy levels with cellular resolution (Matzke et al. 2005) are being developed. Although comprehensive data sets reflecting genome-wide responses at the level of gene expression, the proteome, and various post-translational modifications are becoming available, I posit that as long as these are obtained from, for example, whole tissues, which correspond to mixed populations of cells undertaking different, often opposite responses, they will be confusing and potentially misleading. Fortunately, novel tools and techniques are becoming available that should soon allow the analysis of such genome-wide responses at the cellular level (Birnbaum et al. 2005; Casson et al. 2005; Lee et al. 2006; Mace et al. 2006; Schad et al. 2005).

Finally, a conceptual debate about the most efficient and comprehensive experimental approaches for characterization of growth signaling is also necessary. Recent analysis of large collections of systematically generated knock-out mutants in budding yeast have led to revised views of signaling pathways. Instead of essentially linear pathways with only few lateral inputs, it has been proposed that much larger numbers of genes and their products participate in signaling networks with many products, associated in complexes, contributing quantitatively to signaling in minor ways (Friedman and Perrimon 2007). These conclusions have been drawn on the basis of end-point results, for example the quantitative effect of loss-of-function mutations on a specific trait under investigation. Such approaches are useful for assembly of a collection of cellular components even peripherally involved in signaling. However, the defining feature of signaling networks are that they respond dynamically to constant changes of specific cues to orchestrate desired outcomes at the cellular, organ, or whole-plant level by processing cues and propagating resultant signals. Thus, signaling networks contain two types of components: (i) those that change their activity as they process and transduce signals, and (ii) those that are minor accomplices to assist signal flux. To understand how the environment controls adaptive growth responses, we must focus on those network components that change properties when signaling is active and on their targets by examining the behavior of such networks under conditions of dynamic change.

Acknowledgements The support of the Biotechnology and Biological Sciences Research Council (BBSRC), the Royal Society, The Darwin Trust, and The Samuel Roberts Noble Foundation for work in the Doerner Lab is gratefully acknowledged. I apologize to colleagues whose work was not mentioned due to space restrictions.

References

Anthony RG, Henriques R, Helfer A, Meszaros T, Rios G, Testerink C, Munnik T, Deak M, Koncz C, Bogre L (2004) A protein kinase target of a PDK1 signaling pathway is involved in root hair growth in *Arabidopsis*. EMBO J 23:572–581

Arsham AM, Neufeld TP (2006) Thinking globally and acting locally with TOR. Curr Opin Cell Biol 18:589–597

Bates TR, Lynch JP (1996) Stimulation of root hair elongation in *Arabidopsis thaliana* by low phosphorus availability. Plant Cell Environ 19:529–538

Beemster GT, Baskin TI (1998) Analysis of cell division and elongation underlying the developmental acceleration of root growth in *Arabidopsis thaliana*. Plant Physiol 116:1515–1526

Beemster GT, Baskin TI (2000) *STUNTED PLANT 1* mediates effects of cytokinin, but not of auxin, on cell division and expansion in the root of Arabidopsis. Plant Physiol 124:1718–1727

Beemster GT, De Vusser K, De Tavernier E, De Bock K, Inze D (2002) Variation in growth rate between Arabidopsis ecotypes is correlated with cell division and A-type cyclin-dependent kinase activity. Plant Physiol 129:854–864

Birnbaum K, Jung JW, Wang JY, Lambert GM, Hirst JA, Galbraith DW, Benfey PN (2005) Cell type-specific expression profiling in plants via cell sorting of protoplasts from fluorescent reporter lines. Nat Methods 2:615–619

Bogre L, Okresz L, Henriques R, Anthony RG (2003) Growth signaling pathways in *Arabidopsis* and the AGC protein kinases. Trends Plant Sci 8:424–431

Booker J, Sieberer T, Wright W, Williamson L, Willett B, Stirnberg P, Turnbull C, Srinivasan M, Goddard P, Leyser O (2005) MAX1 encodes a cytochrome P450 family member that acts downstream of MAX3/4 to produce a carotenoid-derived branch-inhibiting hormone. Dev Cell 8:443–449

Boonman A, Prinsen E, Gilmer F, Schurr U, Peeters AJM, Voesenel LACJ, Pons TL (2007) Cytokinin import rate as a signal for photosynthetic acclimation to canopy light gradients. Plant Physiol 143:1841–1852

Carol RJ, Dolan L (2006) The role of reactive oxygen species in cell growth: lessons from root hairs. J Exp Bot 57:1829–1834

Casson S, Spencer M, Walker K, Lindsey K (2005) Laser capture microdissection for the analysis of gene expression during embryogenesis of Arabidopsis. Plant J 42:111–123

Cebolla A, Vinardell JM, Kiss E, Olah B, Roudier F, Kondorosi A, Kondorosi E (1999) The mitotic inhibitor ccs52 is required for endoreduplication and ploidy-dependent cell enlargement in plants. EMBO J 18:4476–4484

Cockcroft CE, den Boer BG, Healy JM, Murray JA (2000) Cyclin D control of growth rate in plants. Nature 405:575–579

Cookson SJ, Van Lijsebettens M, Granier C (2005) Correlation between leaf growth variables suggest intrinsic and early controls of leaf size in *Arabidopsis thaliana*. Plant Cell Environ 28:1355

Cookson SJ, Granier C (2006) A dynamic analysis of the shade-induced plasticity in *Arabidopsis thaliana* rosette leaf development reveals new components of the shade-adaptative response. Ann Bot (Lond) 97:443–452

Cookson SJ, Radziejwoski A, Granier C (2006) Cell and leaf size plasticity in Arabidopsis: what is the role of endoreduplication? Plant Cell Environ 29:1273–1283

Cosgrove DJ (2005) Growth of the plant cell wall. Nat Rev Mol Cell Biol 6:850–861

De Veylder L, Beeckman T, Beemster GT, de Almeida Engler J, Ormenese S, Maes S, Naudts M, Van Der Schueren E, Jacqmard A, Engler G, Inze D (2002) Control of proliferation, endoreduplication and differentiation by the *Arabidopsis* E2Fa-DPa transcription factor. EMBO J 21:1360–1368

del Pozo JC, Boniotti MB, Gutierrez C (2002) Arabidopsis E2Fc functions in cell division and is degraded by the ubiquitin-SCF(AtSKP2) pathway in response to light. Plant Cell 14:3057–3071

del Pozo JC, Diaz-Trivino S, Cisneros N, Gutierrez C (2006) The balance between cell division and endoreplication depends on E2FC-DPB, transcription factors regulated by the ubiquitin–SCFSKP2A pathway in Arabidopsis. Plant Cell 18:2224–2235

Deuschle K, Chaudhuri B, Okumoto S, Lager I, Lalonde S, Frommer WB (2006) Rapid metabolism of glucose detected with FRET glucose nanosensors in epidermal cells and intact roots of *Arabidopsis* RNA-silencing mutants. Plant Cell 18:2314–2325

Disch S, Anastasiou E, Sharma VK, Laux T, Fletcher JC, Lenhard M (2006) The E3 ubiquitin ligase BIG BROTHER controls *Arabidopsis* organ size in a dosage-dependent manner. Curr Biol 16:272–279

Doerner P, Jorgensen JE, You R, Steppuhn J, Lamb C (1996) Control of root growth and development by cyclin expression. Nature 380:520–523

Drew MC, Saker LR (1975) Nutrient supply and growth of seminal root system in barley. 2. Localized, compensatory increases in lateral root growth and rates of nitrate uptake when nitrate supply is restricted to only part of root system. J Exp Bot 26:79–90

Franklin KA, Whitelam GC (2005) Phytochromes and shade-avoidance responses in plants. Ann Bot (Lond) 96:169–175

Friedman A, Perrimon N (2007) Genetic screening for signal transduction in the era of network biology. Cell 128:225–231

Fu X, Harberd NP (2003) Auxin promotes Arabidopsis root growth by modulating gibberellin response. Nature 421:740–743

Granier C, Tardieu F (1999) Leaf expansion and cell division are affected by reducing absorbed light before but not after the decline in cell division rate in the sunflower leaf. Plant Cell Environ 22:1365–1376

Granier C, Tardieu F (1999) Water deficit and spatial pattern of leaf development. Variability in responses can be simulated using a simple model of leaf development. Plant Physiol 119:609–619

Granier C, Inze D, Tardieu F (2000) Spatial distribution of cell division rate can be deduced from that of p34(cdc2) kinase activity in maize leaves grown at contrasting temperatures and soil water conditions. Plant Physiol 124:1393–1402

Gu H, Lalonde S, Okumoto S, Looger LL, Scharff-Poulsen AM, Grossman AR, Kossmann J, Jakobsen I, Frommer WB (2006) A novel analytical method for in vivo phosphate tracking. FEBS Lett 580:5885–5893

Gutierrez C, Ramirez-Parra E, Castellano MM, del Pozo JC (2002) G(1)-to-S transition: more than a cell cycle engine switch. Curr Opin Plant Biol 5:480–486

Hemerly A, Engler Jde A, Bergounioux C, Van Montagu M, Engler G, Inze D, Ferreira P (1995) Dominant negative mutants of the Cdc2 kinase uncouple cell division from iterative plant development. EMBO J 14:3925–3936

Horiguchi G, Ferjani A, Fujikura U, Tsukaya H (2006) Coordination of cell proliferation and cell expansion in the control of leaf size in *Arabidopsis thaliana*. J Plant Res 119:37–42

Horvath BM, Magyar Z, Zhang Y, Hamburger AW, Bako L, Visser RG, Bachem CW, Bogre L (2006) EBP1 regulates organ size through cell growth and proliferation in plants. EMBO J 25:4909–4920

Hu Y, Xie Q, Chua NH (2003) The Arabidopsis auxin-inducible gene *ARGOS* controls lateral organ size. Plant Cell 15:1951–1961

Inze D, Veylder LD (2006) Cell cycle regulation in plant development. Annu Rev Genet 40:77–105

Jorgensen P, Tyers M (2004) How cells coordinate growth and division. Curr Biol 14:R1014–R1027

Keller T, Abbott J, Moritz T, Doerner P (2006) *Arabidopsis REGULATOR OF AXILLARY MERISTEMS1* controls a leaf axil stem cell niche and modulates vegetative development. Plant Cell 18:598–611

Kim GT, Yano S, Kozuka T, Tsukaya H (2005) Photomorphogenesis of leaves: shade-avoidance and differentiation of sun and shade leaves. Photochem Photobiol Sci 4:770–774

Lager I, Looger LL, Hilpert M, Lalonde S, Frommer WB (2006) Conversion of a putative Agrobacterium sugar-binding protein into a FRET sensor with high selectivity for sucrose. J Biol Chem 281:30875–30883

Lai F, Thacker J, Li YY, Doerner P (2007) Cell division activity determines the magnitude of phosphate starvation responses in Arabidopsis. Plant J 50:545–556

Lake JA, Quick WP, Beerling DJ, Woodward FI (2001) Plant development. Signals from mature to new leaves. Nature 411:154

Lee JY, Colinas J, Wang JY, Mace D, Ohler U, Benfey PN (2006) Transcriptional and post-transcriptional regulation of transcription factor expression in *Arabidopsis* roots. Proc Natl Acad Sci USA 103:6055–6060

Li C, Potuschak T, Colon-Carmona A, Gutierrez RA, Doerner P (2005) *Arabidopsis* TCP20 links regulation of growth and cell division control pathways. Proc Natl Acad Sci USA 102:12978–12983

Looger LL, Lalonde S, Frommer WB (2005) Genetically encoded FRET sensors for visualizing metabolites with subcellular resolution in living cells. Plant Physiol 138:555–557

Lopez-Bucio J, Hernandez-Abreu E, Sanchez-Calderon L, Nieto-Jacobo MF, Simpson J, Herrera-Estrella L (2002) Phosphate availability alters architecture and causes changes in hormone sensitivity in the Arabidopsis root system. Plant Physiol 129:244–256

López-Bucio J, Cruz-Ramírez A, Herrera-Estrella L (2003) The role of nutrient availability in regulating root architecture. Curr Opin Plant Biol 6:280–287

Lorberg A, Hall MN (2004) TOR: the first 10 years. Curr Top Microbiol Immunol 279:1–18

Mace DL, Lee JY, Twigg RW, Colinas J, Benfey PN, Ohler U (2006) Quantification of transcription factor expression from Arabidopsis images. Bioinformatics 22:e323–e331

MacKay VL, Mai B, Waters L, Breeden LL (2001) Early cell cycle box-mediated transcription of CLN3 and SWI4 contributes to the proper timing of the G(1)-to-S transition in budding yeast. Mol Cell Biol 21:4140–4148

Maeshima M (2001) Tonoplast transporters: organization and function. Annu Rev Plant Physiol Plant Mol Biol 52:469–497

Mahfouz MM, Kim S, Delauney AJ, Verma DP (2006) *Arabidopsis* TARGET OF RAPAMYCIN interacts with RAPTOR, which regulates the activity of S6 kinase in response to osmotic stress signals. Plant Cell 18:477–490

Marschner H (1995) Mineral nutrition of higher plants. Academic, London

Masle J (2000) The effects of elevated CO_2 concentrations on cell division rates, growth patterns, and blade anatomy in young wheat plants are modulated by factors related to leaf position, vernalization, and genotype. Plant Physiol 122:1399–1415

Matzke AJ, Huettel B, van der Winden J, Matzke M (2005) Use of two-color fluorescence-tagged transgenes to study interphase chromosomes in living plants. Plant Physiol 139:1586–1596

McIntosh K, Bonham-Smith P (2006) Ribosomal protein gene regulation: What about plants? Can J Bot 84:342–362

Melaragno JE, Mehrotra B, Coleman AW (1993) Relationship between endopolyploidy and cell size in epidermal tissue of Arabidopsis. Plant Cell 5:1661–1668

Menand B, Desnos T, Nussaume L, Berger F, Bouchez D, Meyer C, Robaglia C (2002) Expression and disruption of the *Arabidopsis TOR* (target of rapamycin) gene. Proc Natl Acad Sci USA 99:6422–6427

Menges M, Samland AK, Planchais S, Murray JA (2006) The D-type cyclin CYCD3;1 is limiting for the G1-to-S-phase transition in *Arabidopsis*. Plant Cell 18:893–906

Miyawaki K, Matsumoto-Kitano M, Kakimoto T (2004) Expression of cytokinin biosynthetic isopentenyltransferase genes in Arabidopsis: tissue specificity and regulation by auxin, cytokinin, and nitrate. Plant J 37:128–138

Mizukami Y, Fischer RL (2000) Plant organ size control: *AINTEGUMENTA* regulates growth and cell numbers during organogenesis. Proc Natl Acad Sci USA 97:942–947

Morgan D (2007) The cell cycle. Oxford University Press, Corby

Muller M, Schmidt W (2004) Environmentally induced plasticity of root hair development in Arabidopsis. Plant Physiol 134:409–419

Nakaya M, Tsukaya H, Murakami N, Kato M (2002) Brassinosteroids control the proliferation of leaf cells of *Arabidopsis thaliana*. Plant Cell Physiol 43:239–244

Nomura T, Bishop GJ (2006) Cytochrome P450s in plant steroid hormone synthesis and metabolism. Phytochem Rev 5:421–432

Oyama T, Shimura Y, Okada K (2002) The IRE gene encodes a protein kinase homologue and modulates root hair growth in Arabidopsis. Plant J 30:289–299

Planchais S, Samland AK, Murray JA (2004) Differential stability of Arabidopsis D-type cyclins: CYCD3;1 is a highly unstable protein degraded by a proteasome-dependent mechanism. Plant J 38:616–625

Planta RJ (1997) Regulation of ribosome synthesis in yeast. Yeast 13:1505–1518

Polymenis M, Schmidt EV (1997) Coupling of cell division to cell growth by translational control of the G1 cyclin CLN3 in yeast. Genes Dev 11:2522–2531

Potter TI, Rood SB, Zanewich KP (1999) Light intensity, gibberellin content and the resolution of shoot growth in Brassica. Planta 207:505–511

Pritchard SG, Rogers HH, Prior SA, Peterson CM (1999) Elevated CO_2 and plant structure: a review. Glob Change Biol 5:807–837

Rahayu YS, Walch-Liu P, Neumann G, Romheld V, von Wiren N, Bangerth F (2005) Root-derived cytokinins as long-distance signals for NO3-induced stimulation of leaf growth. J Exp Bot 56:1143–1152

Ramirez-Parra E, Lopez-Matas MA, Frundt C, Gutierrez C (2004) Role of an atypical E2F transcription factor in the control of Arabidopsis cell growth and differentiation. Plant Cell 16:2350–2363

Reinhardt D, Mandel T, Kuhlemeier C (2000) Auxin regulates the initiation and radial position of plant lateral organs. Plant Cell 12:507–518

Riou-Khamlichi C, Huntley R, Jacqmard A, Murray JA (1999) Cytokinin activation of Arabidopsis cell division through a D-type cyclin. Science 283:1541–1544

Riou-Khamlichi C, Menges M, Healy JM, Murray JA (2000) Sugar control of the plant cell cycle: differential regulation of Arabidopsis D-type cyclin gene expression. Mol Cell Biol 20:4513–4521

Rymen B, Fiorani F, Kartal F, Vandepoele K, Inze D, Beemster GT (2007) Cold nights impair leaf growth and cell cycle progression in maize through transcriptional changes of cell cycle genes. Plant Physiol 143:1429–1438

Sanchez-Calderon L, Lopez-Bucio J, Chacon-Lopez A, Cruz-Ramirez A, Nieto-Jacobo F, Dubrovsky JG, Herrera-Estrella L (2005) Phosphate starvation induces a determinate developmental program in the roots of *Arabidopsis thaliana.* Plant Cell Physiol 46:174–184

Sanchez-Calderon L, Lopez-Bucio J, Chacon-Lopez A, Gutierrez-Ortega A, Hernandez-Abreu E, Herrera-Estrella L (2006) Characterization of low phosphorus insensitive mutants reveals a crosstalk between low phosphorus-induced determinate root development and the activation of genes involved in the adaptation of Arabidopsis to phosphorus deficiency. Plant Physiol 140:879–889

Schad M, Lipton MS, Giavalisco P, Smith RD, Kehr J (2005) Evaluation of two-dimensional electrophoresis and liquid chromatography–tandem mass spectrometry for tissue-specific protein profiling of laser-microdissected plant samples. Electrophoresis 26:2729–2738

Squatrito M, Mancino M, Donzelli M, Areces LB, Draetta GF (2004) EBP1 is a nucleolar growth-regulating protein that is part of pre-ribosomal ribonucleoprotein complexes. Oncogene 23:4454–4465

Squatrito M, Mancino M, Sala L, Draetta GF (2006) Ebp1 is a dsRNA-binding protein associated with ribosomes that modulates eIF2alpha phosphorylation. Biochem Biophys Res Commun 344:859–868

Strawn D, Doner H, Zavarin M, McHugo S (2002) Microscale investigation into the geochemistry of arsenic, selenium, and iron in soil developed in pyritic shale materials. Geoderma 108:237–257

Teng N, Wang J, Chen T, Wu X, Wang Y, Lin J (2006) Elevated CO_2 induces physiological, biochemical and structural changes in leaves of *Arabidopsis thaliana.* New Phytol 172:92–103

Ticconi CA, Delatorre CA, Lahner B, Salt DE, Abel S (2004) Arabidopsis pdr2 reveals a phosphate-sensitive checkpoint in root development. Plant J 37:801–814

Tocquin P, Ormenese S, Pieltain A, Detry N, Bernier G, Perilleux C (2006) Acclimation of *Arabidopsis thaliana* to long-term CO_2 enrichment and nitrogen supply is basically a matter of growth rate adjustment. Physiol Plantarum 128:677–688

Torrey JG (1950) The induction of lateral roots by indoleacetic acid and root decapitation. Am J Bot 37:257–264

Tsukaya H (2006) Mechanism of leaf shape determination. Annu Rev Plant Biol 57:477–496

Tyson CB, Lord PG, Wheals AE (1979) Dependency of size of *Saccharomyces cerevisiae* cells on growth rate. J Bacteriol 138:92–98

Van Lijsebettens M, Vanderhaeghen R, De Block M, Bauw G, Villarroel R, Van Montagu M (1994) An S18 ribosomal protein gene copy at the Arabidopsis PFL locus affects plant development by its specific expression in meristems. EMBO J 13:3378–3388

Vandenbussche F, Vriezen WH, Smalle J, Laarhoven LJ, Harren FJ, Van Der Straeten D (2003) Ethylene and auxin control the Arabidopsis response to decreased light intensity. Plant Physiol 133:517–527

Vandepoele K, Raes J, De Veylder L, Rouze P, Rombauts S, Inze D (2002) Genome-wide analysis of core cell cycle genes in Arabidopsis. Plant Cell 14:903–916

Vlieghe K, Boudolf V, Beemster GT, Maes S, Magyar Z, Atanassova A, de Almeida Engler J, De Groodt R, Inze D, De Veylder L (2005) The DP-E2F-like gene DEL1 controls the endocycle in *Arabidopsis thaliana.* Curr Biol 15:59–63

Wang D, Harper JF, Gribskov M (2003) Systematic trans-genomic comparison of protein kinases between Arabidopsis and *Saccharomyces cerevisiae.* Plant Physiol 132:2152–2165

Wang H, Qi QG, Schorr P, Cutler AJ, Crosby WL, Fowke LC (1998) ICK1, a cyclin-dependent protein kinase inhibitor from *Arabidopsis thaliana* interacts with both Cdc2a and CycD3, and its expression is induced by abscisic acid. Plant J 15:501–510

Warner JR (1999) The economics of ribosome biosynthesis in yeast. Trends Biochem Sci 24:437–440

Weijers D, Franke-van Dijk M, Vencken RJ, Quint A, Hooykaas P, Offringa R (2001) An *Arabidopsis* Minute-like phenotype caused by a semi-dominant mutation in a *RIBOSOMAL PROTEIN S5* gene. Development 128:4289–4299

Weingartner M, Pelayo HR, Binarova P, Zwerger K, Melikant B, de la Torre C, Heberle-Bors E, Bogre L (2003) A plant cyclin B2 is degraded early in mitosis and its ectopic expression shortens G2-phase and alleviates the DNA-damage checkpoint. J Cell Sci 116:487–498

Werner T, Motyka V, Strnad M, Schmulling T (2001) Regulation of plant growth by cytokinin. Proc Natl Acad Sci USA 98:10487–10492

Werner T, Motyka V, Laucou V, Smets R, Van Onckelen H, Schmulling T (2003) Cytokinin-deficient transgenic Arabidopsis plants show multiple developmental alterations indicating opposite functions of cytokinins in the regulation of shoot and root meristem activity. Plant Cell 15:2532–2550

White DW (2006) *PEAPOD* regulates lamina size and curvature in *Arabidopsis.* Proc Natl Acad Sci USA 103:13238–13243

Williamson LC, Ribrioux SPCP, Fitter AH, Leyser HMO (2001) Phosphate availability regulates root system architecture in Arabidopsis. Plant Physiol 126:875–882

Yano S, Terashima I (2001) Separate localization of light signal perception for sun or shade type chloroplast and palisade tissue differentiation in *Chenopodium album.* Plant Cell Physiol 42:1303–1310

Zhang Y, Woodford N, Xia X, Hamburger AW (2003) Repression of E2F1-mediated transcription by the ErbB3 binding protein Ebp1 involves histone deacetylases. Nucleic Acids Res 31:2168–2177

Plant Cell Monogr (10)
L. Bögre and G. Beemster: Plant Growth Signaling
DOI 10.1007/7089_2007_149/Published online: 26 January 2008

Control of Plant Organ Size

Elena Anastasiou[1] · Michael Lenhard[1,2] (✉)

[1]Institut für Biologie III, Albert-Ludwigs-Universität Freiburg, Schänzlestraße 1, D-79104 Freiburg im Breisgau, Germany

[2]John Innes Centre, Colney Lane, Norwich NR4 7UH, UK
michael.lenhard@bbsrc.ac.uk

Abstract Plant organs grow to highly reproducible sizes that are determined by the plant's genotype and by the identity of the organ. The strong heritability of size differences indicates that organ size is under tight genetic control. The overall increase in size of plant organs is driven by two distinct processes: cell proliferation with the concomitant generation of cytoplasmic mass, and cell expansion due to water uptake into the central vacuole. Molecular genetic analysis has identified a number of genes that influence final organ size by acting on either of these two processes in a promoting or repressing manner. The differences in the sizes and shapes of the various organ types within individuals as well as the size differences among species result from the modification of growth patterns by factors controlling organ identity and by evolutionary change. Genetic analysis in model species is beginning to shed light on how these growth patterns are being controlled and modified. Together, these studies are unraveling how plant organs decide for or against further growth and suggest approaches for manipulating biomass accumulation in plants.

1 Introduction

Lateral plant organs, such as leaves, floral organs and fruits, are produced throughout post-embryonic life. They are ultimately derived from the stem cells of the indeterminate shoot apical meristem (Veit 2006). Yet, while growth of the whole plant does generally not have a fixed endpoint, individual organs only grow to defined sizes and shapes that are determined by the plant's genotype and the identity of the organ; for example, leaves versus floral organs (Ingram and Waites 2006; Mizukami 2001). Our ability to easily distinguish plant species with very similar morphologies but different sizes, for example daisies and marguerites, underlines the strong heritability of organ size, which can only be modified within certain limits by environmental factors. By contrast, evolution has generated an enormous variety of organ sizes, ranging from the giant leaves of some water lilies or bananas, for example, to the more modest dimensions of *Arabidopsis thaliana* (Fig. 1). Despite recent progress, our understanding of the underlying genetic and molecular mechanisms that control plant organ size is still far from comprehensive. Unraveling these regulatory mechanisms is not only of great scientific interest, but is also of substantial applied relevance, as it

Fig. 1 Two extremes of plant organ sizes observed in nature. In contrast to banana trees with their giant leaves (*left*), the diminutive weed *Arabidopsis thaliana* forms only very small lateral organs (*right*). (Image of the banana tree courtesy of Dr. Philippe Vain, John Innes Centre, Norwich, UK.)

would allow the rational manipulation of organ size and thus biomass production, e.g., for generating biofuels. The following chapter will not only discuss the progress that has been made, focusing mainly on studies in *Arabidopsis* and other genetic model organisms, but will also attempt to highlight the gaps to be filled in our current knowledge of how plant organs decide for or against further growth.

Plant organs grow via two fundamentally different, yet closely coordinated processes (Fig. 2; Menand and Robaglia 2004). During a first phase of organ development, cells proliferate mitotically and produce new cytoplasmic mass. After some time, cell proliferation gradually ceases and post-mitotic cells begin to expand by taking up water into their central vacuole. This post-mitotic cell expansion is often accompanied by a ploidy increase due to endoreduplication, for example in leaves and hypocotyls (Sugimoto-Shirasu and Roberts 2003; a process that will be discussed in more detail by Sugimoto-Shirasu in this volume), although in other cases (e.g., *Arabidopsis* floral organs), expanding cells remain diploid. These two phases will be referred to as growth by proliferation and growth by expansion throughout the text, respectively. It is generally assumed that little to no increase in cytoplasmic mass occurs as cells grow by expansion (Sugimoto-Shirasu and Roberts

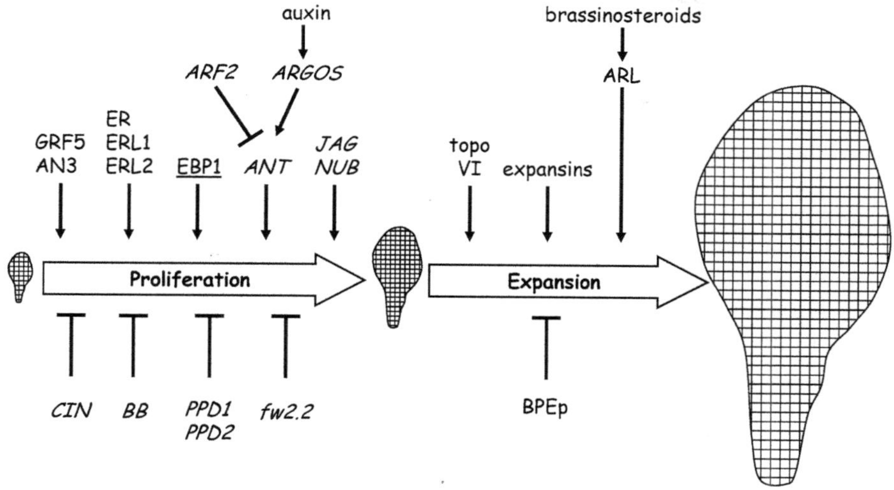

Fig. 2 A schematic overview of the pathways that promote or restrict organ growth. The phases of growth by proliferation and by expansion are shown as temporally separated only for illustrative purposes, although there is a gradual transition between the two phases in planta, with some cells in an organ starting to expand while others still divide. Genes that affect the duration of growth by proliferation are shown in italics, while the underline for EBP1 indicates its effect on the rate of proliferation. For the remaining genes that influence proliferation, the mode of action has not been precisely determined. See text for a detailed discussion of the individual factors

2003). However, it is clear that cells must continue to synthesize plasma and tonoplast membrane and cell wall material to provide for their size increase, implying that growth by expansion is an active process with the potential for regulation.

Thus, several important questions need to be answered with respect to the control of plant organ size. Which factors regulate growth by proliferation and/or expansion, and how do they interact with the basic cell cycle and biosynthetic machinery? How do organs with different identities attain their characteristic size and shape? Is growth of different organs regulated by distinct pathways, or are the same mechanisms active in diverse organ types, with their output being modified by the transcriptional networks that determine organ identity? How has evolution modified size control pathways to generate the amazing variety in organ sizes and shapes that we see in nature? What is actually being measured as an organ grows? Time, size, mass, or something else?

2 Control of Growth by Proliferation

2.1 Promoting Factors

A multitude of *Arabidopsis* mutants show dwarfism because of reduced cell numbers and/or cell size (e.g., Autran et al. 2002; Nelissen et al. 2003). However, for many of them overexpression of the respective gene—where tested—does not lead to extra growth, making it difficult to assess whether these genes fulfill a regulatory role. They will therefore not be discussed further here. Similarly, while activity of the core cell cycle machinery is obviously required for growth by proliferation, overexpression of individual cell cycle stimulators does not generally lead to larger organs, suggesting that organ size is limited by other factors (Dewitte et al. 2003; Doerner et al. 1996). While not increasing final organ size, overexpression of the cell cycle stimulator CyclinD2 in tobacco was reported to accelerate plant growth, both in terms of leaf initiation by the meristem and of growth of individual leaves up to their normal sizes (Cockcroft et al. 2000).

A prominent stimulator of growth by proliferation is the APETALA2 (AP2)-domain protein AINTEGUMENTA (ANT). Loss of function *ant* mutants form smaller leaves and floral organs, whereas *ANT* overexpression increases final organ size (Elliott et al. 1996; Klucher et al. 1996; Krizek 1999; Mizukami and Fischer 2000). The larger organs are due to a prolonged period of cell proliferation, even though the response may vary somewhat amongst different organ types (Krizek 1999; Mizukami and Fischer 2000). The delayed arrest of cell division correlates with the sustained expression of a major cell cycle regulator, *CycD3;1*, although this is unlikely to explain the enlarged organs, as constitutive *CycD3;1* overexpression does not lead to larger organs (Dewitte et al. 2003). ANT is likely to function as a transcription factor, as the protein localizes to the nucleus, can bind to DNA in a sequence-specific manner, and activates gene transcription by virtue of an N-terminal domain (Krizek and Sulli 2006). The identification of ANT target genes that underlie its growth-promoting activity remains an important task for future studies.

The *ARGOS* gene appears to function as an upstream activator of ANT (Hu et al. 2003). *ARGOS* was identified as an auxin-induced gene expressed in developing organs, and its downregulation causes reduced organ growth. By contrast, ARGOS overexpression is sufficient to increase organ size by stimulating excess cell proliferation, yet this effect is abolished in an *ant* mutant background. *ARGOS* encodes a small protein with no discernable functional domains, and how ARGOS acts biochemically is presently unknown.

The *ANGUSTIFOLIA3* (*AN3*)/*GRF-INTERACTING FACTOR1* (*GIF1*) gene encodes a predicted transcriptional co-activator that stimulates organ growth (Horiguchi et al. 2005; Kim and Kende 2004). AN3/GIF1 interacts with the putative transcription factor GROWTH REGULATING FACTOR5 (GRF5), and both genes are expressed in developing organs in the region of maximum cell division activity. Similar to *ANT*, loss of *an3/gif1* or *grf5* function reduces organ size, while overexpression of either gene leads to larger organs, with the changes in size being due to changes in cell numbers (Horiguchi et al. 2005; Kim and Kende 2004). Interestingly, the effects of *an3/gif1* and *grf5* mutations are not isotropic, but affect the leaf width direction more strongly than growth in length, suggesting that growth along the transverse and longitudinal axes of plant organs may be under partly independent control. GRF5 is part of a gene family in *Arabidopsis* (GRF1–GRF9), the members of which are expressed in actively growing organs and tissues of the plant (Horiguchi et al. 2005; Kim et al. 2003). Besides for *GRF5*, a growth promoting effect has also been described for *GRF1*, *GRF2* and *GRF3* (Kim et al. 2003). While overexpression of *GRF1* and *GRF2* is sufficient to enlarge leaf size, triple mutants of *grf1 grf2 grf3* form smaller leaves. However, in contrast to the effect of *GRF5* on cell numbers, the changes in organ size due to altered *GRF1-GRF3* activity result from increased or reduced cell expansion, which may suggest that growth by proliferation and by expansion share common regulatory elements (see also the discussion on *ARGOS-LIKE* below).

Another putative transcriptional regulator with a growth-promoting function is encoded by the *JAGGED* (*JAG*) gene (Dinneny et al. 2004; Ohno et al. 2004). *jag* mutants form smaller floral organs, most prominently petals, because of a premature arrest of cell proliferation in the distal petal blade region. Plants with increased *JAG* activity develop ectopic bracts, i.e., leaves subtending flowers, and leaf lamina tissue on the leaf petiole. However, strong *JAG* overexpression in developing flowers leads to severe developmental defects, including lack of organ differentiation. This suggests that either the main function of *JAG* is to prevent premature differentiation—and the concomitant cell cycle arrest—or that *JAG* has additional functions in tissue patterning and specific differentiation events besides a role in growth control. The recent finding that, together with organ polarity factors, *JAG* is involved in controlling tissue differentiation in the developing gynoecium (Dinneny et al. 2005) can be taken as evidence for the latter (see below). A homolog to *JAG*, called *NUBBIN* (*NUB*), functions redundantly in stimulating tissue growth, most prominently in leaves, stamens and carpels (Dinneny et al. 2006). While *nub* single mutants do not have obvious phenotypes, *jag nub* double mutants form strongly reduced stamens and carpels because of insufficient growth. Both proteins have a C2H2-zinc finger domain and for JAG nuclear localization has been demonstrated (Dinneny et al. 2004; Ohno et al. 2004), suggesting that JAG and NUB regulate transcription.

Work in yeast and animals has demonstrated a crucial role for ribosome biogenesis by the nucleolus in the control of cell and tissue growth (Rudra and Warner 2004). Although *Arabidopsis* plants with a mutation in a ribosomal protein have been reported to show a reduced rate of growth, similar to the classical *Minute* mutants from Drosophila, no detailed measurements of final organ sizes have been reported (Van Lijsebettens et al. 1994; Weijers et al. 2001). Very recently, a potato protein was identified that provides a putative link between ribosome biogenesis and organ size control in plants (Horvath et al. 2006). Plant EBP1 is homologous to the human nucleolar ribosome biogenesis factor EBP1 and localizes to the nucleolus (Pendle et al. 2005). While its downregulation led to smaller organs in both potato and *Arabidopsis* because of a premature arrest of proliferation, its overexpression could increase organ size, although extremely high levels of overexpression were required for this effect. Increased organ size was the result of accelerated proliferation up to wild-type cell numbers followed by enhanced cell expansion (Horvath et al. 2006). Thus, this protein provides initial evidence that the central role of ribosome biogenesis in growth control is not limited to yeast and animals, but also extends to plants. In this respect, it is interesting to note that a class I TCP (TEOSINTE BRANCHED1, CYCLOIDEA, PCF) family protein, TCP20, has been shown to bind to functionally relevant *cis*-regulatory elements in the promoters of important cell cycle regulators and of ribosomal proteins, suggesting a possibility for how cellular growth and cell division could be coordinated (Li et al. 2005).

In addition to the above factors that can be assumed to act cell-autonomously, intercellular signaling via the ERECTA (ER)-family of leucin-rich-repeat receptor-like kinases (LRR-RLKs) appears to play an important role in promoting cell proliferation in developing organs (Shpak et al. 2004). Triple mutants in *er*, *erecta-like1* (*erl1*) and *erecta-like2* (*erl2*) are severely dwarfed because of insufficient cell proliferation. The three genes show overlapping yet distinct expression patterns in young, growing tissues, suggestive of expression-dependent partial redundancy of the three genes. Based on expression analyses, the ER-family kinases appear to act independently of *ANT*. The triple mutants also exhibit defects in stomatal patterning, suggesting that activity of the *ER* kinase family is not only required for the proliferative cell divisions that increase overall organ cell numbers, but that it also influences the asymmetric divisions that lead to the formation of stomata (Shpak et al. 2005). The downstream signaling components and presumed extracellular ligands for the ER-family receptors are still unknown, although signal transduction has been speculated to involve a MAP-kinase cascade (Ingram and Waites 2006). It will be exciting to see whether ER ligands represent the plant-equivalent to animal mitogens and growth factors with the potential to coordinate growth at an organ-wide level. Further details on LRR-RLKs in plant growth control can be found in the chapter by Ingram in this volume.

Several plant hormones influence either cell proliferation or expansion in developing organs. Cytokinins are prominent for their cell cycle stimulating activity, and both reduced cytokinin content due to overexpression of an inactivating enzyme (Werner et al. 2001, 2003) and insensitivity to the hormone (Riefler et al. 2006) cause dwarfism due to reduced cell numbers. However, to our knowledge it has not been unambiguously demonstrated that increased cytokinin levels lead to larger organs. Auxin is also known for promoting cell proliferation and growth, and the reader is referred to the relevant chapter by Offringa in this volume.

2.2 Repressing Factors

In analogy to the previous section, overexpression of many genes has been reported to cause dwarfism because of reduced cell numbers (e.g., ICK1/KRP2, De Veylder et al. 2001; LOB1, Shuai et al. 2002), yet in the absence of a complementary loss-of-function phenotype with increased proliferation, the role of these factors in endogenous proliferation control remains unclear. The following discussion will therefore be limited to those factors that have been shown to be required for limiting proliferation in developing organs.

Several putative transcriptional regulators of the class II TCP family are required for proper arrest of cell proliferation during organ development. Loss of function of the *Antirrhinum* CINCINNATA (CIN) protein leads to larger organs with a wrinkled shape because of excess proliferation at the leaf margins (Nath et al. 2003). The *CIN* expression pattern is intriguing in that expression sweeps across the lamina of the developing leaf in a basipetal manner, preceding the cell cycle arrest front that terminates proliferation in a basipetal direction. Therefore, *CIN* may sensitize leaf margin cells to an unknown cell cycle arrest signal. Downregulation of *Arabidopsis CIN* homologs by overexpression of the *JAW* microRNA (miRNA) leads to a very similar leaf phenotype, suggesting that class II TCP action in limiting proliferation is conserved in plants (Palatnik et al. 2003). However, the picture is somewhat more complicated, as in petals *CIN* has the opposite effect, i.e., it promotes organ growth by proliferation, suggesting organ identity factors may determine the nature of its influence on growth (Crawford et al. 2004).

Recently, the model of how proliferation arrest is achieved in developing leaves has been refined by the identification of the redundant homologs PEAPOD1 (PPD1) and PEAPOD2 (PPD2) (White 2006). *ppd1 ppd2* double mutants form enlarged leaves with a bell shape because of excess growth in the center of the leaf lamina. This excess growth appears to be driven by prolonged proliferation of so-called dispersed meristematic cells, such as stomatal precursors and procambial cells. By contrast, *PPD* overexpression reduces leaf size by causing premature cell cycle arrest of dispersed meristematic cells. *PPD* expression appears to follow the general cell cycle arrest front, i.e., expression is found in the region of organs where most cells have stopped proliferating, but dispersed meristematic cells continue to divide. These observations lead to the proposal that the first, *TCP*-dependent cell cycle arrest front is followed by a second, *PPD*-mediated arrest front that specifically affects the dispersed meristematic cells.

The *BIG BROTHER* (*BB*) gene also limits organ growth by restricting the period of cell proliferation, affecting mainly floral organs and the stem (Disch et al. 2006). Final organ size is tightly correlated with *BB* mRNA expression levels, with a reduction by 50% or a three-fold increase in mRNA amount leading to significant organ enlargement or reduction, respectively. *BB* is expressed in all actively growing tissues. The BB protein has a RING-finger domain, exhibits E3 ubiquitin-ligase activity in vitro, and mutations that abolish this activity also abrogate function *in planta*. Together, these findings suggest that proteasome-mediated degradation of crucial growth stimulators during the phase of cell proliferation determines when proliferation ceases. As *BB* acts independently of *ANT*, *JAG* and phytohormones, the identification of its substrates promises to uncover important additional growth-promoting factors.

Similar to *BB*, loss-of-function mutants for *AUXIN RESPONSE FACTOR2* (*ARF2*) form thicker stems and larger organs, including leaves, integuments and, as a consequence, seeds (Ellis et al. 2005; Okushima et al. 2005; Schruff et al. 2006). The organ enlargement correlates with prolonged expression of *ANT* and its target gene *CycD3;1* (Schruff et al. 2006), suggesting that the excess growth is due to an extended period of cell proliferation. No overexpression phenotype has been reported for *ARF2*. In vitro ARF2 protein has been shown to bind to auxin response elements (AuxREs) and it was able to repress transcription from synthetic AuxRE-containing promoters (Li et al. 2004; Tiwari et al. 2003). However, *arf2* mutants do not exhibit global expression changes in auxin-regulated genes (Okushima et al. 2005), calling into question whether its effects on growth are due to misregulated auxin signaling.

Studies of natural variation in tomato fruit size have identified the *fruit weight2.2* (*fw2.2*) gene as another important repressor of cell proliferation (Frary et al. 2000), as will be discussed in more detail below.

As the preceding discussion has shown, there is a growing inventory of factors that positively or negatively influence cell proliferation and thus contribute to organ size regulation. However, if and how individual factors interact with each other and with the basic cell cycle and growth machinery, how many independent pathways impinge on the control of proliferation, and how the identified factors act molecularly all remain largely unknown.

One generalization that can be drawn from the studies of growth by proliferation is that the factors isolated to date—with the exception of EBP1 and CyclinD2—all seem to influence the duration of cell proliferation, rather than the rate of cell cycling. This could suggest that cells in organ primordia normally cycle at the maximum rate that is possible under the given nutritional and environmental circumstances, and that it may be easier for example to extend the period of proliferation than to speed up cell cycling. Accelerated growth by proliferation that leads to larger organs has only been observed in plants overexpressing the putative ribosome biogenesis factor EBP1, which appears intuitively plausible, as increased ribosome production should allow for a higher rate of macromolecular synthesis by protein translation and thus faster and more overall cell growth. The first glimpses at the molecular evolution of organ size, i.e., the role of *fw2.2* in tomato fruits, seem to support the idea that extending the time of proliferation may be a more accessible path than accelerating growth, as changes in fruit size due to allelic differences at the *fw2.2* locus are the result of altered timing of cell cycle arrest (Cong et al. 2002). Unraveling further cases of evolutionary modification of organ size should indicate whether it is indeed the timing of growth by proliferation that has been the most frequent target of evolutionary change.

3 Control of Growth by Expansion

As mentioned above, after cell proliferation has ceased, cells in plant organs generally undergo a strong expansion that is often coupled with endoreduplication. Although during the proliferation phase cells double in volume between consecutive cell divisions and thus also undergo a limited amount of expansion, the following discussion will exclusively focus on the large-scale expansion after proliferation has arrested.

3.1 Promoting Factors

Compared to proliferation, our knowledge of the molecular control of growth by cell expansion is more limited. Cell expansion in developing organs is driven by water uptake into the central vacuole and involves a massive enlargement of the cell wall through biosynthesis and deposition of new wall material (Menand and Robaglia 2004). During this process, the cell wall must be made extensible in a finely tuned manner to allow it to yield to the turgor pressure while maintaining its integrity. Members of the expansin family of proteins are prime candidates for controlling cell wall extensibility and are presumed to act by breaking noncovalent bonds between cell wall components to allow them to slide relative to each other (Cosgrove 2005). Regulating expansin activity is therefore a plausible mechanism for determining the extent of organ growth through cell expansion. Indeed, downregulating members of the alpha class of expansins in *Arabidopsis*, rice and petunia leads to smaller organs because of reduced cell enlargement. By contrast, overexpression of these genes in *Arabidopsis* and rice is sufficient to increase organ size as a result of larger cells (Cho and Cosgrove 2000; Choi et al. 2003; Zenoni et al. 2004). Thus, expansin activity may indeed be limiting for organ growth by expansion.

Cell expansion in developing organs is promoted by the *ARGOS* homolog *ARGOS-LIKE* (*ARL*) (Hu et al. 2006). Reduced *ARL* expression leads to smaller organs with less expanded cells, while increased *ARL* activity is sufficient to cause organ enlargement due to larger cells. In contrast to *ARGOS* mediating auxin effects on growth, *ARL* appears to function downstream of brassinosteroids, and *ARL* overexpression can partially rescue the dwarfism of brassinosteroid-insensitive mutants. The contrasting effects of *ARGOS* and *ARL* on cell proliferation and expansion, respectively, are reminiscent of the different functions of members of the GRF-family (Horiguchi et al. 2005; Kim et al. 2003; Kim and Kende 2004) (see above). It is intriguing that members of homologous gene pairs or families are involved in the two seemingly very different cellular processes that drive overall organ growth. Unraveling the modes of action of ARGOS and ARL should indicate whether fundamentally similar mechanisms are involved in controlling growth by proliferation and by post-mitotic expansion, with the two genes potentially influencing the process of cellular expansion at different scales, i.e., the limited expansion during the proliferation phase and the large-scale expansion afterwards.

As mentioned before, cell expansion is often accompanied by an increase in ploidy due to endoreduplication. The identification of a topoisomerase VI complex (topo VI) in *Arabidopsis* suggests that endoreduplication is an essential prerequisite for growth by expansion (Sugimoto-Shirasu et al. 2002, 2005). Mutations in any of the presumed topo VI components block endoreduplication beyond a DNA content of 8C (i.e., eight times the DNA amount of the haploid genome) and lead to severe dwarfing, because cells fail to expand. Pharmacological studies with specific inhibitors of topo II and topo VI indicate that they fulfil distinct yet partly overlapping roles in decatenating chromosomes during mitotic cell cycles and endoreduplication (Sugimoto-Shirasu et al. 2005). However, despite the clear requirement for topo VI activity in permitting endoreduplication and cell expansion, it is unlikely that topo VI determines the final ploidy level and thus, due to the correlation of DNA content and cell size, the extent to which organ cells enlarge.

The final ploidy level of hypocotyl and leaf cells as well as the timing of the switch from proliferation to endoreduplication can be influenced by overexpressing or eliminating critical cell cycle regulators (for example De Veylder et al. 2001, 2002; del Pozo et al. 2006). These manipulations can affect final cell size and in some cases also overall organ size, as discussed in more detail by Magyar in this volume.

The phytohormone classes of brassinosteroids and gibberellins are known to promote cell expansion (Clouse and Sasse 1998; Richards et al. 2001; Szekeres et al. 1996). Both brassinosteroid- and gibberellin-insensitive mutants are dwarfed because of insufficient cell expansion, whereas constitutive gibberellin signaling leads to increased cell enlargement, particularly in the stem (Jacobsen and Olszewski 1993). For brassinosteroids, overproduction has been reported to increase overall plant growth, both by proliferation and expansion (Choe et al. 2001), suggesting that brassinosteroids can stimulate both modes of organ growth. The molecular mechanisms of brassinosteroid and gibberellin signaling have been elucidated in considerable detail, as described in recent reviews (Fleet and Sun 2005; Vert et al. 2005).

The extent of cell expansion depends to some degree on the total number of organ cells that have been generated during the preceding proliferation phase: In mutants with reduced cell numbers, such as *ant* or *an3*, organ cells enlarge more strongly than in wild-type, and thus partially compensate for the insufficient cell numbers (Horiguchi et al. 2006; Mizukami and Fischer 2000). This fascinating issue is discussed further by Tsukaya in this volume, whose chapter will also highlight the pathways that determine organ shape.

3.2 Repressing Factors

To date, only a few genes have been isolated that limit cell expansion during organ growth. *BIGPETAL* (*BPE*) encodes a putative transcriptional regulator of the basic helix-loop-helix class, and the *BPE* locus produces two distinct mRNAs through alternative splicing (Szecsi et al. 2006). One of these is present throughout the plant, whereas the other is found specifically in developing petals, and its accumulation is downstream of the known floral homeotic gene activities that specify petal identity. Downregulation of the petal-specific transcript leads to increased petal cell expansion and consequently to larger petals. How the petal-specific form of BPE protein limits cell expansion is currently unknown.

In addition, certain mutant alleles (*rotunda2-1* and *rotunda2-2*) of the presumed transcriptional co-repressor *LEUNIG* (*LUG*, see below) form larger leaves with a wider lamina because of increased cell expansion, indicating that *LUG* is required to prevent excess cell expansion late in leaf growth (Cnops et al. 2004).

4 How Does Organ Identity or Tissue Patterning Modify Growth?

Different plant organs have characteristic sizes and shapes, which are the outcomes of specific growth patterns. For example, cell proliferation in developing leaves of dicotyledonous plants stops first at the leaf tip and sequentially further towards the leaf base, whereas in petals cells continue to divide for the longest period in the central and distal regions that give rise to the petal blade (Dinneny et al. 2004; Disch et al. 2006; Donnelly et al. 1999). Ultimately, these different growth patterns are controlled by the transcriptional networks that determine organ identity (Krizek and Fletcher 2005), but how these modify growth control remains largely unknown. In principle, they could modulate the activity of common growth control pathways, as many of the factors discussed above affect both leaves and floral organs, albeit to different extents or even in opposite directions.

With respect to petals, some initial progress has been achieved in understanding the link between organ identity and growth control. One downstream target that is activated by the B-class homeotic proteins APETALA3 (AP3) and PISTILLATA (PI) in stamens and in petals at the beginning of cell expansion is *NAP*, a homolog of the petunia *NO APICAL MERISTEM* gene (Sablowski and Meyerowitz 1998). Overexpression of *NAP* leads to a failure of petal cells to enlarge properly, suggesting that temporally regulated *NAP* expression is important for normal petal cell expansion. Another downstream target of AP3 and PI, the petal-specific splice form of *BPE*, prevents excess cell enlargement in petals (Szecsi et al. 2006). Thus, the combined activities of *NAP* and the petal-specific form of *BPE* appear to mediate between organ identity factors and patterns of post-mitotic growth.

In contrast to the radially symmetric flowers of *Arabidopsis*, *Antirrhinum* forms zygomorphic (i.e., bilaterally symmetric) flowers with dorsal, lateral and ventral petals of distinct sizes and shapes (Schwarz-Sommer et al. 2003). The class II TCP proteins CYCLOIDEA (CYC) and DICHOTOMA (DICH) are required for dorsal petal identity in a partially redundant manner, and double mutants form radially symmetric flowers (Luo et al. 1996, 1999). Also, in *cyc* mutants, growth of the dorsal stamen, which normally does not develop, is derepressed, correlating with upregulated expression of cell cycle genes (Gaudin et al. 2000). Overexpression of CYC in *Arabidopsis* inhibited cell proliferation and expansion in leaves, but promoted late-stage cell expansion in petals (Costa et al. 2005). Thus, similar to CIN (see above) and in light of the findings on TCP20 (Li et al. 2005), these factors may have a rather direct influence on growth. CYC and DICH activate expression of the small MYB-domain protein RADIALIS in the dorsal region of the flower (Corley et al. 2005), but how this modifies growth patterns in developing organs remains unknown.

The widely accepted model of Waites and Hudson for lateral organ growth, which was derived from the analysis of *Antirrhinum* mutants in the myb-domain protein PHANTASTICA (PHAN) (Waites et al. 1998; see below), states that outgrowth of the lamina requires the juxtaposition of adaxial (towards the shoot, usually the upper side) and abaxial (away from the shoot, usually the lower side) identities (Bowman et al. 2002; Canales et al. 2005; Waites and Hudson 1995). Loss of either adaxial or abaxial identity causes the formation of radialized organs without a lamina, whereas ectopic boundaries between cells with ad- and abaxial identities can induce ectopic lamina outgrowth. How the ad/abaxial boundary induces lamina outgrowth is largely unknown. In addition, recent work has uncovered further complexity in the relation between polarity control and growth regulation, as a number of genes that were originally identified because of growth defects have now been shown to also interact with the pathway determining organ polarity. For example, the growth stimulator ANT was found to influence organ polarity, acting together with the abaxializing YABBY proteins FILAMENTOUS FLOWER (FIL) and

YABBY3 (YAB3) (Nole-Wilson and Krizek 2006). Triple mutants of *ant fil yab3* show very severe polarity defects that correlate with strongly reduced expression of the homeodomain zipper (HD-ZIP) protein PHABULOSA (PHB), normally expressed on the adaxial side. In addition, *JAG* was shown to pattern fruit tissues in cooperation with *FIL* and *YAB3*, and *NUB* is only expressed on the adaxial side of developing organs (Dinneny et al. 2005, 2006). Although the exact mechanistic links are far from clear, these intriguing findings may open the door to detailed studies of the connections between the control of growth and organ polarity.

The putative transcriptional co-repressors LUG and SEUSS (SEU), originally described as repressors of the floral homeotic gene *AGAMOUS* (*AG*) in outer flower whorls, also appear to function in the regulation of floral organ cell proliferation and ad/abaxial patterning, independently of their influence on organ identity (Franks et al. 2002, 2006; Sridhar et al. 2004). Petals of *ag lug seu* triple mutants are strongly reduced in size because of insufficient cell numbers and show defects in ad/abaxial patterning. These are correlated with reduced expression of major regulators of ad/abaxial patterning, PHB and FIL. Although it is not clear from this study whether LUG and SEU act independently on proliferation and on patterning or whether the growth defects result from the polarity defects, the changes in leaf shape and size without apparent polarity defects in the *rotunda2* alleles of *LUG* argue against the latter alternative (Cnops et al. 2004). Thus, ANT, LUG and SEU may both stimulate growth directly and also indirectly by maintaining a robust juxtaposition of the gene expression domains that determine organ polarity.

Patterning of different tissues or domains within organs often involves differential growth. An example is provided by the subdivision of *Arabidopsis* rosette leaves into a bladeless petiole and the distal lamina region. Outgrowth of lamina tissue from the leaf petiole is suppressed by the redundant *BLADE-ON-PETIOLE1* (*BOP1*) and *BOP2* genes, coding for BTB/POZ- and ankyrin-domain proteins that appear to regulate transcription by interacting with additional factors (Ha et al. 2003, 2004; Hepworth et al. 2005; Norberg et al. 2005). Similar to *bop* loss-of-function mutants, *JAG* overexpression causes lamina outgrowth from the petiole. Indeed, *bop* mutants show ectopic expression of *JAG* and its homolog *NUB* in the petiole region, suggesting that exclusion of these growth stimulators from the prospective petiole by *BOP* activity contributes to the petiole's distinct growth pattern.

Lamina outgrowth from the petiole is also suppressed by the *ASYMMETRIC LEAVES1* (*AS1*) and *AS2* genes. *AS1* encodes a myb-domain protein, while *AS2* codes for a protein with a leucin-zipper, which can interact with AS1 (Byrne et al. 2000; Iwakawa et al. 2002). Like mutants in the maize *AS1* ortholog *ROUGH SHEATH2* (*RS2*) and presumably in the *Antirrhinum* ortholog *PHANTASTICA* (*PHAN*), *Arabidopsis as1* and *as2* mutants show ectopic expression of class I *Knotted1*-like homeobox (KNOX) genes in developing leaves (reviewed in Kessler and Sinha 2004). This defect is also found in *bop1* mutants (Ha et al. 2003). Class I KNOX genes act to maintain meristem cells in an undifferentiated state, and their ectopic expression leads to lobed leaves because of prolonged cell proliferation (Scofield and Murray 2006). In addition to repressing KNOX genes in leaves, *AS1/PHAN/RS2* also contribute to ad/abaxial polarity establishment by promoting adaxial cell fate and thus stimulate lamina outgrowth, as postulated by the above model.

Thus, there are interconnected pathways of factors that determine regional and organ identity, e.g., in the case of *KNOX* gene repression by the *BOP* and *AS* genes, leading to different growth patterns that account for the characteristic organ sizes and shapes. Yet, as before, more work will be required to unravel the molecular mechanisms by which these regulators modulate cellular growth and division patterns.

5 Evolution of Plant Organ Size

Just as the factors that regulate organ identity ultimately modulate growth patterns, evolution has brought about an enormous range of different organ sizes and shapes. Over the last few years, we have begun to obtain first insights into how this has been achieved at the molecular level.

Studies of natural variation in organ size and shape and efforts at mapping quantitative trait loci (QTL) that govern this variation have identified numerous QTL for different aspects of leaf and floral organ dimensions (Jünger et al. 2000; Langlade et al. 2005; Perez-Perez et al. 2002). Surprisingly, comparative studies indicate that the QTL that affect leaves and those that influence floral organs are largely distinct, while the sizes and shapes of different floral organs (sepals and petals) are highly correlated and appear to be controlled by the same QTL (Frary et al. 2004; Jünger et al. 2005). Thus, evolution may have acted on distinct leaf and floral growth gene modules, or alternatively the presumed links between organ identity factors and growth control may have been targeted by evolutionary change.

The molecular basis for natural variation in organ size is probably best understood for tomato fruit size and the role of the *fw2.2* gene mentioned above. Allelic variation at this locus has been reported to account for up to 30% of the differences in fruit weight between different tomato cultivars (Frary et al. 2000). The *fw2.2* gene, which encodes a protein with structural similarity to the ras-oncoprotein, limits cell division in the pericarp of developing fruits. The large-fruit allele shows an early peak of *fw2.2* expression in fruit development and produces a lower overall amount of mRNA, whereas expression from the small-fruit allele peaks later and leads to a higher total expression level (Cong et al. 2002; Liu et al. 2003). Thus, a heterochronic change in the promoter of an important growth regulator that also influences overall mRNA abundance seems to underlie the difference in fruit size between certain tomato varieties. It will be interesting to see how widespread such regulatory mutations are in the evolution of organ size differences, as opposed to structural changes in the encoded proteins.

6 Higher-Level Control of Organ Size and Plant-Wide Integration

Tomato fruit size also serves to exemplify the control of plant organ growth by integrative mechanisms that operate on the level of a part of or the entire plant. It has long been known that pruning to reduce the total number of fruits borne by a plant causes an increase in the size of the remaining fruits, indicating that individual organs compete with each other for some limiting factor(s), such as possibly photoassimilates and nutrients. Recent work has shown that a reduction in fruit load leads to altered expression of key cell cycle regulators and of *fw2.2*, suggesting that resource availability may have a rather direct influence on growth and proliferation (Baldet et al. 2006). How this is sensed by individual cells and translated into the appropriate growth response is discussed by Doerner and Tsukaya in the present volume.

Changing not the number of developing sink organs but their sink strength can also influence their final size: apoplastic expression of a yeast invertase, which cleaves sucrose into glucose and fructose and is thought to influence sink strength, was found to increase tuber size in potato tubers (Sonnewald et al. 1997). However, the effect was compartment-specific, and targeting invertase to the cytosol actually had the opposite effect.

7 Concluding Remarks

As we have described in this chapter, there is a growing list of genes that have been shown to influence organ size with the potential for regulation. These factors can affect growth by proliferation and/or expansion, and both positive and negative regulators have been identified. Also, initial insights have been gained into how developmental processes (patterning, cell and organ identity) and evolutionary changes alter growth patterns to make organs larger or smaller. However, one of the most crucial questions for the control of organ size is yet to gain a satisfactory answer: what is it that growing organs are measuring? As discussed above, plants with altered organ sizes mostly show changes in the timing of growth. However, given a constant rate of growth, time and size or mass of the organ are strictly coupled. Thus, experimental manipulation of the rate of growth would be needed to distinguish between whether organs are measuring time or size/mass to decide for or against further growth. Understanding this and additional fundamental growth-related questions in plants can be expected to ultimately lead to novel tools for rationally manipulating the sizes of economically relevant plant organs.

Acknowledgements We thank members of the Lenhard lab for critical reading and helpful comments on the manuscript. Work in the authors' lab is supported by the Deutsche Forschungsgemeinschaft and by a BBSRC David Phillips Fellowship.

References

Autran D, Jonak C, Belcram K, Beemster GT, Kronenberger J, Grandjean O, Inze D, Traas J (2002) Cell numbers and leaf development in Arabidopsis: a functional analysis of the STRUWWELPETER gene. EMBO J 21:6036–6049

Baldet P, Hernould M, Laporte F, Mounet F, Just D, Mouras A, Chevalier C, Rothan C (2006) The expression of cell proliferation-related genes in early developing flowers is affected by a fruit load reduction in tomato plants. J Exp Bot 57:961–970

Bowman JL, Eshed Y, Baum SF (2002) Establishment of polarity in angiosperm lateral organs. Trends Genet 18:134–141

Byrne ME, Barley R, Curtis M, Arroyo JM, Dunham M, Hudson A, Martienssen RA (2000) Asymmetric leaves1 mediates leaf patterning and stem cell function in Arabidopsis. Nature 408:967–971

Canales C, Grigg S, Tsiantis M (2005) The formation and patterning of leaves: recent advances. Planta 221:752–756

Cho HT, Cosgrove DJ (2000) Altered expression of expansin modulates leaf growth and pedicel abscission in Arabidopsis thaliana. Proc Natl Acad Sci USA 97:9783–9788

Choe S, Fujioka S, Noguchi T, Takatsuto S, Yoshida S, Feldmann KA (2001) Overexpression of DWARF4 in the brassinosteroid biosynthetic pathway results in increased vegetative growth and seed yield in Arabidopsis. Plant J 26:573–582

Choi D, Lee Y, Cho HT, Kende H (2003) Regulation of expansin gene expression affects growth and development in transgenic rice plants. Plant Cell 15:1386–1398

Clouse SD, Sasse JM (1998) BRASSINOSTEROIDS: essential regulators of plant growth and development. Annu Rev Plant Physiol Plant Mol Biol 49:427–451

Cnops G, Jover-Gil S, Peters JL, Neyt P, De Block S, Robles P, Ponce MR, Gerats T, Micol JL, Van Lijsebettens M (2004) The rotunda2 mutants identify a role for the LEUNIG gene in vegetative leaf morphogenesis. J Exp Bot 55:1529–1539

Cockcroft CE, den Boer BG, Healy JM, Murray JA (2000) Cyclin D control of growth rate in plants. Nature 405:575–579

Cong B, Liu J, Tanksley SD (2002) Natural alleles at a tomato fruit size quantitative trait locus differ by heterochronic regulatory mutations. Proc Natl Acad Sci USA 99:13606–13611

Corley SB, Carpenter R, Copsey L, Coen E (2005) Floral asymmetry involves an interplay between TCP and MYB transcription factors in Antirrhinum. Proc Natl Acad Sci USA 102:5068–5073

Cosgrove DJ (2005) Growth of the plant cell wall. Nat Rev Mol Cell Biol 6:850–861

Costa MM, Fox S, Hanna AI, Baxter C, Coen E (2005) Evolution of regulatory interactions controlling floral asymmetry. Development 132:5093–5101

Crawford BC, Nath U, Carpenter R, Coen ES (2004) CINCINNATA controls both cell differentiation and growth in petal lobes and leaves of Antirrhinum. Plant Physiol 135:244–253

De Veylder L, Beeckman T, Beemster GT, de Almeida Engler J, Ormenese S, Maes S, Naudts M, Van Der Schueren E, Jacqmard A, Engler G et al. (2002) Control of proliferation, endoreduplication and differentiation by the Arabidopsis E2Fa-DPa transcription factor. EMBO J 21:1360–1368

De Veylder L, Beeckman T, Beemster GT, Krols L, Terras F, Landrieu I, van der Schueren E, Maes S, Naudts M, Inze D (2001) Functional analysis of cyclin-dependent kinase inhibitors of Arabidopsis. Plant Cell 13:1653–1668

del Pozo JC, Diaz-Trivino S, Cisneros N, Gutierrez C (2006) The balance between cell division and endoreplication depends on E2FC-DPB, transcription factors regulated by the ubiquitin-SCFSKP2A pathway in Arabidopsis. Plant Cell 18:2224–2235

Dewitte W, Riou-Khamlichi C, Scofield S, Healy JM, Jacqmard A, Kilby NJ, Murray JA (2003) Altered cell cycle distribution, hyperplasia, and inhibited differentiation in Arabidopsis caused by the D-type cyclin CYCD3. Plant Cell 15:79–92

Dinneny JR, Weigel D, Yanofsky MF (2005) A genetic framework for fruit patterning in *Arabidopsis thaliana*. Development 132:4687–4696

Dinneny JR, Weigel D, Yanofsky MF (2006) NUBBIN and JAGGED define stamen and carpel shape in Arabidopsis. Development 133:1645–1655

Dinneny JR, Yadegari R, Fischer RL, Yanofsky MF, Weigel D (2004) The role of JAGGED in shaping lateral organs. Development 131:1101–1110

Disch S, Anastasiou E, Sharma VK, Laux T, Fletcher JC, Lenhard M (2006) The E3 ubiquitin ligase BIG BROTHER controls Arabidopsis organ size in a dosage-dependent manner. Curr Biol 16:272–279

Doerner P, Jorgensen JE, You R, Steppuhn J, Lamb C (1996) Control of root growth and development by cyclin expression. Nature 380:520–523

Donnelly PM, Bonetta D, Tsukaya H, Dengler RE, Dengler NG (1999) Cell cycling and cell enlargement in developing leaves of Arabidopsis. Dev Biol 215:407–419

Elliott RC, Betzner AS, Huttner E, Oakes MP, Tucker WQ, Gerentes D, Perez P, Smyth DR (1996) AINTEGUMENTA, an APETALA2-like gene of Arabidopsis with pleiotropic roles in ovule development and floral organ growth. Plant Cell 8:155–168

Ellis CM, Nagpal P, Young JC, Hagen G, Guilfoyle TJ, Reed JW (2005) AUXIN RESPONSE FACTOR1 and AUXIN RESPONSE FACTOR2 regulate senescence and floral organ abscission in *Arabidopsis thaliana*. Development 132:4563–4574

Fleet CM, Sun TP (2005) A DELLAcate balance: the role of gibberellin in plant morphogenesis. Curr Opin Plant Biol 8:77–85

Franks RG, Liu Z, Fischer RL (2006) SEUSS and LEUNIG regulate cell proliferation, vascular development and organ polarity in Arabidopsis petals. Planta 224(4):801–811

Franks RG, Wang C, Levin JZ, Liu Z (2002) SEUSS, a member of a novel family of plant regulatory proteins, represses floral homeotic gene expression with LEUNIG. Development 129:253–263

Frary A, Fritz LA, Tanksley SD (2004) A comparative study of the genetic bases of natural variation in tomato leaf, sepal, and petal morphology. Theor Appl Genet 109:523–533

Frary A, Nesbitt TC, Grandillo S, Knaap E, Cong B, Liu J, Meller J, Elber R, Alpert KB, Tanksley SD (2000) fw2.2: a quantitative trait locus key to the evolution of tomato fruit size. Science 289:85–88

Gaudin V, Lunness PA, Fobert PR, Towers M, Riou-Khamlichi C, Murray JA, Coen E, Doonan JH (2000) The expression of D-cyclin genes defines distinct developmental zones in snapdragon apical meristems and is locally regulated by the Cycloidea gene. Plant Physiol 122:1137–1148

Ha CM, Jun JH, Nam HG, Fletcher JC (2004) BLADE-ON-PETIOLE1 encodes a BTB/POZ domain protein required for leaf morphogenesis in *Arabidopsis thaliana*. Plant Cell Physiol 45:1361–1370

Ha CM, Kim GT, Kim BC, Jun JH, Soh MS, Ueno Y, Machida Y, Tsukaya H, Nam HG (2003) The BLADE-ON-PETIOLE 1 gene controls leaf pattern formation through the modulation of meristematic activity in Arabidopsis. Development 130:161–172

Hepworth SR, Zhang Y, McKim S, Li X, Haughn GW (2005) BLADE-ON-PETIOLE-dependent signaling controls leaf and floral patterning in Arabidopsis. Plant Cell 17:1434–1448

Horiguchi G, Ferjani A, Fujikura U, Tsukaya H (2006) Coordination of cell proliferation and cell expansion in the control of leaf size in *Arabidopsis thaliana*. J Plant Res 119:37–42

Horiguchi G, Kim GT, Tsukaya H (2005) The transcription factor AtGRF5 and the transcription coactivator AN3 regulate cell proliferation in leaf primordia of *Arabidopsis thaliana*. Plant J 43:68–78

Horvath BM, Magyar Z, Zhang Y, Hamburger AW, Bako L, Visser RG, Bachem CW, Bogre L (2006) EBP1 regulates organ size through cell growth and proliferation in plants. EMBO J 25:4909–4920

Hu Y, Poh HM, Chua NH (2006) The Arabidopsis ARGOS-LIKE gene regulates cell expansion during organ growth. Plant J 47:1–9

Hu Y, Xie Q, Chua NH (2003) The Arabidopsis auxin-inducible gene ARGOS controls lateral organ size. Plant Cell 15:1951–1961

Ingram GC, Waites R (2006) Keeping it together: co-ordinating plant growth. Curr Opin Plant Biol 9:12–20

Iwakawa H, Ueno Y, Semiarti E, Onouchi H, Kojima S, Tsukaya H, Hasebe M, Soma T, Ikezaki M, Machida C et al. (2002) The ASYMMETRIC LEAVES2 gene of *Arabidopsis thaliana*, required for formation of a symmetric flat leaf lamina, encodes a member of a novel family of proteins characterized by cysteine repeats and a leucine zipper. Plant Cell Physiol 43:467–478

Jacobsen SE, Olszewski NE (1993) Mutations at the SPINDLY locus of Arabidopsis alter gibberellin signal transduction. Plant Cell 5:887–896

Jünger T, Perez-Perez JM, Bernal S, Micol JL (2005) Quantitative trait loci mapping of floral and leaf morphology traits in *Arabidopsis thaliana*: evidence for modular genetic architecture. Evol Dev 7:259–271

Jünger T, Purugganan MD, Mackay TF (2000) Quantitative trait loci for floral morphology in *Arabidopsis thaliana*. Genetics 156:1379–1392

Kessler S, Sinha N (2004) Shaping up: the genetic control of leaf shape. Curr Opin Plant Biol 7:65–72

Kim JH, Choi D, Kende H (2003) The AtGRF family of putative transcription factors is involved in leaf and cotyledon growth in Arabidopsis. Plant J 36:94–104

Kim JH, Kende H (2004) A transcriptional coactivator, AtGIF1, is involved in regulating leaf growth and morphology in Arabidopsis. Proc Natl Acad Sci USA 101:13374–13379

Klucher KM, Chow H, Reiser L, Fischer RL (1996) The AINTEGUMENTA gene of Arabidopsis required for ovule and female gametophyte development is related to the floral homeotic gene APETALA2. Plant Cell 8:137–153

Krizek BA (1999) Ectopic expression of AINTEGUMENTA in Arabidopsis plants results in increased growth of floral organs. Dev Genet 25:224–236

Krizek BA, Fletcher JC (2005) Molecular mechanisms of flower development: an armchair guide. Nat Rev Genet 6:688–698

Krizek BA, Sulli C (2006) Mapping sequences required for nuclear localization and the transcriptional activation function of the Arabidopsis protein AINTEGUMENTA. Planta 224:612–621

Langlade NB, Feng X, Dransfield T, Copsey L, Hanna AI, Thebaud C, Bangham A, Hudson A, Coen E (2005) Evolution through genetically controlled allometry space. Proc Natl Acad Sci USA 102:10221–10226

Li C, Potuschak T, Colon-Carmona A, Gutierrez RA, Doerner P (2005) Arabidopsis TCP20 links regulation of growth and cell division control pathways. Proc Natl Acad Sci USA 102:12978–12983

Li H, Johnson P, Stepanova A, Alonso JM, Ecker JR (2004) Convergence of signaling pathways in the control of differential cell growth in Arabidopsis. Dev Cell 7:193–204

Liu J, Cong B, Tanksley SD (2003) Generation and analysis of an artificial gene dosage series in tomato to study the mechanisms by which the cloned quantitative trait locus fw2.2 controls fruit size. Plant Physiol 132:292–299

Luo D, Carpenter R, Copsey L, Vincent C, Clark J, Coen E (1999) Control of organ asymmetry in flowers of Antirrhinum. Cell 99:367–376

Luo D, Carpenter R, Vincent C, Copsey L, Coen E (1996) Origin of floral asymmetry in Antirrhinum. Nature 383:794–799

Menand B, Robaglia C (2004) Plant cell growth. In: Hall MN (ed) Cell growth: control of cell size. Cold Spring Harbor Laboratory Press, New York, pp 625–637

Mizukami Y (2001) A matter of size: developmental control of organ size in plants. Curr Opin Plant Biol 4:533–539

Mizukami Y, Fischer RL (2000) Plant organ size control: AINTEGUMENTA regulates growth and cell numbers during organogenesis. Proc Natl Acad Sci USA 97:942–947

Nath U, Crawford BC, Carpenter R, Coen E (2003) Genetic control of surface curvature. Science 299:1404–1407

Nelissen H, Clarke JH, De Block M, De Block S, Vanderhaeghen R, Zielinski RE, Dyer T, Lust S, Inze D, Van Lijsebettens M (2003) DRL1, a homolog of the yeast TOT4/KTI12 protein, has a function in meristem activity and organ growth in plants. Plant Cell 15:639–654

Nole-Wilson S, Krizek BA (2006) AINTEGUMENTA contributes to organ polarity and regulates growth of lateral organs in combination with YABBY genes. Plant Physiol 141:977–987

Norberg M, Holmlund M, Nilsson O (2005) The BLADE ON PETIOLE genes act redundantly to control the growth and development of lateral organs. Development 132:2203–2213

Ohno CK, Reddy GV, Heisler MG, Meyerowitz EM (2004) The Arabidopsis JAGGED gene encodes a zinc finger protein that promotes leaf tissue development. Development 131:1111–1122

Okushima Y, Mitina I, Quach HL, Theologis A (2005) AUXIN RESPONSE FACTOR 2 (ARF2): a pleiotropic developmental regulator. Plant J 43:29–46

Palatnik JF, Allen E, Wu X, Schommer C, Schwab R, Carrington JC, Weigel D (2003) Control of leaf morphogenesis by microRNAs. Nature 425:257–263

Pendle AF, Clark GP, Boon R, Lewandowska D, Lam YW, Andersen J, Mann M, Lamond AI, Brown JW, Shaw PJ (2005) Proteomic analysis of the Arabidopsis nucleolus suggests novel nucleolar functions. Mol Biol Cell 16:260–269

Perez-Perez JM, Serrano-Cartagena J, Micol JL (2002) Genetic analysis of natural variations in the architecture of *Arabidopsis thaliana* vegetative leaves. Genetics 162:893–915

Richards DE, King KE, Ait-Ali T, Harberd NP (2001) HOW GIBBERELLIN REGULATES PLANT GROWTH AND DEVELOPMENT: a molecular genetic analysis of gibberellin signaling. Annu Rev Plant Physiol Plant Mol Biol 52:67–88

Riefler M, Novak O, Strnad M, Schmulling T (2006) Arabidopsis cytokinin receptor mutants reveal functions in shoot growth, leaf senescence, seed size, germination, root development, and cytokinin metabolism. Plant Cell 18:40–54

Rudra D, Warner JR (2004) What better measure than ribosome synthesis? Genes Dev 18:2431–2436

Sablowski RW, Meyerowitz EM (1998) A homolog of NO APICAL MERISTEM is an immediate target of the floral homeotic genes APETALA3/PISTILLATA. Cell 92:93–103

Schruff MC, Spielman M, Tiwari S, Adams S, Fenby N, Scott RJ (2006) The AUXIN RESPONSE FACTOR 2 gene of Arabidopsis links auxin signalling, cell division, and the size of seeds and other organs. Development 133:251–261

Schwarz-Sommer Z, Davies B, Hudson A (2003) An everlasting pioneer: the story of Antirrhinum research. Nat Rev Genet 4:657–666

Scofield S, Murray JA (2006) KNOX gene function in plant stem cell niches. Plant Mol Biol 60:929–946

Shpak ED, Berthiaume CT, Hill EJ, Torii KU (2004) Synergistic interaction of three ERECTA-family receptor-like kinases controls Arabidopsis organ growth and flower development by promoting cell proliferation. Development 131:1491–1501

Shpak ED, McAbee JM, Pillitteri LJ, Torii KU (2005) Stomatal patterning and differentiation by synergistic interactions of receptor kinases. Science 309:290–293

Shuai B, Reynaga-Pena CG, Springer PS (2002) The lateral organ boundaries gene defines a novel, plant-specific gene family. Plant Physiol 129:747–761

Sonnewald U, Hajirezaei MR, Kossmann J, Heyer A, Trethewey RN, Willmitzer L (1997) Increased potato tuber size resulting from apoplastic expression of a yeast invertase. Nat Biotechnol 15:794–797

Sridhar VV, Surendrarao A, Gonzalez D, Conlan RS, Liu Z (2004) Transcriptional repression of target genes by LEUNIG and SEUSS, two interacting regulatory proteins for Arabidopsis flower development. Proc Natl Acad Sci USA 101:11494–11499

Sugimoto-Shirasu K, Roberts GR, Stacey NJ, McCann MC, Maxwell A, Roberts K (2005) RHL1 is an essential component of the plant DNA topoisomerase VI complex and is required for ploidy-dependent cell growth. Proc Natl Acad Sci USA 102:18736–18741

Sugimoto-Shirasu K, Roberts K (2003) Big it up: endoreduplication and cell-size control in plants. Curr Opin Plant Biol 6:544–553

Sugimoto-Shirasu K, Stacey NJ, Corsar J, Roberts K, McCann MC (2002) DNA topoisomerase VI is essential for endoreduplication in Arabidopsis. Curr Biol 12:1782–1786

Szecsi J, Joly C, Bordji K, Varaud E, Cock JM, Dumas C, Bendahmane M (2006) BIG-PETALp, a bHLH transcription factor is involved in the control of Arabidopsis petal size. EMBO J 25:3912–3920

Szekeres M, Nemeth K, Koncz-Kalman Z, Mathur J, Kauschmann A, Altmann T, Redei GP, Nagy F, Schell J, Koncz C (1996) Brassinosteroids rescue the deficiency of CYP90, a cytochrome P450, controlling cell elongation and de-etiolation in Arabidopsis. Cell 85:171–182

Tiwari SB, Hagen G, Guilfoyle T (2003) The roles of auxin response factor domains in auxin-responsive transcription. Plant Cell 15:533–543

Van Lijsebettens M, Vanderhaeghen R, De Block M, Bauw G, Villarroel R, Van Montagu M (1994) An S18 ribosomal protein gene copy at the Arabidopsis PFL locus affects plant development by its specific expression in meristems. EMBO J 13:3378–3388

Veit B (2006) Stem cell signalling networks in plants. Plant Mol Biol 60:793–810

Vert G, Nemhauser JL, Geldner N, Hong F, Chory J (2005) Molecular mechanisms of steroid hormone signaling in plants. Annu Rev Cell Dev Biol 21:177–201

Waites R, Hudson A (1995) *Phantastica*: a gene required for dorsoventrality of leaves in *Antirrhinum majus*. Development 121:2143–2154

Waites R, Selvadurai HR, Oliver IR, Hudson A (1998) The PHANTASTICA gene encodes a MYB transcription factor involved in growth and dorsoventrality of lateral organs in Antirrhinum. Cell 93:779–789

Weijers D, Franke-van Dijk M, Vencken RJ, Quint A, Hooykaas P, Offringa R (2001) An Arabidopsis Minute-like phenotype caused by a semi-dominant mutation in a RIBOSOMAL PROTEIN S5 gene. Development 128:4289–4299

Werner T, Motyka V, Laucou V, Smets R, Van Onckelen H, Schmulling T (2003) Cytokinin-deficient transgenic Arabidopsis plants show multiple developmental alterations indicating opposite functions of cytokinins in the regulation of shoot and root meristem activity. Plant Cell 15:2532–2550

Werner T, Motyka V, Strnad M, Schmülling T (2001) Regulation of plant growth by cytokinin. Proc Natl Acad Sci USA 98:10487–10492

White DW (2006) PEAPOD regulates lamina size and curvature in Arabidopsis. Proc Natl Acad Sci USA 103:13238–13243

Zenoni S, Reale L, Tornielli GB, Lanfaloni L, Porceddu A, Ferrarini A, Moretti C, Zamboni A, Speghini A, Ferranti F et al. (2004) Downregulation of the *Petunia hybrida* alpha-expansin gene PhEXP1 reduces the amount of crystalline cellulose in cell walls and leads to phenotypic changes in petal limbs. Plant Cell 16:295–308

Plant Cell Monogr (10)
L. Bögre and G. Beemster: Plant Growth Signaling
DOI 10.1007/7089_2007_148/Published online: 29 January 2008

Control of Leaf Morphogenesis by Long- and Short-Distance Signaling: Differentiation of Leaves Into Sun or Shade Types and Compensated Cell Enlargement

Ali Ferjani[1] · Satoshi Yano[2] · Gorou Horiguchi[1] · Hirokazu Tsukaya[1,2] (✉)

[1]Graduate School of Science, University of Tokyo, 113-0033 Tokyo, Japan
tsukaya@biol.s.u-tokyo.ac.jp

[2]National Institute for Basic Biology/Okazaki Institute for Integrated Bioscience, Okazaki, 444-8585 Aichi, Japan

Abstract The flattened, thin lamina of leaves captures sunlight for photosynthesis and facilitates gas exchange. Therefore, the size and shape of a leaf are fundamentally important features of its integrity and function. Progress in developmental studies has suggested that long- and short-distance signaling pathways are involved in leaf formation. In this chapter, we introduce these signaling pathways, both of which can control final leaf shape and structure: a long-distance signaling pathway that governs the differentiation of leaves into sun and shade types, and a short-distance signaling pathway that appears to be involved in an organ-wide system that integrates cell proliferation and cell enlargement. Although none of the molecules involved in these two pathways have been identified, plausible mechanisms of these pathways are discussed based on present data.

1 Introduction

Leaves are initiated by coordinated changes in the polarity of cell division and the growth of a group of founder cells in the peripheral zone of the shoot apical meristem (SAM) (Esau 1977). After leaf initiation, the lamina is formed by the action of the marginal meristem, which is transiently activated at the leaf margin, followed by the action of the plate meristem, both of which contribute to the flattening and expanding of the developing leaf (Avery 1933; Maksymowych 1963; Donnelly et al. 1999). Recent advances in our understanding of the mechanisms of leaf morphogenesis (Tsukaya 2006; Tsukaya and Beemster 2006) have revealed several key regulatory steps. Among these steps, we have recently found two interesting phenomena that might be regulated by long- or short-distance signaling pathways. Here we present an overview of these phenomena.

Although the size, structure, and shape of leaves of a given species are under robust genetic control, these parameters exhibit a certain degree of flexibility, allowing leaves to tailor their growth based on environmental conditions such as light, water, and nutrient availability (Smith and Hake 1992; Kim et al. 2005; Tsukaya 2006). Light is one of the most important environmental factors for leaves. Irradiance is captured by the photosynthetic organelles, chloroplasts, and the captured energy is converted into photosynthates via a photochemical reaction and the reductive pentose phosphate cycle (Nobel 2005). Because all plant activities require and consume photosynthates, it is crucial that plants produce them in sufficient quantities. Plants can survive under various light conditions by modulating leaf development according to the light environment. Under high-light conditions, leaves thicken and contain large amounts of photosynthetic components per unit leaf area, such as Rubisco and photosystem reaction centers, resulting in high photosynthetic production. High-light-acclimated leaves are called "sun leaves", as compared to "shade leaves", which are formed under low-light conditions. Shade leaves differ from sun leaves in their morphology, which is thinner and broader, and in their physiology, with lower photosynthetic and dark respiration rates, compared to sun leaves. Physiological and ecological studies have revealed details of the differences between these leaf types (for reviews, see Boardman 1977; Björkman 1981; Anderson 1986; Terashima et al. 2001). In addition, the developmental aspects of sun and shade leaves are interesting. Recent studies of the development of sun and shade leaves have revealed that the differentiation of new leaf primordia into sun or shade leaves is controlled remotely by mature leaves. In other words, it is controlled by long-distance signaling. In the first half of this review, we present an overview of how leaf type is systematically controlled by long-distance signaling.

In addition to this long-distance pathway, leaf shape and size also appear to be governed locally in each primordium by a shorter-distance signaling pathway. It is noteworthy that cell proliferation and post-mitotic cell expansion occur simultaneously in separate regions of the same developing leaf (Donnelly et al. 1999; White 2006). Finally, both a precise programmed exit from the mitotic cell cycle and the cessation of post-mitotic cell expansion determine leaf size (White 2006). How are these two processes integrated in leaves? The gross size and shape of leaves are not always the simple sum of the behavior of individual cells (Tsukaya 2002, 2003). In fact, a decrease in the cell number due to mutations or genetic manipulations that decrease cell proliferation activity is often associated with an increase in cell size. We named this phenomenon "compensation" (Tsukaya 2002, 2003). A relationship between decreased cell number and increased cell size has been reported in many transgenic plants and loss-of-function mutants of Arabidopsis. For example, a loss-of-function mutation in the *AN3/GRF-INTERACTING FACTOR1* (*GIF1*) gene (Kim and Kende 2004), which positively regulates cell proliferation in leaf primordia, causes the typical compensation syndrome (Horiguchi et al. 2005). Similarly, several other mutations that affect cell proliferation have been reported to cause compensation, including *aintegumenta* (*ant*), *struwwelpeter* (*swp*), *swellmap*, *G protein a-subunit 1*, and *deformed roots and leaves1* (Mizukami and Fischer 2000; Ullah et al. 2001; Autran et al.

2002; Nelissen et al. 2003; Clay and Nelson 2005). Impaired cell proliferation caused by the reduced activity of cyclin-dependent kinases (CDKs) resulting from the overexpression of either a dominant-negative version of a CDK or a cyclin-dependent kinase inhibitor known as ICK/KRP in tobacco (*Nicotiana tabacum*), Arabidopsis, and rice also induces compensation (Hemerly et al. 1995; Wang et al. 2000; De Veylder et al. 2001; Boudolf et al. 2004; Barrôco et al. 2006). Therefore, compensation is a universal phenomenon in monocot and eudicot species.

In our opinion, this relationship indicates the existence of a short-range signaling within the leaf primordium. In the second half of this review, we will discuss possible mechanisms of leaf-size regulation.

2 Differentiation of Leaves Into Sun and Shade Types

2.1 The Differentiation of Leaves Into Sun and Shade Types Depends on the PPFD

As briefly summarized in the introduction, the differentiation of leaves into sun and shade types is regulated remotely by mature leaves via long-distance signaling. In this section, we focus on the morphology of sun and shade leaves and their acclimation to the environment. Then, we introduce several important findings that indicate unique regulatory mechanism(s) that underlie the formation of sun and shade leaves.

One of the most significant anatomical differences between sun and shade leaves is the thickness of the lamina. Sun leaves are at least 1.5- to 3-fold thicker than shade leaves (Björkman 1981). Sun leaves show increased elongation and/or additional cell layers in the mesophyll, especially in the palisade tissue. The shape of cells in the palisade tissue is cylindrical in sun leaves, but sometimes funnelled or rounded in shade leaves (Haberlandt 1914). The height of the epidermal cells is not significantly different in these leaf types, but the contribution of the epidermis to the leaf thickness is lower in sun leaves because their lamina are thicker (Chabot and Chabot 1977; Dengler 1980; Yano and Terashima 2004). The cuticle of sun leaves is more abundant than that of shade leaves, preventing evaporation under high irradiance (Osborn and Taylor 1990; Ashton and Berlyn 1994). Stomata are also affected by light, showing increased numbers and density (see reviews by Tichá 1982 and Givnish 1988). In addition to light, the stomatal density and/or the stomatal index, which is the proportion of stomata among the epidermal cells, are affected by CO_2 (Woodward 1987; Woodward et al. 2002). These structural differences, which affect photosynthetic performance as determined by the conductivity of CO_2 and optics (Richter and Fukshansky 1998; Terashima et al. 2001), are the result of leaf developmental regulation.

For plants to establish leaves that are adapted to the light environment, the first step is the recognition of the light environment. This raises questions as to whether the light quality or quantity affect the differentiation of sun and shade leaves, which light sensory system is involved, and where does the plant sense light.

There is no direct evidence regarding the issue of whether light quality or quantity is more important in controlling the development of sun and shade leaves. However, several lines of physiological and genetic evidence suggest that light quantity is a major stimulus. White clover (*Trifolium repens* L.) grown under the same photosynthetically-active photon flux density (PPFD) but with various ratios of red and far-red light, which governs the balance between P_{fr} (active) and P_r (inactive) phytochromes, showed the same leaf thickness index (leaf mass per unit area in $mg\,mm^{-2}$, specific leaf area in $mm^2\,mg^{-1}$ leaf DM) but different petiole elongation (Smith 1994; Heraut-Bron et al. 1999, 2001). The idea that light quantity is more important in the stimulation was supported by a genetic analysis. Loss-of-function mutant of *PHYB*, which encodes the main type-II phytochrome in Arabidopsis, also showed leaf thickening with an increase in the light intensity (Kim et al. 2005). On blue light receptors it was revealed that single and double mutants of *CRYTOCHROMES* (*cry1*) and *cry2* and the *PHOTOTROPINS* (*phot1*) mutant showed the same leaf thickening as much as WT plants did (Weston et al. 2000; Lopéz-Juez et al. 2007). Thus, known photoreceptors are probably not involved in the differentiation of sun and shade leaves, although no results have been reported from double mutants of phototropins or disruptants of multiple phytochromes.

If light quantity is the triggering stimulus, does the plant monitor the instantaneous light intensity or the total amount of daily light? Chabot et al. (1979) compared the leaf morphology and photosynthesis efficiency of *Fragaria virginiana* leaves exposed to different light regimes, changing the light intensity and span (maintaining the photoperiod using the background light) but with plants receiving the same quantity of light. Although the instantaneous light intensities among the regimes differed by five-fold, the leaf thickness was not affected. However, the leaf thickness increased with increasing light span while the light intensity was held constant. Thus, the daily light amount regulates differentiation into sun or shade leaves.

2.2 An Unknown Long-Distance Signaling Mechanism Governs Differentiation Into Sun or Shade Leaves

We will now introduce recent studies that indicate the parts of the plant that recognize light and control the differentiation of new leaves, and discuss candidates for light recognition. Lake et al. (2001) showed that the stomatal index of developing Arabidopsis leaves decreased when mature leaves were shaded.

The authors also showed that the stomatal density and index decreased when mature leaves were exposed to CO_2-enriched air, and vice versa. These results indicate that stomatal development in a developing leaf is regulated by long-distance signaling from mature leaves, with leaves sensing the ambient CO_2 concentration and light intensity. Yano and Terashima (2001) reported that in *Chenopodium album*, the leaf thickness, cell-layer number, and cross-sectional area of cells in the palisade tissue increased when mature leaves were exposed to high light, and decreased when mature leaves were exposed to low light. In contrast to the leaf anatomy, chloroplast acclimation was independent of these signals. Therefore, these studies showed that systemic signaling controls leaf development (Fig. 1), although it is unclear whether the information on CO_2 and light are transduced by the same signal. Systemic regulation of leaf development has also been reported in tobacco (Thomas et al. 2004) and poplar (*Populus trichocarpa* × *P. deltoides*, Miyazawa et al. 2006). In tobacco, the leaf area, size, and degree of undulation of epidermal cells are affected by the light environment of mature leaves, as well as the index and density of the stomata. In addition to the systemic signaling, leaf

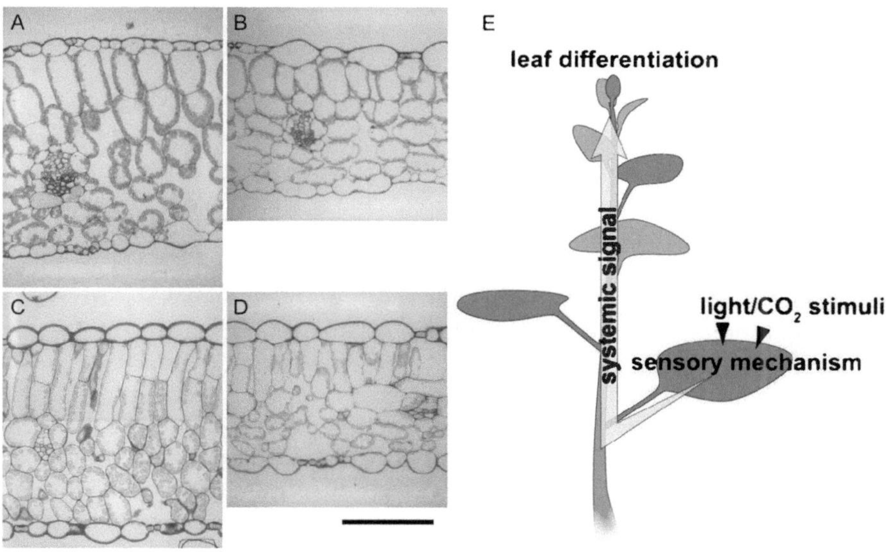

Fig. 1 Anatomy of sun and shade leaves and systemic signals involved in their development. Sun and shade leaves of *Arabidopsis thaliana* and *Chenopodium album* (**A** and **B**, **C** and **D**, respectively). Sun leaves of these species have an additional cell layer in the palisade tissue as compared to shade leaves. Sun and shade leaves were grown under PPFD levels of 360–400 or 60–80 μmol m^{-2} s^{-1}, respectively. The bar represents 100 μm. Information on the light and CO_2 stimuli recognized by mature leaves is transferred to new leaves via systemic signal(s) and determines the differentiation fate of new leaves into sun or shade leaves (**E**)

developmental plasticity is maintained for at least several days, since all of these studies analyzed developing leaves, which were already of visible size at the beginning of the treatments.

Why did plants develop systemic signaling? Let us consider a leaf primordium developing on the SAM. The primordium is insulated from severe environments such as xeric air and strong light. Therefore, the primordium itself cannot sense the light or CO_2 environments. However, mature leaves are exposed to the actual environment, allowing them to sense reliable information. Therefore, plants gather information from mature leaves rather than from the primordium itself to govern the formation of appropriate leaves (Woodward et al. 2002).

As mentioned above, light quantity is more important than light quality in the formation of sun or shade leaves. The light sensory mechanism and the systemic signal are still hypothetical, but should fulfill the following requirements:

1. Have the ability to monitor direct or indirect light and/or CO_2 concentration;
2. Be able to convert environmental information into mobile substance(s) that can be transferred from mature to developing leaves;
3. Be able to induce specific gene expression that regulates cell division and/or cell growth.

Several candidate signals have been considered (Kim et al. 2005; Coupe et al. 2006), including phytohormones, peptides, RNAs (as signals), sugars, and redoxes (as sensors and/or signals). Phytohormones, peptides, and RNAs as signals would require a sensing system, but we are unable to nominate appropriate candidates to date. As mentioned earlier, mutants in major phytochromes and cryptochromes and the *phot1* mutant show the laminar thickening that is observed in sun leaves. In addition, plant behaviors driven by known photoreceptors differ from sun- and shade-leaf formation in their reaction times. Although the involvement of a photoreceptor cannot be ruled out, the probability that known photoreceptors are involved is low. In contrast, sugars and redoxes are simpler explanations than the above candidates. Since sugars are the products of photosynthesis and the PPFD is a limiting factor in natural environments, plants can indirectly monitor the light environment by monitoring the sugar concentration. In addition, sugars are also transferred from a source (such as a photosynthesizing leaf) to a sink (such as the SAM or leaf primordia). Sugar sensing by hexokinase is well organized in plants (Cho et al. 2006). In addition to these features, sugar regulates gene expression (Koch 2000, 2004; Hanson et al. 2001). Thus, sugar is a likely candidate. Redoxes also fulfill the requirements described above. In high- or low-light conditions, the plastoquinone pool in thylakoid membranes is reduced or oxidized, respectively; hence, the redox state represents the light environment. The redox state controls the transcription levels of photosynthetic genes (Escoubas et al. 1995; Pfannschmidt et al. 1999). Karpinski et al. (1999) reported long-distance signaling mediated by a redox system in Arabidopsis. The authors showed that when several parts of rosette leaves were exposed to excess light and their redox state was reduced, the transcription level of *pAPX2::LUC*, a gene composed of the promoter of an antioxidant defense gene and the luciferase coding region, increased in the remaining parts, especially in young leaves around the SAM. The expression level in the remaining parts did not increase when the plastoquinone pool was held in an oxidized state using DCMU, an inhibitor of electron transport from PSII to the plastoquinone pool. At present, the many candidates listed above should be verified. Further physiological and developmental analyses as well as global gene expression profiling, which have been considered by Coupe et al. (2006), should explain these interesting phenomena.

3 Organ-Wide Control of Leaf Size

3.1 Compensation as a Clue to Understanding the Mechanisms of Leaf-Size Control

In this section, we present an overview of our present understanding of the leaf-size control system based on information obtained from studies of Arabidopsis mutants (Tsukaya 2006). Before discussing these topics, let us first re-examine classic and new ideas of organ-size regulation, since this re-examination has led us to studies of a yet unknown, organ-wide control system of leaf size.

To date, two theories, the cell theory and the organismal theory, have been postulated for the mechanisms by which genes control development in plants (Kaplan and Hagemann 1991; Sitte 1992; Hagemann 1992; Kaplan 1992; Tsukaya 2002, 2003). The cell theory posits that the cell is the unit of morphogenesis in leaves. According to this theory, the final leaf size is determined by the sum of behaviors of individual cells, which are under cell-autonomous control. In contrast, the organismal theory proposes that the unit of morphogenesis is the organ, and pre-determined organ space is filled by an increase in cell mass that is achieved through either cell proliferation or cell expansion. Thus, the organismal theory predicts a genetic mechanism that determines leaf shape and size independent of the behaviors of individual cells (Kaplan and Hageman 1991; Sitte 1992; Tsukaya 2002, 2003). In re-examining the validity of these theories by studying leaf morphology mutants, we found that the cell theory is more likely (Tsukaya 2002, 2003). For example, in the *angustifolia* (*an*) mutant, a narrow cell shape due to the inhibition of polar cell expansion directly results in the narrow leaf phenotype (Tsuge et al. 1996;

Kim et al. 2002). In addition, a number of mutants show reduced leaf area due to a specific decrease in the number of cells (Horiguchi et al. 2006a,b). These examples show that changes in the behaviors of cells directly influence leaf shape.

However, leaf size and shape are not always a simple sum of the behavior of individual cells (Tsukaya 2002, 2003), as mentioned in the introduction. This is clear in mutants and transgenic lines that exhibit compensation. Since compensation appears to function to replace the leaf area that is lost due to impaired cell proliferation, one would expect that this phenomenon would support the organismal theory. Organismal theory predicts the existence of a system that maintains a constant leaf size during developmental fluctuations. However, such a regulatory mechanism is highly unlikely since a number of mutants that show decreased cell numbers show no increase in cell size. Likewise, an increase in the cell number caused by the overexpression of *AN3* or *ANT* results in a corresponding increase in the leaf area (Mizukami and Fischer 2000; Horiguchi et al. 2005; reviewed in Tsukaya 2002, 2003). In addition, the leaf area in mutants exhibiting compensation is usually smaller than that in wild-type plants.

In contrast, several possibilities could explain the occurrence of compensation based on the cell theory, with slight modification (Tsukaya 2002, 2003). Namely, compensation can be explained by the "neo cell theory", in which the cell is considered the unit of morphogenesis and the pathway(s) of cell proliferation and cell enlargement are integrated in some way at the organ level (Tsukaya 2002, 2003). In this context, compensation can be regarded as a reflection of certain mechanisms that integrate cell proliferation and cell enlargement. We examine below several possible mechanisms of compensation (or the integration mechanisms) based on available genetic and molecular data. Next, we interpret compensation as the result of a short-range signaling mechanism that operates during leaf morphogenesis.

3.2
Is the Uncoupling of Cell Division from Cell Growth Responsible for Compensation?

In yeast and metazoans, the acceleration or blocking of the progression of the cell cycle by experimental manipulation results in a decrease or increase in cell size, respectively. This uncoupling of cell division from cell growth occurs because growth is not dependent upon cell cycling (Johnston et al. 1977; Jorgensen and Tyers 2004). Does compensation in plants function in a similar manner? The answer is no. First, in a *KRP2* overexpresser (*KRP2* OE), where the mature leaves of which clearly show compensation, proliferating cells are already larger than wild-type cells at the earliest stage of leaf development (De Veylder et al. 2001). This observation appears to suggest, at a glance, that cell division is uncoupled from cell growth, but this is untrue. If the uncoupling of cell division from cell growth occurs, cell size should not be kept, but the cell size is maintained in a certain size also in the *KRP2* OE (Ferjani et al. 2007). Therefore, uncoupling between the cell division and cell growth is not seen in the compensation.

Moreover, one must be cautious of interpreting the contribution of this increase in cytoplasmic volume in early development to the final cell size, since most of the volume of the plant cell is occupied by large vacuoles that develop post-mitotically. In a model that simulates cell growth and cell division in the context of leaf development, the in silico inhibition of cell cycle progression mimics the effects of *KRP2* overexpression, increasing cell size during cell cycling. However, this size difference accounts for only a small proportion of the cell size at the end of post-mitotic cell expansion (Beemster et al. 2006). In fact, a transient increase in cell size due to a delay in cell cycling was observed only in the mitotic phase in triple mutants of cytokinin receptors (Nishimura et al. 2004). In this case, the increase in cell size dissipated in the post-mitotic phase of cell expansion. Moreover, our time-course analyses of cell expansion in compensation-exhibiting mutants revealed no abnormal enhancement of cell enlargement during the cell proliferation stage in most of the mutants examined, but abnormal enhancement did occur, just after entry into the post-mitotic stage (Ferjani et al. 2007). Thus, the uncoupling of cell growth from cell division does not account for the enhanced cell expansion during compensation.

3.3 One-Gene-Two-Functions Model

If a gene has a positive role in cell proliferation and a negative role in post-mitotic cell expansion, the loss of function and overexpression of this gene should produce phenotypes identical to, and opposite, those in compensation-exhibiting mutants, respectively. If this is the case, systems that integrate cell proliferation and cell enlargement are not needed to interpret compensation. However, our present knowledge of the nature of responsive genes in compensation-exhibiting mutants eliminates the possibility of the one-gene-two-functions model. Namely, these genes, such as *AN3* and *ANT*, seem to be involved exclusively in the control of cell proliferation. For example, the overexpression of *AN3* and *ANT* promotes cell proliferation but has no influence on cell expansion (Mizukami and Fischer 2000; Horiguchi et al. 2005). In addition, *AN3*, *ANT*, *FASCIATA1* (*FAS1*), and *SWELLMAP* (*SMP*), which induce compensation when mutated, are expressed strongly in meristematic tissues but at an undetectable level in differentiating tissues (Elliott et al. 1996; Kaya et al. 2001; Autran et al. 2002; Clay and Nelson 2005; Horiguchi et al. 2005; Exner et al. 2006). Thus, the one-gene-two-function model does not hold for at least several instances of compensation.

3.4 Shift in Switching Between Cell Proliferation and Differentiation

The relationship between cell proliferation and post-mitotic cell expansion observed in compensation could be regarded as a simple trade-off between these two developmental processes in leaf primordia. However, such a mechanism is not supported by the time-course data of leaf development in several mutants. For example, the overexpression of *ANT* or loss-of-function of *PEAPOD* (*PPD*) lengthens the cell proliferation phase but does not shorten the duration of post-mitotic cell expansion (Mizukami and Fischer 2000; White 2006). Therefore, compensation cannot be regarded as a simple trade-off between these processes.

However, changes in the timing of the transition from the normal cell cycle to endocycles might be another factor that causes compensation, because the endocycle is a modified cell cycle that lacks a cell division step, and there is a rough positive correlation between endopolyploidy levels and cell size (e.g., Melaragno et al. 1993). Although this correlation seems to exist in some compensation-exhibiting mutants and transgenics (such as *struwwelpeter* (*swp*) and weak lines of *KRP2* overexpressers; Autran et al. 2002; Verkest et al. 2005), however, several lines of evidence have indicated that an increase in the levels of endopolyploidy is not a necessary condition for increased cell size due to compensation. For example, a strong *KRP2* overexpressing line, in which both cell cycling and endocycling are inhibited, clearly shows compensation (De Veylder et al. 2001). In addition, guard cells, which always have a 2C DNA content, also exhibit compensation in transgenic tobacco lines that have reduced expression levels of ribosomal protein L3 genes, and in the Arabidopsis *swp* mutant (Autran et al. 2002; Popescu and Tumer 2004).

Moreover, we recently found more genetic evidence that an increase in endopolyploidy is not an essential condition for the increase in cell size that occurs in compensation. This evidence was found during the isolation and analysis of the *extra-small sisters* (*xs*) mutants, which are defective in cell expansion but have normal numbers of cells in leaves (Fujikura et al. 2007). Among the *xs* mutants, *xs1*, *xs2*, *xs4*, and *xs5* showed suppressed compensation triggered by the *an3* mutation. Importantly, these *xs* mutants differ in the levels of endocycling in leaves, showing normal, increased, or decreased ploidy. This lack of correlation between the level of endocycling and the magnitude of compensation strongly suggests that the level of endopolyploidy and the occurrence of compensation are independent (Fujikura et al. 2007).

3.5 Signaling Systems Linking Cell Proliferation and Cell Expansion

After eliminating the above hypotheses for the mechanisms of compensation and the mechanisms that integrate cell proliferation and cell enlargement, the last possibility is that cell proliferation regulates post-mitotic cell expansion through some unknown linking mechanisms. Such a signal is thought to be produced in mitotic cells and to regulate cell expansion in post-mitotic cells. In this model, four modes of signaling action are possible, as shown in Fig. 2. The hypothetical signal might have a positive (Fig. 2A,C) or negative (Fig. 2B,D) effect on post-mitotic cell expansion and might work either cell-autonomously (Fig. 2A,B) or non-cell-autonomously (Fig. 2C,D).

In all of these models, the strength of the hypothetical signal(s) should correlate with the size of mature cells. If the signal(s) is negative, it should be

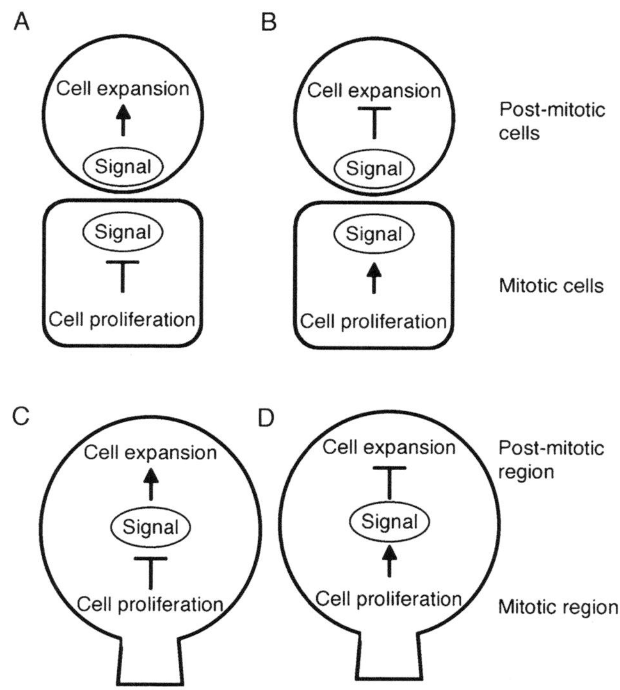

Fig. 2 Models of leaf size control suggested by compensated cell enlargement. To interpret compensation, signals that link cell proliferation and post-mitotic cell expansion are assumed. The hypothetical signals are produced in mitotic cells and function in post-mitotic cells. They exert positive (**A** and **C**) or negative (**B** and **D**) effects on post-mitotic cells and are subjected to negative or positive regulation, respectively, according to the cell proliferation activity. The signals are retained in the cell in the course of leaf development and act cell-autonomously (**A** and **B**) or are released from mitotic cells and act non-cell-autonomously on post-mitotic cells (**C** and **D**). In **A** and **B**, mitotic and post-mitotic cells are shown in the lower and upper rows, respectively. In **C** and **D**, leaf primordia are shown in which mitotic and post-mitotic cells are located in the lower and upper regions of leaf primordia, respectively. In compensation-exhibiting mutants, the strength of positive and negative signals is expected to be stronger and weaker, respectively, leading to enhanced cell expansion in post-mitotic cells

weaker in compensation-exhibiting mutants than in the wild-type, and if it is positive, the signal(s) should be stronger in compensation-exhibiting mutants. This hypothetical signal(s) should have one important feature: its action should be saturated in wild-type leaf primordia if it is negative, or it should be ineffective if it is positive. This conclusion is deduced from the basic nature of compensation: although a reduction in cell numbers due to the loss of function of cell-proliferation activator genes, such as *ANT* or *AN3*, induces accelerated cell expansion, the overproduction of cells caused by the overexpression of the same genes does not affect the cell size (Mizukami and Fischer 2000; Horiguchi et al. 2005). This hypothetical feature of the signal is also predicted by the fact that certain mutants with moderately decreased numbers of leaf cells do not exhibit compensation (Horiguchi et al. 2006a,b). Whether the signal(s) act cell-autonomously or non-cell-autonomously is unknown, but should be revealed through the construction and analysis of chimeric leaf primordia of compensation-exhibiting and wild-type plants. Identification of the above signals should open up a new and exciting field of research on the organ-wide regulation of cell proliferation and cell enlargement.

3.6 Perspectives on the Analysis of Compensated Cell Enlargement

There are no conclusive experimental data as yet on the mechanisms of compensated cell enlargement. However, such mechanisms are beginning to be revealed, as illustrated by the specific manipulation of cell size in compensation-exhibiting mutants (Fujikura et al. 2007). We have identified a wide variety of mutants that enable the genetic manipulation of both cell size and cell number (Horiguchi et al. 2006a, 2006b). Moreover, the recent development of tissue-specific inducible gene expression systems (Brand et al. 2006) and an artificial microRNA expression system (Schwab et al. 2006) will allow local or stage-specific modification of cell proliferation or cell expansion in a more sophisticated manner. Whatever the actual mechanism(s) involved in compensated cell enlargement, characterization of this phenomenon constitutes a challenging undertaking, one that will further understanding of the control of organ size involving a higher-order regulation that links two different developmental processes.

References

Ashton PMS, Berlyn GP (1994) A comparison of leaf physiology and anatomy of *Quercus* (section *Erythrobalanus*-Fagaceae) species in different light environments. Am J Bot 81:589–597

Autran D, Jonak C, Belcram K, Beemster GT, Kronenberger J, Grandjean O, Inzé D, Traas J (2002) Cell numbers and leaf development in *Arabidopsis*: a functional analysis of the *STRUWWELPETER* gene. EMBO J 21:6036–6049

Avery GS (1933) Structure and development of tobacco leaf. Am J Bot 20:565–592

Barrôco RM, Peres A, Droual AM, De Veylder L, le Nguyen SL, De Wolf J, Mironov V, Peerbolte R, Beemster GT, Inzé D, Broekaert WF, Frankard V (2006) The cyclin-dependent kinase inhibitor Orysa; KRP1 plays an important role in seed development of rice. Plant Physiol 142:1053–1064

Beemster GT, Vercruysse S, De Veylder L, Kuiper M, Inze D (2006) The *Arabidopsis* leaf as a model system for investigating the role of cell cycle regulation in organ growth. J Plant Res 119:43–50

Björkman O (1981) In: Encyclopedia of Plant Physiology, Physiological Plant Ecology I. New Series vol 12A. Springer, Berlin Heidelberg, New York, pp 57–107

Boardman NK (1977) Comparative photosynthesis of sun and shade plants. Annu Rev Plant Physiol 28:355–377

Boudolf V, Barrôco R, Engler Jde A, Verkest A, Beeckman T, Naudts M, Inzé D, De Veylder L (2004) B1-type cyclin-dependent kinases are essential for the formation of stomatal complexes in *Arabidopsis thaliana.* Plant Cell 16:945–955

Brand L, Horler M, Nuesch E, Vassalli S, Barrell P, Yang W, Jefferson RA, Grossniklaus U, Curtis MD (2006) A versatile and reliable two-component system for tissue-specific gene induction in Arabidopsis. Plant Physiol 141:1194–1204

Chabot BF, Chabot JF (1977) Effects of light and temperature on leaf anatomy and photosynthesis in *Fragaria vesca.* Oecologia 26:363–377

Chabot BF, Jurik TW, Chabot JF (1979) Influence of instantaneous and integrated light-flux density on leaf anatomy and photosynthesis. Am J Bot 66:940–945

Cho Y-H, Yoo S-D, Sheen J (2006) Regulatory functions of nuclear hexokinase1 complex in glucose signaling. Cell 127:579–589

Clay NK, Nelson T (2005) The recessive epigenetic *swellmap* mutation affects the expression of two step II splicing factors required for the transcription of the cell proliferation gene *STRUWWELPETER* and for the timing of cell cycle arrest in the Arabidopsis leaf. Plant Cell 17:1994–2008

Coupe SA, Palmer BG, Lake JA, Overy SA, Oxborough K, Woodward FI, Gray JE, Quick WP (2006) Systemic signalling of environmental cues in *Arabidopsis* leaves. J Exp Bot 57:329–341

De Veylder L, Beeckman T, Beemster GT, Krols L, Terras F, Landrieu I, van der Schueren E, Maes S, Naudts M, Inzé D (2001) Functional analysis of cyclin-dependent kinase inhibitors of Arabidopsis. Plant Cell 13:1653–1668

Dengler NG (1980) Comparative histological basis of sun and shade leaf dimorphism in *Helianthus annuus.* Can J Bot 58:717–730

Donnelly PM, Bonetta D, Tsukaya H, Dengler RE, Dengler NG (1999) Cell cycling and cell enlargement in developing leaves of *Arabidopsis.* Dev Biol 215:407–419

Elliott RC, Betzner AS, Huttner E, Oakes MP, Tucker WQJ, Gerentes D, Perez P, Smyth DR (1996) *AINTEGUMENTA*, an *APETALA2*-like gene of Arabidopsis with pleiotropic roles in ovule development and floral organ growth. Plant Cell 8:155–168

Esau K (1977) Anatomy of Seed Plants. Wiley, New York

Escoubas J-M, Lmas M, LaRoshe J, Falkowski PG (1995) Light intensity regulation of *cab* gene transcription is signaled by the redox state of the plastoquinone pool. Proc Natl Acad Sci USA 92:10237–10241

Exner V, Taranto P, Schonrock N, Gruissem W, Henning L (2006) Chromatin assembly factor CAF-1 is required for cellular differentiation during plant development. Development 133:4163–4172

Ferjani A, Horiguchi G, Yano S, Tsukaya H (2007) Analysis of leaf development in *fugu* mutants of *Arabidopsis thaliana* reveals three compensation modes that modulate cell expansion in determinate organs. Plant Physiol 144:988–999

Fujikura U, Horiguchi G, Tsukaya H (2007) Dissection of enhanced cell expansion processes in leaves triggered by a defect in cell proliferation, with reference to roles of endoreduplication. Plant Cell Physiol 48:278–286

Givnish TJ (1988) Adaptation to sun and shade: a whole-plant perspective. In: Evans JR, Von Caemmerer S, Adams WW III (eds) Ecology of Photosynthesis in Sun and Shade. CSIRO, Melbourne, pp 63–92

Haberlandt G (1914) Physiological Plant Anatomy. Macmillan, London

Hagemann W (1992) The relationship of anatomy to morphology in plants: a new theoretical perspective. Int J Plant Sci 153:S38–48

Hanson J, Johannesson H, Engström P (2001) Sugar-dependent alterations in cotyledon and leaf development in transgenic plants expressing the HDZhdip gene *ATHB13*. Plant Mol Biol 45:247–262

Hemerly A, Engler Jde A, Bergounioux C, Van Montagu M, Engler G, Inzé D, Ferreira P (1995) Dominant negative mutants of the Cdc2 kinase uncouple cell division from iterative plant development. EMBO J 14:3925–3936

Heraut-Bron V, Robin C, Varlet-Grancher C, Afif D, Guckert A (1999) Light quality (red:far-red ratio): does it affect photosynthetic activity, net CO_2 assimilation, and morphology of young white clover leaves? Can J Bot 77:1425–1431

Héraut-Bron V, Robin C, Varlet-Grancher C, Guckert A (2001) Phytochrome-mediated effects on leaves of white clover: consequences for light interception by the plant under competition for light. Ann Bot 88:737–743

Horiguchi G, Kim G-T, Tsukaya H (2005) The transcription factor AtGRF5 and the transcription coactivator AN3 regulate cell proliferation in leaf primordia of *Arabidopsis thaliana*. Plant J 43:68–78

Horiguchi G, Ferjani A, Fujikura U, Tsukaya H (2006a) Coordination of cell proliferation and cell expansion in the control of leaf size in *Arabidopsis thaliana*. J Plant Res 119:37–42

Horiguchi G, Fujikura U, Ferjani A, Ishikawa N, Tsukaya H (2006b) Large-scale histological analysis of leaf mutants using two simple leaf observation methods: identification of novel genetic pathways governing the size and shape of leaves. Plant J 48:638–644

Johnston GC, Pringle JR, Hartwell LH (1977) Coordination of growth with cell division in the yeast *Saccharomyces cerevisiae*. Exp Cell Res 105:79–98

Jorgensen P, Tyers M (2004) How cells coordinate growth and division. Curr Biol 14:R1014–R1027

Kaplan DR, Hagemann W (1991) The relationship of cell and organism in vascular plants. Bioscience 41:693–703

Kaplan DR (1992) The relationship of organisms to cells in plants: problems and implications of an organismal perspective. Int J Plant Sci 153:S28–S37

Karpinski S, Reynolds H, Karpinska B, Wingsle G, Creissen G, Mullineaux P (1999) Systematic signaling and acclimation in response to excess excitation energy in *Arabidopsis*. Science 284:654–657

Kaya H, Shibahara K, Taoka K, Iwabuchi M, Stillman B, Araki T (2001) *FASCIATA* genes for chromatin assembly factor-1 in *Arabidopsis* maintain the cellular organization of apical meristems. Cell 104:131–142

Kim G-T, Shoda K, Tsuge T, Cho KH, Uchimiya H, Yokoyama R, Nishitani K, Tsukaya H (2002) The *ANGUSTIFOLIA* gene of *Arabidopsis*, a plant *CtBP* gene, regulates leaf-cell expansion, the arrangement of cortical microtubules in leaf cells and expression of a gene involved in cell-wall formation. EMBO J 21:1267–1279

Kim G-T, Yano S, Kozuka T, Tsukaya H (2005) Photomorphogenesis of leaves: shade-avoidance and differentiation of sun and shade leaves. Photochem Photobiol Sci 4:770–774

Kim JH, Kende H (2004) A transcriptional coactivator, AtGIF1, is involved in regulating leaf growth and morphology in Arabidopsis. Proc Natl Acad Sci USA 101:13374–13379

Koch KE, Ying Z, Wu Y, Avigne WT (2000) Multiple paths of sugar-sensing and a sugar/oxygen overlap for genes of sucrose and ethanol metabolism. J Exp Bot 51:417–427

Koch K (2004) Sucrose metabolism: regulatory mechanisms and pivotal roles in sugar sensing and plant development. Curr Opin Plant Biol 7:235–246

Lake JA, Quick QP, Beerling DJ, Woodward FI (2001) Signals from mature to new leaves. Nature 411:154

Lopéz-Juez E, Bowyer JR, Sakai T (2001) Distinct leaf developmental and gene expression responses to light quantity depend on blue-photoreceptor or plastid-derived signals, and can occur in the absence of phototropins. Planta 227:113–123

Maksymowych R (1963) Cell division and cell elongation in leaf development in *Xanthium pennsylvanicum*. Am J Bot 50:891–901

Melaragno JE, Mehrotra B, Coleman AW (1993) Relationship between endopolyploidy and cell size in epidermal tissue of Arabidopsis. Plant Cell 5:1661–1668

Menges M, Samland AK, Planchais S, Murray JA (2006) The D-type cyclin CYCD3;1 is limiting for the G1-to-S-phase transition in Arabidopsis. Plant Cell 18:893–906

Miyazawa S-I, Livingston NJ, Turpin DH (2006) Stomatal development in new leaves is related to the stomatal conductance of mature leaves in poplar (*Populus trichocarpa* × *P deltoides*). J Exp Bot 57:373–380

Mizukami Y, Fischer RL (2000) Plant organ size control: *AINTEGUMENTA* regulates growth and cell numbers during organogenesis. Proc Natl Acad Sci USA 97:942–947

Nelissen H, Clarke JH, De Block M, De Block S, Vanderhaeghen R, Zielinski RE, Dyer T, Lust S, Inzé D, Van Lijsebettens M (2003) DRL1, a homolog of the yeast TOT4/KTI12 protein, has a function in meristem activity and organ growth in plants. Plant Cell 15:639–654

Nishimura C, Ohashi Y, Sato S, Kato T, Tabata S, Ueguchi C (2004) Histidine kinase homologs that act as cytokinin receptors possess overlapping functions in the regulation of shoot and root growth in Arabidopsis. Plant Cell 16:1365–1377

Nobel PS (2005) Physicochemical and Environmental Pant Physiology 3rd edn. Elsevier, USA

Osborn JM, Taylor TN (1990) Morphological and ultrastructural studies on plant cuticular membranes. I. Sun and shade leaves of *Quercus velutina* (Fagaceae). Bot Gaz 151:465–476

Pfannschmidt T, Nilesson A, Allen JF (1999) Photosynthetic control of chloroplast gene expression. Nature 397:625–628

Popescu SC, Tumer NE (2004) Silencing of ribosomal protein L3 genes in *N. tabacum* reveals coordinate expression and significant alterations in plant growth, development and ribosome biogenesis. Plant J 39:29–44

Rhichter T, Fukshansky L (1998) Optics of a bifacial leaf: 3. Implications for photosynthetic performance. Photochem Photobiol 68:337–352

Riou-Khamlichi C, Huntley R, Jacqmard A, Murray JA (1999) Cytokinin activation of *Arabidopsis* cell division through a D-type cyclin. Science 283:1541–1544

Schwab R, Ossowski S, Riester M, Warthmann N, Weigel D (2006) Highly specific gene silencing by artificial microRNAs in *Arabidopsis*. Plant Cell 18:1121–1133

Sitte P (1992) A modern concept of the cell theory: a perspective on competing hypotheses of structure. Int J Plant Sci 153:S1-S6

Smith H (1994) Sensing the light environment: the functions of the phytochrome family. In: Kendrick RE, Kronenberg GHM (eds) Photomorphogenesis in Plants, 2nd edn. Kluwer Academic, Dordrecht, pp 337–416

Smith LG, Hake S (1992) The initiation and determination of leaves. Plant Cell 4:1017–1027

Terashima I, Miyazawa S-I, Hanba YT (2001) Why are sun leaves thicker than shade leaves? Consideration based on analyses of CO_2 diffusion in the leaf. J Plant Res 114:93–105

Thomas PW, Woodward FI, Quick WP (2004) Systemic irradiance signalling in tobacco. New Phytol 161:193–198

Tichá I (1982) Photosynthetic characteristics during ontogenesis of leaves. 7. Stomata density and sizes. Photosynthetica 11:357–471

Tsuge T, Tsukaya H, Uchimiya H (1996) Two independent and polarized processes of cell elongation regulate leaf blade expansion in *Arabidopsis thaliana* (L.) Heynh. Development 122:1589–1600

Tsukaya H (2002) Interpretation of mutants in leaf morphology: genetic evidence for a compensatory system in leaf morphogenesis that provides a new link between cell and organismal theories. Int Rev Cytol 217:1–39

Tsukaya H (2003) Organ shape and size: a lesson from studies of leaf morphogenesis. Curr Opin Plant Biol 6:57–62

Tsukaya H (2006) Mechanism of leaf-shape determination. Annu Rev Plant Biol 57:477–496

Tsukaya H, Beemster GT (2006) Genetics, cell cycle and cell expansion in organogenesis in plants. J Plant Res 119:1–4

Ullah H, Chen JG, Young JC, Im KH, Sussman MR, Jones AM (2001) Modulation of cell proliferation by heterotrimeric G protein in *Arabidopsis*. Science 292:2066–2069

Wang H, Zhou Y, Gilmer S, Whitwill S, Fowke LC (2000) Expression of the plant cyclin-dependent kinase inhibitor ICK1 affects cell division, plant growth and morphology. Plant J 24:613–623

Weston E, Thorogood K, Vinti G, Lopéz-Juez E (2000) Light quantity controls leaf-cell and chloroplast development in *Arabidopsis thaliana* wild-type and blue-light-perception mutants. Planta 211:807–815

White DWR (2006) *PEAPOD* regulates lamina size and curvature in *Arabidopsis*. Proc Natl Acad Sci USA 103:13238–13243

Woodward FI (1987) Stomatal numbers are sensitive to increase in CO_2 from pre-industrial levels. Nature 327:617–618

Woodward FI, Lake JA, Quick WP (2002) Stomatal development and CO_2: ecological consequences. New Phytol 153:477–484

Yano S, Terashima I (2001) Separate localization of light signal perception for sun or shade type chloroplast and palisade tissue differentiation in *Chenopodium album*. Plant Cell Physiol 42:1303–1310

Yano S, Terashima I (2004) Developmental process of sun and shade leaves in *Chenopodium album* L. Plant Cell Env 27:781–793

Plant Cell Monogr (10)
L. Bögre and G. Beemster: Plant Growth Signaling
DOI 10.1007/7089_2007_157/Published online: 29 January 2008

Plant Growth Dynamics: Analysis of Basic Spatial and Temporal Growth Patterns on the Background of Photosynthetic Energy Gain and Interactions with the Environment

Achim Walter

Institute of Chemistry and Dynamics of the Geosphere ICG-III: Phytosphere, Research Center Juelich GmbH, 52425 Juelich, Germany
a.walter@fz-juelich.de

Abstract Leaves and roots show dynamic growth patterns such as diel variations (occurring throughout 24 h) or base-tip gradients that can be investigated non-invasively using image sequence processing methods. These growth patterns are affected by a number of environmental parameters. Light plays a crucial role as it ultimately drives biomass production via photosynthetic energy gain and carbohydrate metabolism. The interaction of growth patterns of different species with light conditions and other environmental factors affecting photosynthesis and carbohydrate metabolism is discussed and put into the context of the physiological framework regulating plant performance.

1 Introduction—The Relevance of Plant Growth

Plant growth is a complex and dynamic process. It is highly regulated and depends on a network of factors. In contrast to most animals, plants grow throughout their entire life. Plant biomass production is driven by photosynthesis which provides energy, raw materials and food for animals. While growth processes in animals are largely deterministic, they are more flexible in plants. This flexibility is an important feature with which plants can dynamically adjust their performance to fluctuating environmental conditions against the background of small-scale intrinsic temporal or spatial organ growth patterns. This dynamic growth potential can be utilized on demand, such as in etiolating, shade-avoiding seedlings, in plants that increase their root-shoot ratio in nutrient-limited situations or in plant organs that show directed growth responses towards or away from vectorial factors such as light, gravity or wind. The temporal dynamics of such acclimation processes; the differential distribution of growth across different plant organs or within single organs or the temporal kinetics with which such growth processes interact with endogenous control mechanisms of the plant (e.g. gene expression, metabolism) are largely unknown. As a crystallizing point for the investigation of growth dynamics, the interaction of growth and photosynthesis will be treated herein.

Plants are photoautotropic organisms, which gain energy by conversion of light to chemical energy. Apart from this gain in energy, plants only depend on uptake of water, mineral nutrients, oxygen and CO_2 to sustain growth. All of the necessary uptake processes take place across plant interfaces that are exposed to the environment. The constant need for the plant to increase its surface to sustain further growth implies that an analysis of its surface expansion processes is a good proxy for a general performance analysis of the dynamic plant in its dynamic environment.

2 Technical Requirements and Optimized Cultivation Systems to Analyze Plant Growth Dynamics

Growth analysis of a plant, organ or tissue region requires knowledge of the size of the investigated system at least at two consecutive points in time. This knowledge can be gained by destructive or non-destructive methods such as determination of fresh or dry weight. Destructive methods do not allow studying growth processes with high resolution. Because of the high variability between different individuals, size differences that are reached within minutes or hours can not be determined with statistical significance by comparing two populations on the basis of destructive measurements. The same argument applies to the destructive analysis of growth differences of spatially neighboring tissue regions. Hence, high spatial or temporal resolution of growth analysis can only be achieved by utilizing non-invasive methods that determine surface or volume of the investigated organ. Often, special cultivation systems have to be established to ensure an exact quantification of plant organ growth and to control environmental parameters with appropriate accuracy.

2.1 Methods for Non-Invasive Determination of Growth

The oldest and easiest method to record growth is to use rulers. This is still an appropriate method for easy and rapid investigation of growth processes within plant populations (Walter and Schurr 1999; Christ et al. 2006).

First descriptions of spatially resolved growth patterns within plant organs date back to the approach of Sachs (1887) for roots and Avery (1933) for leaves. In both cases, ink dots were applied to the organ surfaces and their divergence (increase in distance between dots over time) was recorded. Methods without application of external marks were published decades later (root: Brumfield 1942; leaf: Schmundt et al. 1998). In linearly organized growth zones of leaves of monocotyledonous plants (Ben-Haj-Salah and Tardieu 1995; Beemster et al. 1996) or roots (Goodwin and Stepka 1945; Silk and Erickson 1979), the distribution of cell expansion rate can be quantified elegantly via a so-called "kinematic analysis":

On the basis of the premise that each cell is displaced in a linear way from the end of the cell division zone to the zone of fully differentiated tissue, a plot of average cell length versus distance from the distal end of the meristematic zone can be used to calculate cell elongation along the direction of this line of cells (cell file). First, the velocity distribution along the cell file is calculated by determining, how fast each point along the cell file is displaced away from the meristematic zone. For the elongation zone, this is done by multiplying cell length at each position with "cell flux" (which equals the ratio of the velocity of the root tip and cell length at the end of the growth zone). "Cell flux" is the reciprocal time that is needed to add one new cell to the cell file at the apical interface of the growth zone or to remove one full grown cell at the basal end of the growth zone from the end of the cell file. The elongation rate (or strain rate) is then given by the first derivative of the velocity distribution. For a more detailed explanation of these considerations see Silk et al. (1989).

Cell division rates can also be calculated in such organs based on determination of cell lengths at different positions and on determination of root tip velocity. These measurement parameters are used in the continuity equation (Silk and Erickson 1979; Silk 1984; Beemster and Baskin 1998) to calculate the production of cells and cell division rates within cell files.

Methods with high temporal resolution were initiated in the 1970s when electronic devices began to be applied on a wider scale (Hsiao et al. 1970). Linear Variable Displacement Transducers (LVDTs) register position changes of the leaf tip via a thread that is attached to it, that is kept under tension and that regulates an electric resistance or the inductance of a solenoid. This principle was first applied by Sachs (1887) more than a century ago, when he recorded the diel time series of shoot axis growth rate, but it fell into oblivion for a long time. Herein the term diel is used to depict variations occurring throughout 24 h in a repetitive way. Diel is a synonym of the more commonly found term diurnal that can also depict processes that are only active during the day, but not at night.

Methods providing a combination of high spatial and temporal resolution were elaborated when digital image processing became available (Schmundt et al. 1998; Van der Weele et al. 2003). Again, a pioneering study demonstrating the essential principles on which modern methods are based was leading the way some decades ahead of time: The distribution of growth activity in a maize root growth zone was analyzed throughout 12 h by Erickson and Sax (1956) by continually photographing the ink-dotted root growth zone with a slit camera and a slowly moving film. The transformation of successive temporal events into the x-axis of the image resulted in a so-called "streak image": A group of divergent trajectories was recorded and the inclination of each trajectory depicts the velocity of the cellular element on which the respective ink dot was situated. The difference of the velocity of neighboring elements, normalized by the distance of the elements equals the relative growth rate of the respective length element (Fig. 1). This method was further applied in at least one physiological study (Hejnowicz and Erickson 1968), but fell into oblivion thereafter.

On the basis of the same principle (extended into two dimensions) of so-called "optical flow", digital image sequence processing methods have been established during the last decade (Schmundt et al. 1998; Van der Weele et al. 2003). Images of root growth zones were recorded at time intervals of less than a minute, rendering results for growth distributions in almost cellular resolution (Walter et al. 2002b; Nagel et al. 2006). For leaves, the two-dimensional analysis of divergence of natural patterns like vein crossings or trichomes in image sequences rendered similar distributions of relative (elemental) growth rates (Schmundt et al. 1998; Walter et al. 2002a; Wiese et al. 2007; Fig. 1). Typical intervals between images were two to five minutes. Investigated leaves and roots were illuminated with near-infrared LEDs and were visualized with CCD-cameras that were equipped with appropriate near-infrared filters. Image recording in the near-infrared range has the advantage of achieving constant image brightness throughout day and night. Wavelengths of about 900 nm do not affect any known sensory systems of plants and are not heating up leaves.

Calculation of spatio-temporal growth rate distributions was performed by optimized, custom-made image sequence processing algorithms (Scharr 2004) that are based on the above-mentioned principle of optical flow and that use a so-called structure-tensor approach to calculate motion from orientation in stacks of images (Bigün and Granlund 1987; Haußecker and Spies 1999).

Fig. 1 Calculation of velocity and relative growth rate (RGR) from image sequences of roots and leaves. Root panels: Original images of a *Zea mays* root that was marked with ink dots and that was photographed every 15 min are arranged next to each other (*left*; grid indicates 1 mm). The *black lines* in the *middle panel* depict in the way of streak photographs, how the ink dots were displaced with the extending root. The inclination of each line towards the time-axis corresponds to the velocity of the ink dot. For clarity, those velocities are indicated additionally as colored vectors (*arrows*) in the *middle of each line*. *Colors* are coded via a "*rainbow*" look-up-table shown in the leaf growth panel (*lower right*): *blue* stands for low values, *red* for high values. The right hand root growth panel finally shows the differences of two neighboring velocities divided by the average distance between two neighboring dots throughout the investigated time period. Those differences equal the relative (elemental) growth rate of each segment situated between two neighboring ink dots. They are plotted versus the distance of the corresponding dots from the root tip. *Leaf panels*: Original images of a *Nicotiana tabacum* leaf, taken at $t = 0$ h and $t = 24$ h (*left images*). Images in the *middle* show color-coded distributions of velocity components in *x*- (*top*) and *y*- (*bottom*) directions (equation given above *top panel*) that are projected onto the original image at $t = 0$. The *right hand panel* finally shows the divergence of this velocity distribution projected onto the original image at $t = 0$ which corresponds to the distribution of relative (elemental) growth rates. All calculations follow the same principles as explained for the one-dimensional case of the root, but are performed in two dimensions (*x*- and *y*-direction) ▶

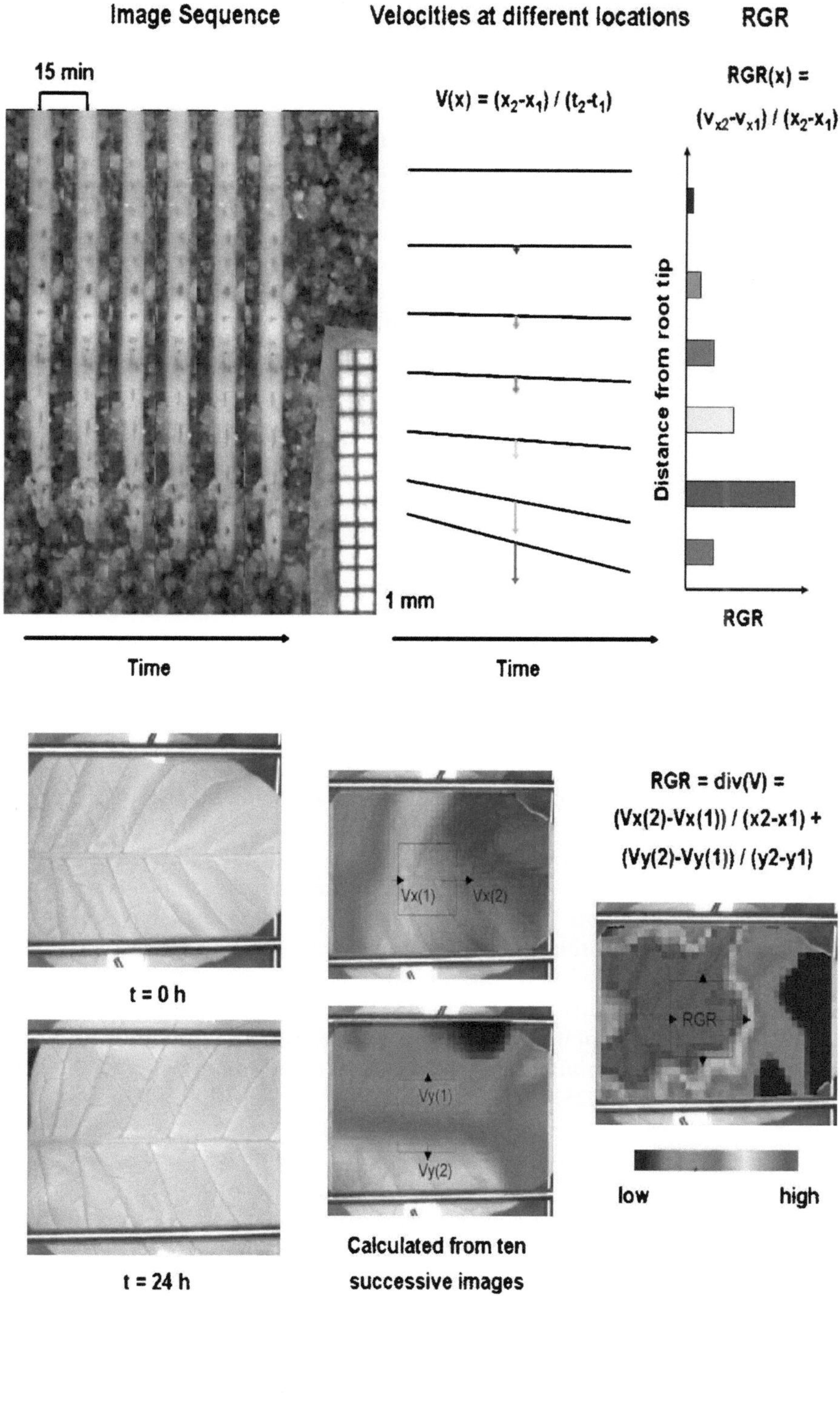
Image Sequence
Velocities at different locations
RGR
15 min
V(x) = (x2-x1) / (t2-t1)
RGR(x) =
(vx2-vx1) / (x2-x1)
Distance from root tip
1 mm
RGR
Time
Time
Vx(1)
Vx(2)
Vy(1)
Vy(2)
RGR = div(V) =
(Vx(2)-Vx(1)) / (x2-x1) +
(Vy(2)-Vy(1)) / (y2-y1)
RGR
t = 0 h
t = 24 h
Calculated from ten
successive images
low
high

Image sequence processing-based methods for determination of growth rates provide both high temporal and spatial resolution. Yet, since smoothing and regularization procedures are a necessary and intrinsic feature of those methods, it has to be pointed out that it is impossible to achieve maximally high spatial and temporal resolution at the same time. Depending on the question to be answered, the focus has to be put either on maximal temporal or on maximal spatial resolution or one has to take deductions of both dimensions into account. Calculation of velocities (and growth rates) takes into account information from spatial regions of about 10×10 pixels and from temporal neighborhoods of about 10–20 image frames. The maximal spatio-temporal resolution that was achieved up to now was reached in a study of root gravitropic curvature processes in *Arabidopsis thaliana* (Chavarría-Krauser et al. 2008). On the basis of the calculation of growth differences in immediate vicinities along the root growth zone, two distinct centers of curvature located about 100 μm apart from each other were identified within the root growth zone which was about 600 μm long. It was shown that the curvature activity within those centers was initiated with a time lag: While the more apical center of curvature reacted towards gravistimulation within minutes, the basal center commenced curvature about 60 minutes past gravistimulus. Pixel resolution of the acquired images was 1.4 μm; images were recorded every 30 s.

2.2 Cultivation Systems to Achieve Optimal Growth Monitoring Situations

An aspect of utmost importance for automated, minimal-invasive analysis of plant growth is to create cultivation systems that enable plant growth conditions that are as "natural" as possible and at the same time facilitate recording of time-lapse movies. Leaves have to be fixed to the focal plane of the camera to facilitate growth recording, because of nyctinastic plant movements (Bünning 1948, Bünning and Moser 1966) that otherwise lead to movements of the leaf surface out of the focal plane. This is achieved by straining the leaf via threads that are fixed to its margins and that are guided by a solid frame, ensuring that the weights fixed to the other end of each thread lead to a constantly horizontal orientation of the leaf surface (Fig. 2). It was shown for the example of *Ricinus communis* that the choice of an optimal strain force is possible (Walter et al. 2002a). Application of this force neither altered the diel pattern of leaf growth nor the base-tip gradient. Forces that were either too small or too large, however, did alter spatial and temporal leaf growth patterns. It has to be noted that not only optical, but also mechano-electrical procedures of automated leaf growth recording, such as LVDTs require the application of a strain force. Optimal strain forces have to be established for each new species to be investigated by comparing growth of strain-force treated leaves with growth of control leaves.

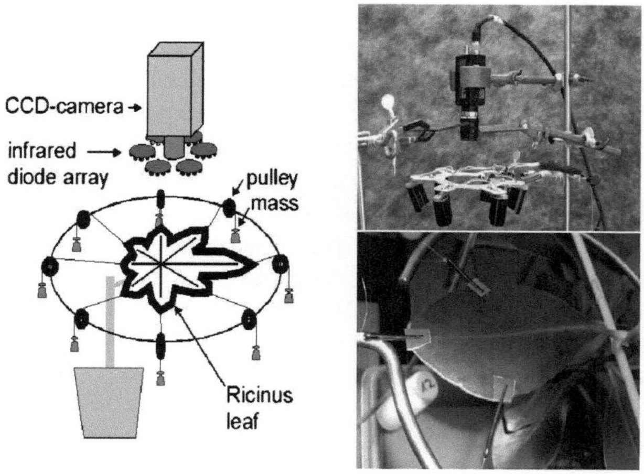

Fig. 2 Setup for leaf growth monitoring. Schematic drawing (*left*), CCD-camera and infrared LED arrays (*upper right*), and leaf of *Clusia minor* mechanically fixed to the focal plane of a camera by three weights of 12 g each

To enable a controlled variation of environmental parameters, growth monitoring systems have to be put in growth chambers or in controlled conditions of greenhouse facilities such as the Biosphere 2 Center (Walter and Lambrecht 2004) where the setup is protected from advert conditions like wind, rain or dew and where for example the effect of altered atmospheric CO_2-conditions can be studied in detail. For root growth, image acquisitions have been performed either in agarose-filled Petri dishes (Beemster and Baskin 1998; Nagel et al. 2006) or in hydroponic cultivation systems using an inclined base plate (Walter et al. 2002b, 2003a). In both systems, roots are situated in a translucent medium allowing optical recording, are well supplied with nutrients and forced to grow in two dimensions only.

3 Basic Spatial and Temporal Patterns of Organ Growth Distributions

Spatial and temporal patterns of growth distributions are often intimately connected to each other, since neighboring regions of an expanding organ often consist of tissue of successive developmental stages. For monocotyledonous leaves and for root growth zones, which are organized linearly with clearly distinguished zones of cell division and expansion, development is deterministic to a high extent. Hence, the principal distribution of growth rates along the axial direction of those organs is very similar and has been well investigated throughout the last decades. Much less attention has been directed towards distributions of growth rates in leaves of dicotyledonous plants.

The temporal sequence of cell elongation also differs strongly between root and leaf tissue. While in roots, cell elongation is typically finished within between 6 h (*Arabidopsis thaliana*; Beemster and Baskin 1998) and 24 h (*Zea mays*; Silk et al. 1989), cells of leaves typically expand for 7 d (*Arabidopsis thaliana*; De Veylder et al. 2001) or 14 d (*Nicotiana tabacum*; Walter et al. 2003b). Correspondingly, maximal relative growth rates are much higher in root cells, reaching values of up to 40% h^{-1}, while in leaf tissue, peak values hardly ever exceed 4% h^{-1}. Another fundamental difference between growth dynamics of roots and leaves is, that root growth activity is typically not fluctuating throughout the diel cycle (Head 1965; Iijima et al. 1998; Walter et al. 2002b; Walter and Schurr 2005), while it has long been known that leaf growth intensity is not constant throughout 24 h.

3.1 Spatial Growth Distribution in Leaves of Dicotyledonous Plants

In contrast to roots and monocot leaves, leaves of dicotyledonous plants possess growth zones that extend with comparable intensity in two dimensions and that do not show a clear distinction between dividing and expanding tissue. For *Arabidopsis* leaves, it was recently shown that the transition from the "proliferative" growth phase, in which the organ grows by cell division only, to the "expansive" growth phase, in which the individual cell area increases is accompanied by the onset of endoreduplication by which the ploidy level is increased (Beemster et al. 2005). It has been a central axiom of dicot leaf growth research for decades that cells of the leaf tip are initiated earlier than cells located at the leaf base (Foster 1936). This developmental lag is then retained until the latest stages of leaf development, leading to pronounced base-tip gradients of growth distribution that reflect the differential developmental status of the two regions (VanVolkenburgh 1987). In *Nicotiana tabacum* for example, a developmental time lag of 4 days was identified (Walter et al. 2003b). *Nicotiana tabacum* has been the model system for dicotyledonous leaf growth for a long time (Avery 1933; Hannam 1968). Similar base-tip gradients were identified for *Ricinus communis* (Walter et al. 2002a) and for the molecular model system *Arabidopsis thaliana* (Donnelly et al. 1999; Wiese et al. 2007). Yet, compelling evidence has been reported recently that base-tip gradients of dicot leaf growth and connected cellular developmental lags between base and tip are not a uniform pattern within dicotyledonous plants. Other plants such as *Glycine max* (Ainsworth et al. 2005) and *Populus deltoides* (Walter et al. 2005; Matsubara et al. 2006) do not show pronounced base-tip gradients but an almost homogeneous distribution of growth rates within the leaf lamina, connected to uniform cell sizes during the entire post-emergent phase of leaf development. This leads to the conclusion that either there is no developmental lag in the initiation of leaves from the different tunica layers of the shoot apical meristem or that development of those tissue regions is synchronized in a later stage of leaf development, possibly under the influence of external signals such as light.

3.2
Characteristic Diel Cycles of Leaf Growth Activity

Leaves of all investigated species show characteristic diel cycles of growth activity under all reported environmental conditions (for a review see Walter and Schurr 2005; or Matsubara and Walter 2006). The intensity and the finestructure of the diel variation varies from day to day and the phasing clearly differs between species (Fig. 3): Leaves of *Ricinus communis* (Walter et al. 2002a), *Nicotiana tabacum* (Walter and Schurr 2005) and *Arabidopsis thaliana* (Wiese et al. 2007) for example showed maximal growth activity in the beginning of the day. These species were also following the classical pattern of base-tip distribution of growth rate with higher growth rates at the leaf base and lower growth rates at the tip. An exactly contrasting phase behavior was displayed by leaves of *Populus deltoides* (Walter et al. 2005; Matsubara et al. 2006). Here, maximal growth activity was found at the end of the day; growth rate decreased throughout the entire night and

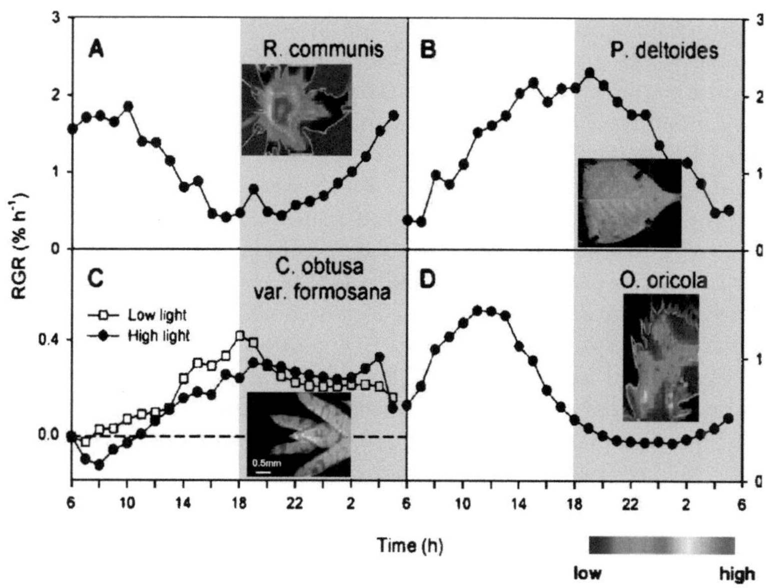

Fig. 3 Diel leaf growth cycles for some dicotyledonous plant species and typical examples for color-coded distribution of relative growth rate (RGR) across the leaf lamina. In high light conditions, *Chamecyparis obtusa* leaves shrink in the morning. This figure is a reproduction of Fig. 1 in Matsubara and Walter (2006) and is printed with kind permission of Springer Science and Business Media

increased continually throughout the day. The growth behavior of *Glycine max* is largely comparable, with a somewhat later occurrence of the growth rate maximum (Ainsworth et al. 2005). Interestingly, those two species did not show a base-tip gradient. For leaves of *Populus deltoides* it was shown by microarray analysis that the nocturnal decrease of growth rates was correlated with a downregulation of genes encoding ribosomal proteins and histones, indicating a decrease of cytoplasmic growth (Matsubara et al. 2006). The investigated species of *Populus* retains cell division activity within leaves practically until they reach full size (Van Volkenburgh and Talor 1996), while leaves of *Ricinus communis* (Roggatz et al. 1999) and *Nicotiana tabacum* (Walter et al. 2003b) merely show cell expansion in post-emergent growth stages. These results support the hypothesis that the difference in phasing of the two growth types might be induced by the differential extent to which leaf growth is driven by cytoplasmic (connected to cell division) and vacuolar (connected to cell expansion) growth.

Diel leaf growth cycles have also been monitored in scale leaves of the gymnosperm species *Chamaecyparis obtusa* and *Chamaecyparis formosensis*. Here, maxima of growth activity were recorded at the end of the day or beginning of the night—as in *Populus deltoides* and *Glycine max*. Yet, in the two *Chamaecyparis* species, growth activity was almost constant throughout the night and a base-tip gradient of growth was recorded (Lai et al. 2005). Moreover, significant shrinking of the foliage was recorded in *Chamaecyparis* when light intensity was increased at the beginning of the day, demonstrating that growth in this gymnosperm genus might be restricted by other factors as in the above mentioned angiosperm species.

A third mode of diel leaf growth activity was recorded in four different species of CAM-plants (Gouws et al. 2005): Leaves of *Kalanchoe beharensis*, *Opuntia oricola*, *Opuntia phaeacantha* and *Opuntia engelmanii* showed growth maxima in the middle of the day, if plants were exposed to low water availability. The maximum of growth activity was correlated with phase III of the crassulacean acid metabolism, in which CO_2-fixation via RubisCO is taking place with stomata closed and CO_2 gained from malate metabolized at night and stored in the vacuole. The cellular status in this phase of CAM is providing excellent conditions for growth since the availability of carbohydrates is high, cytoplasmic pH is low and turgor is maximal. In C_4-plants, the middle of the day also is indicated to be the preferred phase of growth (Watts 1974; Christ 1978; Seneweera et al. 1995).

4
Utilization of Light Energy as a Basic Requirement of Plant Growth

Light is an essential source of energy for photoautotropic organisms. Plants are adapted to use this energy source as efficiently as possible and are acclimating towards alterations of light climate during their entire development. Phases of low light availability such as the night or shade periods can be "buffered" in terms of sustained growth activity via remobilization of energy from transient storage compounds such as starch, but they can also lead to photomorphogenetic growth reactions (Matsubara and Walter 2006). Light intensity directly affects assimilation rate and is hence the most important environmental factor with which plant growth rates are controlled over periods of weeks to seasons (Kruger and Volin 2006; Shipley 2006).

The spatial and temporal heterogeneity of the light climate induces a large number of responses on different levels of organization; especially in the shoot (Björkman 1981). Among them are state transitions in thylacoid membranes (Demmig-Adams and Adams 1992; Allen and Nilsson 1997; Niyogi 1999; Wollman 2001), ontogenetic modifications in sun and shade leaves of some plant species (Sack et al. 2003; Terashima et al. 2006) and acclimation responses of plant canopy architecture (Terashima and Hikosaka 1995; Evans and Poorter 2001; Frak et al. 2002; Niinemets et al. 2004a,b). This wide spectrum of phenotypic plasticity facilitates efficient acclimation of photosynthesis, resource allocation and biomass production to dynamically fluctuating light climates and habitat conditions (Schurr et al. 2006).

4.1 Heterogeneity of Light Use Across the Leaf Lamina

Light intensity reaching a leaf can fluctuate rapidly and can vary strongly within a short distance, for example in forest understorey situations, where often light flecks penetrating canopy gaps are the main light source. Moreover, leaf tissue which is utilizing light energy for growth processes also shows a high degree of heterogeneity. How does a plant manage to coordinate growth processes of such different tissues as vein tissue consisting of largely differentiated sclerenchymatic elements providing mechanical stability and fully functional xylem and phloem elements and largely undifferentiated parenchymatic tissue of intercostal leaf regions?

To address the connection between heterogeneity of light use and growth distribution, analyses of the spatial distribution of growth and photosynthetic efficiency were performed in developing leaves of the tropical understorey species *Coccoloba uvifera* and *Sanchezia nobilis* (Walter et al. 2004). Although base-tip growth gradients and spatial heterogeneities of growth were observed across the leaf lamina, photosynthetic efficiency was distributed homogeneously between base and tip. Yet, differences in the distribution of potential quantum yield F_V/F_m were found between veins and intercostal tissue, indicating differences in the development of the photosynthetic apparatus in those tissues that might reflect a differential demand for locally produced photosynthates. Immediately after leaf unfolding, F_V/F_m was higher in vein tissue compared to intercostal tissue. Gradually, this difference reversed and in full grown leaves, F_V/F_m was higher in intercostal tissue than in vein tissue. This indicates that in early phases of post-emergent leaf development, an increased amount of photosynthates might be required for energy-consuming growth processes within the leaf vein tissue, while differentiation of the photosynthetic machinery can proceed more slowly in intercostal tissue. There, carbohydrates can be supplied in sufficient amount via phloem import.

4.2 Adaptation Towards Different Temporal Variability of Light Intensity

Light climate can change extremely fast. In experiments with two congener gymnosperm tree species of the Taiwanese cloud forest (*Chamaecyparis obtusa* and *Chamaecyparis formosensis*), it was shown that plant growth can react rapidly towards alterations of light climate (Lai et al. 2005). Differences in abundance and ecological niche of the two species were reflected in growth dynamics: While *Chamaecyparis obtusa* grew faster in low light intensity and adjusted the amplitude of its diel leaf growth cycle faster towards light intensity changes, *Chamaecyparis formosensis* profited more strongly from higher light intensity and showed higher growth rates. This correlates with the abundance of *Chamaecyparis obtusa* under closed canopies of mature forest patches and with the high abundance of *Chamaecyparis formosensis* in canopy gaps and at the forest edge.

4.3 Acclimation of Total Leaf Area Towards Alterations of Light Intensity

The dynamics of growth reactions towards alterations of light intensity were also shown in a study of seedlings of *Nicotiana tabacum* that were exposed to varying daylength and light intensity (Walter et al. 2007). It was shown for example that within 24 h of increasing light intensity by a factor of two, significant changes of relative growth rate occur. The plant reaction could only be quantified by monitoring a sufficient number of plants (about 25 per population) and by comparing relative growth rates instead of absolute leaf areas. The degree of automation and reliability of image analysis distinguished the applied phenotyping procedure from that of other studies, in which total leaf area of rosette stage plants has also been quantified (Leister et al. 1999; Barbagallo et al. 2003; El-Lithy et al. 2004; Granier et al. 2006). Not only leaf area, but also shoot fresh and dry weight increased as a reaction towards inclining light quantity.

4.4 Acclimation of Root Growth Towards Alterations of Light Intensity

Root growth can react even more strongly than leaf growth to increased shoot light perception. In another study of seedlings of *Nicotiana tabacum*, it was shown that a very pronounced and characteristic reaction of root growth occurs when light intensity is increased (Nagel et al. 2006): During the first three hours after increase of light intensity by a factor of five, a characteristic fluctuation of root tip growth velocity was observed that was connected to parallel decreases and increases of expansion within the meristematic zone and the zone of cell elongation. The fluctuations were caused by a superposition of a transient, hydraulic decrease of growth activity due to increased transpiration and an accelerating increase of growth activity induced by sucrose import. Experiments with tobacco plants that had a decreased activity of sucrose-phosphate phosphatase (Chen et al. 2005) showed far less pronounced growth reactions. As the total increase of root growth activity by a factor of four within four days exceeded the increase of shoot growth acceleration by far, a significant shift of root-shoot-ratio was detected as a consequence of dynamic growth alterations in response to increased light intensity (Walter and Nagel 2006). Whether a shift of root-shoot-ratio is a general consequence of increased light intensity is controversial in the literature; the reaction will probably depend strongly on the species and on interaction with other factors such as nutrient availability (Lambers and Posthumus 1980; Hodge et al. 1997). Yet, an increase of root growth activity in response to increasing light intensity has been reported in a number of studies before, using different species and observing reactions on longer time scales (Webb 1976; Vincent and Gregory 1989; Aguirrezabal et al. 1994; Bingham et al. 1997).

5
The Carbohydrate Metabolism Mediates Between the Gain of Light Energy and the Production of Biomass

A close investigation of the connection between carbohydrate metabolism and growth dynamics is necessary to move towards a mechanistic understanding of the spatial and temporal relations between energy gain in photosynthesis and biomass production during growth processes. Carbohydrates are the transportable currency unit that is produced in photosynthesis. The majority of plant dry matter consists of carbohydrates, mainly cellulose, hemicellulose and starch. Carbohydrate metabolism is governed by diel or circadian rhythms and concentrations of carbohydrates in leaves typically reach maxima at day (Kemp and Blacklow 1980; Matt et al. 1998; Geiger et al. 2000; Chia et al. 2004; Walter and Schurr 2005). The circadian clock not only governs carbohydrate metabolism but a large number of metabolic, physiological or ontogenetic processes in plants (Somers 1999; McClung 2001; Staiger 2002; Schultz and Kay 2003; Lüttge 2003). The performance of a plant is clearly increased by optimized temporal fine-tuning of circadian processes with the 24-h-cycle of light and dark (Ouyang et al. 1998; Green et al. 2002; Dodd et al. 2005).

5.1 Variation of Diel Leaf Growth Cycles

For extreme situations it has been shown that alterations of the diel variation of carbohydrate metabolism can induce changes of the diel leaf growth pattern. In leaves of *Populus deltoides* that were exposed to a three-times elevated CO_2-content, a strong, transient decay of growth rate was observed during the afternoon, which temporally coincided with a transient decrease of glucose content (Walter et al. 2005). An amplification of nocturnal growth activity was found in leaves of transgenic potato plants that showed an increased starch content at the end of the day (Kehr et al. 1998), indicating that the pool of transitory starch is of utmost relevance for nocturnal growth activity. A clear indication how carbohydrate metabolism affects diel leaf growth cycles was recently observed in *Arabidopsis thaliana*: Starch-free-mutant plants (stf1, Kofler et al. 2000) that do not possess a relevant pool of transitory starch showed much lower growth rates than wild-type plants at night, but were able to grow with comparable intensity to wild-type plants during the day (Wiese et al. 2007). Another hint at the important role of temporal availability of carbohydrates in the regulation of growth dynamics is given by the observation that CAM- and C_4-plants mainly grow during the day when availability of carbohydrates is high due to the decarboxylation of malate, while C_3-plants show growth maxima at night-day transition phases, when metabolites coming from transitory starch probably play an important role. The fact that the investigated C_3-plants do not show an unequivocal diel leaf growth cycle demonstrates that a large number of factors affects the basic temporal pattern of leaf growth and shows the dimension of the challenge to come to a conclusive picture of the regulation of dicot leaf growth dynamics.

5.2 Small-Scale Variation of Leaf Growth

Carbohydrate production of the growing leaf itself might affect the homogeneity of growth distribution within the lamina (Walter et al. 2005): When comparing leaves of *Populus deltoides* that were exposed to the sun with leaves that were completely shaded, but growing on sunlit trees, it was observed that relative growth rate distribution across the lamina was more patchy and temporal fluctuation of average relative leaf growth rate was much stronger in leaves that grew in the shade. At night, leaves of both populations showed comparable temporal fluctuations of relative growth rate. This leads to the speculation that in shaded leaves the fluctuating but strong import of carbohydrates from source leaves led to spatial and temporal heterogeneities of growth that were dampened in the case of sunlit growing leaves by the stabilizing effect of carbohydrates produced "on-site".

Transient variations of leaf growth are almost always seen in laboratory experiments immediately after switching lights on or off (Walter and Schurr 2005). These variations have to be considered in the context of immediate alterations of the hydraulic properties of the plant (closing or opening of stomates, turgor changes) and of a change in apoplastic pH: Switching lights off leads to a transient acidification of the apoplast (Mühling et al. 1995) that is connected with an amplification of cell-wall extensibility (Cosgrove 1999).

5.3
Variation of Root Growth

Finally, the enormous relevance of carbohydrate metabolism for short-term growth variations is also seen in roots. As mentioned above, plants with decreased activity of sucrose-phosphate phosphatase reacted much slower than wild-type plants towards an increase of light intensity (Nagel et al. 2006). Experiments with excised root systems, that continued growth—without being provided with photosynthates from the shoot—as long as sucrose was available from the growth medium, supported the hypothesis that carbohydrate availability is crucial for the intensity of root growth. A direct correlation of root growth intensity with carbohydrate concentration of the growth zone has also been described for *Arabidopsis thaliana* (Freixes et al. 2002). Yet, the small-scale distribution of growth across the root growth zone is not connected to carbohydrate concentrations, but probably regulated by carbohydrate deposition rates and deposition rates of other growth substrates such as mineral nutrients (Walter et al. 2003a).

6
The Effect of Altered Environmental Factors Interfering with the Connection Between Light, Photosynthesis and Leaf Growth Dynamics

A wide range of environmental factors affects leaf and root growth dynamics via direct or indirect alteration of photosynthetic energy gain. Two factors that are investigated in the context of global climate change and its effect on managed plant systems are atmospheric CO_2- and O_3 concentration. While increased CO_2-levels generally ameliorate photosynthesis and primary production, increased amounts of ozone in lower atmospheric layers decrease plant performance via oxidative stress.

6.1
Increased Content of Atmospheric CO_2

The increase of atmospheric CO_2-concentration from pre-industrial values around 280 ppm to currently more than 350 ppm is strongly accelerating the "greenhouse effect" and is hence the main driver for global warming. Several studies performed at the Biosphere 2 Center investigated in which way growth behavior of a large stand of the agroforestry model species *Populus deltoides* was affected by two- and threefold (most extreme scenario envisaged for the year 2100, IPCC 2001) increased concentration of atmospheric CO_2 (Murthy et al. 2005; Barron-Gafford et al. 2005; Walter et al. 2005). Plants were cultivated for four years in completely enclosed and climate-controlled biomes of 600 m^2 area, 10 m height and 1 m soil depth.

For a number of species, results have shown that photosynthesis is increased by an elevation of atmospheric CO_2 (Poorter and Navas 2003; Long et al. 2004; Ainsworth and Long 2005). The degree of photosynthetic amelioration depends on interaction with other factors such as temperature (Turnbull et al. 2002) and nutrient availability (Kruse et al. 2003). In growing leaves, assimilation is often stimulated more strongly than in fully differentiated leaves (Pearson and Brooks 1995; Miller et al. 1997; Wait et al. 1999), which leads to increasing contents of starch, sucrose, glucose and fructose (Poorter et al. 1997). If plants are exposed to elevated CO_2 for a longer time, acclimation is observed and assimilation decreases gradually (Stitt 1991; Ainsworth et al. 2003). Growth usually reacts in a less-pronounced way towards elevated CO_2, but data reported in the literature shows an enormous variability (Poorter and Navas 2003). Similar to biomass growth, crop yield from studies under elevated CO_2 shows huge variability which always depends on interaction of elevated CO_2 with other environmental factors. Hence, it is extremely difficult to predict the effect of elevated CO_2 on future crop yields (Long et al. 2006).

Investigation of leaf growth of *Populus deltoides* in the Biosphere 2 Center during the 2002 growth period showed that a three-times elevated CO_2-concentration led to an increase of final leaf area of 22% (Walter et al. 2005). This value correlated closely with the total increase of aboveground biomass of 27% in this season (Barron-Gafford et al. 2005) and confirmed results of studies performed with other species of Populus (Ferris et al. 2001). The increase of total leaf area was realized predominantly in later phases of individual leaf development and not by enlargement of meristematic leaf initials, which is also supported by findings from the literature (Taylor et al. 2003). For the first time, it could be shown in the study of Walter et al. (2005) that the difference in size of leaves from ambient and elevated CO_2 was increasing throughout the season and that it was correlated with a transient decrease of growth activity under elevated CO_2 in the afternoon which lasted approximately 3 h. This growth decrease coincided temporally with a reduced content of glucose in the growing leaves of the elevated CO_2-treatment, indicating that at this time of the day, carbohydrate partitioning guides photosynthates away from metabolic paths directly feeding growth processes (e.g. glucose) towards increased storage (e.g. starch) or export into the root. This transient afternoon growth decrease explains on a phenotypic level, why the overall increase in leaf growth at elevated CO_2 does not reach the level of enhancement of photosynthesis at elevated CO_2.

Genes that control this acclimation have been investigated in a field experiment at the SoyFACE-facility in Illinois (Ainsworth et al. 2006). There, the effect of a 1.5-times elevated atmospheric CO_2-content on growth, performance and gene transcription in *Glycine max*, one of the most important dicotyledonous crop plants, was investigated. CO_2 was released in a computer-controlled way from a tube system surrounding a field plot of about 20 m diameter (for more details of experimental design see Miglietta et al. 2001). Leaves of *Glycine max* were harvested in the middle of the night when they showed strongest growth activity (Ainsworth et al. 2005). In a cDNA-microarray analysis, 1146 transcripts were identified showing significant differences between developing and full grown leaves. 139 transcripts showed a significant interaction between development and CO_2 content. Both groups comprised a number of transcripts that encoded for ribosomal proteins and for genes involved in the cell cycle and in cell-wall loosening. 327 genes were identified showing differential expression at different external CO_2 concentrations. In contrast to other studies, not only genes leading to increased photosynthesis and carbohydrate production were found, but also a stimulation of respiratory breakdown of starch was reflected at the transcript level.

6.2
Increased Ozone Concentration in Lower Troposphere Layers

The effect of elevated ozone on leaf growth was investigated at the same experimental field site in 2004 (Christ et al. 2006). Ozone concentration was set 20% above ambient concentration; this increase is expected to be seen by 2050. Ozone concentration of the lower tropospheric layer has increased from pre-industrial values of around 10 ppb to a current summer average of about 40 ppb (Pritchard and Amthor 2005; Morgan et al. 2006). Ozone is a powerful oxidizing agent that is already expected to be leading to crop yield losses in the range of some billion $ in the US (Murphy et al. 1999; Lorenzini and Saitanis 2003). The sensitivity for ozone damage differs between plant species; *Glycine max* is known as a generally very sensitive species (Lesser et al. 1990).

In contrast to expectations based on laboratory experiments and "open-top-chamber-studies" (Reid and Fiscus 1998; Rogers et al. 2004; Fiscus et al. 2005) no decrease in crop yield of the investigated cultivar Spencer was observed. This might have been caused by optimal growing conditions for *Glycine max* in Illinois leading to extremely high crop yield in 2004 (Leakey et al. 2006). Yet, the results of Christ et al. (2006) showed that growth, photosynthesis and carbohydrate content of leaves from the upper canopy were reduced markedly in the treated plants. Leaves of the upper canopy are developing during the life phase of pod filling; a developmental stage of the plant when it is especially sensitive to ozone damage (Morgan et al. 2004). Those leaves are the prime source for assimilates going into the pods and developing seeds (Thrower 1962; Gifford and Evans 1981). The fact that those leaves remained smaller than leaves of control plants shows that the plants were able to divert their reserves towards generative organs at the cost of the vegetative growth of the leaves of the upper canopy. This implicates that at least the investigated cultivar still has high potential for optimizing crop yield. In future studies, it will be important to compare the vegetative growth performance of upper canopy leaves in different environmental situations to optimize the selection of lines or cultivars for certain regions and climatic scenarios.

7 Conclusions

Plants possess an enormous phenotypic plasticity and can divert growth substrates and hence growth activity between growing plant modules in a very flexible way, allowing them to react efficiently to fluctuations in environmental parameters. They can increase the root:shoot ratio when light intensity reaching the shoot increases (Walter and Nagel 2006) or when nutrient availability decreases (Scheible et al. 1997; Walter et al. 2003a). The ability of plants to alter the direction of growth by differential redistribution of growth activity across the growth zone of an organ when reacting towards gradients of nutrient concentration, light or gravity has fascinated plant biologists for a long time (Darwin 1880; Perrin 2005). Since the fitness of a plant is strongly increased by dynamic growth reactions towards dynamically changing environmental parameters, plants and organs with diverse growth reactions have evolved in different ecological niches.

When comparing the reaction patterns of leaves and roots, one has to consider that the heterogeneity of environmental conditions to which those organs are exposed, differs strongly. This difference has led to different basic patterns of growth dynamics in leaves and roots. The steady rotation of day and night phases accompanied by strong differences in temperature and humidity during 24 h has led to a huge diel variation of relative growth rate in leaves, even if environmental factors other than light are kept constant. In contrast to this, roots of a wide variety of species grow continually throughout 24 h. Yet, since roots are not used to buffer strong variations of environmental factors, they modulate their growth rates strongly and rapidly in reaction to singular changes of temperature (Pahlavanian and Silk 1988; Walter et al. 2002b), water availability (Fan and Neumann 2004), nutrient availability (Walter et al. 2003a) or light intensity (Nagel et al. 2006).

The control of growth processes is regulated on a number of system levels, ranging from biomechanical constraints (Niklas 1999) via transcriptional control in roots (Birnbaum et al. 2003; Bassani et al. 2004) and leaves (Traintotti et al. 2004; Matsubara et al. 2006; Ainsworth et al. 2006) to regulation by long-distance signals (Heckenberger et al. 1998). In which way environmental stimuli are affecting this regulatory network has to be investigated more intensely in future studies to understand plant performance in fluctuating environmental situations. Models of cellular behavior in the context of growing organs and of plant architecture will help to gain insight into mechanisms of plant development. Supported by such models, the connection between patterns of gene expression and plant architecture is currently being revealed (Prusinkiewicz 2004; Coen et al. 2004).

The investigation of the interaction of heterogeneities of different environmental parameters with dynamic growth patterns will lead to an improved understanding of past, present and future plant behavior. This in turn will help us to understand evolutionary processes, to breed and design optimal crop plants for different environmental scenarios and to assess, how plant ecosystems will react to global climate change.

References

Aguirrezabal LAN, Deleens E, Tardieu F (1994) Root elongation rate is accounted for by intercepted PPFD and source-sink relations in field and laboratory-grown sunflower. Plant Cell Environ 17:443–450

Ainsworth EA, Davey PA, Hymus GJ, Osborne CP, Rogers A, Blum H, Noesberger J, Long SP (2003) Is stimulation of leaf photosynthesis by elevated carbon dioxide concentration maintained in the long term? A test with *Lolium perenne* grown for 10 years at two nitrogen fertilization levels under Free Air CO_2 Enrichment (FACE). Plant Cell Environ 26:705–714

Ainsworth EA, Long SP (2005) What have we learned from 15 years of Free-Air CO_2 Enrichment (FACE)? A meta-analytic review of the responses of photosynthesis, canopy properties and plant production to rising CO_2. New Phytol 165:351–372

Ainsworth EA, Walter A, Schurr U (2005) *Glycine max* leaflets lack a base-tip gradient in growth rate. J Plant Res 118:343–346

Ainsworth EA, Rogers A, Vodkin LO, Walter A, Schurr U (2006) The effects of elevated CO_2 concentration on soybean gene expression. An analysis of growing and mature leaves. Plant Physiol 142:135–147

Allen JF, Nilsson A (1997) Redox signalling and the structural basis of regulation of photosynthesis by protein phosphorylation. Physiol Plant 100:863–868

Avery GS (1933) Structure and development of tobacco leaves. Am J Bot 20:565–592

Barbagallo RP, Oxborough K, Pallett KE, Baker NR (2003) Rapid, non-invasive screening for perturbations of metabolism and plant growth using chlorophyll fluorescence imaging. Plant Physiol 132:485–493

Barron-Gafford G, Martens D, Grieve K, Bil K, Kudeyarov V, McLain JET, Lipson D, Murthy R (2005) Growth of Eastern Cottonwoods (*Populus deltoides*) in elevated [CO_2] stimulates stand-level respiration and rhizodeposition of carbohydrates, accelerates soil nutrient depletion, yet stimulates above- and belowground biomass production. Global Change Biol 11:1220–1233

Bassani M, Neumann PM, Gepstein S (2004) Differential expression profiles of growth-related genes in the elongation zone of maize primary roots. Plant Mol Biol 56:367–380

Beemster GTS, Masle J, Williamson RE, Farquhar GD (1996) Effects of soil resistance to root penetration on leaf expansion in wheat (*Triticum aestivum* L.): kinematic analysis of leaf elongation. J Exp Bot 47:1663–1678

Beemster GTS, Baskin TI (1998) Analysis of cell division and elongation underlying the develop-mental acceleration of root growth in *Arabidopsis thaliana*. Plant Physiol 116:1515–1526

Beemster GTS, De Veylder L, Vercruysse S, West G, Rombaut D, Van Hummelen P, Galichet A, Gruissem W, Inzé D, Vuylsteke M (2005) Genome-wide analysis of gene expression profiles associated with cell cycle transitions in growing organs of *Arabidopsis*. Plant Physiol 138:734–743

Ben-Haj-Salah H, Tardieu F (1995) Temperature affects expansion rate of maize leaves without change in spatial distribution of cell length. Analysis of the coordination between cell division and cell expansion. Plant Physiol 109:861–870

Bigün J, Granlund GH (1987) Optimal orientation detection of linear symmetry. Proc 1st Int Conf Computer Vision. ICCV, London, June 8–11, 1987

Bingham IJ, Blackwood JM, Stevenson EA (1997) Site, scale and time-course for adjustments in lateral root initiation in wheat following changes in C and N supply. Annals Bot 80:97–106

Birnbaum K, Shasha DE, Wang JY, Jung JW, Lambert GM, Galbraith DW, Benfey PN (2003) A gene expression map of the *Arabidopsis* root. Science 302:1956–1960

Björkman O (1981) Responses to different quantum flux densities. In: Lange OL, Nobel PS, Osmond CB, Ziegler H (eds) Encyclopaedia of Plant Physiology vol 12A. Physiological Plant Ecology I Responses to the Physical Environment. Springer, Berlin Heidelberg New York, pp 96–117

Brumfield RT (1942) Cell growth and division in living root meristems. Am J Bot 29:533–543

Bünning E (1948) Entwicklungs- und Bewegungsphysiologie der Pflanze. Springer, Berlin Heidelberg New York

Bünning E, Moser I (1966) Response-Kurven bei der circadianen Rhythmik von *Phaseolus*. Planta 69:101–110

Chavarría-Krauser A, Nagel KA, Palme K, Schurr U, Walter A, Scharr H (2008) Spatio-temporal quantification of differential growth processes in root growth zones based on a novel combination of image sequence processing and refined concepts describing curvature production. New Phytol 175:811–821

Chen S, Hajirezaei M, Peisker M, Tschiersch H, Sonnewald U, Börnke F (2005) Decreased sucrose-6-phosphate phosphatase level in transgenic tobacco inhibits photosynthesis, alters carbohydrate partitioning, and reduces growth. Planta 221:479–492

Chia T, Thorneycroft D, Chapple A, Messerli G, Chen J, Zeeman SC, Smith SM, Smith AM (2004) A cytosolic glucosyltransferase is required for conversion of starch to sucrose in *Arabidopsis* leaves at night. The Plant J 37:853–863

Christ MM, Ainsworth EA, Nelson R Schurr U, Walter A (2006) Anticipated yield loss in field-grown soybean under elevated ozone can be avoided at the expense of leaf growth during early reproductive growth stages in favourable environmental conditions. J Exp Bot 57:2267–2275

Christ RA (1978) The elongation rate of wheat leaves. I. The elongation rates during night and day. J Exp Bot 29:603–610

Coen E, Rolland-Lagan A-G, Matthews M, Bangham JA, Prusinkiewicz P (2004) The genetics of geometry. Proc Natl Acad Sci USA 101:4728–4735

Cosgrove DJ (1999) Enzymes and other agents that enhance cell wall extensibility. Ann Rev Plant Physiol Plant Mol Biol 50:391–417

Darwin C (1880) The power of movements in plants. Murray, London

Demmig-Adams B, Adams WW III (1992) Photoprotection and other responses of plants to high light stress. Annu Rev Plant Physiol Plant Mol Biol 43:599–626

De Veylder L, Beeckman T, Beemster GTS, Krols L, Terras P, Landrieu I, Van der Schueren E, Maes S, Naudts M, Inze D (2001) Functional analysis of cyclin-dependent kinase inhibitors of *Arabidopsis*. Plant Cell 13:1653–1667

Dodd AN, Salathia N, Hall A, Kévei E, Tóth R, Nagy F, Hibberd JM, Millar AJ, Webb AAR (2005) Plant circadian clocks increase photosynthesis, growth, survival, and competitive advantage. Science 309:630–633

Donnelly PM, Bonetta D, Tsukaya H, Dengler RE, Dengler NG (1999) Cell cycling and cell enlargement in developing leaves of *Arabidopsis*. Develop Biol 215:407–419

El-Lithy ME, Clerkx EJM, Ruys GJ, Koornneef M, Vreugdenhil D (2004) Quantitative trait locus analysis of growth-related traits in a new *Arabidopsis* recombinant. Plant Physiol 135:444–458

Erickson RO, Sax KW (1956) Experimental growth rate of primary root of *Zea mays*. Proc Am Philosoph Soc 100:487–498

Evans JR, Poorter H (2001) Photosynthetic acclimation of plants to growth irradiance: the relative importance of specific leaf area and nitrogen partitioning in maximizing carbon gain. Plant Cell Environ 24:755–767

Fan L, Neumann PM (2004) The spatially variable inhibition by water deficit of maize root growth correlates with altered profiles of proton flux and cell wall pH. Plant Physiol 135:2291–2300

Ferris R, Sabatti M, Miglietta F, Mills RF, Taylor G (2001) Leaf area is stimulated in *Populus* by free air CO_2 enrichment (POPFACE), through increased cell expansion and production. Plant Cell Environ 24:305–315

Fiscus EL, Booker FL, Burkey KO (2005) Crop responses to ozone: uptake, models of action, carbon assimilation and partitioning. Plant Cell Environ 28:997–1011

Foster AS (1936) Leaf Differentiation in Angiosperms. The Botanical Rev 2:349–372

Frak E, Le Roux X, Millard P, Adam B, Dreyer E, Escuit C, Sinoquet H, Vandame M, Variet-Grancher C (2002) Spatial distribution of leaf nitrogen and photosynthetic capacity within the foliage of individual trees: disentangling the effects of local light quality, leaf irradiance, and transpiration. J Exp Bot 53:2207–2216

Freixes S, Thibaud MC, Tardieu F, Muller B (2002) Root elongation and branching is related to local hexose concentration in *Arabidopsis thaliana* seedlings. Plant Cell Environ 25:1357–1366

Geiger DR, Servaites JC, Fuchs MA (2000) Role of starch in carbon translocation and partitioning at the plant level. Aust J Plant Physiol 27:571–582

Gifford RM, Evans LT (1981) Photosynthesis, carbon partitioning and yield. Ann Rev Plant Physiol 32:485–509

Goodwin RH, Stepka W (1945) Growth and differentiation in the root tip of *Phleum pratense*. Am J Bot 32:36–46

Gouws LM, Osmond CB, Schurr U, Walter A (2005) Distinctive diel growth cycles in leaves and cladodes of CAM plants: differences from C_3-plants and putative interactions with substrate availability, turgor and cytoplasmic pH. Funct Plant Biol 32:421–428

Granier C, Aguirrezabal L, Chenu K, Cookson SJ, Dauzat M, Hamard P, Thioux JJ, Rolland G, Bouchier-Combaud S, Lebaudy A, Muller B, Simonneau T, Tardieu F (2006) PHENOPSIS, an automated platform for reproducible phenotyping of plant responses to soil water deficit in *Arabidopsis thaliana* permitted the identification of an accession with low sensitivity to soil water deficit. New Phytol 169:623–625

Green RM, Tingay S, Wang ZY, Tobin EM (2002) Circadian rhythms confer a higher level of fitness to *Arabidopsis* plants. Plant Physiol 129:576–584

Hannam RV (1968) Leaf growth and development in the young tobacco plant. Aust J Biol Sci 21:855–870

Haußecker H, Spies H (1999) Motion. In: Jähne B, Haußecker H, Geißler P (eds) Handbook of Computer Vision and Applications. Academic Press, San Diego CA, USA, pp 310–369

Head GC (1965) Studies of diurnal changes in cherry root growth and nutational movements of apple root tips by time-lapse cinematography. Annals Bot 29:219–224

Heckenberger U, Roggatz U, Schurr U (1998) Effect of drought stress on the cytological status in *Ricinus communis*. J Exp Bot 49:181–189

Hejnowicz Z, Erickson RO (1968) Growth inhibition and recovery in roots following temporary treatment with auxin. Physiol Plantarum 21:302–313

Hodge A, Paterson E, Thornton B, Millard P, Killham K (1997) Effects of photon flux density on carbon partitioning and rhizosphere carbon flow of *Lolium perenne*. J Exp Bot 48:1797–1805

Hsiao TC, Acevedo E, Henderson DW (1970) Maize leaf elongation: Continuous measurements and close dependence on plant-water status. Science 168:590–591

Iijima M, Oribe Y, Horibe Y, Kono Y (1998) Time lapse analysis of root elongation rates of rice and *sorghum* during the day and night. Annals Bot 81:603–607

IPCC (2001) In: Houghton JT, Ding Y, Griggs DJ, Noguer M, van der Linden PJ, Dai X, Maskell K, Johnson CA (eds) Climate Change 2001: The Scientific Basis. Cambridge University Press, Cambridge, UK

Kehr J, Hustiak F, Walz C, Willmitzer L, Fisahn J (1998) Transgenic plants changed in carbon allocation pattern display a shift in diurnal growth pattern. The Plant J 16:497–503

Kemp DR, Blacklow WM (1980) Diurnal extension rates of wheat leaves in relation to temperatures and carbohydrate concentrations of the extension zone. J Exp Bot 31:821–828

Kofler H, Häusler RE, Schulz B, Gröner F, Flügge UI, Weber A (2000) Molecular characterisation of a new mutant allele of the plastid phosphoglucomutase in *Arabidopsis*, and complementation of the mutant with the wild-type cDNA. Mol Gen Genet 263:978–986

Kruger EL, Volin JC (2006) Reexamining the empirical relation between plant growth and leaf photosynthesis. Funct Plant Biol 33:421–429

Kruse J, Hetzger I, Mai C, Polle A, Rennenberg H (2003) Elevated pCO_2 affects *N*-metabolism of young poplar plants (*Populus tremolo* XP. alba) differently at deficient and sufficient N-supply. New Phytol 157:65–81

Lai IL, Scharr H, Chavarria-Krauser A, Küsters R, Wu JT, Chou CH, Schurr U, Walter A (2005) Leaf growth dynamics of two congener gymnosperm tree species reflect the heterogeneity of light intensities given in their natural ecological niche. Plant Cell Environ 28:1496–1505

Lambers H, Posthumus F (1980) The effect of light intensity and relative humidity on growth rate and root respiration of *Plantago lanceolota* and *Zea mays*. J Exp Bot 31:1621–1630

Leakey ADB, Uribelarrea M, Ainsworth EA, Naidu SL, Rogers A, Ort DR, Long SP (2006) Photosynthesis, productivity and yield of maize are not affected by open-air elevation of CO_2 concentration in the absence of drought. Plant Physiol 140:779–790

Leister D, Varotto C, Pesaresi P, Niwergall A, Salamini F (1999) Large-scale evaluation of plant growth in *Arabidopsis thaliana* in non-invasive image analysis. Plant Physiol Biochem 37:671–678

Lesser VM, Rawlings JO, Spruill SE, Somerville MC (1990) Ozone effects on agricultural crops: Statistical methodologies and estimated dose-response relationships. Crop Sci 30:148–155

Long SP, Ainsworth EA, Rogers A, Ort DR (2004) Rising atmospheric carbon dioxide: Plants face the future. Ann Rev Plant Biol 55:591–628

Long SP, Ainsworth EA, Leakey ADB, Nösberger J, Ort DR (2006) Food for thought: Lower-Than-Expected Crop Yield Stimulation with Rising CO_2 Concentrations. Science 312:1918–1921

Lorenzini G, Saitanis C (2003) Ozone: a novel plant pathogen. In: Sanita di Toppi L, Pawlik-Skowronska B (eds) Abiotic stresses in plants. Kluwer Academic Press, Dordrecht, The Netherlands, pp 205–229

Lüttge U (2003) Circadian rhythmicity: Is the biological clock hardware or software? Prog Bot 64:277–319

Matsubara S, Hurry V, Druart N, Benedict C, Janzik I, Chavarría-Krauser A, Walter A, Schurr U (2006) Nocturnal changes in leaf growth of *Populus deltoides* are controlled by cytoplasmic growth. Planta 223:1315–1328

Matsubara S, Walter A (2006) In: Esser K, Lüttge U, Beyschlag W, Murata J (eds) Living in day-night cycles—Specific diel leaf growth patterns and the circadian control of photomorphogenesis. Progress in Botany 68. Springer, Berlin Heidelberg New York, pp 288–314

Matt P, Schurr U, Klein D, Krapp A, Stitt M (1998) Growth of tobacco in short-day conditions leads to high starch, low sugars, altered diurnal changes in the Nia transcript and low nitrate reductase activity and inhibition of amino acid synthesis. Planta 207:27–41

McClung CR (2001) Circadian rhythms in plants. Annu Rev Plant Physiol Plant Mol Biol 52:139–162

Miglietta F, Peressotti A, Vaccari FP, Zaldei A, deAngelis P, Scarascia-Mugnozza G (2001) Free-air CO_2 enrichment (FACE) of a poplar plantation: the POPFACE fumigation system. New Phytol 150:465–476

Miller A, Tsai CH, Hemphill D, Endres M, Rodermel S, Spalding M (1997) Elevated CO_2 effects during leaf ontogeny (A new perspective on acclimation). Plant Physiol 115:1195–1200

Morgan PB, Bernacchi DJ, Ort DR, Long SP (2004) An in vivo analysis of the effect of season-long open-air elevation of ozone to anticipated 2050 levels on photosynthesis in soybean. Plant Physiol 135:2348–2357

Morgan PB, Mies TA, Bollero GA, Nelson RL, Long SP (2006) Season-long elevation of ozone concentration to projected 2050 levels under fully open-air conditions substantially decreases the growth and production of soybean. New Phytol 170:333–343

Mühling KH, Plieth C, Hansen UP, Sattelmacher B (1995) Apoplastic pH of intact leaves of *Vicia faba* as influenced by light. J Exp Bot 46:377–382

Murphy JJ, Delucchi MA, McCubbin DR, Kim HJ (1999) The cost of crop damage caused by ozone air pollution from motor vehicles. J Environ Manage 55:273–289

Murthy R, Barron-Gafford G, Dougherty PM, Engel VC, Grieve K, Handley L, Klimas C, Potosnak MJ, Zarnoch SJ, Zhang J (2005) Increased leaf area dominates carbon flux response to elevated CO_2 in stands of *Populus deltoides* (Bartr.) and underlies a switch from canopy light-limited CO_2 influx in well-watered treatments to individual leaf, stomatally-limited influx under water stress. Global Change Biol 11:716–731

Nagel KA, Schurr U, Walter A (2006) Dynamics of root growth stimulation in *Nicotiana tabacum* in increasing light intensity. Plant Cell Environ 29:1936–1945

Niinemets Ü, Kull O, Tenhunen JD (2004a) Within canopy variation in the rate of development of photosynthetic capacity is proportional to integrated quantum flux density in temperate deciduous trees. Plant Cell Environ 27:293–313

Niinemets Ü, Tenhunen JD, Beyschlag W (2004b) Spatial and age-dependent modifications of photosynthetic capacity in four Mediterranean oak species. Funct Plant Biol 31:1179–1193

Niklas KJ (1999) A mechanical perspective on foliage leaf form and function. New Phytol 143:19–31

Niyogi KK (1999) Photoprotection revisited: Genetic and molecular approaches. Ann Rev Plant Physiol Plant Mol Biol 50:333–359

Ouyang Y, Andersson CR, Kondo T, Golden SS, Johnson CH (1998) Resonating circadian clocks enhance fitness in cyanobacteria. Proc Natl Acad Sci USA 95:8660–8664

Pahlavanian AM, Silk WK (1988) Effect of temperature on spatial and temporal aspects of growth in the primary maize root. Plant Physiol 87:529–532

Pearson M, Brooks GL (1995) The influence of elevated CO_2 on growth and age-related-changes in leaf gas-exchange. J Exp Bot 46:1651–1659

Perrin RM, Young LS, Murthy N, Harrison BR, Wang Y, Will JL, Masson PH (2005) Gravity signal transduction in primary roots. Annals Bot 96:737–743

Poorter H, Navas ML (2003) Plant growth and competition at elevated CO_2: on winners, losers and functional groups. New Phytol 157:175–198

Poorter H, Van Berkel Y, Baxter R, Den Hertog J, Dijkstra P, Gifford RM, Griffin KL, Roumet C, Roy J, Wong SC (1997) The effect of elevated CO_2 on the chemical composition and construction costs of leaves of 27 C-3 species. Plant Cell Environ 20:472–482

Pritchard SG, Amthor JS (2005) Crops and environmental change. An introduction to effects of global warming, increasing atmospheric CO_2 and O_3 concentrations, and soil salinization on crop physiology and yield. Food Products Press, New York, pp 1–37

Prusinkiewicz P (2004) Modeling plant growth and development. Curr Opin Plant Biol 7:79–83

Reid CD, Fiscus EL (1998) Effects of elevated [CO_2] and/or ozone on limitations to CO_2 assimilation in soybean (*Glycine max*). J Exp Bot 49:885–895

Rogers A, Allen DJ, Davey PA, Morgan PB, Ainsworth EA, Bernacchi CJ, Cornic G, Dermody O, Dohleman FG, Heaton EA, Mahoney J, Zhu X-G, Delucia EH, Ort DR, Long SP (2004) Leaf photosynthesis and carbohydrate dynamics of soybeans grown throughout their life-cycle under free-air carbon dioxide enrichment. Plant Cell Environ 27:449–458

Roggatz U, McDonald AJS, Stadenberg I, Schurr U (1999) Effect of nitrogen deprivation on cell division and expansion of *Ricinus communis* L. Plant Cell Environ 22:81–90

Sachs J (1887) Vorlesung über Pflanzen-Physiologie. Wilhelm Engelmann, Leipzig

Sack L, Cowan PD, Jaikumar N, Holbrook NM (2003) The "hydrology" of leaves: coordination of structure and function in temperate woody species. Plant Cell Environ 26:1343–1356

Scharr H (2004) Optimal filters for extended optical flow. Int Workshop on Complex Motion. Springer, Heidelberg, Günzberg, Germany

Scheible WR, Lauerer M, Schulze ED, Caboche M, Stitt M (1997) Accumulation of nitrate in the shoot acts as a signal to regulate shoot-root allocation in tobacco. The Plant J 11:671–691

Schmundt D, Stitt M, Jähne B, Schurr U (1998) Quantitative analysis of the local rates of growth of dicot leaves at a high temporal and spatial resolution, using image sequence analysis. The Plant J 16:505–514

Schultz TF, Kay SA (2003) Circadian clocks in daily and seasonal control of development. Science 301:326–328

Schurr U, Walter A, Rascher U (2006) Functional dynamics of plant growth and photosynthesis – from steady-state to dynamics – from homogeneity to heterogeneity. Plant Cell Environ 29:340–352

Seneweera SP, Basra AS, Barlow EW, Conroy JP (1995) Diurnal regulation of leaf blade elongation in rice by CO_2. Plant Physiol 108:1471–1477

Shipley B (2006) Net assimilation rate, specific leaf area and leaf mass ratio: which is most closely correlated with relative growth rate? A meta-analysis. Funct Ecol 20:565–574

Silk WK, Erickson RO (1979) Kinematics of plant growth. J Theor Biol 76:481–501

Silk WK (1984) Quantitative descriptions of development. Ann Rev Plant Physiol 35:479–518

Silk WK, Lord EM, Eckard KJ (1989) Growth patterns inferred from anatomical records. Plant Physiol 90:708–713

Somers DE (1999) The physiology and molecular bases of the plant circadian clock. Plant Physiol 121:9–19

Staiger D (2002) Circadian rhythms in *Arabidopsis*: time for nuclear proteins. Planta 214:334–344

Stitt M (1991) Rising carbon dioxide levels and their potential significance for carbon flow in photosynthetic cells. Plant Cell Environ 14:741–762

Taylor G, Tricker PJ, Zhang FZ, Alston VJ, Miglietta F, Kuzminsky E (2003) Spatial and temporal effects of free-air CO_2 enrichment (POPFACE) on leaf growth, cell expansion, and cell production in a closed canopy of poplar. Plant Physiol 131:177–185

Terashima I, Hikosaka K (1995) Comparative ecophysiology of leaf and canopy photosynthesis. Plant Cell Environ 18:1111–1128

Terashima I, Hanba YT, Tazoe Y, Vyas P, Yano S (2006) Irradiance and phenotype: comparative eco-development of sun and shade leaves in relation to photosynthetic CO_2 diffusion. J Exp Bot 57:343–354

Thrower SL (1962) Translocation of labeled assimilates in the soybean. II. The pattern of translocation in intact and defoliated plants. Aust J Biol Sci 15:629–649

Trainotti L, Pavanello A, Casadoro G (2004) Differential expression of genes in apical and basal tissues of expanding tobacco leaves. Plant Sci 167:679–686

Turnbull MH, Murthy R, Griffin KL (2002) The relative impacts of daytime and nighttime warming on photosynthetic capacity in *Populus deltoides*. Plant Cell Environ 25:1729–1737

van der Weele CM, Jiang HS, Palaniappan KK, Ivanov VB, Palaniappan K, Baskin TI (2003) A new algorithm for computational image analysis of deformable motion at high spatial and temporal resolution applied to root growth. Roughly uniform elongation in the meristem and also, after an abrupt acceleration, in the elongation zone. Plant Physiol 132:1138–1148

Van Volkenburgh E (1987) Regulation of dicotyledonous leaf growth. In: Cosgrove DJ, Knievel DP (eds) Physiology of cell expansion during plant growth. Am Soc Plant Physiol, 193–201

Van Volkenburgh E, Taylor G (1996) Leaf growth physiology. In: Stettler RF, Bradshaw HD Jr, Heilman PE, Hinckley TM (eds) Biology of Populus. NRC Research Press, Ottawa, pp 283–299

Vincent CD, Gregory PJ (1989) Effects of temperature on the development and growth of winter wheat roots. I. Controlled glasshouse studies of temperature, nitrogen and irradiance. Plant Soil 119:87–97

Wait DA, Jones CG, Wynn J, Woodward FI (1999) The fraction of expanding to expanded leaves determines the biomass response of *Populus* to elevated CO_2. Oecologia 121:193–200

Walter A, Schurr U (1999) The modular character of growth in *Nicotiana tabacum* plants under steady-state nutrition. J Exp Bot 50:1169–1177

Walter A, Feil R, Schurr U (2002a) Restriction of nyctinastic movements and application of tensile forces to leaves affects diurnal patterns of expansion growth. Function Plant Biol 29:1247–1258

Walter A, Spies H, Terjung S, Kuesters R, Kirchgeßner N, Schurr U (2002b) Spatio-temporal dynamics of expansion growth in roots: automatic quantification of diurnal course and temperature response by digital image sequence processing. J Exp Bot 53:689–698

Walter A, Feil R, Schurr U (2003a) Expansion dynamics, metabolite composition and substance transfer of the primary root growth zone of *Zea mays* L. grown in different external nutrient availabilities. Plant Cell Environ 26:1451–1466

Walter A, Roggatz U, Schurr U (2003b) Expansion kinematics are an intrinsic property of leaf development and are scaled from cell to leaf level at different nutrient availability. Plant Biol 5:642–650

Walter A, Lambrecht SC (2004) Biosphere 2 Center as a unique tool for environmental studies. J Environ Monitor 6:267–277

Walter A, Rascher U, Osmond B (2004) Transitions in photosynthetic parameters of midvein and interveinal regions of leaves and their importance during leaf growth and development. Plant Biol 6:184–191

Walter A, Schurr U (2005) Dynamics of leaf and root growth— endogenous control versus environmental impact. Annals Bot 95:891–900

Walter A, Christ MM, Barron-Gafford GA, Grieve KA, Murthy R, Rascher U (2005) The effect of elevated CO_2 on diel leaf growth cycle, leaf carbohydrate content and canopy growth performance of *Populus deltoides*. Global Change Biol 11:1207–1219

Walter A, Nagel KA (2006) Root growth reacts rapidly and more pronounced than shoot growth towards increasing light intensity in tobacco seedlings. Plant Signal Behav 1:225–226

Walter A, Scharr H, Gilmer F, Zierer R, Nagel KA, Ernst M, Wiese A, Virnich O, Christ MM, Uhlig B, Jünger S, Schurr U (2007) The dynamics of seedling growth acclimation towards altered light conditions can be quantified via GROWSCREEN – a setup and procedure designed for rapid optical phenotyping of different plant species. New Phytol 174:447–455

Watts WR (1974) Leaf extension in *Zea mays.* J Exp Bot 25:1085–1096

Watts WR (1974) Leaf extension in *Zea mays.* J Exp Bot 25:1085–1096

Webb DP (1976) Root growth in acer saccharum marsh. Seedlings: Effects of light intensity and photoperiod on root elongation rates. Bot Gaz 137:211–217

Wiese A, Christ MM, Virnich O, Schurr U, Walter A (2007) Spatio-temporal leaf growth patterns of *Arabidopsis thaliana* and evidence for sugar control of the diel leaf growth cycle. New Phytol 174:752–761

Wollman F-A (2001) State transitions reveal the dynamics and flexibility of the photosynthetic apparatus. EMBO J 20:3623–3630

Plant Cell Monogr (10)
L. Bögre and G. Beemster: Plant Growth Signaling
DOI 10.1007/7089_2007_156/Published online: 18 January 2008

Keeping the Balance Between Proliferation and Differentiation by the E2F Transcriptional Regulatory Network is Central to Plant Growth and Development

Zoltan Magyar

School of Biological Sciences, Royal Holloway, University of London, Egham Hill, Egham TW20 0EX, UK
Zoltan.Magyar@rhul.ac.uk

Abstract The canonical role for E2F transcription factors is to regulate G1 to S transition, but it is becoming apparent in many systems that E2Fs have broader functions and that, besides the regulation of cell cycle transitions, they coordinate cell proliferation with cell growth and differentiation. The current model is that E2Fs can work both as positive and negative regulators of transcription, dependent on their structure and on the function of the retinoblastoma (RB) tumour suppressor protein. Here we discuss the current understanding of how the plant RB-E2F pathway works and its central role in plant growth and development.

1 Introduction

In higher plants, organ formation occurs throughout their life via the maintenance of a reservoir of stem cells in the shoot and root apical regions called meristems (Doerner 2007; Scofield and Murray 2006; Sharma et al. 2003). Plant growth regulation is tightly tuned with the genetic setup and environmental conditions. In meristems undifferentiated cells are produced by cell proliferation, and when these cells stop dividing, as they leave the meristematic region, they differentiate into specific tissues. During differentiation, plant cells frequently increase their DNA content by a modified mitotic cycle called endoreduplication, a process of continuous DNA synthesis without intervening mitosis (Inze and De Veylder 2006). Cell division and differentiation are well co-ordinated events during plant life, but the molecular mechanisms that maintain the balance between these processes are still not well understood.

The plant cell cycle is regulated by conserved molecular elements; pivotal among them are cyclins and cyclin-dependent kinases (CDKs) that drive the cell through the cell cycle control points. The canonical CDK, CDKA1 is an essential gene in *Arabidopsis* and is required both for gamete and somatic cell proliferation (Dissmeyer et al. 2007). The CDK family of genes have largely expanded with around 30 members in *Arabidopsis*. CDKBs are particularly specialized in the regulation of mitosis, while others might work outside the cell cycle, such as in the regulation of transcription. The family of cyclins is also expanded in plants where they are similarly grouped to D-, A-, and B-types as in animals (Inze and De Veylder 2006). CDK activity is regulated by inhibitors of CDK (ICKs), also known as Kip-related proteins (KRPs) in plants (De Veylder et al. 2001; Wang et al. 1997). Recently a new class of CDK inhibitor genes was discovered, SIAMESE (SIM), a regulator of trichome development through endoreduplication (Churchman et al. 2006). CDK is also regulated through activating phosphorylation at the T-loop by a CDK activation kinase (CAK) (Umeda et al. 2005) and through inhibitory phosphorylation by the WEE1 kinase (Inze 2005). The identity of the phosphatase that removes this inhibitory phosphorylation from CDKA is currently debated (Boudolf et al. 2006).

This chapter focusses on the role of the RB-E2F pathway in plant growth control. Readers are directed to excellent recent reviews on general description of the plant cell cycle control (Dewitte and Murray 2003; Inze 2005; Inze and De Veylder 2006; Ramirez-Parra et al. 2005). Our knowledge about E2Fs and RBs comes from animal studies. Models of how the RB-E2F pathway controls the G1–S transition can be found in every textbook of molecular biology. However, recent discoveries have put these models under serious challenge (Rowland and Bernards 2006). Therefore first I will summarize the available data on the RB-E2F pathway in animal cells and later focus on the study of this pathway in plants.

2 Animal E2Fs

In the majority of eukaryotic organisms, the decision to enter or leave the cell division cycle is taken in the G1 phase. The E2F-RB pathway plays an important role in this regulatory process. RB was the first tumour suppressor gene cloned from mammalian cells, while E2F1 was identified through its ability to form a complex with the RB protein (Bagchi et al. 1991; Bandara and La Thangue 1991; Du and Pogoriler 2006).

More than 100 proteins are potentially able to interact with the animal RB, indicating that RB function is far more complex than it was initially suggested, but still the best studied binding partners of RB are the E2F transcription factors (Du and Pogoriler 2006). Structural relatives of this pathway have been identified in the unicellular green algae *Chlamydomonas reinhardtii* as well as in multicellular organisms, indicating that these regulators might play a role in an evolutionary conserved mechanism. The structural relatives of RB and E2Fs are missing from yeast, but the regulatory logic also applies; the G1 to S transition is regulated by the removal of a transcriptional repressor Whi5 of G1 regulation (Cooper 2006).

The widely accepted model for E2F function is the formation of heterodimers between E2F and its dimerization partner (DP) that will activate the expression of genes required for entering the cell cycle. RB inhibits this event by physically binding to E2Fs at their carboxyl terminal RB binding motif and inhibiting its activity. Upon mitogen stimulation, this repression is suppressed by hyperphosphorylation of RB by specific CDK–cyclin D complexes (Sherr and Roberts 1999), leading to the activation of genes required for DNA synthesis. Further studies, however, revealed that the function of E2Fs is much more complex, since animal E2Fs can either activate or repress transcription (Du and Pogoriler, 2006; Rowland and Bernards 2006). For instance, the majority of the eight mammalian E2Fs (E2F1–8) are repressors (E2F4–8) that could either inhibit transcription in RB-dependent (E2F4 and E2F5) or RB-independent manner (E2F6, E2F7, E2F8).

The basis of the RB-dependent transcriptional repression through E2Fs is the ability of RB, and its pocket protein relatives, p107 and p130, to simultaneously bind to E2Fs and chromatin remodelling enzymes such as histone deacetylases (HDACs) (Rayman et al. 2002) or histone methyl transferases (e.g. SUV39H1) (Liu et al. 2005). In addition, there are examples that activator E2Fs are also able to repress transcription (Aslanian et al. 2004). Nevertheless, loss of activator E2Fs in mammalian cells reduces gene expression of E2F target genes and inhibits cell division (Wu et al. 2001). On the other hand, mutation of repressor E2Fs resulted in an increase in E2F-dependent gene expression in quiescent cells (Attwooll et al. 2004). However, whether the repressor/activator function of cell proliferation is the most important role of the mammalian E2Fs is still not clear, since blocking all E2F activities by overexpression of a dominant negative E2F mutant form lacking the C-terminal transactivation and RB binding domains, rather than inhibiting cell proliferation, results in blocking of cell cycle exit and differentiation (Rowland et al. 2002; Zhang et al. 1999).

In *Drosophila*, the interplay between E2Fs is simplified as there are only two E2F transcription factors, dE2F1 and dE2F2, that have antagonistic effects on cell division during larval development: dE2F1 is an activator while dE2F2 is a repressor (Frolov et al. 2001). Loss of dE2F1 function resulted in a serious proliferation defect in the mutant fly, which surprisingly was restored by the simultaneous loss of the repressing dE2F2. This observation indicates that dE2F1 activates transcription by replacing the repressor dE2F2 from promoter sequences of target genes containing E2F-binding sites. According to this model, E2F-mediated repression limits the rates of cell proliferation. Further studies also revealed that dE2F2 and the retinoblastoma family (RBF) of proteins provide a repressor activity that is uncoupled from cell cycle progression, and that loss of E2F-mediated repression results in the inappropriate expression of tissue-specific genes and markers for differentiation (Dimova et al. 2003) (Fig. 1). Similarly loss of repressor E2F4 in mouse embryonic fibroblasts allows cells to undergo spontaneous differentiation (Landsberg

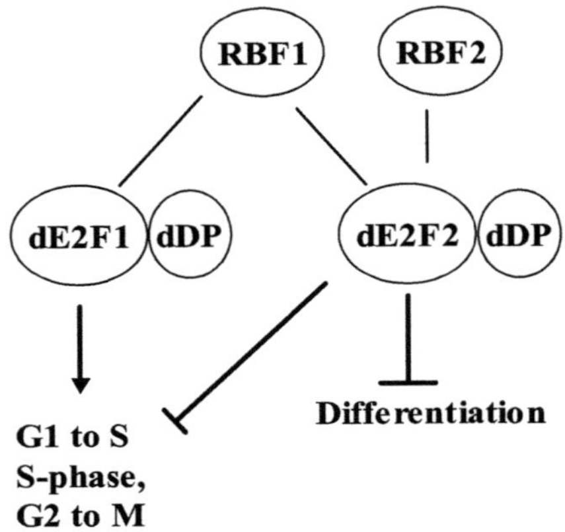

Fig. 1 Model of E2F transcriptional regulation in *Drosophila*. dE2F1 and dE2F2 are the only E2Fs in *Drosophila* and both need to interact with the single dDP protein for efficient DNA-binding, but they show different preferences for the retinoblastoma protein. dE2F1 interacts only with RBF1, while dE2F2 associates both with RBF1 and RBF2. dE2F1 and dE2F2 are functionally antagonistic transcription factors. dE2F1 activates the expression of key cell cycle regulators for both G1–S and G2–M transitions such as cyclin E and Cdc25 genes, respectively (Neufeld and Edgar 1998), by preventing the recruitment of repressor dE2F2 to the DNA on cell cycle gene promoters. dE2F2, on the other hand, binds and inhibits the expression of cell cycle genes as well as of a variety of tissue-specific genes involved in differentiation that are not activated by dE2F1 (Dimova et al. 2003; Frolov et al. 2001)

et al. 2003). Although it has been suggested that dE2Fs are not essential for cell proliferation, since an E2F-independent mechanism is sufficient for high basal level of gene expression in the absence of dE2Fs, it is likely, however, that the balance of E2F-dependent activation and repression is involved in the coordination of cell proliferation and differentiation.

Even though initially the animal E2F function was associated mainly with the control of G1–S transition, further studies clearly show that they are involved in the regulation of G2–M transition as well. There are a number of mitotic target genes for mammalian E2Fs including cyclin B1 (Zhu et al. 2005) and Mad2, a spindle checkpoint gene (Hernando et al. 2004). In *Drosophila*, the key target gene of dE2F1 to control G2–M transition is the phosphatase, CDC25, an activator of the mitotic CDK1 (Neufeld and Edgar 1998).

3
The *Chlamidomonas* E2F-RB Pathway; the Simplest Scenario

The simplest extensively studied photosynthesizing organism that contains the conserved elements of the RB-E2F pathway is the unicellular green alga, *Chlamydomonas reinhardtii*. Recent genome sequencing analysis of this alga revealed that it has single copy genes encoding homologues of RB/MAT3, E2F and DP; therefore it provides the simplest model for the analysis of the RB-E2F pathway (Bisova et al. 2005; and another chapter in this volume). RB (mat3) mutant algae cells undergo supernumerous divisions, while mutations in dp1 or in e2f1 suppress this ability (Fang et al. 2006; Umen and Goodenough 2001). These data suggest a similar function for RB/MAT3 as its animal counterparts in repressing E2F-DP function. Since rb/mat3/e2f1/dp1mutant algae could divide more or less normally, the RB-E2F pathway might not be essential for cell division in *Chlamydomonas*. However, cell size control was abrogated in the mat3-e2f-dp mutant; rb/mat3 divide at an extremely small size, while dp1 and certain e2f1 mutants became larger than wild-type, but the molecular mechanism behind these observations is not well understood.

4 The *Arabidopsis* E2F Family

Arabidopsis contains a family of six E2F related proteins (Vandepoele et al. 2002). Structurally they could be divided into two distinct subgroups: the first one contains E2FA, E2FB and E2FC. These three E2Fs have all the domains conserved that are present in the animal E2F1–3, including the DNA-binding, dimerization, transactivation and RB-binding domains (Fig. 2). Members of the second group in *Arabidopsis* are DEL1, DEL2 and DEL3 (DP-E2F-like proteins). These are structurally related to mammalian E2F7 and E2F8 that lack all the E2F-specific domains except the DNA-binding domain, which is present in tandem duplication. Unlike E2Fs, *Arabidopsis* DEL DNA-binding function does not require heterodimer formation with the two known dimerization partner proteins, DPA and DPB (Fig. 2). The single *Arabidopsis* retinoblastoma-related protein (RBR1) interacts with all the three E2Fs by physically associating within their C-terminal region, but there is no known binding site for RBR1 in DELs.

Although E2Fs and DELs are clearly different in structural organization, in vitro they bind to the same DNA motif, indicating that they might compete with each other for binding to target promoter sequences. In agreement, all the E2Fs are able to transactivate on promoters containing an E2F-binding site with variable efficiency, while DELs antagonize their transcriptional activity (Kosugi and Ohashi 2002a,b; Mariconti et al. 2002). Therefore the two subgroups are functionally different; E2Fs are potentially activators while DELs are repressors. However, similarly to animal E2Fs that become repressors when in complex with RB proteins, due to binding to chromatin remodelling enzymes, (Rowland and Bernards 2006) *Arabidopsis* E2Fs are also able to interact with RBR1 protein although the function of this interaction is still unknown (de Jager et al. 2001; del Pozo et al. 2002; Magyar et al. 2005). Therefore, *Arabidopsis* E2Fs might work either as activators or

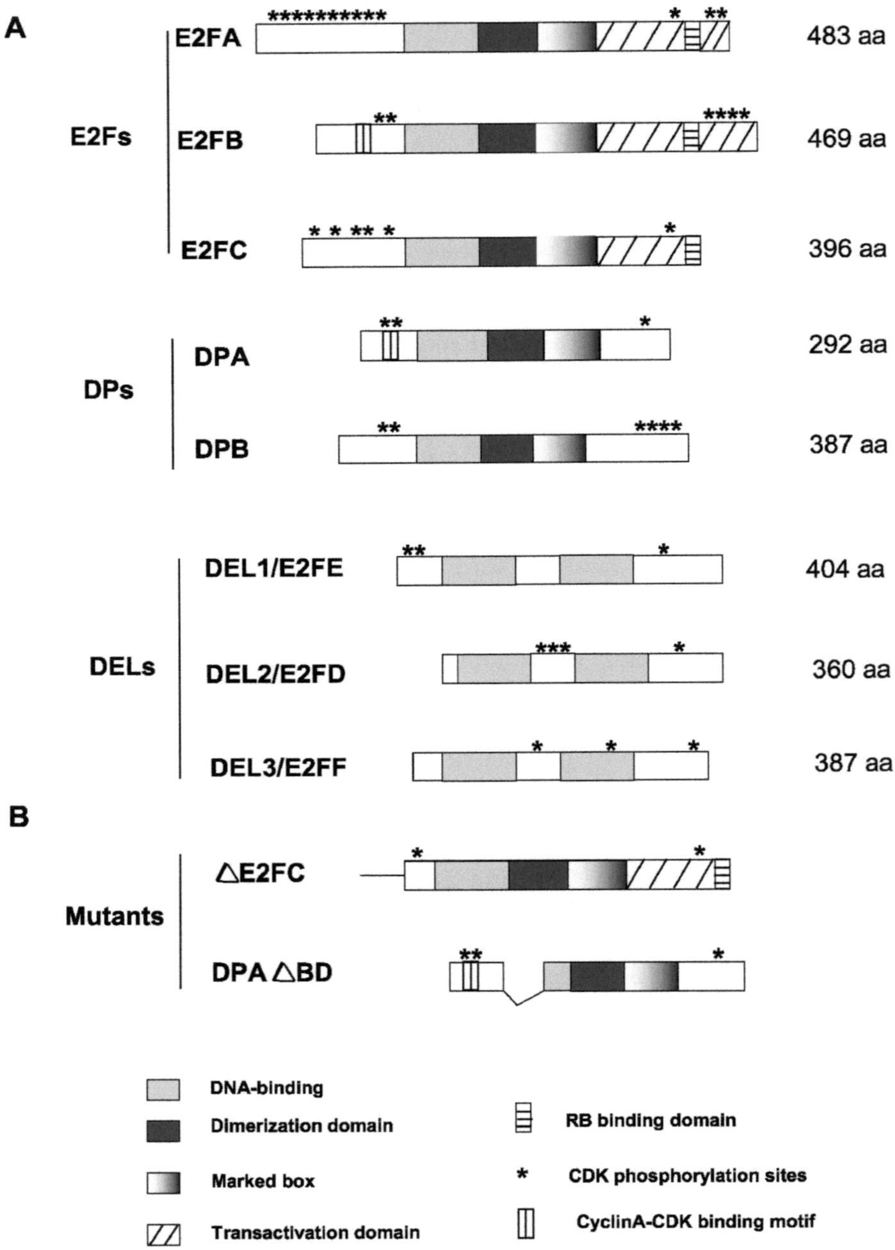

Fig. 2 Family of *Arabidopsis* E2F transcription factors. *A* Based on their different structural organization, the eight *Arabidopsis* proteins are classified into subgroup of E2Fs, DPs and DELs (Vandepoele et al. 2002). *B* Deletion mutants of E2FC and DPA used as molecular tools for studying E2F functions (del Pozo et al. 2002, 2006; Ramirez-Parra et al. 2003; Magyar and DeVeylder, unpublished results)

repressors depending on their association with RB, while DELs could be repressors or "anti-repressors" antagonizing with both types of E2F complexes on target promoters. Consequently, one would expect that modulating the expression of DELs in plants could change the activity of cell proliferation. However, the currently available data, rather than supporting a role for regulation of cell division, indicates that DELs might have specific roles in cell growth and expansion since loss of function or ectopic expression of DEL1 or DEL3 resulted in fairly normal-looking *Arabidopsis* plants (Ramirez-Parra et al. 2004; Vlieghe et al. 2005). Furthermore, variation of DEL3 levels lead to misexpression of E2F target genes unrelated to cell cycle but involved in cell wall biogenesis (Ramirez-Parra et al. 2004). DEL1 was found to influence ploidy level; loss of function *del1* increased the DNA amount while ectopic DEL1 repressed it, indicating a specific role for this transcription factor in the regulation of endoreduplication. Correspondingly, there was a correlation between the expressions of some S-phase regulated E2F target genes (e.g. CDC6, MCM3) and the level of DEL1 in the cotyledons (Vlieghe et al. 2005). Altogether, DEL1 could be an "anti-repressor" with a function to antagonize an E2F-dependent repression of mitotic cell cycle and thus act on the developmentally regulated switch from mitosis to endocycle. However, the idea that DELs could antagonize all the E2F complexes on target promoters is questionable, since the ectopic expression of DEL1 could not suppress the ectopic cell divisions caused by the simultaneous overexpression of both E2FA and DPA (Vlieghe et al. 2005).

On the basis of these data it was suggested that DELs and E2Fs could have different target specificities and/or that they compete with each other for binding to promoter sequences on only a subset of E2F target genes. In this respect it is worth mentioning that the mechanism governing the selectivity of E2F-promoter interaction is more complex than a simple E2F–DNA recognition as we know in animal cells (Zhu et al. 2005). The DNA-binding specificity of E2F has been attributed to protein interactions mediated by the marked box domain (Black et al. 2005). Plant E2Fs contain a similar domain, but DELs are missing it (Fig. 2), indicating that the target specificity of *Arabidopsis* E2Fs and DELs could be directed via protein interactions to distinct promoters.

4.1 Plant E2Fs and RBR Proteins; Regulators of Cell Division and Differentiation

In contrast to DELs, manipulating E2F activities by gain or loss of function mutations strongly affects cell proliferation and the pattern of cell differentiation, strongly supporting the idea that E2Fs are regulators of the G1–S control point (De Veylder et al. 2002; del Pozo et al. 2002, 2006; Magyar et al. 2005; Sozzani et al. 2006). In agreement with their expected G1–S function *Arabidopsis* E2Fs are up-regulated at an early stage during re-entry into cell division (de Jager et al. 2001). On the basis of their cell cycle-dependent expression, E2FA function was linked to S-phase control, while both E2FB and E2FC might have overlapping or distinct roles outside of S-phase. *Arabidopsis* E2Fs work similarly to their animal counterparts; they must form heterodimers with one of the two DP proteins to efficiently bind to E2F-targeted promoters (Magyar et al. 2000; Mariconti et al. 2002; Kosugi and Ohashi 2002b). Furthermore, heterodimerization with DPA, but not with DPB, results in nuclear localization of E2FA and E2FB, suggesting that E2F–DPA and E2F–DPB heterodimers are functionally different (Kosugi and Ohashi 2002b). Interestingly, neither DPA nor DPB could enhance the nuclear translocation of E2FC, probably because another post-translational modification is required to fully activate E2FC-dependent transcription (Kosugi and Ohashi 2002b).

In *Arabidopsis* all the three E2Fs are controlled by the single retinoblastoma related protein (RBR1). Interestingly, cereals such as maize and rice contain more than one RB-related gene and on the basis of their expression patterns it was suggested that they have different regulatory roles; maize RBR1 controlling cell differentiation and RBR3 regulating mitosis (Sabelli and Larkins 2006). Genetic analysis suggests that the *Arabidopsis* RBR1 has essential functions early in plant development since the knock out *rbr1* mutant is gametophytic lethal. Inactivation of the RBR1 gene in *Arabidopsis* endosperm of female megagametophyte results in over-proliferation due to failure in blocking mitosis, indicating a negative regulatory role for RBR1 in the mitotic cell division cycle (Ebel et al. 2004). In addition, decreasing the level or activity of RBR1, either by virus-induced gene silencing (VIGS) in tobacco, or by inducible expression of a viral RBR-binding protein (geminivirus RepA) in *Arabidopsis*, caused abnormal leaf development, probably due to prolonged cell proliferation (Desvoyes et al. 2006; Park et al. 2005). Interestingly, in both cases, leaf cells show increased ploidy levels later in development. Further studies revealed that RBR1 could control differentiation as well; ectopic expression in the shoot or root apical meristems stimulates in early differentiation (Wildwater et al. 2005; Wyrzykowska et al. 2006). Suppressing its expression in the root apical meristem by using *RBR1*-RNAi resulted in the production of several extra layers of undifferentiated cells in the columella root cap (Wildwater et al. 2005). Moreover, *Arabidopsis* RBR1 could work according to the canonical CycD/RBR/E2F pathway model to regulate stem cell maintenance: increasing RBR1 phosphorylation by the ectopic expression of cyclin D led to supernumerous stem cell layers while hypophosphorylation of RBR1 by overexpression of KRP2, a cyclin-dependent kinase inhibitor from *Arabidopsis*, resulted in the loss of stem cells similar to that seen with RBR overexpression (Wildwater et al. 2006). In agreement with the model, increasing the amount of RB-free E2F complexes by ectopic coexpression of E2FA and DPA resulted in an excess of stem cells. However, these results raise several questions, e.g. how the single *Arabidopsis* RBR1 could regulate mitosis and differentiation, and how and which E2Fs are part of these RBR1-regulated processes during plant development?

4.2
Activator and Repressor E2Fs; Who is Who?

As mentioned above, RBR1 function is essential in *Arabidopsis*, suggesting that plant E2Fs are required for the regulation of cell division and differentiation. However, overexpression of a dominant negative (DN) mutant form of DPA, which affected DNA binding but maintaining the dimerization ability, challenged this notion (Ramirez-Parra et al. 2003; Magyar and De Veylder, unpublished data) (Fig. 2). In these transgenic cells where E2F transcriptional activity is blocked, DELs should be the only remaining E2Fs due to their DP-independent binding to E2F-targeted promoters. Although a similar mutation in mammalian DP-1 induces G1 cell cycle arrest (Wu et al. 1996), in these DPA-DN *Arabidopsis* plants there were no macroscopic phenotypes or cell cycle defects, just a minor reduction in the expression levels of certain E2F target genes (e.g. RNRII, ORC1) (Ramirez-Parra et al. 2003; Magyar and DeVeylder, unpublished data). These data further support the idea that, in plants, DELs are unlikely to be the major repressors of cell division that antagonize E2F functions. It is possible, however, that the mutant DPA eliminates both the repressor and the activator DP-dependent E2F complexes from the DNA, which resulted in normal development of the transgenic *Arabidopsis* plants. Mutations of the activator and repressor *Drosophila* dE2Fs resulted in only a minor defect in cell proliferation and growth during larval development (Frolov et al. 2001). Interestingly, the effects of inactivating the single *Drosophila* dDP appeared indistinguishable from the effects of inactivating both dE2F1 and dE2F2 (Frolov et al. 2001, 2003). Therefore, the individual *Arabidopsis* E2Fs might work antagonistically on cell division and growth in a similar way to that demonstrated in *Drosophila*. Since cell proliferation did not change significantly in the DP mutant plants, it is likely that plant E2F functions are not vital for the transcriptional activation of cell cycle genes, in agreement to what was found with E2F and DP mutants in *Drosophila* and in *Chlamydomonas* (Fang et al. 2006; Frolov et al. 2001). However, it cannot be excluded that the mutant DPA protein is not able to eliminate all the DP-dependent E2F functions in the cells, and thus some remaining E2F activities could still drive cells through the cell cycle. Further studies of individual E2F mutants and the analysis of their combinations are required to understand the functions and interactions of E2Fs during plant development, and to fully address whether they are indispensable for regulation of cell cycle transitions.

Functional characterization of the individual members of *Arabidopsis* E2Fs have already revealed differences among them: ectopic expression of E2FA with DPA resulted in strong activation of both mitotic cell cycle and endocycle (De Veylder et al. 2002; Kosugi and Ohashi 2003); overexpression of E2FB was also able to activate mitosis but it repressed the endocycle (Magyar et al. 2005; Sozzani et al. 2006), whereas reduction in the level of E2FC confirmed its negative regulatory function in mitosis but a positive one in endoreduplication (del Pozo et al. 2006). According to these data, E2FB and E2FC are antagonistic transcription factors, while E2FA has a dual functionality. We do not know much about the target gene specificities of E2FB and E2FC, but there is a clear antagonistic effect of E2FB and E2FC on the regulation of the mitotic cyclin B1;1 gene (del Pozo et al. 2006; Sozzani et al. 2006). Although the promoter of this cyclin gene does not contain consensus E2F binding elements, decreasing the amount of E2FC in transgenic *Arabidopsis* plants, by using *E2FC*-specific RNAi, dramatically activated its expression (90 times up), even to a much higher level than the expression of known E2F target genes such as CDC6, EXP3. This indicates that this mitotic cyclin could be a direct target for E2FC-dependent repression. Since overexpression of E2FB, or co-expression of E2FA with DPA, significantly activates the expression of cyclin B1;1, it is also possible that it has an alternative E2F-binding site; however, this needs to be tested experimentally. In contrast, these data strongly support a G2–M regulatory function for plant E2Fs, similarly to animal E2Fs (Hernando et al. 2004; Neufeld and Edgar 1998). Furthermore, increased E2FB levels led to shortened cell cycle duration, and it was suggested that E2FB function in plant cells is comparable to that of *Drosophila* dE2F1, which simultaneously increases the expression of critical S- and M-phase regulators (Magyar et al. 2005).

Our recent data show that E2FB can directly induce the promoter of the *Arabidopsis* CDKB1;1 gene, a plant-specific regulator of the G2 to M transition (Magyar and Bogre, unpublished results). It is important to note that reduction in the level of E2FC caused significant changes in gene expression in mature leaves but not in young leaves, suggesting that E2FC acts by repressing E2F-regulated genes in mature differentiated cells. As the ploidy level decreased in these transgenic plants, it can be argued that E2FC has an important regulatory role in the switch from mitotic cell cycle to endocycle by repressing the expression of mitotic genes such as cyclin B1;1. Interestingly, in switching from mitotic to endocycles, cells in *Dorosophila* embryo terminate the expression of mRNAs encoding the mitotic regulators such as cyclin B1 and B3 genes (Edgar and Orr-Weaver 2001). Ectopic expression of mitotic cyclin B1;2 in *Arabidopsis* trichome cells switch the endocycle into mitosis, indicating that reducing the expression of mitotic cyclins is an important regulatory step towards the activation of endocycle (Schnittger et al. 2002). How E2FC regulates the transcription of cyclin B1;1 is not known yet, but previous studies indicated that E2FC could work as a direct transcriptional repressor on E2F target genes.

Ectopic expression of a stabilized mutant ΔE2FC lacking the amino terminal domain in dark-grown *Arabidopsis* plants resulted in reduced expression of CDC6, an *S*-phase regulatory gene (del Pozo et al. 2002) (Fig. 2). Intriguingly, the two alternative splicing variants of the mouse E2F3, the natural full length (E2F3a) and the amino terminally deleted (E2F3b) forms displayed opposite transcriptional functions where the activator role of E2F3a is converted into transcriptional repressor in E2F3b (Aslanian et al. 2004; Leone et al. 2000). Possibly the analogous deletion on *Arabidopsis* E2FC might have similar functional impacts (Fig. 2). Although the precise role of the N-terminal region of mammalian E2F1–3 is still being uncovered, it is required for ubiquitin-mediated degradation of E2F1 protein (Marti et al. 1999). The N-terminal extension of plant E2Fs contains a number of CDK phosphorylation sites (Fig. 2) and the E2FC protein was targeted by the ubiquitin-mediated proteosome pathway in a CDK-dependent manner. Furthermore, deletion of the N-terminal part stabilized the E2FC as well as E2FA protein (del Pozo et al. 2002; Magyar 2005; Magyar and Bogre, unpublished). Although clear evidence that E2FC works as a direct transcriptional repressor on E2F target genes is still missing. The strong growth-arrested phenotype of the N-terminal deletion ΔE2FC mutant co-expressed with DPB dimerization partner was interpreted as a consequence of reduced cell proliferation due to abundance of E2FC-DPB transcriptional repressor complex (del Pozo et al. 2006). The higher ploidy level observed in the double transgenic mutant ΔE2FC-DPB further supported the hypothesis that E2FC is a negative regulator for mitosis but an activator of endocycle.

It would be interesting to know whether E2FC has a role in meristems or in young leaves, since its mitosis inhibitory function seems to be restricted to mature leaves, while E2FC expression was found to be high in actively dividing tissues (del Pozo et al. 2002, 2006). Microarray studies indicate constitutive expression in both leaves and roots (Beemster et al. 2005). Ectopic expression of E2FA with DPA also resulted in a strong growth-arrested phenotype in transgenic *Arabidopsis* plants (De Veylder et al. 2002) or inhibited tobacco growth in a concentration-dependent manner (Kosugi and Ohashi 2003). Cell proliferation and endoreduplication, however, were both strongly up-regulated in E2FA/DPA overexpressors. How E2FA could control in parallel these different, spatially and temporally separated processes has not yet been addressed. *E2FA* transcripts were detected both from mitotic and endoreduplicating tissues, indicating that E2FA could have a dual regulatory role in vivo in both of these events. Moreover, it was found that ectopic E2FA has an opposite effect on cell proliferation during *Arabidopsis* development. It increased cell number in the cotyledons (De Veylder et al. 2002) but resulted in fewer cells in mature leaves (He et al. 2004). Probably, E2FA could activate or repress cell division depending on the developmental stage. In addition, increased E2FA activity in E2FA/DPA co-expressors resulted in almost complete inhibition of growth early after germination, probably due to arrest of cell cycle exit (De Veylder et al. 2002).

Genome-wide expression analysis of these transgenic *Arabidopsis* plants revealed that genes encoding proteins required for DNA synthesis are highly up-regulated (e.g. CDC6, ORC1, CDC45, RNRII, MCM3), strongly supporting a regulatory role for E2FA during DNA synthesis. Moreover, G2–M regulatory genes such as cyclin B1;1, and CDKB1;1 were also found among the potential

E2FA target genes (Vandepoele et al. 2005; Vlieghe et al. 2003). Promoter analysis of CDKB1;1 further supported the notion that the E2FA–DPA heterodimer could activate this G2–M specific promoter through its E2F binding element (Boudolf et al. 2004). However, these results raise the questions on how E2FA could stimulate G2 to M transition when its own expression is restricted to the S-phase of the cell cycle, and how E2FA could activate endocycle if it is a positive transcriptional regulator of genes required for G2 to M transition.

Intriguingly, E2FA seems to negatively regulate cell proliferation in mature leaves, as observed for E2FC, but the molecular mechanism behind this ob-

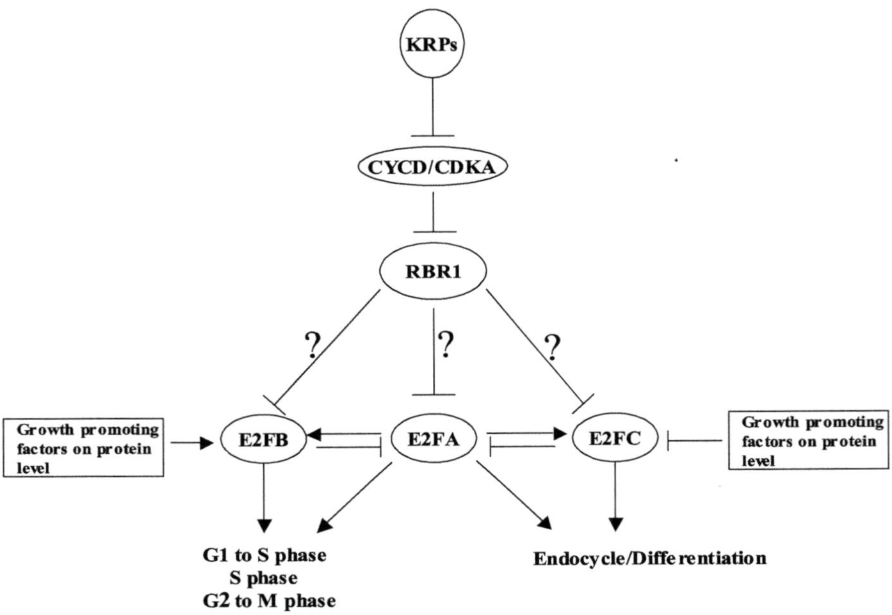

Fig. 3 Model of *Arabidopsis* RBR1-E2F pathway. The model reflects our current understanding of how RBR1 and E2Fs control cell proliferation and differentiation. Growth-promoting factors such as light and auxin activate RBR–kinase complexes, consisting of cyclin D or cyclin A and CDKA;1 that inactivate RBR1 function by hyperphosphorylation. CDK inhibitors (KRPs) suppress RBR1 phosphorylation resulting in hypophosphorylated RBR1. The single *Arabidopsis* RBR1 protein is able to interact with all the three E2Fs and thus repress their transactivation functions, although the mechanism of this interaction is still unknown. E2FB and E2FC have opposing functions: E2FB activates both G1–S and G2–M cell cycle transitions, while it inhibits endoreduplication. In contrast, E2FC stimulates the switch from mitosis to endocycle by repressing G2–M transition. Growth-promoting factors such as light and auxin stabilize E2FB, but destabilize E2FC proteins. E2FA is a strong activator for the S-phase of the cell cycle, but it also could stimulate mitosis and endocycle through the up-regulation of E2FB and E2FC expression. E2FC and E2FB could negatively regulate E2FA expression and protein accumulation, respectively, by unknown feed-back mechanisms (del Pozo et al. 2006; Sozzani et al. 2006)

servation is not known (He et al. 2004). It has been reported that E2FA is able to activate the expression of both E2FB and E2FC, the two potentially antagonistic E2F transcription factors on the regulation of mitosis and endocycle, respectively (Vandepoele et al. 2005). Therefore, it is possible that the balance between E2FB and E2FC levels will determine whether the cells continue in mitosis or switch to endocycle. E2FA has an effect on both of these factors through the regulation of E2FB and E2FC. How the ratio between E2FC and E2FB is set is not known, but an interesting hypothesis suggests that auxin distribution plays a role in this process.

Auxin regulates cell division and elongation in a concentration-dependent manner; elevated auxin levels activate cell division in the meristems, while reduced amounts repress mitosis as cells leave the meristematic regions, and in parallel it enhances cell growth (Scheres and Xu 2006). Auxin increases the stability of E2FB, and co-expression of E2FB with DPA in plant cells could maintain cell proliferation even in the absence of auxin (Magyar et al. 2005). Moreover, elevated levels of E2FB–DPA heterodimer in plant cells resulted in extremely small cell size, indicating that E2FB inhibits growth. Ectopic expression of E2FB in *Arabidopsis* plants also stimulates cell division and results in smaller cells both in leaf and in roots that became significantly shorter (Sozzani et al. 2006). It was suggested that auxin could influence cell proliferation and growth through the modulation of the level of E2FB protein; a high auxin level would stabilize E2FB, which stimulates cell division. In contrast, E2FC stability was shown to be oppositely regulated, destabilized in growth-promoting physiological conditions (e.g. in plants grown in light), and regulated by the ubiquitin–SCF pathway (del Pozo et al. 2002). Therefore, an elevated level of E2FB could specifically keep cells dividing in the meristems and in young tissues, while the E2FC protein level would increase above E2FB in mature leaves and thus would stimulate the switch from mitotic cell cycle to endocycle by repressing mitotic genes (Fig. 3).

5 Conclusions

Growth of organs is controlled by two processes: cell division and cell expansion. The timing of transition from proliferative growth to cell expansion largely determines the cell number in organs and thereby their growth rate potential. In this chapter we have discussed the central role of the RBR1–E2F pathway in this process. RBR1 can potentially affect all E2F functions, but the relative concentrations of two E2F factors, E2FB and E2FC, could determine whether cells maintain their proliferation potential or exit the mitotic cycle and begin the cell differentiation accompanied by cell enlargement and endoreduplication (Fig. 3).

Acknowledgements Thanks to Prof. László Bögre for critical reading of the manuscript and Dr. Rossana Henriques and Dr. Virginia Balbi for helping with the preparation of the manuscript. The work was supported by BBSRC.

References

Aslanian A, Iaquinta PJ, Verona R, Lees JA (2004) Repression of the Arf tumor suppressor by E2F3 is required for normal cell cycle kinetics. Genes Dev 18:1413–1422

Attwooll C, Lazzerini Denchi E, Helin K (2004) The E2F family: specific functions and overlapping interests. EMBO J 23:4709–4716

Bagchi S, Weinmann R, Raychaudhuri P (1991) The retinoblastoma protein copurifies with E2F-I, an E1A-regulated inhibitor of the transcription factor E2F. Cell 65:1063–1072

Bandara LR, La Thangue NB (1991) Adenovirus E1a prevents the retinoblastoma gene product from complexing with a cellular transcription factor. Nature 351:494–497

Beemster GT, De Veylder L, Vercruysse S, West G, Rombaut D, Van Hummelen P, Galichet A, Gruissem W, Inzé D, Vuylsteke M (2005) Genome-wide analysis of gene expression profiles associated with cell cycle transitions in growing organs of Arabidopsis. Plant Physiol 138:743–743

Bisova K, Krylov DM, Umen JG (2005) Genome-wide annotation and expression profiling of cell cycle regulatory genes in *Chlamydomonas reinhardtii*. Plant Physiol 137:475–491

Black EP, Hallstrom T, Dressman HK, West M, Nevins JR (2005) Distinctions in the specificity of E2F function revealed by gene expression signatures. Proc Natl Acad Sci USA 102:15948–15953

Boudolf V, Inze D, De Veylder L (2006) What if higher plants lack a CDC25 phosphatase? Trends Plant Sci 11:474–479

Boudolf V, Vlieghe K, Beemster GT, Magyar Z, Torres Acosta JA, Maes S, Van Der Schueren E, Inze D, De Veylder L (2004) The plant-specific cyclin-dependent kinase CDKB1;1 and transcription factor E2Fa-DPa control the balance of mitotically dividing and endoreduplicating cells in *Arabidopsis*. Plant Cell 16:2683–2692

Churchman ML et al. (2006) SIAMESE, a plant-specific cell cycle regulator, controls endoreplication onset in *Arabidopsis thaliana*. Plant Cell 18:3145–3157

Cooper K (2006) Rb, whi it's not just for metazoans anymore. Oncogene 25:5228–5232

de Jager SM, Menges M, Bauer UM, Murra JA (2001) *Arabidopsis* E2F1 binds a sequence present in the promoter of S-phase-regulated gene AtCDC6 and is a member of a multigene family with differential activities. Plant Mol Biol 47:555–568

De Veylder L et al. (2002) Control of proliferation, endoreduplication and differentiation by the *Arabidopsis* E2Fa-DPa transcription factor. EMBO J 21:1360–1368

De Veylder L et al. (2001) Functional analysis of cyclin-dependent kinase inhibitors of *Arabidopsis*. Plant Cell 13:1653–1668

del Pozo JC, Boniotti MB, Gutierrez C (2002) *Arabidopsis* E2Fc functions in cell division and is degraded by the ubiquitin-SCF(AtSKP2) pathway in response to light. Plant Cell 14:3057–3071

del Pozo JC, Diaz-Trivino S, Cisneros N, Gutierrez C (2006) The balance between cell division and endoreplication depends on E2FC-DPB, transcription factors regulated by the ubiquitin-SCFSKP2A pathway in *Arabidopsis*. Plant Cell 18:2224–2235

Desvoyes B, Ramirez-Parra E, Xie Q, Chua NH, Gutierrez C (2006) Cell type-specific role of the retinoblastoma/E2F pathway during *Arabidopsis* leaf development. Plant Physiol 140:67–80

Dewitte W, Murray JA (2003) The plant cell cycle. Annu Rev Plant Biol 54:235–264

Dimova DK, Stevaux O, Frolov MV, Dyson NJ (2003) Cell cycle-dependent and cell cycle-independent control of transcription by the Drosophila E2F/RB pathway. Genes Dev 17:2308–2320

Dissmeyer N, Nowack MK, Pusch S, Stals H, Inze D, Grini PE, Schnittger A (2007) T-loop phosphorylation of *Arabidopsis* CDKA;1 is required for its function and can be partially substituted by an aspartate residue. Plant Cell 19:972–985

Doerner P (2007) Plant meristems: cytokinins – the alpha and omega of the meristem. Curr Biol 17:R321–R323

Du W, Pogoriler J (2006) Retinoblastoma family genes. Oncogene 25:5190–5200

Ebel C, Mariconti L, Gruissem W (2004) Plant retinoblastoma homologues control nuclear proliferation in the female gametophyte. Nature 429:776–780

Edgar BA, Orr-Weaver TL (2001) Endoreplication cell cycles: more for less. Cell 105:297–306

Fang SC, de los Reyes C, Umen JG (2006) Cell size checkpoint control by the retinoblastoma tumor suppressor pathway. PLoS Genet 2:e167

Frolov MV, Huen DS, Stevaux O, Dimova D, Balczarek-Strang K, Elsdon M, Dyson NJ (2001) Functional antagonism between E2F family members. Genes Dev 15:2146–2160

Frolov MV, Stevaux O, Moon NS, Dimova D, Kwon EJ, Morris EJ, Dyson NJ (2003) G1 cyclin-dependent kinases are insufficient to reverse dE2F2-mediated repression. Genes Dev 17:723–728

He SS, Liu J, Xie Z, O'Neill D, Dotson S (2004) *Arabidopsis* E2Fa plays a bimodal role in regulating cell division and cell growth. Plant Mol Biol 56:171–184

Hernando E et al. (2004) Rb inactivation promotes genomic instability by uncoupling cell cycle progression from mitotic control. Nature 430:797–802

Inze D (2005) Green light for the cell cycle. EMBO J 24:657–662

Inze D, De Veylder L (2006) Cell cycle regulation in plant development. Annu Rev Genet 40:77–105

Kosugi S, Ohashi Y (2002a) E2Ls, E2F-like repressors of *Arabidopsis* that bind to E2F sites in a monomeric form. J Biol Chem 277:16553–16558

Kosugi S, Ohashi Y (2002b) Interaction of the *Arabidopsis* E2F and DP proteins confers their concomitant nuclear translocation and transactivation. Plant Physiol 128:833–843

Kosugi S, Ohashi Y (2003) Constitutive E2F expression in tobacco plants exhibits altered cell cycle control and morphological change in a cell type-specific manner. Plant Physiol 132:2012–2022

Landsberg RL, Sero JE, Danielian PS, Yuan TL, Lee EY, Lees JA (2003) The role of E2F4 in adipogenesis is independent of its cell cycle regulatory activity. Proc Natl Acad Sci USA 100:2456–2461

Leone G, Nuckolls F, Ishida S, Adams M, Sears R, Jakoi L, Miron A, Nevins JR (2000) Identification of a novel E2F3 product suggests a mechanism for determining specificity of repression by Rb proteins. Mol Cell Biol 20:3626–3632

Liu DX, Nath N, Chellappan SP, Greene LA (2005) Regulation of neuron survival and death by p130 and associated chromatin modifiers. Genes Dev 19:719–732

Magyar Z, Atanassova A, De Veylder L, Rombauts S, Inze D (2000) Characterization of two distinct DP-related genes from *Arabidopsis thaliana*. FEBS Lett 486:79–87

Magyar Z, De Veylder L, Atanassova A, Bako L, Inze D, Bogre L (2005) The role of the *Arabidopsis* E2FB transcription factor in regulating auxin-dependent cell division. Plant Cell 17:2527–2541

Mariconti L, Pellegrini B, Cantoni R, Stevens R, Bergounioux C, Cella R, Albani D (2002) The E2F family of transcription factors from *Arabidopsis thaliana*. Novel and conserved components of the retinoblastoma/E2F pathway in plants. J Biol Chem 277:9911–9919

Marti A, Wirbelauer C, Scheffner M, Krek W (1999) Interaction between ubiquitin-protein ligase SCFSKP2 and E2F-1 underlies the regulation of E2F-1 degradation. Nat Cell Biol 1:14–19

Neufeld TP, Edgar BA (1998) Connections between growth and the cell cycle. Curr Opin Cell Biol 10:784–790

Park JA, Ahn JW, Kim YK, Kim SJ, Kim JK, Kim WT, Pai HS (2005) Retinoblastoma protein regulates cell proliferation, differentiation, and endoreduplication in plants. Plant J 42:153–163

Ramirez-Parra E, Desvoyes B, Gutierrez C (2005) Balance between cell division and differentiation during plant development. Int J Dev Biol 49:467–477

Ramirez-Parra E, Frundt C, Gutierrez C (2003) A genome-wide identification of E2F-regulated genes in *Arabidopsis*. Plant J 33:801–811

Ramirez-Parra E, Lopez-Matas MA, Frundt C, Gutierrez C (2004) Role of an atypical E2F transcription factor in the control of *Arabidopsis* cell growth and differentiation. Plant Cell 16:2350–2363

Rayman JB, Takahashi Y, Indjeian VB, Dannenberg JH, Catchpole S, Watson RJ, te Riele H, Dynlacht BD (2002) E2F mediates cell cycle-dependent transcriptional repression in vivo by recruitment of an HDAC1/mSin3B corepressor complex. Genes Dev 16:933–947

Rowland BD, Bernards R (2006) Re-evaluating cell-cycle regulation by E2Fs. Cell 127:871–874

Rowland BD, Denissov SG, Douma S, Stunnenberg HG, Bernards R, Peeper DS (2002) E2F transcriptional repressor complexes are critical downstream targets of p19(ARF)/p53-induced proliferative arrest. Cancer Cell 2:55–65

Sabelli PA, Larkins BA (2006) Grasses like mammals? Redundancy and compensatory regulation within the retinoblastoma protein family. Cell Cycle 5:352–355

Scheres B, Xu J (2006) Polar auxin transport and patterning: grow with the flow. Genes Dev 20:922–926

Schnittger A, Schobinger U, Stierhof YD, Hulskamp M (2002) Ectopic B-type cyclin expression induces mitotic cycles in endoreduplicating *Arabidopsis* trichomes. Curr Biol 12:415–420

Scofield S, Murray JA (2006) The evolving concept of the meristem. Plant Mol Biol 60:V-VII

Sharma VK, Carles C, Fletcher JC (2003) Maintenance of stem cell populations in plants. Proc Natl Acad Sci USA 100(1):11823–11829

Sherr CJ, Roberts JM (1999) CDK inhibitors: positive and negative regulators of G1-phase progression. Genes Dev 13:1501–1512

Sozzani R, Maggio C, Varotto S, Canova S, Bergounioux C, Albani D, Cella R (2006) Interplay between *Arabidopsis* activating factors E2Fb and E2Fa in cell cycle progression and development. Plant Physiol 140:1355–1366

Umeda M, Shimotohno A, Yamaguchi M (2005) Control of cell division and transcription by cyclin-dependent kinase-activating kinases in plants. Plant Cell Physiol 46:1437–1442

Umen JG, Goodenough UW (2001) Control of cell division by a retinoblastoma protein homolog in *Chlamydomonas*. Genes Dev 15:1652–1661

Vandepoele K, Raes J, De Veylder L, Rouze P, Rombauts S, Inze D (2002) Genome-wide analysis of core cell cycle genes in *Arabidopsis*. Plant Cell 14:903–916

Vandepoele K, Vlieghe K, Florquin K, Hennig L, Beemster GT, Gruissem W, Van de Peer Y, Inze D, De Veylder L (2005) Genome-wide identification of potential plant E2F target genes. Plant Physiol 139:316–328

Vlieghe K et al. (2005) The DP-E2F-like gene DEL1 controls the endocycle in *Arabidopsis thaliana*. Curr Biol 15:59–63

Vlieghe K et al. (2003) Microarray analysis of E2Fa-DPa-overexpressing plants uncovers a cross-talking genetic network between DNA replication and nitrogen assimilation. J Cell Sci 116:4249–4259

Wang H, Fowke LC, Crosby WL (1997) A plant cyclin-dependent kinase inhibitor gene. Nature 386:451–452

Wildwater M et al. (2005) The RETINOBLASTOMA-RELATED gene regulates stem cell maintenance in *Arabidopsis* roots. Cell 123:1337–1349

Wu CL, Classon M, Dyson N, Harlow E (1996) Expression of dominant-negative mutant DP-1 blocks cell cycle progression in G1. Mol Cell Biol 16:3698–3706

Wu L et al. (2001) The E2F1–3 transcription factors are essential for cellular proliferation. Nature 414:457–462

Wyrzykowska J, Schorderet M, Pien S, Gruissem W and Fleming AJ (2006) Induction of differentiation in the shoot apical meristem by transient overexpression of a retinoblastoma-related protein. Plant Physiol 141:1338–1348

Zhang HS, Postigo AA, Dean DC (1999) Active transcriptional repression by the Rb-E2F complex mediates G1 arrest triggered by p16INK4a, TGFbeta, and contact inhibition. Cell 97:53–61

Zhu W, Giangrande PH, Nevins JR (2005) Temporal control of cell cycle gene expression mediated by E2F transcription factors. Cell Cycle 4:633–636

Plant Cell Monogr (10)
L. Bögre and G. Beemster: Plant Growth Signaling
DOI 10.1007/7089_2007_158/Published online: 29 January 2008

Plant Cell Growth Signalling and Its Link to Ploidy

Takeshi Yoshizumi[1] · Christian Breuer[1,2] · Minami Matsui[1] ·
Keiko Sugimoto-Shirasu[1,2] (✉)

[1]Plant Science Center, RIKEN, Yokohama, 230-0045 Kanagawa, Japan
sugimoto@psc.riken.jp

[2]Department of Cell and Developmental Biology, John Innes Centre, Norwich NR4 7UH, UK

Abstract What determines the final size of plant cells is a fundamental question in plant growth and development but the cellular mechanism that mediates this control remains largely unknown. Recent genetic studies using model plants *Arabidopsis*, maize and legume *Medicago* demonstrate that increasing DNA content or ploidy by a process called endoreduplication contributes to the post-mitotic cell expansion in higher plants. During successive rounds of endoreduplication or endocycle, cells replicate chromosomal DNA in the absence of mitosis, and the progression of the endocycle is both positively and negatively regulated by developmental and/or environmental signals. Plants also possess some ploidy-independent mechanism to control cell size and at least some part of this control involves a global feedback mechanism called organ-size checkpoint that balances cell proliferation and cell expansion within an organ to maintain its size homeostasis.

1 Introduction

One of the features that make plant growth distinct from that of animals and yeasts is the variety of the final size that individual cells reach in plant tissues. Most yeast and animal cells only double their size during their development but many plant cells often expand at least ten times their original size and some cells even expand more than 1000 times. Such massive increase in cell size is supported by a unique growth mechanism in plants that utilizes both macromolecular production in cytoplasm and water uptake into vacuoles. Given that cell size is fairly constant among different plant species, the determination of final cell size must be under genetic control. Elucidating these genetic mechanisms underlying plant cell growth has been challenging but thanks to the recent rapid accumulation of genetic, genomic and bioinformatics resources available in model plants, we have begun to uncover the complex pathways that determine cell size in plants. One of the key factors that often correlates with plant cell size is the nuclear DNA content or ploidy, and this correlation can be found either at a single cell level as the endoreduplicated cell (Fig. 1) or at the whole plant level as tetraploid or polyploidy plants (Fig. 2). Herein we will first describe our current knowledge on how endoreduplication is controlled in higher plants. We will then examine how

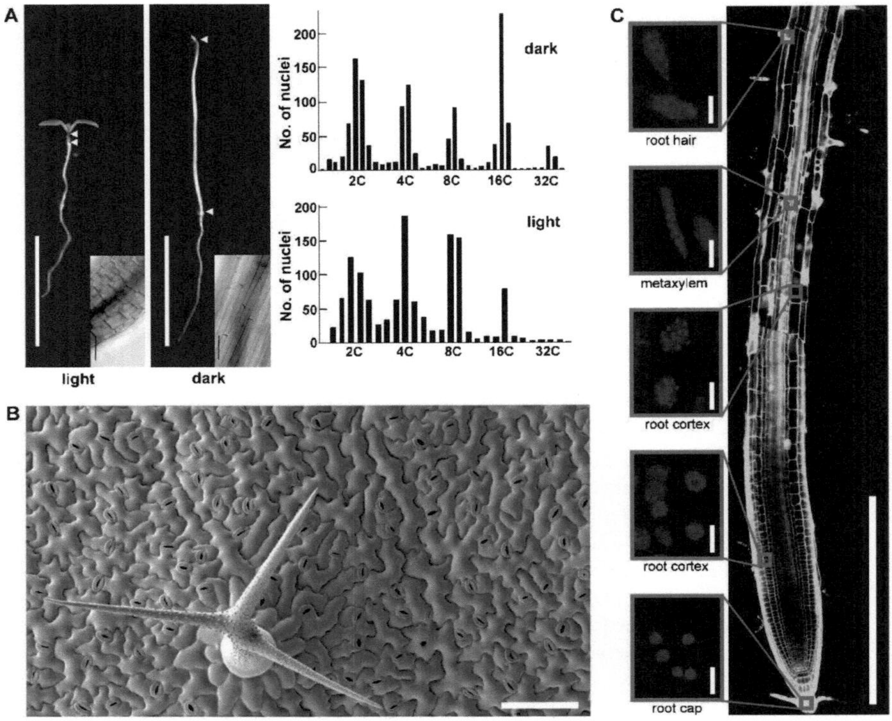

Fig. 1 Various cell types support several cycles of endoreduplication in *Arabidopsis* and resulting increase in ploidy contributes to an increase in cell size. **A** *Arabidopsis* hypocotyls endoreduplicate up to 32C in the dark and up to 16C in the light. The increased ploidy positively correlates with the size of epidermal cells (*insets*). *White triangles* indicate the top and bottom of hypocotyls. *Scale* 500 μm (seedlings), 50 μm (insets). **B** The size of *Arabidopsis* leaf epidermal cells varies more than 500-fold and the variation in cell size is linked to ploidy. For example, the ploidy level of nuclei in a pair of guard cells remains 2C whereas ploidy in a fully developed trichome goes up to 32C. *Scale* 100 μm. **C** *Arabidopsis* root cells enter the endocycle after they exit from the mitotic cell cycle and the progression of the endocycle coincides with the post-mitotic cell expansion. Increased ploidy also contributes to the formation of large cell types including root hairs and vascular metaxylem cells. DAPI-stained nuclei from root cap cells (2C), root cortex cells (4C and 8C), metaxylem cells (16C) and root hair cells (16C) (*Insets*). *Scale* 10 μm (DAPI-stained nuclei), 500 μm (root tip)

endoreduplication and resulting increase in ploidy influence post-mitotic cell expansion in plants. Finally, we will also discuss recent findings that address how the control of cell size may be governed by overall developmental programmes.

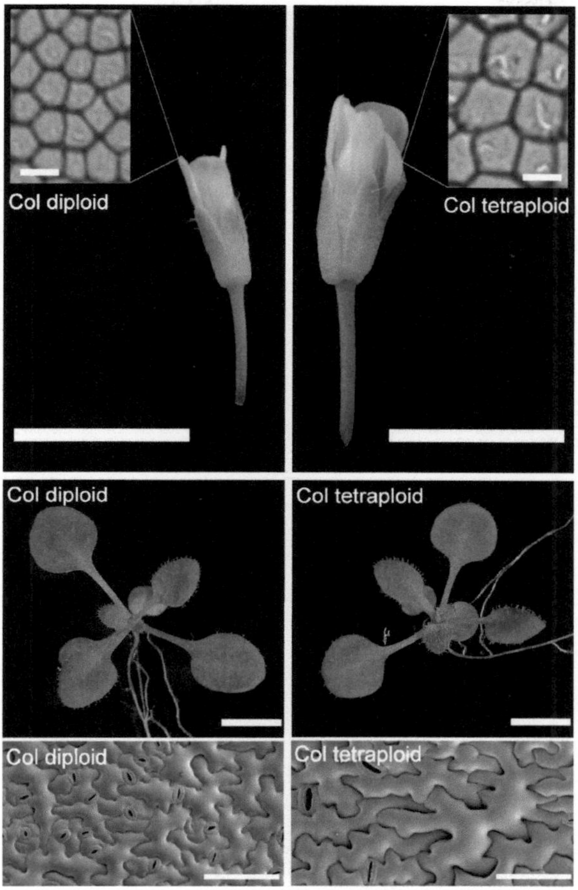

Fig. 2 Cell and organ size correlate with the nuclear DNA content in *Arabidopsis* tetraploid plants. The nuclear DNA content in tetraploid plants is doubled compared to diploid plants, and this leads to the formation of larger cells in both flowers and leaves. An increase in cell size also contributes to an increase in flower size but does not appear to affect the overall leaf size. *Scale* 4 mm (flower), 10 μm (flower epidermis), 5 mm (seedlings), 100 μm (leaf epidermis). Light micrographs of flower epidermis are courtesy of Hirokazu Tsukaya (University of Tokyo)

2
How is Endoreduplication Controlled in Plants?

2.1
Control of the Endocycle by Cell Cycle-Related Genes

Endoreduplication is often viewed as a short-cut of the mitotic cell cycle that skips mitosis and re-enters the S-phase. During the normal mitotic cell cycle, cells have a mechanism to prevent entry into the S-phase without going through the M-phase. Cells that initiate the endocycle, however, are able to break this rule and re-enter the S phase in the absence of mitosis (Fig. 3). Despite these key differences between the mitotic cell cycle and endocycle, at least some of the molecular machineries that trigger the S-phase appear to be commonly shared between these two types of cell cycles and only subsets of these controls are specific to the endocycle (Table 1). E2F is a transcription factor that regulates entry into the S-phase during the mitotic cell cycle. E2F

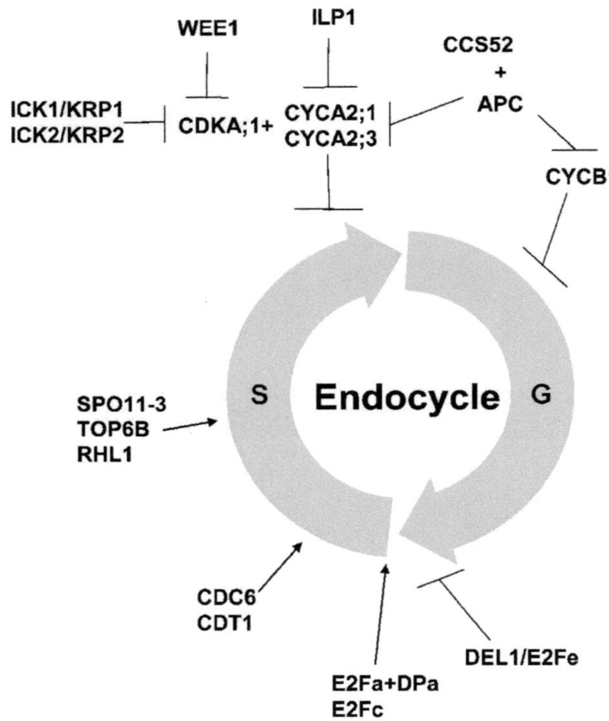

Fig. 3 A schematic diagram that describes how endoreduplication may be controlled in plants. During the endocycle, cells re-enter the S-phase without going through the M-phase. Whether there is a gap phase (G phase) between successive S-phases is not currently known. A key step to switch from the mitotic cell cycle to endocycle is to drop the M-phase-specific CDK activity to inhibit the induction of mitosis and this process is mediated through the down-regulation of M-phase-specific cyclins such as CYCAs and CYCBs. Re-entry into the S-phase during the endocycle utilizes several E2F transcription factors E2Fa, E2Fc and DEL1/E2Fe that also regulate S-phase entry in the mitotic cell cycle. DNA replication during the endocycle uses mostly the same molecular machinery as that used for the S-phase in the mitotic cell cycle, e.g. components of a pre-replication complex such as CDC6 and CDT1, but at least some including components of the plant DNA topoisomerase VI, SPO11-3, TOP6B and RHL1, appear to have a specific role in the endocycle

Table 1 Mutants and transgenic plants that show ploidy phenotypes

Gene	Expected function	General phenotypes	Ploidy phenotypes	Refs.
(A) Genes identified from gain-of-function studies				
Cell cycle signalling				
E2Fa and DPa	G1/S transition, transcription factor	Dwarf[1]	Increased in seedlings and trichomes	De Veylder et al. 2002
CDC6a	Component of DNA replication licensing complex	Over-branched trichomes, increased stomatal density[1]	Increased in leaves and trichomes	Castellano et al. 2001, 2004
CDT1a	Component of DNA replication licensing complex	Over-branched trichomes, increased stomatal density[1]	Increased in leaves and trichomes	Castellano et al. 2004
CYCB1;2	M cyclin	Multicellular trichomes[2]	Reduced in trichomes	Schnittger et al. 2003
ICK1/KRP1	CDK inhibitor	Small/under-branched trichomes[2]	Reduced in trichomes, increased in neighboring cells	Schnittger et al. 2003; Weinl et al. 2005
ICK2/KRP2	CDK inhibitor	Small and serrated leaves, large leaf cells[1]	Reduced in leaves	De Veylder et al. 2001
ILP1	Transcriptional repressor	Thick hypocotyls, large cotyledons, increased cell size in seedlings[3]	Increased in seedlings and leaves	Yoshizumi et al. 2006
Light signalling				
IPD1	CUE domain protein	Slightly longer hypocotyls in the dark[3]	Increased in dark-grown hypocotyls	Tsumoto et al. 2006
(B) Genes identified from loss-of-function studies				
Cell cycle signalling				
E2Fc	G1/S transition, transcription factor	Small curled leaves, reduced cell size[4]	Reduced in leaves	del Pozo et al. 2006
DEL1	DP-E2F-like, transcription factor	Not known[5]	Increased in seedlings and true leaves	Vlieghe et al. 2005

Table 1 (continued)

Gene	Expected function	General phenotypes	Ploidy phenotypes	Refs.
SPO11-3	DNA topoisomerase VI subunit A	Dwarf, reduced root hairs, small/under-branched trichomes[5]	Reduced in hypocotyls, leaves, and trichomes	Sugimoto-Shirasu et al. 2002; Hartung et al. 2002
TOP6B	DNA topoisomerase VI subunit B	Dwarf, reduced root hairs, small/under-branched trichomes[5]	Reduced in hypocotyls, leaves, and trichomes	Sugimoto-Shirasu et al. 2002; Hartung et al. 2002
RHL1	A component of DNA topoisomerase VI	Dwarf, reduced root hairs, small/under-branched trichomes[5]	Reduced in hypocotyls, leaves, and trichomes	Sugimoto-Shirasu et al. 2005
ccs52	An activator of APC	Small petioles, hypocotyls, and roots[6]	Reduced in petioles, hypocotyls, and roots	Cebolla et al. 1999
HBT	Subunit of APC	Cell division arrest, increased cell size[5]	Increased in leaves	Serralbo et al. 2006
CYCA2;3	M cyclin	Not known[5]	Increased in leaves	Imai et al. 2006
CYCA2;1	M cyclin	Not known[5]	Increased in seedlings	Yoshizumi et al. 2006
SIM	ICK/KRP homolog	Multicellular trichomes[5]	Reduced in trichomes and hypocotyls	Walker et al. 2000; Churchman et al. 2006
Developmental signalling				
TRY	MYB transcription factor	Large and over-branched trichomes[5]	Increased in trichomes	Perazza et al. 1999; Schellmann et al. 2002
KAK	E3 ubiquitin ligase	Large and over-branched trichomes, long hypocotyls[5]	Increased in trichomes and hypocotyls	Perazza et al. 1999; El Refy et al. 2003
RFI	Not identified	Large and over-branched trichomes[5]	Increased in trichomes	Perazza et al. 1999
UVI4/PYM	Unknown protein	Large and over-branched trichomes, resistant to UV-B[5]	Increased in trichomes and hypocotyls	Perazza et al. 1999; Hase et al. 2006
Light signalling				
PHYB	Photo receptor	Long hypocotyls[5]	Increased in light-grown hypocotyls	Gendreau et al. 1998

Table 1 (continued)

Gene	Expected function	General phenotypes	Ploidy phenotypes	Refs.
COP1	E3 ubiquitin ligase	Constitutive photomorphogenesis[5]	Reduced in dark-grown hypocotyls	Gendreau et al. 1998
DNA repair signalling				
BRU1/TSK/ MGO3	TPR + LRR protein	Disorganized meristem, hypersensitive to DNA damage[5]	Increased in seedlings and flowers	Takeda et al. 2004; Suzuki et al. 2005
FAS1	Subunit of chromatin assembly factor	Disorganized meristem, increased cell size, hypersensitive to DNA damage[5]	Increased in leaves and trichomes	Kaya et al 2001; Exner et al. 2006; Kirik et al. 2006
FAS2	Subunit of chromatin assembly factor	Disorganized meristem, increased cell size, hypersensitive to DNA damage[5]	Increased in leaves and trichomes	Kaya et al 2001; Exner et al. 2006
MSI1	Subunit of chromatin assembly factor	Disorganized meristem, increased cell size, hypersensitive to DNA damage[6]	Increased in leaves and trichomes	Kaya et al 2001; Exner et al. 2006
Plant hormone signalling				
SPY	O-GlcNAc transferase	Large and over-branched trichomes[5]	Increased in leaves and trichomes	Jacobsen et al. 1996; Perazza et al. 1999
CTR1	Raf family of MAPKKK	Ethylene triple response[5]	Increased in hypocotyls	Gendreau et al. 1999
GA1	Ent-kaurene synthase (GA-biosynthetic enzyme)	Short hypocotyls[5]	Reduced in hypocotyls	Gendreau et al. 1999
CPM17	P450	Short hypocotyls, brassinoride deficient[5]	Reduced in dark-grown hypocotyls	Gendreau et al. 1998
FRL1	Sterol methyltransferase	Dwarf, abnormal petal shape, increased petal cell size[5]	Increased in petals	Hase et al. 2000, 2005

[1] Phenotypes observed by 35S over-expression, [2] *pGL2* expression, [3] activation tagging and 35S over-expression, [4] RNA interference, [5] EMS mutation or T-DNA/tranposon insertion, [6] antisense

interacts with DIMERIZATION PARTNER (DP) protein to activate the transcription of S-phase genes. There are three typical and three atypical E2Fs in *Arabidopsis* (Mariconti et al. 2002; Kosugi and Ohashi 2002). Co-expression of E2Fa, one of the typical E2Fs, and its interacting protein DPa, under 35S promoter promotes both cell division and endoreduplication (De Veylder et al. 2002), suggesting that E2Fa and DPa positively regulate both the mitotic cell cycle and endocycle (Fig. 3). In contrast, E2Fc, which also belongs to the typical E2F family (del Pozo et al. 2002), appears to promote the endocycle but inhibit the mitotic cell cycle since down-regulation of *E2Fc* expression by RNA interference leads to the formation of small and twisted leaves that have many small cells with reduced ploidy (del Pozo et al. 2006, Fig. 3). E2Fe/DP-E2F-like protein 1 (DEL1), one of the atypical E2Fs in *Arabidopsis* (Kosugi and Ohashi 2002), on the other hand, represses the endocycle since the *del1-1* mutation leads to increased ploidy and the ectopic expression of E2Fe/DEL1 reduces endoreduplication (Vlieghe et al. 2005, Fig. 3). All of these *Arabidopsis* E2Fs can bind to the consensus *cis*-element to which authentic E2F bind (Mariconti et al. 2002), and the promotion or repression of the endocycle in these mutants and transgenic plants correlate with the altered expression of E2F target genes required for DNA replication (Vlieghe et al. 2005). DNA replication is licensed to occur when a pre-replication complex, composed of the origin recognition complex (ORC), CELL DIVISION CYCLE6 (CDC6), cyclin10 target 1 (CDT1) and the minichromosome maintenance (MCM) complex proteins, assembles at the replication origins. The transcription of these genes is up-regulated in *E2Fa-DPa* co-expressing lines (De Veylder et al. 2002), and even over-expression of *CDC6* (Castellano et al. 2001) or *CDT1* (Castellano et al. 2004) alone can drive extra cell division and the endocycle. Interestingly, promotion of these two different cell cycles is cell-type specific, i.e. cells that possess meristematic competence undergo further proliferation and cells that are committed to endoreduplication such as trichomes undergo another round of endocycle (Castellano et al. 2004), suggesting that the activity of the pre-replication complex is required for the progression of the endocycle but it is not involved in the mechanism that switches from the mitotic cell cycle to the endocycle (Fig. 3). Recent genetic studies show that the DNA topoisomerase VI (topo VI) complex has a specific role in endoreduplication (Fig. 3). *Arabidopsis* mutants in the components of the plant topo VI complex such as *HYPOCOTYL 6 (HYP6)*, *ROOT HAIRLESS 2 (RHL2)* and *RHL1* (Sugimoto-Shirasu et al. 2002; Hartung et al. 2002; Sugimoto-Shirasu et al. 2005) are all viable and when they are induced to proliferate in callus induction media, they are capable of undergoing cell division at a similar rate to wild type. However, these mutants cannot complete the endocycle to reach 32C and instead they stall at 8C (Sugimoto-Shirasu et al. 2002, 2005; Hartung et al. 2002; Fig. 4). The plant topo VI complex is not likely to have a regulatory role in the endocycle. Instead, it probably acts during the S-phase in the endocycle to prevent the entanglement of replicated chromosomes.

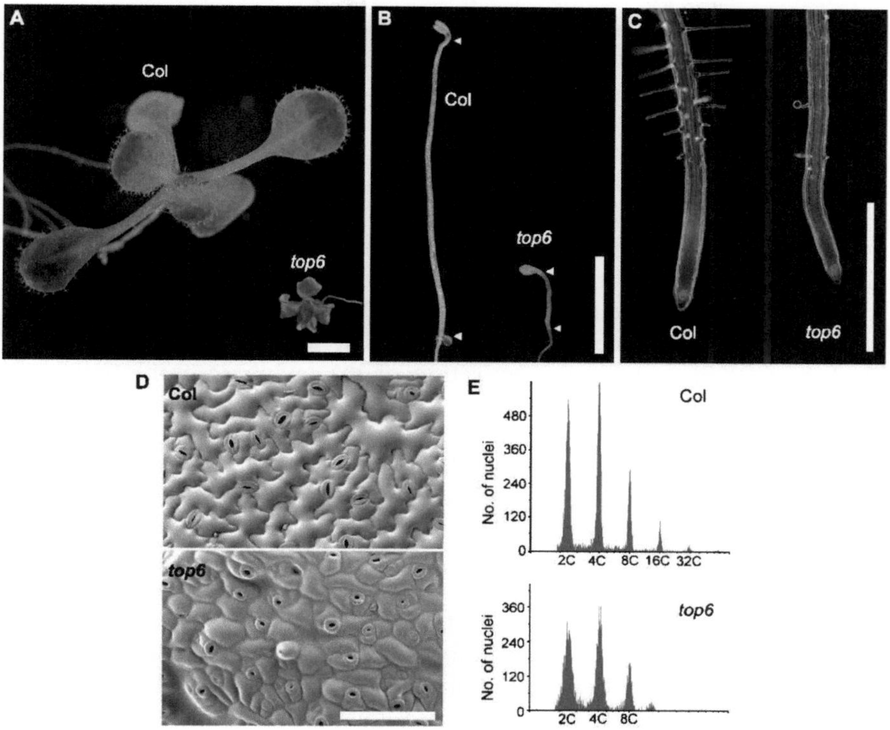

Fig. 4 *Arabidopsis* mutants that display ploidy phenotypes help us identify molecular components required for the endocycle. **A** A light micrograph of 2-week-old, light-grown wild-type (Col) and DNA topoisomerase VI mutant (*top6*) plants. The *top6* mutants display extreme dwarf phenotypes. *Scale* 2 mm. **B** A light micrograph of 4-day-old, dark-grown wild-type (Col) and *top6* hypocotyls. The growth of *top6* hypocotyls is severely compromised. *White triangles* indicate the top and bottom of hypocotyls. *Scale* 4 mm. **C** A light micrograph of wild-type (Col) and *top6* mutant roots. *top6* is defective in the initiation and subsequent outgrowth of root hairs. *Scale* 1 mm. **D** Scanning electron micrographs of wild-type and *top6* leaf epidermis demonstrate that the final cell size is reduced in *top6*. *Scale* 100 μm. **E** Flow cytometric analysis shows the ploidy of wild-type leaves ranges from 2C to 32C whereas the ploidy in *top6* reaches only 8C, indicating that *top6* has defects in the progression of successive endocycles beyond 8C

Another key process for cells to enter the endocycle is to drop the M-phase-specific CDK activity to inhibit the induction of mitosis. This process is primarily mediated through the down-regulation of M-phase-specific cyclins, i.e. A-type and B-type cyclins (*CYCA*, *CYCB*, respectively), and several upstream regulators of these cyclins have been recently characterized (Fig. 3). The *ilp1-1D* mutant was originally identified as an increased ploidy mutant from *Arabidopsis* activation tagging lines (Yoshizumi et al. 2006). Various phenotypes including thick hypocotyls, large cotyledons, and long roots are observed at the early seedling stage in *ilp1-1D*, although its adult plants do not exhibit any obvious morphological change. *ilp1-1D* undergoes an extra round of endocycle and this phenotype is associated with the down-regulation of *CYCA2* genes including *CYCA2;1* and *CYCA2;3*, suggesting that ILP1 negatively regulates the expression of these *CYCA2* genes (Fig. 3). Loss of function of *CYCA2;1* and *CYCA2;3* also induces an extra round of endocycle in young seedlings (Yoshizumi et al. 2006) and mature leaves (Imai et al. 2006), respectively. CYCA2;3 physically interacts with CDKA;1 and *CDKA;1* is expressed in developed trichomes (Imai et al. 2006). Therefore, the CYCA2/CDKA;1 complex appears to act as a break for endoreduplication and increasing its activity is sufficient to terminate the endocycle (Fig. 3).

Another well-documented mechanism that allows down-regulation of M-phase CDK activity is the activation of the anaphase-promoting complex (APC), a ubiquitin ligase that targets CYCA and CYCB for degradation (Fig. 3). HOBITT (HBT1) is a subunit of the plant APC complex, and loss of *HBT1* in *hbt*^{2311} inhibits cell proliferation and promotes the endocycle (Serralbo et al. 2006). Furthermore, a study of the legume *Medicago truncatula* has demonstrated that the APC complex is activated by its up-stream regulator CCS52/FIZZY-RELATED (FZR) protein, because suppression of the *CCS52/FZR* gene by the antisense gene causes reduced ploidy levels and small cells (Cebolla et al. 1999; Fig. 3). In *Arabidopsis* CYCA2;3 is one of the APC targets since over-expression of the stable CYCA2;3 that has a mutation in the destruction box strongly reduces ploidy level (Imai et al. 2006; Fig. 3).

M-phase-specific CDK activity is also regulated by inhibitors of CYC/CDK complexes such as ICK1/KRP1 and ICK2/KRP2 (Verkest et al. 2005, Weinl et al. 2005). High expression of these inhibitors blocks both the mitotic cell cycle and endocycle in *Arabidopsis* but their moderate expression appears to interfere only with the activity of mitotic CDKA, thus leading to an early entry into the endocycle (Verkest et al. 2005; Weinl et al. 2005). Another putative CDK inhibitor encoded by *SIAMESE (SIM)* interacts with D-type cyclin and CDKA;1 (Churchman et al. 2006). Loss of *SIM* function results in multicellular trichomes with individual cells having reduced ploidy (Walker et al. 2000), and this is associated with the ectopic expression of *CYCB1;1* in trichomes (Churchman et al. 2006), suggesting that SIM inhibits the mitotic cell cycle in trichomes by down-regulating the expression of CYCB1. The WEE1 kinase, a negative regulator of CDK activity, is also suggested to have a role at the transition from the mitotic cell cycle to endocycle since its transcription is upregulated at the onset of endoreduplication in maize endosperm (Sun 1999; Fig. 3).

2.2
Control of the Endocycle by Developmental Signals

Endoreduplication is often associated with cell differentiation in plant development, and most cell types except those in floral organs undergo several rounds of endocycle. It is well established that, in *Arabidopsis*, cells in leaves, hypocotyls and roots endoreduplicate up to 32C (Fig. 1), and recent genetic studies have provided key insights into the regulation of endoreduplication in the developmental context. Leaf trichome, a branched hair cell that is formed on the epidermal layer, is comprised of a single cell in *Arabidopsis*, and its nuclei endoreduplicate up to 32C (Hülskamp 2000). Many mutations that have increased or decreased branching have been identified and in almost all cases, these defects are associated with enhanced or reduced levels of endoreduplication, respectively (Hülskamp 2000; Table 1). For example, *triptychon (try)* shows increased ploidy in over-branched trichomes, suggesting that TRY protein is a negative regulator of endoreduplication (Perazza et al. 1999). TRY is a truncated MYB transcription factor that affects trichome initiation (Schellmann et al. 2002). It is thus likely that the signalling pathways that specify cell fate during trichome development also act as an upstream regulator of the endocycle. Another group of mutants such as *kaktus (kak)*, *rastifari (rfi)* and *polychome (pym)* also produces over-branched trichomes with increased ploidy (Perazza et al. 1999). The double mutants *pym kak* and *pym rfi*, but not *kak rfi*, show strong additive phenotypes, suggesting that at least two independent genetic pathways, one governed by PYM and the other by KAK and RFI, negatively regulate the endocycle. *KAK* encodes a HECT-type ubiquitin E3 ligase that promotes ubiquitin transfer to appropriate targets for proteolysis (El Refy et al. 2003). How KAK represses the endocycle is not currently known but it is reasonable to predict that KAK targets some cell cycle-related protein for degradation. The *pym* mutation has been recently shown to be allelic to the *uvi4* mutant that is isolated as a new mutation that gives resistance to UV-B irradiation (Hase et al. 2006). *UVI4/PYM* encodes an unknown protein that has a weak similarity to ferredoxin hydrogenase. Understanding how UVI4/PYM suppresses the endocycle requires further characterization but one possible mechanism is that UVI4/PYM is involved in maintaining the mitotic status within a cell because the *UVI4/PYM* expression pattern is similar to that of *CYCB2;2*, one of the M-phase specific cyclins (Hase et al. 2006).

2.3
Control of the Endocycle by Light Signals

Endoreduplication is controlled by both external and internal signals, and one of the most established environmental signals that affect the endocycle in plants is light. The growth of *Arabidopsis* hypocotyls is light-dependent, i.e. in the dark they etiolate and form long and thin hypocotyls whereas in the light they form short and thick hypocotyls (Fig. 1A). Regardless of the light condition where seedlings are grown, these hypocotyls are composed of approximately 22 cells along their longitudinal axis and their growth is primarily driven by cell expansion rather than cell division. This growth is associated with an increase in ploidy up to 16C in the light and to 32C in the dark (Gendreau et al. 1997, Fig. 1A). *PHYTOCHROMEB (PHYB)* encodes a photoreceptor to absorb red light, and hypocotyls from its loss-of-function mutants display etiolated growth in the light. This phenotype is accompanied by an increase in ploidy (Gendreau et al. 1998), suggesting that PHYB-mediated red light signalling is involved in the control of endoreduplication. Interestingly, mutations in a blue light receptor CRYPTOCHROME 1 (CRY1) also lead to the etiolated hypocotyl phenotype in the light but these mutations have only minor effects on the endocycle (Gendreau et al. 1998). Therefore, the mechanism that suppresses hypocotyl elongation in the light appears to involve at least two genetic pathways, i.e. one that is endocycle dependent and mediated by red light and another that is endocycle independent and mediated by blue light.

What are the downstream signals that transduce these light signals to control the endocycle? One strong candidate that might be involved in this light signalling is CONSTITUTIVELY PHOTOMORPHOGENESIC 1 (COP1), a negative regulator of general light signalling in *Arabidopsis* because both hypocotyl elongation and endoreduplication is suppressed in *cop1* mutants (Gendreau et al. 1998). Another candidate gene that is likely to be involved in these signalling cascades is *INCREASED POLYPLOIDY LEVEL IN DARKNESS 1-1D (IPD1-1D)* that encodes a plant-specific protein with unknown function (Tsumoto et al. 2006). The *ipd1-1D* mutant was identified as an over-endoreduplicated hypocotyl mutant from activation tagging lines. Interestingly, *ipd1-1D* mutants only display the ploidy phenotype in the dark and in agreement with this, *IPD1* is transcriptionally down regulated by light. Part of the IPD1 protein is homologous to the CUE domain which is required for binding to mono ubiquitin, suggesting that IPD1 may target some negative regulator of the endocycle for protein modification. However, the CUE domain within IPD1 lacks a well-conserved amino acid that is essential for its binding to mono ubiquitin. Therefore, it is possible that IPD1 has some unrelated function in regulating the endocycle and hypocotyl elongation through light signalling.

2.4 Control of the Endocycle by Plant Growth Regulators

Various plant growth regulators are also involved in the control of the endocycle. A brassinosteroid (BR) deficient mutant *cpm17* displays short hypocotyl phenotypes in the dark and this is associated with reduced ploidy (Gendreau et al. 1998), suggesting that BR positively regulates the endocycle in the dark. In contrast, other BR signalling mutants such as *de-etiolated2* (*det2*) and *brassinosteroid-insensitive1* (*bri1*) that display a severe dwarf phenotype in the light have a normal ploidy level (Stacey and Sugimoto-Shirasu, unpublished results), suggesting that BR may act downstream of light signalling to modulate endoreduplication.

Gibberellin (GA) also appears to promote the endocycle since *ga1* mutants, deficient in GA biosynthesis, show a strongly reduced ploidy phenotype in the dark (Gendreau et al. 1998). Consistent with this, the *spindly* (*spy*) mutation, which exhibits a constitutive GA response (Jacobsen et al. 1996), shows an increased ploidy phenotype in light-grown leaves. The *spy* mutation causes an over-branched trichome phenotype and this is enhanced in the *try* mutant background (Perraza et al. 1999), indicating that GA promotes an additional round of endocycle in a pathway independent from the TRY-mediated genetic pathway.

The involvement of ethylene in the endocycle control is demonstrated from studies on the *constitutive triple response1* (*ctr1*) mutants that show constitutive ethylene-response phenotypes. When *ctr1* mutants are grown in the dark, they develop short and thick hypocotyls, and microscopic analyses revealed that this is associated with increased ploidy, suggesting that ethylene acts as a positive regulator of the endocycle (Gendreau et al. 1999).

It is also likely that the endocycle is under the control of many other small signalling molecules. One potential modulator of these signals that has been recently identified is *FRILL1 (FRL1)* which encodes SMT2, a sterol metyltransferase (Hase et al. 2005). Most cells in floral organs such as sepals and petals do not normally undergo endoreduplication in *Arabidopsis*, but in *frl1* these cells endoreduplicate ectopically, leading to the development of serrated sepals and petals (Hase et al. 2000). Further studies revealed that the ploidy level is also increased in other tissues and that the composition of endogenous sterol is altered in *frl1* (Hase et al. 2005). The application of BR and several other known sterols does not rescue the *frl1* mutant phenotypes (Hase et al. 2005), suggesting that some novel sterol compound controls the switch from the mitotic cell cycle to the endocycle.

2.5 Links Between DNA Repair and Endoreduplication

In addition to the genetic and environmental control of endoreduplication described above, increasing evidence points to an intriguing link between the maintenance of genome integrity and endoreduplication. FASCIATA1 (FAS1), FAS2 and MICROSATELLITE INSTABILITY1 (MSI1) are subunits of chromatin assembly factor CAF1 that is involved in the nucleosome assembly on newly replicated DNA (Kaya et al. 2001). Loss of this protein complex in *fas1, fas2, and msi1* mutants leads to pleiotropic developmental phenotypes including fasciation, disorganized apical meristems and dwarfism (Kaya et al. 2001; Kirik et al. 2006; Exner 2006). These defects are associated with various cellular and subcellular abnormalities such as delayed progression of the mitotic cell cycle, impaired heterochromatic and euchromatic conformation and stimulation of homologous recombination and other DNA repair response (Schonrock et al. 2006; Exner et al. 2006; Kirik et al. 2006). Interestingly, *fas1, fas2, and msi1* mutants also have higher ploidy than wild type, suggesting that the endocycle is promoted in these mutant backgrounds (Schonrock et al. 2006; Exner et al. 2006; Kirik et al. 2006). Because of the complex phenotypes observed in these mutants, how ploidy is increased in these mutants is not clear. However, one interesting possibility is that the suboptimal genome integrity and/or resulting arrest of the mitotic cell cycle somehow triggers the endocycle as an alternative mechanism to protect genetic information and thus to sustain the life of cells. Another example that links DNA repair and endocycle control is the *brushy1/tonsoku/mgoun3 (bru1/tsk/mgo3)* mutants in *Arabidopsis* that exhibit the faciation phenotype similar to *fas* mutants (Takeda et al. 2004). The *bru1/tsk/mgo3* mutants also display increased homologous recombination and elevated DNA damage response as well as stochastic release of transcriptional gene silencing and abnormal heterochromatin structure (Takeda et al. 2004). Furthermore, these phenotypes also associate with the delayed progression of mitotic cell cycle from G2 to M-phase and an increased level of ploidy (Suzuki et al. 2005), further supporting the functional link between chromosome replication, DNA repair and endocycle control. The *BRU1/TSK/MGO3* gene encodes a novel nuclear protein that contains TPR and LRR repeats (Takeda et al. 2004), and its homolog in *Nicotiana tobaccum* is expressed specifically at the S-phase (Suzuki et al. 2005). Thus, it is likely that BRU1/TSK/MGO3 has a role during and/or just after DNA replication.

3 How Is Cell Size Controlled in Plants?

3.1 Ploidy-Dependent Control of Cell Size

Cell size is determined by highly dynamic and probably flexible signalling pathways in plants and our understanding of its control is still limited. Nevertheless, recent studies on various mutants and transgenic plants mainly in *Arabidopsis*, maize and legume *Medicago* show clear correlation between the level of ploidy and cell size (Table 1, Fig. 4), providing strong genetic evidence to support the long-standing "karyoplasmic ratio" or "nuclear-cytoplasmic ratio" theory. A question still remains as to which, an increase in ploidy or an increase in cell size, comes first, since deducing causal relationships based on the observed mutant phenotypes is not always straight-forward. However, our recent study using *Arabidopsis* tetraploid plants clearly demonstrates that increasing ploidy first does have a positive impact on cell size (Breuer et al. 2007; Fig. 2), suggesting that increased DNA content can indeed support further growth of cells. It is worth noting that increasing ploidy also allows further growth of some plant organs, e.g. floral petals and sepals, but not of others, e.g. leaves, which appear to have some compensatory mechanism to maintain the overall organ size (Fig. 2).

Whether plant cells possess some active mechanism to judge the level of ploidy and to determine the extent of cell growth accordingly still remains elusive. It is generally thought that increasing ploidy multiplies the DNA template available for gene expression and this, in turn, enhances the metabolic activity of cells to support further growth. However, this model has been challenged by a recent finding in maize endosperm where down-regulation of the endocycle by over-expression of the dominant-negative *CDKA* does not change the overall level of starch and other storage proteins (Leiva-Neto et al. 2004).

3.2 Ploidy-Independent Control of Cell Size

Rapidly accumulating results from various genetic studies also have revealed that cell size and ploidy are not always tightly coupled and that in some cases modification of the endocycle leads to only minor or opposite changes in cell size (Table 1). For example, the size of root cells from different *Arabidopsis* ecotypes varies considerably but those differences in cell size have very little correlation with ploidy (Beemster et al. 2002). Likewise, mis-expression of CDK inhibitors or dominant-negative CDKs in *Arabidopsis* and tobacco results in the drastic reduction of ploidy while at the same time leads to an increase in cell size (Hemerly et al. 1995; Wang et al. 2000; De Veylder et al. 2001). These results suggest that ploidy is not the final determinant of cell size and some other ploidy-independent signalling pathways contribute to the determination of final cell size in plants. Possible candidates that may be involved in these pathways include signals that are imposed from the total organ-checkpoint control (Tsukaya 2006).

3.3 Upstream Regulators that Link Cell-Size Control and Plant Development

Cellular processes that underlie post-mitotic cell expansion must be tightly regulated by overall developmental programmes but how such upstream signals coordinate the timing and extent of cell expansion is not well understood. A recent study suggests that the *BIGPETALp (BPEp)* gene encoding a basic helix-loop-helix (bHLH) transcription factor might be a part of these signalling cascades since BPEp acts downstream of petal organ identity genes such as *APETALA3, PISTILLATA, APETALA1* and *SEPALLATA3*, and it regulates petal growth by restricting its cell expansion (Szecsi et al. 2006). Plant growth regulators such as auxin, GA and BR are also likely to play a role in transducing such overall developmental cues to individual cells (Belkhadir and Chory 2006). An involvement of some regulatory genes such as *ARGOS*-

LIKE (ARL) has been suggested in BR-related cell expansion (Hu et al. 2006). In addition, an intriguing link between auxin signalling and ribosome-mediated post-mitotic cell expansion has been revealed by a study on the *Arabidopsis* and potato EBP1 protein (Horvath 2006).

4 Conclusion

Over the last couple of years we have witnessed considerable progress in our understanding of how cells endoreduplicate in plants and how this process is influenced by various internal and external factors. It is now evident that increasing ploidy has a positive consequence on cell size, the next big question is how this process is mediated in plants. More systematic analyses of already existing ploidy/cell size mutants and transgenic plants by combinations of transcriptome, metabolome and computer-assisted imaging techniques should help our further understanding of this very intriguing connection. Future research should also uncover what contributes to the ploidy-independent plant cell growth and how these size controls at the cellular level are linked to overall plant growth and development.

Acknowledgements This work is supported by a BBSRC David Phillips Fellowship to K.S-S. C.B is a recipient of an EU Marie-Curie Early Stage Training studentship.

References

Beemster GT, De Vusser K, De Tavernier E, De Bock K, Inze D (2002) Variation in growth rate between Arabidopsis ecotypes is correlated with cell division and A-type cyclin-dependent kinase activity. Plant Physiol 129:854–864

Belkhadir Y, Chory J (2006) Brassinosteroid signaling: a paradigm for steroid hormone signaling from the cell surface. Science 314:1410–1411

Breuer C, Stacey NJ, West CE, Zhao Y, Chory J, Tsukaya H, Azumi Y, Maxwell A, Roberts K, Sugimoto-Shirasu K (2007) BIN4, a novel component of the plant DNA topoisomerase VI complex, is required for endoreduplication in *Arabidopsis*. Plant Cell 19:3655–3668

Castellano MM, del Pozo JC, Ramirez-Parra E, Brown S, Gutierrez C (2001) Expression and stability of Arabidopsis CDC6 are associated with endoreplication. Plant Cell 13:2671–2686

Castellano Mdel M, Boniotti MB, Caro E, Schnittger A, Gutierrez C (2004) DNA replication licensing affects cell proliferation or endoreplication in a cell type-specific manner. Plant Cell 16:2380–2393

Cebolla A, Vinardell J, Kiss E, Olah B, Roudier F, Kondorosi A, Kondorosi E (1999) The mitotic inhibitor *ccs52* is required for endoreduplication and ploidy-dependent cell enlargement in plants. EMBO J 18:4476–4484

Churchman ML, Brown ML, Kato N, Kirik V, Hulskamp M, Inze D, De Veylder L, Walker JD, Zheng Z, Oppenheimer DG, Gwin T, Churchman J, Larkin JC (2006) SIAMESE, a Plant-Specific Cell Cycle Regulator, Controls Endoreplication Onset in *Arabidopsis thaliana*. Plant Cell 18:3145–3157

del Pozo JC, Boniotti MB, Gutierrez C (2002) Arabidopsis E2Fc functions in cell division and is degraded by the ubiquitin-SCF(AtSKP2) pathway in response to light. Plant Cell 14:3057–3071

del Pozo JC, Diaz-Trivino S, Cisneros N, Gutierrez C (2006) The balance between cell division and endoreplication depends on E2FC-DPB, transcription factors regulated by the ubiquitin-SCF^{SKP2A} pathway in *Arabidopsis*. Plant Cell 18:2224–2235

De Veylder L, Beeckman T, Beemster GT, Krols L, Terras F, Landrieu I, van der Schueren E, Maes S, Naudts M, Inze D (2001) Functional analysis of cyclin-dependent kinase inhibitors of *Arabidopsis*. Plant Cell 13:1653–1668

De Veylder L, Beeckman T, Beemster GT, de Almeida Engler J, Ormenese S, Maes S, Naudts M, Van Der Schueren E, Jacqmard A, Engler G, Inze D (2002) Control of proliferation, endoreduplication and differentiation by the *Arabidopsis* E2Fa-DPa transcription factor. EMBO J 21:1360–1368

El Refy A, Perazza D, Zekraoui L, Valay JG, Bechtold N, Brown S, Hulskamp M, Herzog M, Bonneville JM (2003) The *Arabidopsis KAKTUS* gene encodes a HECT protein and controls the number of endoreduplication cycles. Mol Genet Genomics 270:403–414

Exner V, Taranto P, Schonrock N, Gruissem W, Hennig L (2006) Chromatin assembly factor CAF-1 is required for cellular differentiation during plant development. Development 133:4163–4172

Gendreau E, Traas J, Desnos T, Grandjean O, Caboche M, Höfte H (1997) Cellular basis of hypocotyl growth in *Arabidopsis thaliana*. Plant Physiol 114:295–305

Gendreau E, Hofte H, Grandjean O, Brown S, Traas J (1998) Phytochrome controls the number of endoreduplication cycles in the *Arabidopsis thaliana* hypocotyl. Plant J 13:221–230

Gendreau E, Orbovic V, Hofte H, Traas J (1999) Gibberellin and ethylene control endoreduplication levels in the *Arabidopsis thaliana* hypocotyl. Planta 209:513–516

Hartung F, Angelis KJ, Meister A, Schubert I, Melzer M, Puchta H (2002) An archaebacterial topoisomerase homolog not present in other eukaryotes is indispensable for cell proliferation of plants. Curr Biol 12:1787–1791

Hase Y, Tanaka A, Baba T, Watanabe H (2000) *FRL1* is required for petal and sepal development in *Arabidopsis*. Plant J 24:21–32

Hase Y, Fujioka S, Yoshida S, Sun G, Umeda M, Tanaka A (2005) Ectopic endoreduplication caused by sterol alteration results in serrated petals in *Arabidopsis*. J Exp Bot 56:1263–1268

Hase Y, Trung KH, Matsunaga T, Tanaka A (2006) A mutation in the *uvi4* gene promotes progression of endo-reduplication and confers increased tolerance towards ultraviolet B light. Plant J 46:317–326

Hemerly A, Engler Jde A, Bergounioux C, Van Montagu M, Engler G, Inze D, Ferreira P (1995) Dominant negative mutants of the Cdc2 kinase uncouple cell division from iterative plant development. EMBO J 14:3925–3936

Horvath BM, Magyar Z, Zhang Y, Hamburger AW, Bako L, Visser RG, Bachem CW, Bogre L (2006) EBP1 regulates organ size through cell growth and proliferation in plants. EMBO J 25:4909–4920

Hu Y, Poh HM, Chua NH (2006) The Arabidopsis *ARGOS-LIKE* gene regulates cell expansion during organ growth. Plant J 47:1–9

Hülskamp M (2000) How plants split hairs. Curr Biol 10:R308–310

Imai KK, Ohashi Y, Tsuge T, Yoshizumi T, Matsui M, Oka A, Aoyama T (2006) The A-Type Cyclin CYCA2;3 Is a Key Regulator of Ploidy Levels in *Arabidopsis* Endoreduplication. Plant Cell 18:382–396

Jacobsen SE, Binkowski KA, Olszewski NE (1996) SPINDLY, a tetratricopeptide repeat protein involved in gibberellin signal transduction in *Arabidopsis*. Proc Natl Acad Sci USA 93:9292–9296

Kaya H, Shibahara KI, Taoka KI, Iwabuchi M, Stillman B, Araki T (2001) *FASCIATA* genes for chromatin assembly factor-1 in *Arabidopsis* maintain the cellular organization of apical meristems. Cell 104:131–142

Kirik A, Pecinka A, Wendeler E, Reiss B (2006) The Chromatin Assembly Factor Subunit FASCIATA1 Is Involved in Homologous Recombination in Plants. Plant Cell 18:2431–2442

Kosugi S, Ohashi Y (2002) E2Ls, E2F-like repressors of Arabidopsis that bind to E2F sites in a monomeric form. J Biol Chem 277:16553–16558

Leiva-Neto JT, Grafi G, Sabelli PA, Dante RA, Woo YM, Maddock S, Gordon-Kamm WJ, Larkins BA (2004) A dominant negative mutant of cyclin-dependent kinase A reduces endoreduplication but not cell size or gene expression in maize endosperm. Plant Cell 16:1854–1869

Mariconti L, Pellegrini B, Cantoni R, Stevens R, Bergounioux C, Cella R, Albani D (2002) The E2F family of transcription factors from *Arabidopsis thaliana*. Novel and conserved components of the retinoblastoma/E2F pathway in plants. J Biol Chem 277:9911–9919

Perazza D, Herzog M, Hülskamp M, Brown S, Dorne AM, Bonneville JM (1999) Trichome cell growth in *Arabidopsis thaliana* can be derepressed by mutations in at least five genes. Genetics 152:461–476

Schellmann S, Schnittger A, Kirik V, Wada T, Okada K, Beermann A, Thumfahrt J, Jurgens G, Hulskamp M (2002) *TRIPTYCHON* and *CAPRICE* mediate lateral inhibition during trichome and root hair patterning in *Arabidopsis*. EMBO J 21:5036–5046

Schnittger A, Weinl C, Bouyer D, Schobinger U, Hülskamp M (2003) Misexpression of the cyclin-dependent kinase inhibitor ICK1/KRP1 in single-celled *Arabidopsis* trichomes reduces endoreduplication and cell size and induces cell death. Plant Cell 15:303–315

Schonrock N, Exner V, Probst A, Gruissem W, Hennig L (2006) Functional genomic analysis of CAF-1 mutants in *Arabidopsis thaliana*. J Biol Chem 281(14):9560–9568

Serralbo O, Perez-Perez JM, Heidstra R, Scheres B (2006) Non-cell-autonomous rescue of anaphase-promoting complex function revealed by mosaic analysis of *HOBBIT*, an *Arabidopsis CDC27* homolog. Proc Natl Acad Sci USA 103:13250–13255

Sugimoto-Shirasu K, Stacey NJ, Corsar J, Roberts K, McCann MC (2002) DNA topoisomerase VI is essential for endoreduplication in *Arabidopsis*. Curr Biol 12:1782–1786

Sugimoto-Shirasu K, Roberts GR, Stacey NJ, McCann MC, Maxwell A, Roberts K (2005) RHL1 is an essential component of the plant DNA topoisomerase VI complex and is required for ploidy-dependent cell growth. Proc Natl Acad Sci USA 102:18736–18741

Sun Y, Dilkes BP, Zhang C, Dante RA, Carneiro NP, Lowe KS, Jung R, Gordon-Kamm WJ, Larkins BA (1999) Characterization of maize (Zea mays L.) Wee1 and its activity in developing endosperm. Proc Natl Acad Sci USA 96:4180–4185

Suzuki T, Nakajima S, Inagaki S, Hirano-Nakakita M, Matsuoka K, Demura T, Fukuda H, Morikami A, Nakamura K (2005) *TONSOKU* is expressed in S phase of the cell cycle and its defect delays cell cycle progression in *Arabidopsis*. Plant Cell Physiol 46:736–742

Szecsi J, Joly C, Bordji K, Varaud E, Cock JM, Dumas C, Bendahmane M (2006) *BIGPETALp*, a *bHLH* transcription factor is involved in the control of *Arabidopsis* petal size. EMBO J 25:3912–3920

Takeda S, Tadele Z, Hofmann I, Probst AV, Angelis KJ, Kaya H, Araki T, Mengiste T, Mittelsten Scheid O, Shibahara K, Scheel D, Paszkowski J (2004) *BRU1*, a novel link between responses to DNA damage and epigenetic gene silencing in *Arabidopsis*. Genes Dev 18:782–793

Tsukaya H (2006) Mechanism of leaf-shape determination. Annu Rev Plant Biol 57:477–496

Tsumoto Y, Yoshizumi T, Kuroda H, Kawashima M, Ichikawa T, Nakazawa M, Yamamoto N, Matsui M (2006) Light-dependent polyploidy control by a CUE protein variant in *Arabidopsis*. Plant Mol Biol 61:817–828

Verkest A, Manes CL, Vercruysse S, Maes S, Van Der Schueren E, Beeckman T, Genschik P, Kuiper M, Inze D, De Veylder L (2005) The cyclin-dependent kinase inhibitor KRP2 controls the onset of the endoreduplication cycle during Arabidopsis leaf development through inhibition of mitotic CDKA;1 kinase complexes. Plant Cell 17:1723–1736

Vlieghe K, Boudolf V, Beemster GT, Maes S, Magyar Z, Atanassova A, de Almeida Engler J, De Groodt R, Inze D, De Veylder L (2005) The DP-E2F-like gene *DEL1* controls the endocycle in *Arabidopsis thaliana*. Curr Biol 15:59–63

Wang H, Zhou Y, Gilmer S, Whitwill S, Fowke LC (2000) Expression of the plant cyclin-dependent kinase inhibitor ICK1 affects cell division, plant growth and morphology. Plant J 24:613–623

Walker JD, Oppenheimer DG, Concienne J, Larkin JC (2000) *SIAMESE*, a gene controlling the endoreduplication cell cycle in *Arabidopsis thaliana* trichomes. Development 127:3931–3940

Weinl C, Marquardt S, Kuijt SJ, Nowack MK, Jakoby MJ, Hulskamp M, Schnittger A (2005) Novel functions of plant cyclin-dependent kinase inhibitors, ICK1/KRP1, can act non-cell-autonomously and inhibit entry into mitosis. Plant Cell 17:1704–1722

Yoshizumi T, Tsumoto Y, Takiguchi T, Nagata N, Yamamoto YY, Kawashima M, Ichikawa T, Nakazawa M, Yamamoto N, Matsui M (2006) INCREASED LEVEL OF POLYPLOIDY1, a conserved repressor of *CYCLINA2* transcription, controls endoreduplication in *Arabidopsis*. Plant Cell 18:2452–2468

Plant Cell Monogr (10)
L. Bögre and G. Beemster: Plant Growth Signaling
DOI 10.1007/7089_2007_140/Published online: 22 August 2007

Epidermal Signalling and the Control of Plant Shoot Growth

Gwyneth C. Ingram

Institute of Molecular Plant Science, Daniel Rutherford Building, Kings Buildings, University of Edinburgh, Edinburgh EH9 3JR, UK
gwyneth.ingram@ed.ac.uk

Abstract Although the epidermal cell-layer is of undisputed physiological importance to all angiosperms, its potential role in controlling the growth of plant organs has remained a subject of research and contention for more than a century. Recent advances in our understanding of the molecular control of plant cell proliferation, growth and specification, in combination with novel biotechnological techniques, have provided new tools for addressing the control of organ growth control. New data have also elucidated the mechanisms via which plant cells communicate and, taken with classical studies, provide tantalising vistas into the mechanisms controlling co-ordinated organ growth. However, despite this explosion in knowledge, a clear picture of the contribution of specific cell-types, including the epidermis, to the regulation of growth and morphogenesis in plant organs remains elusive.

1 Introduction

The correct development of the plant epidermis is of the utmost importance for plant survival. Epidermal development is finely honed to provide a compromise between protective roles against pathogen attack, radiation damage, water-loss and other environmental aggressions, and roles as an interactive interface with the environment, vital for gas and nutrient exchange, water uptake and light penetration. In line with its multiple roles, the plant epidermis can differentiate a wide range of morphologically dramatic specialised cell types, including root hair cells, trichomes and stomatal guard cells. Because of their accessibility, and the fact that disrupting their development in the laboratory context does not lead to inviability, a relatively large amount is known about how epidermal cell-types are specified from a basic pavement of epidermal precursors (protoderm). In contrast, loosing or even compromising basic epidermal identity usually leads to plant lethality. Correspondingly, much less is known about how basic epidermal identity is specified, and to what extent specification and growth of the epidermis is necessary for and/or co-ordinated with development and growth of underlying cell-layers during development. Herein I will attempt to analyse evidence for the importance of the epidermis for normal plant growth, and will present current knowledge regarding how signalling between epidermal cells and their neighbours controls the development of this vitally important tissue.

2 The Ontogeny of the Shoot Epidermal Cell Layer

2.1 The Embryonic Protoderm

At what point during plant development the epidermal cell layer is specified remains an open question. In many angiosperms, including the model species *Arabidopsis thaliana*, embryonic cytokinesis commences with an asymmetric division partitioning the fertilized zygote into a small, highly cytoplasmic, apical cell and a larger basal cell. In *Arabidopsis*, the progeny of the apical cell are destined to give rise to most of the cells in the mature plant, including all epidermal cells, whilst one descendent from the basal cell forms the organising centre (quiescent centre) of the root meristem (Jurgens 2001). In *Arabidopsis* an outer cell layer (or protoderm/dermatogen) is demarcated after just four rounds of cell division in the apical cell. After this so-called dermatogen stage, protodermal cells undergo principally anticlinal cell divisions and thus only give rise to more protodermal cells (Goldberg 1994; Willemsen and Scheres 2004). The point at which protodermal demarcation occurs during embryogenesis in different plant species depends largely upon patterns of cell division early in embryogenesis. The division of the embryos in other dicotyledonous species, such as cotton (Pollock 1964), *Citrus* (Bruck 1985a) and apple (Meyer 1958), and of monocotyledonous species, such as maize (Randolph 1936) and barley (Merry 1941), lack the pleasing predictability of those observed in *Arabidopsis*. Despite this they lead to the production of an organised protodermal cell layer, albeit in an embryo comprising many more than 16 cells. Moreover, disruption of the cell division patterns of the early *Arabidopsis* embryos in some developmental mutants, so that dermatogen demarcation occurs later or asynchronously in different embryonic regions, does not appear to affect the propensity of embryos to produce a discrete protoderm and epidermis later in development (Torres-Ruiz and Jurgens 1994). Thus, acquisition of protodermal identity appears solely a function of "outside" cell position. The question of the positional information required for the specification of protodermal identity is complex, and will be addressed later.

2.2 The Endosperm Aleurone Layer

Interesting developmental similarities exist between the embryonic protoderm and the outside cell layer of the second product of double fertilization in angiosperms, the endosperm. Although some debate still exists regarding the evolutionary origin of the angiosperm endosperm (Friedman 2001; Friedman and Floyd 2001; Baroux et al. 2002), the most generally accepted current hypothesis is that it is a sexualised homologue of nutritive tissues which develop from the megagametophyte, for example in the gymnosperms. In the endosperm of many angiosperms, the outside cell-layer has specific cellular modifications, and a different developmental fate to other endosperm cells (Olsen 2004). In *Arabidopsis* the outer cell layer of the endosperm is maintained intact until seed maturity, whilst internal cells are broken down and used to support the developing embryo. The outer endosperm cells may play a role in taking up nutrients from the neighbouring endodermis of the seed coat. In monocotyledonous plants, for example the Gramineae, this specialisation of the outer endosperm cell layer is more extreme with the development of the highly specialised aleurone cell-layer surrounding the persistent starchy endosperm. The developmental ontogeny of the outer layer of the endosperm is rather different from that of the protoderm, in that the endosperm initially develops as a syncitial monolayer of nuclei lining the endosperm cavity. Upon cellularisation these nuclei become the progenitors of the aleurone, but also of all internal endosperm cells, to which they give rise by periclinal divisions (Olsen 2001). Several recent studies have shown that, as for the embryonic protoderm, the specification of aleurone cell fate is dictated largely by outside position (Geisler-Lee and Gallie 2005; Gruis et al. 2006).

2.3 The L1 Layer of the Shoot Apical Meristem

The embryonic protoderm gives rise, via anticlinal cell divisions, to the epidermal cells of embryonic organs such as the hypocotyl and cotyledons (in dicots) and the scutellum and coleoptile in monocots. In addition to strictly embryonic structures, the seedling shoot apical meristem (SAM), which will give rise to all shoot structures, is also specified during embryogenesis. In *Arabidopsis*, the SAM is composed of three layers of cells denoted L1 (epidermal), L2 (subepidermal) and L3. The L1 and L2 are a single cell thick and undergo principally anticlinal cell divisions. Together they form the tunica. The underlying L3 cells do not show particular restrictions in their division planes and form the corpus. In *Arabidopsis*, embryonic protoderm cells give rise to the L1 or outer cell layer of the shoot apical meristem, with underlying cell layers in the meristem originating from hypodermal embryonic cell layers (Barton 1993). A similar developmental ontogeny is probably applicable in the maize embryonic SAM (Poethig 1986), although there is often no easily distinguishable subepidermal tunica cell layer in maize vegetative tissues, and in maize "L2" is often used to denote the meristematic corpus.

3
Epidermal Contribution to Plant Growth Control

The question of whether the growth and morphogenesis of plant organs is controlled at the whole organ level (the organismal theory), the level of the combined action of single cells (cell theory), or by a less simplistic mechanism governed by cell–cell communication and co-ordination (the neo-cell theory) has been of interest for decades (Tsukaya 2002). Integrally linked to this question, is whether the mechanisms regulating organ growth and morphogenesis are active in any particular cell-layer of the developing organ. In 1933 George Avery carried out a detailed study of leaf development in tobacco (Avery, 1933). He suggested that in young leaf primordia the mesophyll provides the "impetus for development", placing the epidermis under tension at "the marginal meristem" [similar to the situation proposed by Wegner (below)]. However, he also remarked that although the epidermis ceases to divide before the mesophyll, its cells expand parallel to the lamina surface for much longer than underlying cells and are responsible for "pulling" mesophyll cells apart and thus play a potentially important role in determining final leaf shape. His study serves to elegantly illustrate the intuitively attractive idea that as a continuous planar monolayer of relatively uniform cells which encases every organ, the epidermis appears the simplest, most logical tissue in which to express the information regulating organ growth, at least in organs with an extended lamina. In the last 70 years however, many studies have sought to clarify the roles of different tissue layers in organ development by analysis of periclinal chimeras or clonal sectors in which different cell-layers have different developmental characteristics, studies of the effects of genetically or physically ablating the epidermis, and the study of developmental mutants. Despite this considerable body of work, it still remains unclear whether any one cell-layer can be thought of as the prime expression site for information regulating general organ growth and form. However, in organs with an extended lamina (such as leaves, petals and sepals), a certain body of evidence has been accrued suggesting that the epidermis can play a very important role in regulating organ shape.

3.1
Contribution of the L1 Lineage to Plant Organs

The fact that cells in the embryonic protoderm and meristematic L1 layer undergo predominantly anticlinal divisions can give the impression that L1 cell-layer identity is maintained in a clonal fashion throughout plant growth. Indeed it is true that in many higher plants shoot epidermal cells are all clonally related to a few protodermal precursors in the developing embryo. The study of periclinal mosaics in which the cell-layers of the shoot meristem differ for visible or histological markers, and clonal analysis using similar markers has been widely used to follow cell fates during plant development (reviewed in Neilson-Jones 1969; Tilney-Basset 1986; Poethig 1987; Marcotrigiano 2001). Such studies have confirmed that several structures that appear to be composed only of epidermal cells, for example the ligules of maize leaves and the sepal margins in *Arabidopsis* (Jenik and Irish 2000) are indeed entirely derived from the meristematic L1-layer. Such studies have also, however, shown that in some dicotyledonous species (such as privet) the mesophyll, which in the leaves of most dicots is predominantly L2-derived, can be L1-derived at leaf margins due to periclinal cell divisions in the L1 during organ development (Stewart 1975). Indeed in monocotyledonous species such analyses have shown that leaf marginal tissue is almost always L1 derived, and that periclinal divisions of the L1 cell layer are a routine feature of organ inception. Moreover, in maize, periclinal divisions of the meristematic L1 layer have been documented during development (Sharman 1940), and in rare cases epidermal cells derived from L2 lineages have even been noted (Poethig 1986). In every case where incursions between lineages have been noted, cells develop according to their positions rather than to their lineage, leading to the realisation that plant cells talk constantly to one another, and that their developmental fate, at least in the meristem and during early organ development is dependent on their physical position (Stewart 1975). Another discovery made during the study of periclinal mosaics, which is of particular importance to this discussion, is the fact that making mosaics in which different cell layers have different developmental vigour can markedly change the contribution of different meristematic cell-layers to organs without changing the overall shape of the organ (Stewart 1974). This serves to underline the degree of plasticity, and compensatory mechanisms manifest during plant organ development.

3.2 Control of Division Planes in the Meristematic and Primordium L1 Cell Layer

Epidermal cells in the shoot meristematic regions of many plants undergo predominantly anticlinal divisions to give more epidermal cells. However, clonal analysis has led to periclinal divisions being observed in the epidermal lineage of organ primordia, and sometimes also in shoot apical meristems. The degree to which L1 cells divide periclinally appears to be dependent upon both species and developmental stage. For example, incursions between the tunica cell-layers of *Arabidopsis* and *Antirrhinum* meristems appear to be relatively rare although incursions in later organ development of many dicots are observed especially at organ margins. In maize it is clear that the L1 can divide periclinally during both meristem and organ development (Sharman 1942), and in later nodes of the maize plant, most tissues can be derived from cells situated in the L1 layer of the embryonic SAM.

In a seductive series of mathematical models Jens Wegner proposes that the propensity of the L1 layer of meristems to divide periclinally is a function of the tangential strain placed on the outer cell layers of the tissue in question. This in turn is dependent on both cell size and the radius of the meristem or organ margin in question (Wegner 2000a,b). Thus, a corpus containing many cells and forming a wide meristematic dome, would generate a relatively large tangential strain within the outermost cell layers. This would have two predicted consequences; firstly it would encourage the occurrence of anticlinal divisions at the expense of periclinal divisions in the outermost (L1) cell layers and secondly, it would cause the formation of more tunica layers (i.e. layers of cells undergoing predominantly anticlinal divisions). In support of this theory maize vegetative meristems, which have only one tunica layer showing periclinal division, have a radius of only a few cell lengths. In comparison, dicot meristems, such as those of *Antirrhinum* are broader with a radius of many more cell widths and in agreement with Wegner's model contain two anticlinally dividing and clonally distinct tunica layers.

Wagner's models are based on the assumption that all meristematic cells divide at the same rate and are the same size at division, and that there is no genetic control over cell division planes. The first of these assumptions is not met, as several studies have demonstrated that the cells at the centre/apex of the shoot meristem, even in angiosperms, divide more slowly than those on the flanks (Grandjean et al. 2004) (reviewed in Carraro et al. 2006; Fleming 2006), although this discrepancy appears small enough in angiosperms that it does not completely destroy the predictions of the model. In contrast, the apical meristems of many gymnosperms (for example *Cupressus* species) do not possess a discrete tunica layer. Instead they have a population of one to three enlarged and histologically distinct apical initial cells which undergo both anticlinal and periclinal divisions to give rise to both epidermal and underlying cells (Pillai 1963). However, there is a tendency for periclinal divisions in superficial cell-layers to be lost in some gymnosperms [for example Monkey Puzzle (*Araucaria*)] in favour of the production of a more uniform anticlinally dividing tunica layer. In a study of the seasonal variation in structure of the shoot apices of *Araucaria columnaris*, Pillai (1964) notes that in the summer, when the meristem is relatively narrow, it contains only one tunica layer, whereas winter meristems, which are much broader, contain two distinct tunica layers, as predicted by Wegner's model (Pillai 1964). The second assumption underlying these models (that there is no genetic control over cell division planes) is controversial, and will be dealt with in more detail later.

Wegner also notes a correlation between the frequency of L1 incursion into underlying cell-layers at leaf borders and the form of these borders in transverse sections. In general, species or cultivars with "narrow" (in terms of cell radii) leaf margins are more likely to develop L1 derived mesophyll at leaf margin (Wegner 2003a,b). Although the assumptions underlying Wagner's models are open to discussion, the fact remains that L1 cells can and do undergo periclinal divisions and contribute to underlying tissues in some angiosperm species, an observation that complicates analysis of both how epidermal growth is controlled, and to what extent the epidermis is responsible for controlling the growth of underlying tissue.

The control of cell division planes of cells which are not at the organ margin is obviously also an important issue for plant growth and development. In particular maintenance of predominantly anticlinal division planes in the developing organ surface is predicted to be crucial in allowing laminar outgrowth due to the apparent role of epidermal cells in blade expansion. Mutants where epidermal cell specification is compromised (see later sections) tend to loose their ability to regulate epidermal cell division planes, suggesting that restriction of epidermal cell divisions to the anticlinal plane is an intrinsic consequence of the acquisition of epidermal identity. However, interpreting the phenotypes of such mutants in terms of growth regulation can be difficult. One mutant in which cell division planes appear to be uncoupled without noticeable loss of epidermal identity is the Extra Cell Layer1 (*Xcl1*) mutant of maize, in which late oblique divisions in the developing epidermis lead to the production of two epidermal cell-layers. In support of the proposed role of the epidermis in lamina expansion, the blade width of *Xcl1* mutant leaves is considerably reduced (Kessler et al. 2002). The division defect in this mutant is proposed to occur late enough in development that cells are irreversibly committed to an epidermal fate. However, the mutation is semi-dominant and the gene responsible has not yet been cloned making it difficult to draw mechanistic conclusions from this particular mutant.

3.3
Periclinal Mosaics and Clonal Analysis of the Role of the Epidermis

As described above, the use of periclinal mosaics has been invaluable in ascertaining the contributions of various cell-layers to organ development. However, they have also been used in addressing the potential roles of specific cell layers/tissue types in the control of organ shape (reviewed in Szymkowiak and Sussex 1996; Marcotrigiano 2001). The generation and study of periclinal mosaics (either full periclinal chimeras or clones induced during development), where cell layers carry mutations in developmental genes or are derived from different species, has been an important approach. The use of periclinal chimeras in such experiments has the associated disadvantage that certain chimeric combinations can affect the contributions of different cell layers to organs, giving a false impression of normal development. An alternative has been to analyse clones, usually marked by irradiating plants heterozygous for a recessive developmental marker at a known point in development. Because this provides unequivocally clonal groups of marked cells (mericlinal chimeras) from progenitors at defined developmental stages it can give more accurate information about cellular contributions, and has been used for multiple studies.

Relatively large numbers of both types of study have not provided a consensus as to what extent a given cell layer controls organ shape and growth, as the results vary between organs, species, and according to the developmental marker studied. Interestingly however, several studies on leaves and other organs with an extended lamina (such as sepals and petals) using plants chimeric at just one locus (rather than intraspecific combinations with significant degrees of genetic divergence), have shown that the presence of a wild-type allele in the L1 cell layer can play a greater role in restoring wild-type phenotypes than expression in underlying cell-layers (Hantke et al. 1995; Perbal et al. 1996; Vincent et al. 2003). However, this is not always the case as some genes appear to be required in underlying cell layers for normal development, including the *Nicotiana LAM1* gene (McHale and Marcotrigiano 1998), the *OKRA* gene from cotton (Dolan 1998) and the *LIGULELESS-1(LG1)* gene from maize (Becraft et al. 1990).

How can the differences in behaviour of different developmental genes be explained? It may be that in organs with extended laminas, the epidermal cell-layer plays a significant role in expressing the information which determines final organ size and shape, signalling to underlying cell-layers to control their development. Developmental genes such as *LAM1* or *LG1*, which are needed in underlying cell-layers for normal development, may be involved in these communication processes. Indeed, some indication that *LG1* may be required to perpetuate positional signalling across the leaf blade has been gained from analysing the developmental effects of *lg1* mutant sectors (Becraft and Freeling 1991).

Another possible explanation for the observed differences lies in the ability of adjacent cell-layers to accommodate discrepancies in growth, and ultimately co-ordinate their development. The molecular mechanisms underlying this accommodation probably include the symplastic movement of developmentally important molecules, such as transcription factors and regulators of the cell-cycle, between cell-layers in the meristem and in young organ primordia via plasmodesmatal connections (reviewed in Lucas and Lee 2004). The products of both the different developmental genes used in clonal studies, and downstream effectors of the markers, may move from layer-to-layer to different extents during development. Such differential capacity for movement has been observed for several proteins including the transcription factors *LEAFY* and *APETALA1* in *Arabidopsis* (Sessions et al. 2000; Wu et al. 2003).

Additionally, as pointed out by Vincent et al., chimeras are affected not only in the localisation of developmental gene expression, but also in the absolute amounts of their product (Vincent et al. 2003). Perception of protein concentrations may be more important for the actions of some developmentally important genes than for others.

Interestingly, in organs lacking laminas, such as stamens and ovaries, the importance of the L1 appears more often to be overridden by that of the L2 and L3 (Szymkowiak and Sussex 1992; Vincent et al. 2003). One explanation for this is that the L1 cell layer may contribute proportionally more cells to the total volume of an organ with an extended lamina than to an effectively cylindrical organ such as a stamen. Protein dosage could therefore also be an important factor in the perceived differences in cell-layer contribution to organ development in mosaic laminar and non-laminar organs.

3.4 Chemical and Physical Ablation of Epidermal Cells

Cell ablation has been widely used to study plant development, particularly development of the accessible epidermal cell layer. Although these techniques are often extreme and cause obvious physical damage and stress, they have nonetheless provided interesting information about the roles of epidermal tissues at various stages in development. Genetic ablation of protodermal cells in young *Arabidopsis* embryos by expression of Diphtheria Toxin A under an epidermis-specific promoter leads to developmental arrest of the embryo proper at the early globular stage, suggesting a role for the protoderm in promoting continuing embryogenesis (Weijers et al. 2003). Ablation of protodermal cells in cotyledons during later embryogenesis seriously disrupts cotyledon development, consistent with a role in cotyledon morphogenesis and expansion. Likewise post-embryonic ablation of L1 cells in leaves using BARNASE expression causes major disruption in leaf development, characterised by the production of small unexpanded organs which tend to be thicker (in terms of cell numbers) than wild-type organs (Baroux et al. 2001). Thus, in agreement with the findings of chimera studies, it appears possible that the L1 is important for expressing information required for the correct outgrowth of the lamina as well as for organ expansion. However, when expressing toxins within developing tissues it is always possible that they may move to adjacent cells causing them to behave aberrantly. In addition the presence of the protoderm might be needed for the correct specification of underlying cell-layers, and thus growth defects could be a secondary effect of cell-specification defects.

Physical ablation of the epidermal cells of plant embryos is technically very challenging, although studies using cultured embryos of *Citrus jambhiri* (Bruck 1985a,b) have provided several interesting observations which will be discussed in more depth later. In general, physical ablation of epidermal cells in organ primordia is difficult to carry out over large enough areas to allow the examination of epidermal roles in growth control, and the physical properties of the scar tissues which form at wound sites also complicate interpretation. However, laser ablation studies in meristems of tomato have suggested a role of the meristematic L1 layer in controlling meristem growth. For example, ablating the meristematic L1 layer causes a dramatic change in the behaviour of underlying (L2) cells which cease to divide anticlinally and commence periclinal divisions, suggesting that the presence of the L1 cell layer restricts division planes in underlying cell layers (Reinhardt et al. 2003a). This observation is interesting in light of the models proposed by Wegner, although the assumption that epidermal cells all have developmentally similar properties becomes important when interpreting this type of study. Differences between concentric zones of the meristem might also be relevant. In the centre of the angiosperm meristem (central zone) a group of self-maintaining, slowly dividing stem cells produce progeny which enter a more rapidly proliferating peripheral zone, prior to being incorporated into organ primordia or stem tissues (Grandjean et al. 2004; Carraro et al. 2006). The molecular mechanism underlying the maintenance of the stem cell population in the central zone is relatively well characterised and hinges on expression of a putative peptide ligand (CLAVATA3) in the central zone tunica cells in response to the activity of the transcription factor WUSCHEL in the lower corpus. Perception of CLV3 is thought to negatively regulate WUS expression via the Receptor-Like Kinase CLV1 and associated factors, restricting the size of the central zone (reviewed in Doerner 2003; Williams and Fletcher 2005). Bearing this in mind, the question then arises whether the phenotype caused by ablating the L1 of the CZ is specifically due to removal of the epidermal layer, or to loss of CLV3 expressing cells which could cause a developmental reprogramming of underlying cells. Interestingly, changes in the orientation of cell divisions from anticlinal to periclinal in underlying cell layers of root meristems upon chemical ablation of the root epidermis have also been noted (Baroux et al. 2001), suggesting that one of the intrinsic properties of the L1 is indeed to restrict cell division planes in underlying cell layers. Additionally, loss of L1 cells either through ablation or in mutants with decreased epidermal specification leads to premature differentiation of meristem tissue, thus, although CLV3 signalling from the tunica acts to repress the size of the stem cell population, the presence of the L1 appears to be intrinsically necessary for the very existence of stem cells (Baroux et al. 2001; Abe et al. 2003; Johnson et al. 2005).

In addition to its role in meristem activity, the L1 appears to be necessary for the inception of organ primordia at the shoot apex. Thus, physical removal of the L1 layer causes an immediate cessation in organ initiation at the site of removal, even though underlying cells are still able to divide and expand normally (Reinhardt et al. 2003a). Reinhard et al. argue that this is unlikely to be due to an immediate loss of meristem identity as meristematic markers are lost only slowly from "skinned" meristems. One possible explanation is that removing the epidermis perturbs auxin transport. Polar auxin transport, which is now widely accepted to cause localised increases in auxin activity that promote organ formation, appears to be largely restricted to the L1 cell layer in the shoot meristem (discussed below). Another possible explanation, is that the L1 cell layer is key in the perception of signals required for organ initiation (again, possibly auxin, but equally possibly other signals).

The role of the L1-specific WOX proteins in recruiting subsets of meristematic cells to organ primordia (again discussed below) may support this view.

3.5 The Epidermis as a Physical Constraint to Organ Growth

In the models proposed by Jens Wegner, the plane of epidermal division is regulated by the force exerted upon it by the proliferation of underlying cells. In this model it is tempting to envisage the epidermis as a type of "bag" which is stretched over underlying cells and exerts an inward pressure on them, thus restricting their growth. The switch from anticlinal to periclinal cell divisions observed in the meristematic L2 after L1 removal in the experiments of Reinhardt et al. appears to support this view. Interestingly, similar regular periclinal divisions are induced when pressure is applied to *Coleus* internodes and removed one day later (Lintilhac 1981), and in underlying parenchyma when epidermal cells are excised from epicotyl in *Citrus* (Bruck 1985b). This could be interpreted, then, as a universal reaction to release of pressure, but it could equally well reflect a universal response to wounding, and the two possibilities are very difficult to separate.

Another observation which supports a restrictive role for the epidermis in organ growth is that, at least at the early seedling stage, mutants where epidermal integrity is compromised often appear bigger, and grow faster than wild-type plants in tissue culture conditions where humidity is very high (Ingram, unpublished results). Whether this apparent increased growth is due to increased cell division, or simply to abnormal cell expansion has not been accurately ascertained.

3.6 Mutants in Interpreting the Role of the Epidermis

In theory mutants affecting the development of specific tissue types within plant organs should provide a useful source of information regarding the relative contributions of different tissues to final leaf shape. In this scenario it is a cell-type rather than a cell lineage which is affected and thus, although some accommodation may take place between cell layers, incursions from different cell lineages should have no effect on the conclusions. A good example is the *Arabidopsis reticulata* mutant which shows a strong reduction in mesophyll cell density, apparently due to a defect in mesophyll cell proliferation during development (Gonzalez-Bayon et al. 2006). The paucity of mesophyll cells in the lamina gives mutant leaves a pale aspect, with veins showing up as darker lines. The mutant epidermal cell-layer appears phenotypically wild-type and, moreover, the overall form of the lamina is comparable to that in wild-type plants although slightly reduced in size. The ability of this mutant to produce a normal lamina despite its lack of mesophyll cells does provide an intriguing indication that the leaf epidermis could be the primary determinant of leaf shape, although, since the exact role of the RETICULATA protein is unclear, it is perhaps unwise to draw too many conclusions from this mutant. Additionally, the role of L3-derived vascular tissue in regulating leaf shape should not be underestimated.

Theoretically mutants which are thought to be directly compromised in specification of L1 identity might provide a more direct source of information regarding the role of the L1 in growth control. The HDZipIV-class homeodomain proteins ATML1 and PDF2 are expressed in the L1 in Arabidopsis meristems and organ primordia during shoot development (Lu et al. 1996; Sessions et al. 1999; Abe et al. 2003). They act redundantly to specify epidermal identity in shoot organs and maintain the meristematic L1 cell layer (Abe et al. 2003). They are thought to act by binding elements called L1 boxes in the promoters of targets, which include their own genes and genes involved in cuticle biosynthesis and epidermal differentiation such as FIDDLEHEAD (Abe et al. 2001). Double *atml1 pdf* mutants show characteristic defects which start during embryogenesis with abnormal cell division patterns in the apical part of the embryo including the meristem L1 layer and epidermis of the cotyledon primordia. Cotyledons frequently fail to form properly and are commonly fused. Although one or two leaf-like organs can form they are small, misshapen and usually consume the poorly maintained shoot apical meristem. If the meristem is not consumed it differentiates as vacuolarised cells. The surfaces of shoot organs are covered with cells which resemble mesophyll, with the exception of occasional stomatal clusters, which indicate that double mutants do retain some epidermal identity (Abe et al. 2003; Ingram, unpublished data). This may be provided by another related protein from the same family (Nakamura et al. 2006; Ingram, unpublished data). The inability to form morphologically normal organs when epidermal identity is compromised is an important indication of the role of the epidermis in organ growth and morphogenesis, and correlates with the observations obtained from ablation studies discussed above. However, the extreme nature of the organ phenotypes shown by these mutants, and the accompanying meristem degeneration, preclude their use in quantitative studies.

The fact that the defects observed in *atml1/pdf2* double mutants do not affect early embryogenesis suggests three things. (1) Either epidermal/protodermal identity is not entirely lost in these mutants, and enough is left to permit normal protoderm specification during early embryogenesis or (2) ATML1, PDF2 and related genes are required for epidermal differentiation in late embryogenesis and post germination, but not for the specification of protoderm identity during early embryogenesis. (3) That conclusions from chemical ablation studies suggesting that specification of the protoderm is necessary for early embryogenesis to proceed normally are wrong. Ongoing studies will distinguish between the three possibilities, although the phenotypes associated with reduced function of the *Arabidopsis AtDEK1* gene provides a tantalising indication that the first explanation could be correct. *AtDEK1* is a likely membrane-bound cysteine protease that was first isolated by homology to the maize *DEFECTIVE KERNEL1 (DEK1)* gene (Lid et al. 2002, 2005; Johnson et al. 2005). Strong mutants in *DEK1* in maize show defects in the specification of aleurone cell identity in the endosperm, and early embryo arrest. Weaker alleles show defects in leaf development, including aberrant specification of specific epidermal cell-types (Becraft et al. 2002). Null mutants in *ADEK1* show early embryonic arrest and defects in the patterning of endosperm cell divisions in *Arabidopsis*. Null mutant embryos show aberrant cell division patterns and, notably, do not form a morphologically distinct protodermal cell layer (Johnson et al. 2005; Lid et al. 2005) and do not express epidermal markers, including *ATML1* (Johnson et al. 2005). The extreme nature of the null mutant phenotype, and the unrestricted expression pattern of *ADEK1* make it difficult to conclude whether the role of *ADEK1* in young embryos is restricted to the protoderm. However, RNAi knockdown of *ADEK1* in all cell-layers during late embryogenesis leads to the differentiation of mesophyll in the place of epidermal cells on the cotyledons, and a loss of meristematic activity in the SAM; phenotypes that are strongly reminiscent of those shown by *atml1/pdf2* double mutants. This indicates that at least one of the roles of *ADEK1* is the maintenance of epidermal identity in embryonic cells (Johnson et al. 2005). In plants with reduced *ADEK1* expression due to RNAi, cotyledons appear narrower than in the wild-type, possibly due to aberrant lateral expansion of the lamina, and the SAM is usually lost so that no post-embryonic organs are produced.

3.7 Genetic Manipulation of the Growth Properties of the L1 Cell Layer

The identification of L1-expressed genes, and the subsequent characterisation of their promoters, coupled to the identification of key growth regulators in plants, has recently smoothed the way for a new generation of experiments aimed at understanding the contribution of the L1 to plant growth. Two independent studies have used expression of cognate growth regulators under the *ATML1* promoter to analyse how manipulating growth specifically in the L1 of the SAM and young leaf primordia affects organ development.

In the first study two cyclin-dependent kinase inhibitors, KRP1/ICK1 and KRP4, were expressed in the L1, leading to a decrease in L1 proliferation. The resulting epidermal cells demonstrated a compensatory increase in size, but overall plant and organ size were none-the-less decreased. Surprisingly however, cell numbers in the underlying mesophyll did not differ significantly from those in wild-type leaves leading the authors to conclude that the control of proliferation in mesophyll and epidermal cell layers are regulated by autonomous programmes during development (Bemis and Torii 2007). This conclusion is in agreement with those reached during the study of the *reticulata* mutant (Sect. 3.6) (Gonzalez-Bayon et al. 2006). However, during wild-type development, cessation of cell division in the epidermis occurs later than that in underlying tissues (Sect. 3.3). Moreover, no effect on the SAM was observed in this study, suggesting that the transgenic KRP activity may be context dependent. It is thus possible that the KRP-induced block in cell division in this study could function predominantly at stages in primordium development when mesophyll proliferation is more advanced than that of the epidermis, leading, in part, to the observed imbalance in cell-numbers in the mature leaves.

The second study used a similar strategy to reintroduce perception of the growth-promoting molecule brassinosteroid (br) specifically in the L1 layer of dwarf brassinosteroid insensitive mutants. This was sufficient to rescue normal organ growth, despite the fact that br perception appeared to occur largely cell autonomously. Moreover, depletion of br in the L1 was sufficient to cause dwarfing. It was concluded, in line with several clonal studies discussed previously, that a non-cell-autonomous signal from the epidermis controls the growth of underlying cells allowing the epidermis to both drive and restrict the growth of shoots (Savaldi-Goldstein et al. 2007).

When considering the apparently contradictory conclusions drawn from these studies, it should be borne in mind that the control of growth mediated by br acts predominantly (although not exclusively) at the level of cell expansion, whereas KRPs exert a direct control of cell proliferation. This not only serves to make the two sets of observations difficult to compare directly, but also underlines the potential complexity, as well as the utility, of analysing growth control using such approaches.

3.8 L1-Expressed WOX-Class Homeobox Genes and Recruitment of Cells into Organs

Another class of mutants which may have some bearing on the genetic regulation of cell division planes in the L1 cell layer of meristems and organ primordia is that composed of the *pressed flower (prs)* and *pretty few seeds2 (pfs2)* mutants of *Arabidopsis* (Matsumoto and Okada 2001; Park et al. 2005) and the double *narrow sheath1 (ns1)/narrow sheath2 (ns2)* mutant of maize (Scanlon et al. 1996; Scanlon 2000). These mutants all affect the recruitment of cells into developing organs, and the cells affected are principally (but possibly not exclusively) epidermal. Mutants in *PRS* affect recruitment of cells into the lateral sepals during early flower primordium development, and later recruitment into the margins of the remaining adaxial and abaxial sepals (Matsumoto and Okada 2001). Additionally, *prs* mutants fail to recruit most of the cells which would form the stipules and other minor basal structures of the wild-type leaf (Nardmann et al. 2004). The *ns1/ns2* mutant of maize, shows an analogous deletion of lateral regions of the basal leaf domain, leading to a marked reduction in the width of the leaf sheath. *PRS*, *NS1* and *NS2* all encode proteins of the WUSCHEL-related homeobox (WOX) class (Haecker et al. 2004) and are expressed in restricted region of the L1 in meristems where cells are to be recruited to the lateral domains of flower or organ primordia, and in the marginal domains of young organ primordia (Nardmann et al. 2004). Ectopic expression of *PRS* leads to the ectopic outgrowth of epidermal bilayers on organ surfaces (Matsumoto and Okada 2001).

PFS, another WOX protein, may play a similar role to that of NS1/2 and PRS, but in recruiting L1 cells from the ovule chalaza into incipient integument primordia, a role that it appears to share with WUSCHEL itself (Matsumoto and Okada 2001; Gross-Hardt et al. 2002). WUS expression, like that of *NS1/2*, is largely restricted to the L1 cell layer of the developing ovule primordium prior to integument initiation (J. Goodrich, unpublished results). Local ectopic expression of WUS in the zone of integument initiation under the control of the *AINTEGUMENTA (ANT)* promoter is reported to lead to the initiation of ectopic integuments, an effect reminiscent of that following the ectopic expression of *PRS* (Gross-Hardt et al. 2002).

Interestingly, clonal analysis of *ns* mutants indicated that *NS* function was required not in the L1, where *NS1 and NS2* are expressed, but in a small population of underlying L2 cells. *NS* function was also found to be non-cell autonomous within the NS domain (that is deleted in *ns* mutants) (Scanlon 2000). It seems possible then, that the functions of WOX genes in allowing L1 cells to enter organ primordia may be complex and involve interactions between L1 and underlying cells. One possible explanation is that WOX function is somehow required either to block production or perception of an L2 derived signal, which normally prevents L1 cells from proliferating independently of L2 growth. This interpretation is consistent with the effects of ectopically expressing *PRS* and *WUS*. How this function interacts with that of other genes involved in organ initiation, such as *ANT*, remains to be clarified.

3.9 The Epidermal Monolayer as an Information Superhighway

One of the major stumbling blocks when analysing the contribution of the epidermis to plant growth is undoubtedly the fact that, at least in *Arabidopsis*, much of the extra-vascular movement of auxin to the shoot meristem and to the margins of young organ primordia (termed acropetal transport for the purpose of this review) is restricted to the L1 cell layer. Acropetal Polar Auxin transport (PAT) in L1 cells is absolutely required for phyllotaxis and the initiation and outgrowth of defined organ primordia, as well as the subsequent patterning and development of vascular strands within organs (mediated by basipetal PAT in internal cells) (Scarpella et al. 2006) (reviewed in Tanaka et al. 2006). At least two of the key proteins required for auxin transport in the apical meristem, the PIN1 efflux carrier and the AUX1 influx carrier, are primarily localized in the L1 cell layer in the shoot meristem and leaf primordia (Reinhardt et al. 2003b). Thus, disrupting L1 integrity should disrupt both organ initiation and vascular development. Indeed defects in these processes are observed in some L1 identity mutants (above), although it is difficult to decide with certainty the causes of such phenotypes when meristem maintenance and leaf morphology are also compromised.

Relatively little work has been carried out to assess why acropetal PAT is localized to the shoot L1. This type of study is non-trivial given the fact that auxin and PAT themselves control expression of auxin transporters (reviewed in Tanaka et al. 2006). However, one obvious possibility is that the genes encoding the proteins responsible for PAT such as AUX1, PIN1 and the MDR/PGP subfamily of ABC transporters are predisposed to L1 expression due to the presence of tissue specific elements in their promoters. No elements (such as L1-boxes) which confer L1-specific expression in shoots are obvious in the promoters of these genes (GI unpublished data). Moreover, during the establishment of PAT during embryogenesis, expression of transporters such as PIN1 is not restricted to L1 cells until well after protodermal identity has been established (Friml et al. 2003). In addition, the same transporters required for acropetal PAT in the L1 are also required internally for basipetal PAT in the developing vasculature. However, auxin-mediated transcriptional regulation, at least of PIN protein expression appears to occur via an AUX/IAA dependent pathway. Because of the potentially auxin independent, or at least tissue/organ specific expression of components of this pathway (for example AUXIN RESPONSE FACTOR transcription factors) this may add a layer of cell-type or tissue-type specificity to the process (Vieten et al. 2005).

Another explanation of the L1-localization of acropetal PAT is that the intrinsic physical properties of the L1-layer somehow favour PAT in the L1 over PAT in underlying cells. As PAT positively regulates the expression of at least some the transporters responsible, this would lead to an effective restriction of auxin movement to the L1-cell layer without the need to invoke tissue specific expression. The epidermis is organized as a planar monolayer composed of cells with a precisely defined apical-basal polarity with respect to the plant surface, and tight lateral cell–cell junctions without nascent air spaces. This may play an important role in facilitating the efficient movement of auxin and subsequent up-regulation of transporter expression. The cell walls between neighbouring L1 cells may even have properties which facilitate the efficient apoplastic movement of auxin. The plant surface could also serve to effectively corral auxin in the epidermal layer giving rise to higher auxin concentrations, and thus PAT in epidermal cells. However, testing whether such factors are important is again technically difficult. Interestingly, expression of a maize homologue of the *Arabidopsis* auxin efflux carrier PIN1 is down regulated in maize *crinkly4* mutants (see below), where L1 identity and organization appear to be compromised at the shoot apex (Kessler et al. 2006).

4 Specifying and Maintaining a Functional Epidermal Cell Layer

The signals underlying epidermal specification during embryogenesis and the maintenance of L1 identity in meristems and organ primordia are largely unknown. This lack of information is probably attributable to two factors. Firstly, abnormal epidermal specification leads to the production of sub-viable embryos, making the isolation and characterisation of mutants difficult. Secondly, there may be a high level of genetic and functional redundancy between some of the factors involved in these critical developmental processes. Despite these problems some indications regarding the types of signalling molecules which could be involved, and how they may be perceived have emerged from recent research.

4.1 Signalling and Positional Information in Protodermal Specification

The nature of the developmental signals required for early embryo patterning have long been a source of speculation to plant scientists. When considering protodermal specification, it is first interesting to ask when cells first acquire protodermal characteristics. There is a certain amount of evidence to suggest that protodermal identity is a default during the development of the embryo-proper and that in at least some species cells have protodermal identity well before a defined dermatogen/protoderm is set aside by restrictions in cell division planes. Bruck and Walker, in their study of embryogenesis in *Citrus jambhiri*, conclude that even at the very early globular stage, (well before protoderm demarcation) a cuticle has been secreted on the outside of the developing embryo (Bruck 1985a). Cuticle secretion is considered to be a strictly epidermal function, and this suggests that cells at the surface possess epidermal characteristics which are then presumably lost from the more internal progeny of a periclinal division. Interestingly, this hypothesis is borne out at the gene expression level in *Arabidopsis*. At least two protodermal markers (*ATML1* and *ACR4)* are expressed in all cells of the embryo proper at the eight cell stage (when all eight cells are in contact with the periphery), and their expression is then lost in "internal" cells subsequent to the periclinal divisions which give rise to the dermatogen embryo (Lu et al. 1996; Gifford et al. 2003). It is fair then, to suggest that the signals required for protoderm development are present from very early on in embryogenesis.

The "loss" of protodermal characteristics in embryonic cells which loose contact with the periphery, is very reminiscent of the process of aleurone proliferation where most starch endosperm (internal) cells are derived by periclinal divisions from cells on the endosperm periphery, which will eventually differentiate as aleurone. These similarities are backed up by gene expression patterns which are common to the two tissues. For example some of the Outside Cell Layer (OCL) genes from maize, which are homologous to the HDZipIV class of transcription factors to which ATML1 and PDF2 belong, are expressed in both the developing embryonic protoderm and the incipient aleurone layer during caryopsis development (Ingram et al. 2000). Although the precise function of these genes in maize has not yet been ascertained, they do appear to play a role in the growth of the developing caryopsis (Khaled et al. 2005), and, extrapolating from the role of their *Arabidopsis* homologues in protoderm development it is possible that aleurone development is also specified by these genes. The agronomic importance of the cereal endosperm, coupled to the characterisation of a wide range of easily visualised aleurone specific pigment markers, has meant that many loci involved in the specification of the aleurone layer have been extensively characterised by maize geneticists over the past decades. Several of these genes appear to play parallel roles in the specification of embryonic protoderm, including DEK1, which is required for aleurone specification and normal embryogenesis in maize (Becraft et al. 2002). As mentioned previously, the *Arabidopsis DEK1* orthologue *AtDEK1* also appears to be necessary for embryonic protoderm specification and endosperm development (Johnson et al. 2005; Lid et al. 2005). *DEK1* encodes putative membrane localised calpain protease, suggesting that it acts by cleaving other protein molecules (Lid et al. 2002). However, to date, no confirmed targets of this cleavage activity have been identified. Additionally, the non cell-layer specific expression pattern and severe phenotypes associated with *DEK1* and orthologues in *Arabidopsis* and tobacco (Ahn et al. 2004), has led to the suggestion that it may play a general role in plant cell differentiation rather than a specific role in endosperm development. The maize *Extra cell layers (Xcl)* mutant, previously discussed for its defects in the regulation of epidermal cell division planes in leaves, also shows an analogous increase in aleurone cell layer number (Kessler et al. 2002). Similarly the *disorganised aleurone layer 1* and *2* mutants of maize both show disorganised epidermal cell arrangement in rescued homozygous plants (Lid et al. 2004). Finally, the maize *supernumerary aleurone layer 1* mutant also affects embryo development, although in this case the embryonic defects have not been characterised in depth (Shen et al. 2003).

Another maize gene, *CRINKLY4 (CR4)*, has also been implicated both in aleurone cell fate specification and the development of the embryo, with weak alleles showing defects in leaf epidermal fate (Becraft et al. 1996, 2001). *CR4* encodes an RLK and again *Arabidopsis* homologues, including the suspected orthologue *ACR4*, have been identified (Tanaka et al. 2002; Gifford et al. 2003; Watanabe et al. 2004; Cao et al. 2005). Unlike *CR4, ACR4* expression is restricted to the L1 cells layer of the embryo and all shoot meristems and organ primordia. The null *acr4* mutant phenotype and the sub cellular localisation of ACR4 protein are consistent with a role in signalling between L1 cells during development, and/or between L1 cells and surrounding tissue (Gifford et al. 2003; Watanabe et al. 2004). The *acr4* mutant phenotype is restricted to ovule integuments and sepal margins, with slight defects in leaf surface formation reported. Although no ligands for CR4 or ACR4 have been identified, it has been postulated that the *ABNORMAL LEAF SHAPE1 (ALE1)* encoding a subtilisin-like serine protease is involved in generating an *ACR4* ligand from a peptide precursor, due to the genetic interaction between *acr4* and *ale1* mutant alleles (Watanabe et al. 2004). If they are not protected from desiccation, mutants in *ALE1* die after germination due to aberrant cuticle formation on the seedling cotyledons and early leaves. *ALE1* is expressed in the Embryo-Surrounding Region (ESR) of the endosperm during embryogenesis, and may also be expressed weakly in the very young embryo. Double mutants between *ALE1* and *ACR4* show a dramatically exacerbated embryo/seedling phenotype. Recent results have shown that ACR4 may act with a second unrelated RLK, ALE2, during the development of the embryonic epidermis (Tanaka et al. 2007).

The analysis of *ALE1* has fuelled the ongoing debate regarding whether the correct specification and differentiation of the embryonic protoderm is dependent on signals from the surrounding endosperm. Interestingly, a parallel debate exists regarding the role of maternal tissues in the differentiation of the endosperm aleurone layer in cereals. As described previously, considerable evidence that "outside" or "surface" position is critical both for protoderm and for aleurone cell specification has been published. As a result it is natural that the finger should be pointed at surrounding tissues (the ESR for the protoderm, and the maternal nucellus for the aleurone) as a source of developmental signals. Indeed studies in maize have shown that several genes including small secreted peptides, including those encoding the secreted ESR-proteins (Opsahl-Ferstad et al. 1997; Bonello et al. 2000, 2002), are expressed in the embryo surrounding region of maize. ESR proteins belong to the CLE (CLV3/ESR) family of secreted peptides which includes CLV3 a putative ligand of the CLAVATA1 RLK (Cock and McCormick 2001; Sharma et al. 2003). Moreover, considerable evidence now exists that these peptides could be cleaved as part of their activation, potentially by secreted proteases such as that encoded by ALE1 (Ito et al. 2006; Kondo et al. 2006; Ni and Clark 2006). However, debate still exists as to whether embryogenesis (either zygotic or somatic) can occur in plants in the absence of endosperm-like cells. In many studies in vitro, embryogenesis appears to depend upon the presence of non-embryogenic cells with some endosperm-like characteristics (Magnard et al. 2000, and references therein). However, it is difficult to distinguish between the requirements for initiation of embryogenesis, and those for cell-fate specification in many of these systems. A recent publication addressed the question of whether aleurone specification can occur in the absence of maternal nucellus tissue, and concluded that it could, although sucrose concentrations used in the culture systems were abnormally high (Gruis et al. 2006). Interestingly, recent research in *Arabidopsis* has shown the probable presence of specific sucrose transporters in both the ESR (Baud et al. 2005), and the maternal endothelium (the inner-most maternal cell layer which directly juxtaposes the developing aleurone) (Lauterbach 2007).

If positional signals are not derived from the tissue surrounding the developing protoderm and aleurone, then two further possibilities exist. The first is that a signal present and immobilized within the zygote, egg or central cell, possibly in the cell wall or membrane, is perceived by cells at the embryo/endosperm periphery. Such signals are known to play roles in the development of algae, such as *Fucus* (Kropf et al. 1988; Quatrano et al. 1991; Shaw and Quatrano 1996). The second possibility is that embryo and/or endosperm cells have a mechanism for sensing the presence of their neighbours, and those cells which sense that they lack neighbours on one face, respond by expressing protoderm/aleurone specific genes. For the moment neither of these possibilities can be either proven or disproved.

4.2 Maintaining L1 Identity in Meristems and Organ Primordia

It seems likely that the molecular mechanisms involved in specification of protoderm cell fate during early embryogenesis may be closely related to those involved in maintaining that fate later in embryogenesis and during post germination growth. The maintenance of epidermal identity via an active signalling pathway is crucial to the maintenance of an organised epidermal monolayer, which in turn is of fundamental importance to proper plant function, especially in leaves. It is probably almost as physiologically important to prevent cells which leave the L1 layer via periclinal division during organ development from maintaining their epidermal fate, as it is to ensure that the cells on the outside of the plant maintain theirs.

Two possible mechanisms can again be envisaged for the maintenance of epidermal cell fate. The first is a constant perception of "outside" (the environment, or a lack of neighbours) which permits only externally situated cells to develop with L1 fate. The second is constant "lateral" signalling between epidermal cells, so that if cells leave the monolayer they lose their epidermal fate. Compelling evidence that a variation on the latter rather than the former mechanism may be the case comes from the observation that after removal of the epidermal cell layer, cells at the wound site never re-assume an epidermal fate, although they can redifferentiate as parenchyma tissues. Thus, it would seem that the positional information required for epidermal differentiation is irretrievably lost. This type of observation led Bruck and Walker to describe "epidermal differentiation as a one-time event" during their studies of *Citrus jambhiri* (Bruck 1985b). What form this lateral signalling takes is still an open question. One intriguing possibility is that signals deposited in the external cell wall or cuticle (which is present in a rudimentary form even in meristems) are responsible for part of this signalling, echoing possible mechanisms of cell-fate specification in the developing embryo.

The phenotypes of weak alleles of *CR4* and *DEK1* in maize, and the *acr4* and *ADEK1* RNAi phenotypes in *Arabidopsis* all indicate that both classes of protein are likely to be involved in interpretation of the positional signals required for maintaining epidermal identity and organisation. Ongoing studies of these proteins should thus allow the isolation of other potential signalling components in the pathways involved both in L1 maintenance and protodermal specification during embryogenesis.

4.3 Co-ordination of L1 Growth with that of Underlying Cells

Several pieces of evidence have been presented herein to support the view that the plant epidermal cell layer may play an important role in regulating plant growth and morphology. However, it is very likely that mechanisms also exist to tailor the growth of the epidermal cell layer to that of underlying cells. The observations made regarding the potential roles of WOX proteins in organ initiation, and particularly the fact that their over expression can lead to ectopic growth of the epidermis, are important indicators to the existence of such mechanisms.

The molecular basis for the co-ordination of plant epidermal growth with that of underlying tissues may, as previously proposed, involve the movement of developmentally important transcriptional or cell-cycle regulators between cell-layers. A good example of this involves the well-documented movement of the transcription factor KNOTTED1 (KN1) from the L2 (where it is expressed) to the L1 cell layer of the maize shoot apical meristem. Studies of the regulation of the movement of KN1 in maize, and that of related proteins in *Arabidopsis*, suggest that it is mediated by sequences within the proteins and may be necessary for endogenous protein function (Lucas et al. 1995; Kim et al. 2003).

In addition to the movement of proteins through plasmodesmata, it is also possible that the growth of cell layers may be co-ordinated by apoplastic signalling. This possibility is supported by the observation that three Leucine Rich Repeat RLKs, ERECTA (ER), ERECTA-LIKE1 (ERL1) and ERL2 act synergistically to control organ growth in *Arabidopsis* (Shpak et al. 2003, 2004). Triple mutants between the three genes are severely dwarfed and show patterning defects in floral organs. Shpak and colleagues propose that ER, ERL1 and ERL2 are responsible in part for linking cell proliferation to the growth and morphogenesis of organs. This is an interesting hypothesis especially in light of recent work from the same group showing that in the epidermis, the synergistic activity of these three proteins is apparently modified by the superposition of epidermis-specific apoplastic signalling, involving the TOO MANY MOUTHS receptor-like protein, involved in deciding the developmental fate of cells in the stomatal lineage (Shpak et al. 2005). This is an interesting example of how a proliferative vs. differentiation decision can involve local modifications of globally expressed signalling pathways and may provide a paradigm for another mechanism by which proliferation in different tissues/cell layers could be co-ordinated.

In conclusion, the plant epidermis, in addition to its obvious and critically important physiological roles, also appears to play an important role in controlling the growth and morphogenesis of shoot organs. Because loss of epidermal identity or integrity is usually fatal, investigating this role presents serious technical challenges. However, as our knowledge of the types of molecules potentially involved both in epidermal specification/organisation, and in the regulation of plant growth increases, so does the number of tools available to investigate how these developmental processes interact.

Acknowledgements I would like to thank the Royal Society of London and the EPSRC (RCUK fellowships) for financial support, Dr. Justin Goodrich for access to his data, and Prof. Andrew Hudson for helpful comments.

References

Abe M, Katsumata H, Komeda Y, Takahashi T (2003) Regulation of shoot epidermal cell differentiation by a pair of homeodomain proteins in Arabidopsis. Development 130:635–643

Abe M, Takahashi T, Komeda Y (2001) Identification of a cis-regulatory element for L1 layer-specific gene expression, which is targeted by an L1-specific homeodomain protein. Plant J 26:487–494

Ahn JW, Kim M, Lim JH, Kim GT, Pai HS (2004) Phytocalpain controls the proliferation and differentiation fates of cells in plant organ development. Plant J 38:969–981

Avery GS (1933) Structure and Development of the Tobacco Leaf. Am J Botany 20:565–592

Baroux C, Blanvillain R, Moore IR, Gallois P (2001) Transactivation of BARNASE under the AtLTP1 promoter affects the basal pole of the embryo and shoot development of the adult plant in *Arabidopsis*. Plant J 28:503–515

Baroux C, Spillane C, Grossniklaus U (2002) Evolutionary origins of the endosperm in flowering plants. Genome Biol 3:reviews 1026

Barton MK, Poethig RS (1993) Formation of the shoot apical meristem in *Arabidopsis thaliana*: an analysis of development in the wild type and in the *shoot meristemless* mutant. Development 119:823–831

Baud S, Wuilleme S, Lemoine R, Kronenberger J, Caboche M, Lepiniec L, Rochat C (2005) The AtSUC5 sucrose transporter specifically expressed in the endosperm is involved in early seed development in Arabidopsis. Plant J 43:824–836

Becraft PW, Bongard-Pierce DK, Sylvester AW, Poethig RS, Freeling M (1990) The liguleless-1 gene acts tissue specifically in maize leaf development. Dev Biol 141:220–232

Becraft PW, Freeling M (1991) Sectors of liguleless-1 tissue interrupt an inductive signal during maize leaf development. Plant Cell 3:801–807

Becraft PW, Kang SH, Suh SG (2001) The maize CRINKLY4 receptor kinase controls a cell-autonomous differentiation response. Plant Physiol 127:486–496

Becraft PW, Li K, Dey N, Asuncion-Crabb Y (2002) The maize dek1 gene functions in embryonic pattern formation and cell fate specification. Development 129:5217–5225

Becraft PW, Stinard PS, McCarty DR (1996) CRINKLY4: A TNFR-like receptor kinase involved in maize epidermal differentiation. Science 273:1406–1409

Bemis SM, Torii KU (2007) Autonomy of cell proliferation and developmental programs during Arabidopsis aboveground organ morphogenesis. Dev Biol 304:367–381

Bonello JF, Opsahl-Ferstad HG, Perez P, Dumas C, Rogowsky PM (2000) Esr genes show different levels of expression in the same region of maize endosperm. Gene 246:219–227

Bonello JF, Sevilla-Lecoq S, Berne A, Risueno MC, Dumas C, Rogowsky PM (2002) Esr proteins are secreted by the cells of the embryo surrounding region. J Exp Bot 53:1559–1568

Bruck DK, Walker DB (1985a) Cell Determination During Embryogenesis in Citrus jambhiri. I. Ontogeny of the Epidermis. Botanical Gazette 146:188–195

Bruck DK, Walker DB (1985b) Cell Determination during Embryogenesis in Citrus jambhiri. II. Epidermal Differentiation as a One-Time Event. Am J Botany 72:1602–1609

Cao X, Li K, Suh SG, Guo T, Becraft PW (2005) Molecular analysis of the CRINKLY4 gene family in *Arabidopsis thaliana*. Planta 220:645–657

Carraro N, Peaucelle A, Laufs P, Traas J (2006) Cell differentiation and organ initiation at the shoot apical meristem. Plant Mol Biol 60:811–826

Cock JM, McCormick S (2001) A large family of genes that share homology with CLAVATA3. Plant Physiol 126:939–942

Doerner P (2003) Plant meristems: a merry-go-round of signals. Curr Biol 13:R368–374

Dolan L, Poethig RS (1998) The Okra Leaf Shape Mutation in Cotton is Active in all Cell Layers of the Leaf. Am J Botany 85:322–327

Fleming AJ (2006) The co-ordination of cell division, differentiation and morphogenesis in the shoot apical meristem: a perspective. J Exp Bot 57:25–32

Friedman WE (2001) Developmental and evolutionary hypotheses for the origin of double fertilization and endosperm. CR Acad Sci III 324:559–567

Friedman WE, Floyd SK (2001) Perspective: the origin of flowering plants and their reproductive biology – a tale of two phylogenies. Evolut Int J Org Evolution 55:217–231

Friml J, Vieten A, Sauer M, Weijers D, Schwarz H, Hamann T, Offringa R, Jurgens G (2003) Efflux-dependent auxin gradients establish the apical-basal axis of Arabidopsis. Nature 426:147–153

Geisler-Lee J, Gallie DR (2005) Aleurone cell identity is suppressed following connation in maize kernels. Plant Physiol 139:204–212

Gifford ML, Dean S, Ingram GC (2003) The Arabidopsis ACR4 gene plays a role in cell layer organisation during ovule integument and sepal margin development. Development 130:4249–4258

Goldberg RB, de Paiva G, Yadegari R (1994) Plant Embryogenesis: Zygote to Seed. Science 266:605–614

Gonzalez-Bayon R, Kinsman EA, Quesada V, Vera A, Robles P, Ponce MR, Pyke KA, Micol JL (2006) Mutations in the RETICULATA gene dramatically alter internal architecture but have little effect on overall organ shape in Arabidopsis leaves. J Exp Bot 57:3019–3031

Grandjean O, Vernoux T, Laufs P, Belcram K, Mizukami Y, Traas J (2004) In vivo analysis of cell division, cell growth, and differentiation at the shoot apical meristem in Arabidopsis. Plant Cell 16:74–87

Gross-Hardt R, Lenhard M, Laux T (2002) WUSCHEL signaling functions in interregional communication during Arabidopsis ovule development. Genes Dev 16:1129–1138

Gruis DF, Guo H, Selinger D, Tian Q, Olsen OA (2006) Surface position, not signaling from surrounding maternal tissues, specifies aleurone epidermal cell fate in maize. Plant Physiol 141:898–909

Haecker A, Gross-Hardt R, Geiges B, Sarkar A, Breuninger H, Herrmann M, Laux T (2004) Expression dynamics of WOX genes mark cell fate decisions during early embryonic patterning in *Arabidopsis thaliana*. Development 131:657–668

Hantke SS, Carpenter R, Coen ES (1995) Expression of floricaula in single cell layers of periclinal chimeras activates downstream homeotic genes in all layers of floral meristems. Development 121:27–35

Ingram GC, Boisnard-Lorig C, Dumas C, Rogowsky PM (2000) Expression patterns of genes encoding HD-ZipIV homeo domain proteins define specific domains in maize embryos and meristems. Plant J 22:401–414

Ito Y, Nakanomyo I, Motose H, Iwamoto K, Sawa S, Dohmae N, Fukuda H (2006) Dodeca-CLE peptides as suppressors of plant stem cell differentiation. Science 313:842–845

Jenik PD, Irish VF (2000) Regulation of cell proliferation patterns by homeotic genes during Arabidopsis floral development. Development 127:1267–1276

Johnson KL, Degnan KA, Ross Walker J, Ingram GC (2005) AtDEK1 is essential for specification of embryonic epidermal cell fate. Plant J 44:114–127

Jurgens G (2001) Apical-basal pattern formation in Arabidopsis embryogenesis. EMBO J 20:3609–3616

Kessler S, Seiki S, Sinha N (2002) Xcl1 causes delayed oblique periclinal cell divisions in developing maize leaves, leading to cellular differentiation by lineage instead of position. Development 129:1859–1869

Kessler S, Townsley B, Sinha N (2006) L1 division and differentiation patterns influence shoot apical meristem maintenance. Plant Physiol 141:1349–1362

Khaled AS, Vernoud V, Ingram GC, Perez P, Sarda X, Rogowsky PM (2005) Engrailed-ZmOCL1 fusions cause a transient reduction of kernel size in maize. Plant Mol Biol 58:123–139

Kim JY, Yuan Z, Jackson D (2003) Developmental regulation and significance of KNOX protein trafficking in Arabidopsis. Development 130:4351–4362

Kondo T, Sawa S, Kinoshita A, Mizuno S, Kakimoto T, Fukuda H, Sakagami Y (2006) A plant peptide encoded by CLV3 identified by in situ MALDI-TOF MS analysis. Science 313:845–848

Kropf DL, Kloareg B, Quatrano RS (1988) Cell wall is required for fixation of the embryonic axis in Fucus zygotes. Science 239:187–190

Lauterbach C, Niedermeier M, Besenbeck R, Stadler R, Sauer N (2007) Immunolocalization of the PmSUC1 Sucrose Transporter in *Plantago major* Flowers and Reporter-Gene Analyses of the PmSUC1 Promoter Suggest a Role in Sucrose Release from the Inner Integument. Plant Biol (Stuttg) 9:357–365

Lid SE, Al RH, Krekling T, Meeley RB, Ranch J, Opsahl-Ferstad HG, Olsen OA (2004) The maize disorganized aleurone layer 1 and 2 (dil1, dil2) mutants lack control of the mitotic division plane in the aleurone layer of developing endosperm. Planta 218:370–378

Lid SE, Gruis D, Jung R, Lorentzen JA, Ananiev E, Chamberlin M, Niu X, Meeley R, Nichols S, Olsen OA (2002) The defective kernel 1 (dek1) gene required for aleurone cell development in the endosperm of maize grains encodes a membrane protein of the calpain gene superfamily. Proc Natl Acad Sci USA 99:5460–5465

Lid SE, Olsen L, Nestestog R, Aukerman M, Brown RC, Lemmon B, Mucha M, Opsahl-Sorteberg HG, Olsen OA (2005) Mutation in the *Arabidopisis thaliana* DEK1 calpain gene perturbs endosperm and embryo development while over-expression affects organ development globally. Planta 221:339–351

Lintilhac PM, Vesecky TB (1981) Mechanical Stress and Cell Wall Orientation in Plants. II. The Application of Controlled Directional Stress to Growing Plants; with a Discussion on the Nature of the Wound Reaction. Am J Botany 68:1222–1230

Lu P, Porat R, Nadeau JA, O'Neill SD (1996) Identification of a meristem L1 layer-specific gene in Arabidopsis that is expressed during embryonic pattern formation and defines a new class of homeobox genes. Plant Cell 8:2155–2168

Lucas WJ, Bouche-Pillon S, Jackson DP, Nguyen L, Baker L, Ding B, Hake S (1995) Selective trafficking of KNOTTED1 homeodomain protein and its mRNA through plasmodesmata. Science 270:1980–1983

Lucas WJ, Lee JY (2004) Plasmodesmata as a supracellular control network in plants. Nat Rev Mol Cell Biol 5:712–726

Magnard JL, Le Deunff E, Domenech J, Rogowsky PM, Testillano PS, Rougier M, Risueno MC, Vergne P, Dumas C (2000) Genes normally expressed in the endosperm are expressed at early stages of microspore embryogenesis in maize. Plant Mol Biol 44:559–574

Marcotrigiano M (2001) Genetic mosaics and the analysis of leaf development. Int J Plant Sci 162:513–525

Matsumoto N, Okada K (2001) A homeobox gene, PRESSED FLOWER, regulates lateral axis-dependent development of Arabidopsis flowers. Genes Dev 15:3355–3364

McHale NA, Marcotrigiano M (1998) LAM1 is required for dorsoventrality and lateral growth of the leaf blade in *Nicotiana*. Development 125:4235–4243

Merry J (1941) Studies on the Embryo of Hordeum sativum-I. The Development of the Embryo. Bull Torrey Bot Club 68:585–598

Meyer CF (1958) Cell Patterns in Early Embryogeny of the McIntosh Apple. Am J Botany 45:341–349

Nakamura M, Katsumata H, Abe M, Yabe N, Komeda Y, Yamamoto KT, Takahashi T (2006) Characterization of the class IV homeodomain-Leucine Zipper gene family in Arabidopsis. Plant Physiol 141:1363–1375

Nardmann J, Ji J, Werr W, Scanlon MJ (2004) The maize duplicate genes narrow sheath1 and narrow sheath2 encode a conserved homeobox gene function in a lateral domain of shoot apical meristems. Development 131:2827–2839

Neilson-Jones W (1969) Plant Chimeras. Methuen & Co., London

Ni J, Clark SE (2006) Evidence for functional conservation, sufficiency, and proteolytic processing of the CLAVATA3 CLE domain. Plant Physiol 140:726–733

Olsen OA (2001) ENDOSPERM DEVELOPMENT: Cellularization and Cell Fate Specification. Annu Rev Plant Physiol Plant Mol Biol 52:233–267

Olsen OA (2004) Nuclear endosperm development in cereals and *Arabidopsis thaliana*. Plant Cell 16(Suppl):S214–227

Opsahl-Ferstad HG, Le Deunff E, Dumas C, Rogowsky PM (1997) ZmEsr, a novel endosperm-specific gene expressed in a restricted region around the maize embryo. Plant J 12:235–246

Park SO, Zheng Z, Oppenheimer DG, Hauser BA (2005) The PRETTY FEW SEEDS2 gene encodes an Arabidopsis homeodomain protein that regulates ovule development. Development 132:841–849

Perbal MC, Haughn G, Saedler H, Schwarz-Sommer Z (1996) Non-cell-autonomous function of the Antirrhinum floral homeotic proteins DEFICIENS and GLOBOSA is exerted by their polar cell-to-cell trafficking. Development 122:3433–3441

Pillai SK (1963) Structure and seasonal study of the shoot apex of some *Cupressus* species. New Phytol 62:335–341

Pillai SK (1964) Structure and Seasonal Study of the Shoot Apex of two species of *Araucaria*. Plant System Evolut 111:273–284

Poethig RS (1987) Clonal analysis of cell lineage patterns in plant development. Am J Botany 74:581–594

Poethig RS, Coe EH, Johri MM (1986) Cell lineage patterns in maize embryogenesis: a clonal analysis. Develop Biol 117:392–404
Pollock EG, Jensen WA (1964) Cell Development During Early Embryogenesis in Capsella and Gossypium. Am J Botany 51:915–921
Quatrano RS, Brian L, Aldridge J, Schultz T (1991) Polar axis fixation in Fucus zygotes: components of the cytoskeleton and extracellular matrix. Dev Suppl 1:11–16
Randolph LF (1936) Developmental morphology of the caryopsis in maize. J Agric Res 53:881
Reinhardt D, Frenz M, Mandel T, Kuhlemeier C (2003a) Microsurgical and laser ablation analysis of interactions between the zones and layers of the tomato shoot apical meristem. Development 130:4073–4083
Reinhardt D, Pesce ER, Stieger P, Mandel T, Baltensperger K, Bennett M, Traas J, Friml J, Kuhlemeier C (2003b) Regulation of phyllotaxis by polar auxin transport. Nature 426:255–260
Savaldi-Goldstein S, Peto C, Chory J (2007) The epidermis both drives and restricts plant shoot growth. Nature 446:199–202
Scanlon MJ (2000) NARROW SHEATH1 functions from two meristematic foci during founder-cell recruitment in maize leaf development. Development 127:4573–4585
Scanlon MJ, Schneeberger RG, Freeling M (1996) The maize mutant narrow sheath fails to establish leaf margin identity in a meristematic domain. Development 122:1683–1691
Scarpella E, Marcos D, Friml J, Berleth T (2006) Control of leaf vascular patterning by polar auxin transport. Genes Dev 20:1015–1027
Sessions A, Weigel D, Yanofsky MF (1999) The *Arabidopsis thaliana* MERISTEM LAYER 1 promoter specifies epidermal expression in meristems and young primordia. Plant J 20:259–263
Sessions A, Yanofsky MF, Weigel D (2000) Cell-cell signaling and movement by the floral transcription factors LEAFY and APETALA1. Science 289:779–782
Sharma VK, Ramirez J, Fletcher JC (2003) The Arabidopsis CLV3-like (CLE) genes are expressed in diverse tissues and encode secreted proteins. Plant Mol Biol 51:415–425
Sharman BC (1940) A periclinal division in the dematogen at the tip of the maize growing point. Nature 146:778
Sharman BC (1942) Shoot apex in grasses and cereals. Nature 149:82
Shaw SL, Quatrano RS (1996) The role of targeted secretion in the establishment of cell polarity and the orientation of the division plane in Fucus zygotes. Development 122:2623–2630
Shen B, Li C, Min Z, Meeley RB, Tarczynski MC, Olsen OA (2003) sal1 determines the number of aleurone cell layers in maize endosperm and encodes a class E vacuolar sorting protein. Proc Natl Acad Sci USA 100:6552–6557
Shpak ED, Berthiaume CT, Hill EJ, Torii KU (2004) Synergistic interaction of three ERECTA-family receptor-like kinases controls Arabidopsis organ growth and flower development by promoting cell proliferation. Development 131:1491–1501
Shpak ED, Lakeman MB, Torii KU (2003) Dominant-negative receptor uncovers redundancy in the Arabidopsis ERECTA Leucine-rich repeat receptor-like kinase signaling pathway that regulates organ shape. Plant Cell 15:1095–1110
Shpak ED, McAbee JM, Pillitteri LJ, Torii KU (2005) Stomatal patterning and differentiation by synergistic interactions of receptor kinases. Science 309:290–293
Stewart RN, Dermen H (1975) Flexibility in Ontogeny as Shown by the Contribution of the Shoot Apical Layers to Leaves of Periclinal Chimeras. Am J Botany 62:935–947

Stewart RN, Semeniuk P, Dermen H (1974) Competition and Accommodation between Apical Layers and their Derivatives in the Ontogeny of Chimeral Shoots of *Pelargonium* x Hortorum. Am J Botany 61:54–67

Szymkowiak EJ, Sussex IM (1992) The internal meristem layer (L3) determines floral meristem size and carpel number in tomato periclinal chimeras. Plant Cell 4:1089–1100

Szymkowiak EJ, Sussex IM (1996) What Chimeras Can Tell Us About Plant Development. Annu Rev Plant Physiol Plant Mol Biol 47:351–376

Tanaka H, Dhonukshe P, Brewer PB, Friml J (2006) Spatiotemporal asymmetric auxin distribution: a means to coordinate plant development. Cell Mol Life Sci 63:2738–2754

Tanaka H, Watanabe M, Sasabe M, Hiroe T, Tanaka T, Tsukaya H, Ikezaki M, Machida C, Machida Y (2007) Novel receptor-like kinase ALE2 controls shoot development by specifying epidermis in Arabidopsis. Development 134:1643–1652

Tanaka H, Watanabe M, Watanabe D, Tanaka T, Machida C, Machida Y (2002) ACR4, a putative receptor kinase gene of *Arabidopsis thaliana*, that is expressed in the outer cell layers of embryos and plants, is involved in proper embryogenesis. Plant Cell Physiol 43:419–428

Tilney-Basset RAE (1986) Plant Chimeras. Edward Arnold, London

Torres-Ruiz RA, Jurgens G (1994) Mutations in the FASS gene uncouple pattern formation and morphogenesis in Arabidopsis development. Development 120:2967–2978

Tsukaya H (2002) Interpretation of mutants in leaf morphology: genetic evidence for a compensatory system in leaf morphogenesis that provides a new link between cell and organismal theories. Int Rev Cytol 217:1–39

Vieten A, Vanneste S, Wisniewska J, Benkova E, Benjamins R, Beeckman T, Luschnig C, Friml J (2005) Functional redundancy of PIN proteins is accompanied by auxin-dependent cross-regulation of PIN expression. Development 132:4521–4531

Vincent CA, Carpenter R, Coen ES (2003) Interactions between gene activity and cell layers during floral development. Plant J 33:765–774

Watanabe M, Tanaka H, Watanabe D, Machida C, Machida Y (2004) The ACR4 receptor-like kinase is required for surface formation of epidermis-related tissues in *Arabidopsis thaliana*. Plant J 39:298–308

Wegner J (2000a) Empirical investigations into the tunica structuring point of the shoot apex of Pelargonium zonale. Plant Sci 153:171–175

Wegner J (2000b) A theoretical approach to the genesis of cell layer arrangements in undifferentiated tissues. Plant Sci 153:177–183

Wegner J (2003a) On the developmental analogy of the shoot apex and the leaf margin and the genesis of L1-derived mesophyll areas in the leaf margin. Plant Sci 164:565–569

Wegner J (2003b) On the histogenesis of L1-derived tissue in the leaf - empirical investigations into the mantle structuring point. Plant Sci 164:571–578

Weijers D, Van Hamburg JP, Van Rijn E, Hooykaas PJ, Offringa R (2003) Diphtheria toxin-mediated cell ablation reveals interregional communication during Arabidopsis seed development. Plant Physiol 133:1882–1892

Willemsen V, Scheres B (2004) Mechanisms of pattern formation in plant embryogenesis. Annu Rev Genet 38:587–614

Williams L, Fletcher JC (2005) Stem cell regulation in the Arabidopsis shoot apical meristem. Curr Opin Plant Biol 8:582–586

Wu X, Dinneny JR, Crawford KM, Rhee Y, Citovsky V, Zambryski PC, Weigel D (2003) Modes of intercellular transcription factor movement in the Arabidopsis apex. Development 130:3735–3745

Plant Cell Monogr (10)
L. Bögre and G. Beemster: Plant Growth Signaling
DOI 10.1007/7089_2007_150/Published online: 26 January 2008

Signaling in Auxin-Dependent Plant Development

Marcelo Kemel Zago · Carlos S. Galvan-Ampudia · Remko Offringa (✉)

Institute of Biology Leiden, Leiden University,
Wassenaarseweg 64, 2333 AL Leiden, The Netherlands
r.offringa@biology.leidenuniv.nl

Abstract Plants are versatile organisms that, in order to compensate for their sessile life style, employ significant developmental plasticity to respond to environmental changes. To support this flexibility, plants have developed a plethora of signaling pathways that, in comparison to other organisms, involve a relatively large number of molecular switches, e.g. the protein kinases and phosphoprotein phosphatases. The plant hormone auxin plays a central role in these adaptation processes, and its action is not only determined by the perception and signal transduction of this signaling molecule, but also by its polar transport-mediated differential distribution. Herein we will review the role of the main signaling components in auxin-dependent plant development. Interestingly, auxin signaling is very unusual, as it does not involve a canonical signaling pathway comprising a receptor and a downstream kinase cascade. Instead, auxin employs the TIR1/AFB F-box proteins as receptors to initiate targeted proteolysis of Aux/IAA repressors leading to an increase in auxin-responsive gene expression. However, protein kinases play a crucial role in regulating and directing differential auxin distribution. Of particular interest is a group of serine/threonine kinases belonging to the plant specific AGCVIII subfamily. Of this group, the PINOID kinase has been shown to direct polar auxin transport by controlling the basal-to-apical deployment of the PIN auxin efflux carriers, a process that is reversible by PP2A phosphatases.

1 Introduction

Land plants are sessile organisms, and as a consequence they have acquired a flexible developmental program that allows them to adapt to sudden changes in their environment. One of the well-studied examples of plant developmental plasticity is the growth of seedling shoots towards the light. Already by the end of the 19th century, Darwin's observations on the bending of canary grass coleoptiles to unidirectional light led him to conclude that some matter in the upper part of the coleoptile is acted on by light, and then transmits its effects to the lower part of this tissue (Darwin and Darwin 1881). Around 1930, this matter was identified as indole-3-acetic acid (IAA) and named after the Greek word "auxein" meaning "to grow" (Went 1926; Kogl and Haagen-Smit 1931). Detailed observations by Went and Cholodny on the auxin-mediated orientation of plant growth to unidirectional light (phototropism) or gravity (gravitropism) led to the Cholodny and Went hypothesis (Went 1926; Cholodny 1927; Went F. and Thimann 1937). This model states that tropic growth is the result of predominant distribution of auxin to the dark or lower side upon light or gravity stimulation, and that, due to differences in sensitivity to auxin, these elevated auxin concentrations enhance cell elongation in the shoot, and inhibit cell growth in the root, ultimately leading to bending of the organ. In support of this hypothesis, more recent experiments demonstrated asymmetric auxin distribution leading to expression of auxin responsive genes in light and gravity-induced shoots (Li et al. 1991; Friml et al. 2002; Esmon et al. 2006) and roots (Larkin et al. 1996; Luschnig et al. 1998; Ottenschlager et al. 2003).

The tropic growth experiments clearly demonstrate that auxin action is a result of the interplay between the local auxin concentration—which is determined by biosynthesis, transport, and inactivation—and the sensitivity or responsiveness of cells to this plant hormone. Currently, auxin is recognized as a central regulator of plant development that not only controls elementary processes such as cell division and elongation, but also directs complex developmental and patterning processes such as embryogenesis, vascular differentiation, phyllotaxis and fruit development (Tanaka et al. 2006). Herein we will analyze the signaling pathways that regulate auxin-dependent plant development.

A canonical signaling pathway consists of a phosphorelay system of transmembrane receptor-activated protein kinases, enzymes that mediate reversible phosphorylation of substrate proteins thereby altering the activity, stability and/or subcellular localization of these proteins. Protein kinases control a great variety of cellular functions, and their catalytic domain is extremely conserved among different organisms. Database searches show that in bacteria, in general, less than 10 genes encode protein kinases, whereas 130 have been identified in yeast, 251 in flies, 411 in worms, approximately 520 in humans and more than 1000 in *Arabidopsis thaliana* (Leonard et al. 1998; Manning et al. 2002; Champion et al. 2004; Milanesi et al. 2005) (PlantsP Database: http://plantsp.sdsc.edu/). Considering that the complexity of the cellular signaling in part reflects the extent to which an organism senses and reacts to environmental changes, the greater number of kinases coded by plants reflects their need for developmental plasticity and robust defense mechanisms against pathogens and other stresses. As we will discuss below, in spite of their abundance in plants, kinases unexpectedly do not play a central, but rather an accessory role in auxin-signaling. In contrast, a particular class of plant specific kinases appears to have an important role in regulating the transport, thereby controlling local levels of auxin.

2
Auxin Signaling

Several processes are known to occur within a few minutes after auxin application to plant tissues. These vary from rapid changes in enzymatic activities

(Morre et al. 2003; Droog et al. 1995; Bilang and Sturm 1995) and gene expression (Theologis 1986; Guilfoyle 1986; Abel and Theologis 1996) to changes in transporter activities, leading to acidification of the cell wall (Grebe 2005), and a rapid increase in the membrane potential (Karcz and Burdach 2002) or the cytosolic calcium levels (Gehring et al. 1990). For most of these primary responses the signaling pathways are yet unknown, but in the last few years the events leading to auxin responsive gene expression have been uncovered.

2.1 Auxin Responsive Gene Expression

Differential screens of cDNA libraries in the 1980s led to the identification of the first auxin responsive genes (Theologis et al. 1985; Hagen and Guilfoyle 1985; Key et al. 1986; van der Zaal et al. 1987). Most of these genes are activated within minutes after auxin stimulation in a process independent of de novo synthesis of proteins. Several auxin responsive elements (AuxREs) have been identified in the promoters of these primary auxin response genes (Ulmasov et al. 1997; Guilfoyle et al. 1998a; Guilfoyle et al. 1998b), and Auxin Response Factors (ARFs) were shown to bind to these elements and to activate or to repress transcription (Ulmasov et al. 1999a).

ARFs in general contain three well-defined domains: a DNA binding domain (DBD) that binds AuxREs, a middle region domain, and a third domain that was found to mediate homo- or heterodimerization. Whether an ARF is an activator or repressor depends on the structure of its middle region domain. For example, ARFs with Q-rich middle regions activate transcription, while ARFs with P/S/T-rich middle region repress transcription (Ulmasov et al. 1997, 1999a).

Some of the primary auxin response genes were found to encode small short-lived proteins, named Aux/IAA proteins, that resemble bacterial repressors (Abel and Theologis 1996; Abel et al. 1994). Like ARFs, also Aux/IAA proteins contain three distinct domains, of which domain I has been shown to have transcription repression activity (Tiwari et al. 2004), domain II is involved in Aux/IAA protein stability and may be a target for ubiquitination (Ramos et al. 2001), and the third domain that allows Aux/IAA proteins to homo- or heterodimerize with ARFs or other Aux/IAA proteins (Ulmasov et al. 1999b). The interaction between Aux/IAAs and ARFs was shown to result in repression of the ARF-stimulated gene expression (Ulmasov et al. 1997), and the repression of ARF action can only be released upon degradation of the Aux/IAAs (Fig. 1) (Weijers and Jürgens 2004).

Apart from being identified in screens for auxin responsive genes, the Aux/IAA encoding genes have also been identified through gain-of-function mutations that lead to auxin insensitivity. Most Aux/IAA proteins are extremely short lived, and all gain-of-function mutations in the *Aux/IAA* genes result in a specific amino acid change in domain II that stabilizes the encoded protein, and thus represses auxin responses leading to auxin insensitivity phenotypes (Liscum and Reed 2002).

Unfortunately, loss-of-function mutants in the *Aux/IAA* genes provide very little information about the functions of the individual genes. All of the mutant plant lines analyzed to date display no or only very subtle phenotypes, suggesting that there is functional redundancy between Aux/IAAs. By contrast, loss-of-function mutations have been very informative for the ARF family of transcription factors. For example, based on mutant phenotypes, ARF3/ETTIN has been associated with gynoecium development (Sessions et al. 1997), ARF5/MONOPTEROS with early embryogenesis and vascular development (Hardtke and Berleth 1998), ARF7/NON PHOTOTROPIC HYPOCOTYL4 with differential growth responses in aerial tissues (Liscum and Reed 2002; Liscum and Briggs 1995) and ARF8/FRUIT WITHOUT FERTILIZATION with fruit initiation (Goetz et al. 2006).

The *Arabidopsis* genome encodes 29 AUX/IAA proteins and 23 ARFs, which can combine to translate the auxin signal into a gene expression response. For example, it has been shown in yeast two-hybrid assays that specific combinations of ARFs and Aux/IAAs are preferred interaction partners (Kim et al. 1997; Hardtke et al. 2004). The specificity of these interactions seems essential to differentiate auxin responses in different cell types, and is further supported by the observation that both ARFs and Aux/IAAs are at best only partially interchangeable in each others expression domain (Weijers and Jürgens 2004; Weijers et al. 2005a).

2.2 Auxin Perception Leads to SCF E3 Ligase-Mediated Degradation of Aux/IAA Proteins

The identification of Aux/IAAs as repressors of auxin responsive gene expression (Ulmasov et al. 1997) and the observation that they are very instable, especially upon auxin stimulation (Abel et al. 1994; Gray et al. 2001), placed the proteolysis machinery at a central position in auxin signaling. Moreover, the isolation and characterization of several *Arabidopsis* auxin-resistant mutants revealed that a ubiquitination-mediated degradation pathway is likely to be involved. Ubiquitin is a regulatory protein that is first activated by a ubiquitin-activating enzyme E1, which transfers the ubiquitin component to a ubiquitin-conjugating enzyme, E2. E2 subsequently acts in concert with an E3 ubiquitin-ligase, to link the ubiquitin to lysine residues in the target protein (Fig. 1). A final poly-ubiquitylation step is generally thought to label these target proteins for degradation by the 26S proteasome (Moon et al. 2004).

Arabidopsis mutants revealed that the E3 ligase that participates in auxin signaling is the SKP1/Cullin/F-box (SCF) complex comprising the *Arabidopsis* SKP1-like protein ASK1, CULLIN1 (CUL1), the F-Box protein Transport Inhibitor Response 1 (TIR1), and the E2-interacting RING domain protein RBX1 (SCF^{TIR1}; Fig. 1) (Hellmann et al. 2003; Gray et al. 1999, 2002; Rueg-

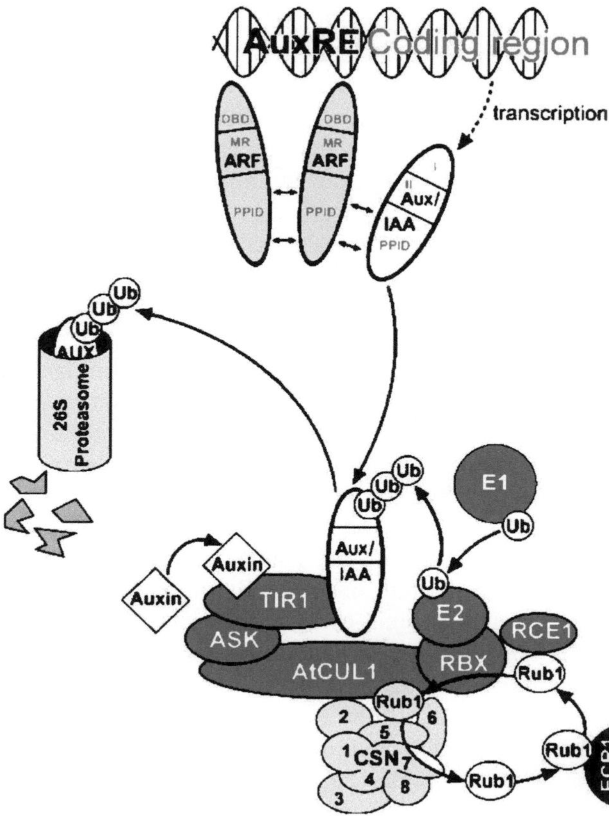

Fig. 1 Auxin signaling in *Arabidopsis thaliana* is regulated by proteolysis of Aux/IAA proteins mediated by the SCF^{TIR1} E3 ubiquitin ligase. Aux/IAAs, which inhibit auxin responsive gene expression (AuxRE: Auxin Responsive Element) through binding ARFs (through their Protein–Protein Interaction Domain—PPID), are likely labelled for proteolysis by ubiquitination. This process is mediated by the ubiquitin activating enzyme E1, the ubiquitin conjugating enzyme E2 and the ubiquitin ligase E3. E1 transfers the ubiquitin component to E2 that in turn binds to and acts in concert with E3 to ubiquitinate the substrate protein. Once targets are ubiquitinated, they are degraded by the 26S proteasome. SCF^{TIR1}, the E3 ligase complex that participates in auxin signaling, consists of CULLIN1 (AtCUL1), the *Arabidopsis* SKP1 homolog ASK1, and the F-box protein TIR1. Auxin binding to TIR1 enhances its interaction with the Aux/IAAs which leads to enhanced degradation of these proteins. Regulatory subunits of SCF^{TIR1} include RCE1, which modifies AtCUL1 by adding RUB1 that is previously activated by the AXR1-ECR1 sub-complex. The COP9 signalosome (CSN) removes RUB1 from AtCUL1, leading to subsequent dissociation of the SCF^{TIR1} complex

ger et al. 1998). Aux/IAA proteins were shown to specifically interact with the F-box protein TIR1 (Gray et al. 2001), and the recent finding that auxin-binding to TIR1 enhances this interaction and thus leads to enhanced degradation of Aux/IAAs (Dharmasiri et al. 2003a) uncovered TIR1, together with the closely related proteins AFB1, 2 and 3, as the long-sought auxin receptors (Fig. 1) (Kepinski and Leyser 2005; Dharmasiri et al. 2005a,b). In view of these findings, it is likely that auxin-promoted poly-ubiquitination by the SCF^{TIR1} E3 ligase marks Aux/IAA proteins for degradation. Remarkably, however, auxin-induced poly-ubiquitination has not yet been demonstrated for Aux/IAA proteins.

Several regulatory components of SCF E3 ligases have been identified. For example, it has been found that the CUL1 subunit of the SCF complex is modified by the addition of the ubiquitin-like protein RUB1/NEDD8 (del Pozo and Estelle 1999). RUB1/NEDD8 is activated by the E1 complex AXR1-ECR1, which catalyzes the transfer of RUB1 to the RUB conjugating enzyme RCE1 that acts together with RBX1 in the RUB modification of CUL1 (Fig. 1) (Gray et al. 2002; Dharmasiri et al. 2003b). Knock-out mutations in most of these regulatory components lead to auxin-resistant phenotypes, and the double mutant *axr1/rce1* causes embryonic defects similar to *mp*, leading to the hypothesis that RUB modification positively regulates SCF activity (Lincoln et al. 1990; Bostick et al. 2004; Larsen and Cancel 2004; Dharmasiri et al. 2003b). The RUB-conjugated state of the SCF complex is regulated by the COP9 Signalosome (CSN), a protein complex that shares reasonable similarity to the lid of the 26S proteasome (Fu et al. 2001). CSN action has been demonstrated to be necessary for both auxin response and RUB1 removal from CUL1 (Schwechheimer et al. 2001), which probably destabilizes the SCF complex after its function so that new complexes can be formed (Fig. 1) (Cope and Deshaies 2003; Schwechheimer 2004). The CSN is also known to interact with other types of E3 ligases, such as the photomorphogenesis related COP1, and to be required for the nuclear import of this RING finger protein (Chamovitz et al. 1996; von Arnim et al. 1997; Seo et al. 2003). COP1 and the CSN have been shown to promote degradation of HY5 (Osterlund et al. 2000a,b), a transcription factor that positively regulates photomorphogenesis, explaining why loss-of-function mutations in both *COP1* and the CSN-subunit encoding genes cause constitutive photomorphogenesis (*cop*) rather than auxin-related phenotypes (Schwechheimer and Deng 2000).

2.3 Receptors and Phosphorylation in Auxin Signaling

With the molecular basis of the central auxin signaling pathway largely uncovered, it is intriguing to note that there is limited evidence for the involvement of canonical pathways, comprising a membrane-bound receptor and a protein kinase cascade, in auxin signaling. The most likely candidate for a membrane-bound auxin receptor is still Auxin Binding Protein 1 (ABP1), a protein that is retained in the ER and is only present in low levels at the cell surface. Several lines of evidence indicate that plasma membrane surface-localized ABP1 is involved in signaling cascades leading to rapid, TIR1/AFB-independent cell expansion and early electrophysiological auxin responses (Fig. 2) (Napier et al. 2002).

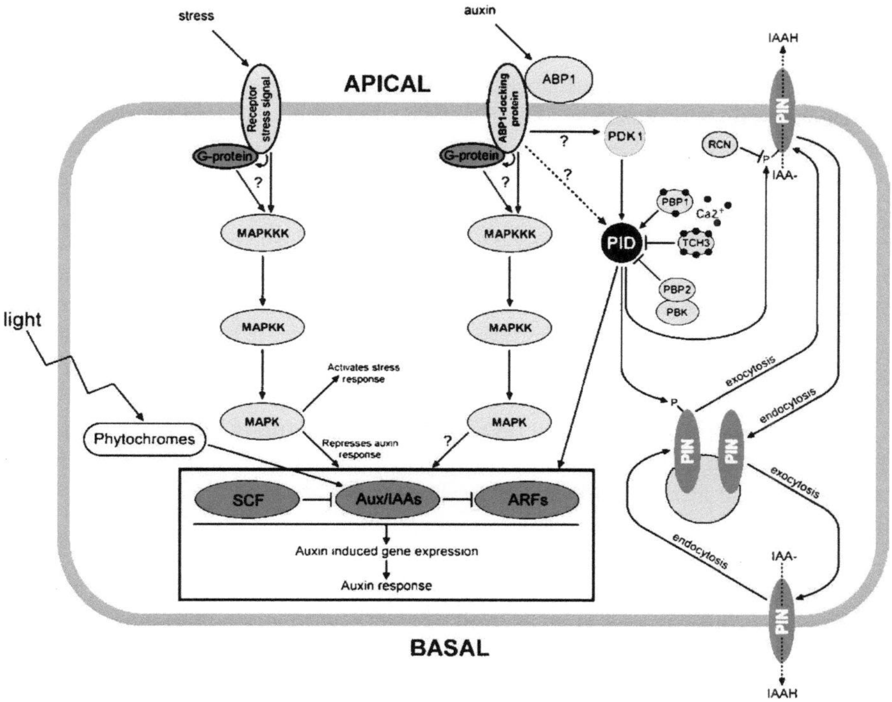

Fig. 2 Hypothetical auxin signaling pathway involving ABP1 and the PDK1/PID pathway regulating PAT. A hypothetical signal transduction pathway for auxin starts with its reception by apoplastic ABP1, which modifies its transmembrane ABP1-docking protein. The latter could transduce the signal in two different ways: through a MAPK cascade or through a hypothetical PDK1 pathway. In the first case, the ABP1-docking protein either activates a G-protein or directly activates a MAPK cascade ultimately leading to the induction of auxin response. In the second case, the ABP1-docking protein modifies PDK1, which starts a signaling cascade passing through PID or other AGC VIII subfamily kinase and ending in the regulation of auxin response. Alternatively, the ABP1-docking protein could, through unidentified factors (*dashed arrow*), induce transport of PID to the plasma membrane region where PDK1 is anchored; once in this position, PID would be activated by PDK1 and continue to transduce the signal. The acknowledged role of PID as regulator of PINs is also shown: PID activity, which can be repressed by the PBP2-PBK complex, or regulated in response to cytosolic calcium levels by PBP1 (upregulation) or TCH3 (downregulation), results in PIN phosphorylation either in endosomes or at the plasma membrane, in which case PINs are respectively induced for exocytosis or to remain at the PM at the apical cellular pole. Other putative pathways are also shown: stress signals activate MAPKKK NPK1 signal transduction leading to repression of auxin response; phytochromes may phosphorylate Aux/IAA proteins to regulate light and/or auxin induced gene expression

The ultimate data supporting this hypothesis, however, remain to be obtained, since the phosphorylation events occasionally reported to be involved in auxin signaling have not been linked to ABP1. For example, Aux/IAA proteins have been shown to be phosphorylated by phytochromes in vitro, suggesting that light signaling acts on auxin responsive gene expression by influencing the stability of Aux/IAA proteins (Colon-Carmona et al. 2000). Also the Mitogen Activated Protein Kinase (MAPK) cascade has been implicated in the modulation of auxin response. Roots of *Arabidopsis* seedlings treated with auxin showed an increase in MAPK activity and this activation was inhibited in the auxin resistant *axr4* mutant (Mockaitis and Howell 2000). However, it has been recently shown that AXR4 is involved in the regulation of auxin import by regulating the localization of the putative auxin influx carrier AUX1 at the plasma membrane (Dharmasiri et al. 2006). Therefore, the lack of auxin-induced MAPK activity in *axr4* is likely to be the consequence of reduced auxin uptake. This does not exclude, however, that the MAPK signaling pathway is involved in the modulation of auxin response. In fact, it has been shown that the MAP3K NPK1 activates stress responses and represses auxin-induced gene expression (Kovtun et al. 1998). Finally, the serine/threonine kinase PINOID (PID) has been proposed to be a regulator of auxin signaling (Christensen et al. 2000). Although more recent data have clearly related PID activity to the modulation of polar auxin transport (PAT) (Benjamins et al. 2001; Friml et al. 2004), a role for this kinase in auxin signaling can not yet be excluded.

In conclusion, although not part of the central auxin signaling pathway, phosphorylation events seem to modulate auxin responses, and to serve at least to integrate other signals, such as light or stress, with auxin signaling (Fig. 2).

3 Signaling Directing Auxin Distribution

3.1 Directional Transport Generates Dynamic Gradients of Auxin

The observations by Darwin on tropisms (Darwin and Darwin 1881) and subsequent experiments by plant biologists such as Went (Went and Thimann 1937) not only led to the identification of auxin, but also revealed directional transport of this hormone from source (biosynthesis) tissues to sites of action. Transport measurements using radio-labeled auxin showed the existence of two types of auxin transport systems: fast and non-directional transport through the phloem, and slow and directional cell-to-cell transport that is referred to as polar auxin transport (PAT). The transport through the phloem was first detected by Morris and Thomas (Morris and Thomas 1978) and occurs in both basi- and acropetal directions at approximately 5–20 cm/h (Nowacki and Bandurski 1980). Experiments performed by Baker

(Baker 2000) indeed revealed significant presence of IAA in the phloem. A connection between the fast transport of auxin and PAT was demonstrated in experiments performed in pea, in which radio-labeled IAA initially present in the phloem was detected in the polar transport system at a later time point (Cambridge and Morris 1996).

In contrast to the phloem-mediated auxin transport, PAT is restricted to free IAA, is unidirectional and occurs in a cell-to-cell manner. The velocity is much slower, and has been estimated at approximately 5–20 mm/h. In dicotyledon plants, polar auxin transport is known to travel long-distance from the shoot apex to the root via the vascular cambium and xylem parenchyma and differentiating xylem vessels. On its way down, lateral transport of auxin from the vascular cells to the outer cell layers is needed to stimulate shoot elongation, inhibit lateral bud outgrowth and initiate lateral root primordia. At the root tip, PAT is redirected upwards, proceeding basipetally through the root epidermis towards the root elongation zone. Polar auxin transport inhibitors, such as 1-*N*-naphphthlylphtalamic acid (NPA) have been used to show that this cell-to-cell transport of auxin directs plant development (Morris et al. 2004). Auxin induces cell division and cell expansion, and the dynamic auxin gradients and maxima established by its polar transport not only direct tropic growth responses, as described by the Cholodny and Went hypothesis (Went 1926; Cholodny 1927; Went and Thimann 1937), but are also instructive for embryo patterning (Friml et al. 2003; Weijers et al. 2005b), root meristem maintenance (Sabatini et al. 1999) and the initiation and positioning of lateral root and aerial organ primordia (phyllotaxis) (Benkova et al. 2003; Reinhardt et al. 2003).

3.2
Efflux Carriers as Drivers of Polar Auxin Transport

The chemiosmotic hypothesis proposed for the mechanism of polar auxin transport in the early 1970s indicates that polar cell-to-cell transport of auxin involves both influx and efflux carriers and that the asymmetric subcellular localization of the efflux carriers determines the direction of transport (Rubery and Sheldrake 1974; Raven 1975). Candidates for these different transport proteins were identified through characterization of different auxin-related *Arabidopsis* mutants. The auxin-resistant mutant *aux1* led to the identification of the AUX1/LAX family of permease-like membrane proteins as likely auxin influx carriers (Bennett et al. 1996; Parry et al. 2001). Recently, expression of AUX1 in the heterologous Xenopus oocyte system confirmed its function as an auxin influx transporter (Yang et al. 2006). Moreover, characterization of the *Arabidopsis ethylene insensitive root 1/agravitropic root 1* (*eir1-1/agr1*) and *pin formed 1* (*pin1*) mutants that mimic growth of wild-type plants on auxin transport inhibitors led to the identification of the PIN family of membrane proteins. PIN proteins have now been recognized as rate-limiting components of auxin efflux (Paponov et al. 2005; Petrasek et al. 2006). Both AUX1 and PIN proteins show asymmetric sub-cellular distribution and, in accordance with the chemiosmotic hypothesis, the polar localization of PIN proteins correlates perfectly with and in fact was found to determine the direction of auxin transport (Friml et al. 2003; Benkova et al. 2003; Paponov et al. 2005; Wisniewska et al. 2006).

In addition, some plant homologs of the animal multi-drug resistance/ P-glycoprotein (MDR/PGP) ABC transporters have been identified as binding to the auxin transport inhibitor NPA, and these proteins were found to transport auxin when expressed in heterologous host cells. *Arabidopsis* loss-of-function mutants in the corresponding genes show reduced polar auxin transport, suggesting a role for these ABC transporters in auxin transport. At this moment, however, the exact role of the MDR/PGP-dependent auxin transport pathway is still unclear, and MDR/PGP sub-cellular localization is not as well correlated with the direction of auxin transport as that of the PIN proteins (Geisler and Murphy 2006).

3.3 Regulators of Polar Auxin Transport

The role of PIN proteins as rate-limiting components directing polar auxin transport is well described. However, not much is known about the determinants and regulators of the polar localization of these membrane proteins. In fact, very little is known about polarity establishment in plant cells, let alone the polar localization of PIN proteins. Experiments by Geldner and coworkers using the exocytotic vesicle trafficking inhibitor Brefeldin A and cytoskeleton depolymerizing drugs indicated that, in interphase cells, PIN1 proteins cycle between the plasma membrane and endosome-like compartments in an actin-dependent and microtubule-independent manner (Geldner et al. 2001). This PIN recycling is likely to be important for both maintaining and facilitating rapid changes in PIN polar localization. However, in dividing cells, PIN deposition was found to occur at both sides of the cell plate in a microtubule-dependent manner (Geldner et al. 2001; Boutte et al. 2006), indicating that polar localization of PIN proteins is established via post-mitotic redistribution. Following their initial observations, Geldner and coworkers identified the ADP-ribosylation factor–GDP/GTP exchange factor (ARF-GEF) EMB30/GNOM as the Brefeldin A sensitive component in the trafficking of PIN1 vesicles from endosomal compartments to the plasma membrane (Geldner et al. 2003). In accordance with this role, *emb30/gnom* loss-of-function mutants are defective in proper subcellular localization of PIN1 and show developmental defects that resemble those caused by inhibition of polar auxin transport (Steinmann et al. 1999; Geldner et al. 2004). Although the activity of EMB30/GNOM and of other vesicle trafficking components, such as the small GTP binding protein ADP-ribosylation factor 1

(ARF 1) and the ARF-GTPase activating protein (ARF-GAP) SCF/VAN3, has been correlated with polar auxin transport and PIN vesicle trafficking (Xu and Scheres 2005; Sieburth et al. 2006), no clear evidence has yet been provided that these proteins determine PIN polarity.

3.4 Signaling Pathways Regulating Polar Auxin Transport

In contrast to the accessory role of kinases in the modulation of auxin responses, several kinases have been found to have a direct involvement in the control of polar auxin transport. For example, Dai and co-workers (Dai et al. 2006) provided evidence that PAT is regulated by a MAPK cascade, as both loss-of-function and overexpression of the MAPKK BUD1/MKK7 resulted in phenotypes that suggest a role of MKK7 as a negative regulator of PAT (Dai et al. 2006). Moreover, the first and to date only identified PIN polarity determinant is the protein serine/threonine kinase PINOID (PID), as above-threshold activity of this kinase induces PIN sub-cellular re-targeting from the basal (root tip-facing) to the apical (shoot apex-facing) cell pole (Friml et al. 2004). *PID* was identified as a primary auxin-response gene (Benjamins et al. 2001), and therefore the kinase is likely to be part of a feedback loop by which auxin regulates the direction of its own transport in accordance with the canalization hypothesis, which describes the self-organizing role of auxin transport in tissue patterning (Sachs 1991). Below we will further elaborate on the role of PID, how its activity is regulated, and whether closely related protein kinases are involved in the polar targeting of the PIN auxin efflux carriers.

3.4.1 The PINOID Kinase Orienting Polar Auxin Transport

The first indication that PID has a role in polar auxin transport came from the phenotypic similarities between the *pinoid* and the *pin1* loss-of-function mutants, as both mutants show cotyledon defects and develop pin-like inflorescences (Bennett et al. 1995). Subsequent experiments showed that overexpression of *PID* results in phenotypes such as agravitropic growth and collapse of the main root meristem. The root meristem collapse could be rescued by the PAT inhibitor NPA, suggesting that the PID protein kinase is a regulator of PAT (Benjamins et al. 2001). More recently, it was shown that PID determines the apico-basal subcellular polarity of PIN proteins. PID overexpression in the root tip causes basally localized PINs (PIN1, 2 and 4) to be re-localized to the apical cell side. This results in apicalized auxin transport that deprives the root meristem of auxin, and leads to meristem collapse. Conversely, *pid* loss-of-function causes apically localized PIN1 to be re-localized to the basal side of epidermis cells in the inflorescence apex, preventing the formation of auxin maxima needed for organ initiation, thus resulting in pin-like inflorescences

(Friml et al. 2004). Thus, cellular PID levels modulate the direction of PAT by determining the subcellular targeting of PIN proteins.

Several lines of evidence point to the fact that phosphorylation of PINs is important in the regulation of their subcellular localization. Firstly, modification of a serine residue in one of the short cytoplasmic loops of PIN2 to a glycine resulted in failure of PIN2 deployment to the plasma membrane (Petrasek et al. 2006). Moreover, large-scale identification of phosphorylation sites in plasma membrane proteins by mass spectrometry showed that the large central hydrophilic loop of several *Arabidopsis thaliana* PIN proteins is phosphorylated at multiple positions in vivo (Nuhse et al. 2004; Benschop et al. 2007). Recently, we have obtained evidence that PID may be (one of) the kinase(s) responsible for phosphorylating the PIN proteins in their large hydrophilic loop (Michniewicz et al. 2007).

3.4.2
PINOID and Protein Phosphatase 2A Activity Antagonistically Mediate PIN Polar Targeting

Interestingly, a mutant has been identified in a gene encoding the regulatory A subunit of a trimeric protein phosphatase 2A (PP2A), that displays root curling in response to NPA (*rcn1*) (Garbers et al. 1996). The *rcn1* mutant shows increased root basipetal auxin transport, reduced gravitropic response and a delay in the establishment of differential auxin-induced gene expression across a gravity-stimulated root tip, aspects that were restored to normal upon NPA treatment (Rashotte et al. 2001). Michniewicz and co-workers showed that PP2A activity, of which RCN1 is one of the three regulatory A subunits (PP2AA) in *Arabidopsis*, antagonizes the action of PID on PINs. PP2AA loss-of-function resulted in the same basal-to-apical switches in PIN polarity, leading to similar embryo and root defects as PID overexpression, whereas the *pid* mutation partially suppressed these *pp2aa* loss-of-function phenotypes. Moreover, both PID and PP2AA partially co-localize with PIN proteins on the plasma membrane, and in vivo and semi in vivo phosphorylation assays showed that they act antagonistically on the phosphorylation status of PIN proteins (Michniewicz et al. 2007). This beautifully exemplifies a dual molecular switch in which PID-mediated phosphorylation of PINs induces apical localization of these efflux carriers, resulting in upward PAT, while PP2A-mediated dephosphorylation of PINs reverses this situation, consequently resulting in downward PAT (Fig. 2).

3.4.3
The PID Regulatory Complex

Several PID regulatory proteins have been identified in a yeast two-hybrid screen as interactors of PID. Two of them, PINOID Binding Protein 1 (PBP1) and TOUCH3 (TCH3), are calmodulins that show calcium-dependent binding to PID and respectively enhance and downregulate PID activity in vitro (Fig. 2). Neither PBP1 nor TCH3 seem to be phosphorylation targets of PID, and assays in which *35S::PID* seedlings were treated with calcium transporter and calmodulin inhibitors showed enhanced root meristem collapse (Benjamins et al. 2003). In addition, we have obtained further data that support a role for PBP1 and TCH3 as regulators of the activity and subcellular localization of the PID kinase in vivo (H. Robert et al., unpublished data).

A third interactor and regulator of PID is the PINOID Binding Protein 2 (PBP2). PBP2 contains two protein–protein interaction domains, the BTB/POZ- (Bric-a-brac, Tramtrack and Broad Complex/Pox virus and Zinc finger) and the TAZ (Transcriptional Adaptor putative Zinc Finger) domain. PBP2 has previously been identified as a calmodulin binding transcriptional regulator AtBT1 (Du and Poovaiah 2004). Our observations suggest that PBP2 acts as an inhibitor of PID, since it represses PID auto- and trans-phosphorylation activity in vitro (Benjamins 2004; Zago 2006). Moreover, the fact that the GFP-PBP2 fusion protein shows a cytoskeleton-like localization in onion cells (Benjamins 2004), suggests that PBP2 provides a possible link between the established roles of PID and the cytoskeleton in regulating PAT. Although the subcellular localization of PBP2-GFP in *Arabidopsis* protoplasts and plants showed nuclear and general cytosolic rather than cytoskeleton localization, a yeast two-hybrid screen identified two closely related microtubule motors (kinesins) as interacting partners of PBP2, which were named PBP2 Binding Kinesin 1 and 2 (PBK1 and 2). Preliminary data from both in vitro and in vivo experiments corroborate the involvement of the PBKs in the PID signaling pathway. On the basis of these observations, we speculate that the PBKs, together with PBP2, are involved in repression, and possibly determine the subcellular localization of PID activity (Zago 2006).

3.5
Other Plant AGC Kinases Involved in Auxin-Related Processes

In *Arabidopsis*, PID belongs to the plant specific AGC- (after the cAMP-dependent protein kinase A, cGMP-dependent protein kinase G and phospholipids-dependent protein kinase C) VIII subfamily of kinases (Hanks and Hunter 1995). Members of this subfamily show sequence signatures and functional data suggesting that they represent the plant orthologs of the PKA and PKC in animals (Trewavas et al. 2002; Bogre et al. 2003). They do contain peculiar characteristics which distinguish them from their animal counterparts. For example, the classical DFG motif present in subdomain VII of the catalytic kinase domain is replaced by DFD. Furthermore, kinases of the AGCVIII group show an unusual insertion of about 50–80 aa in the activation loop of the catalytic domain. Previous sequence comparisons that were based on

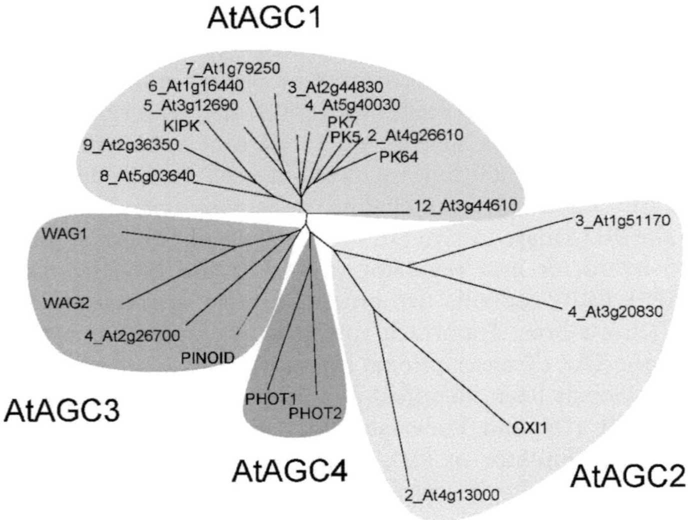

Fig. 3 Phylogram of the AGCVIII family of protein kinases in *Arabidopsis*, based on an alignment of amino acid sequence comprising the catalytic domain of these proteins

alignment of complete protein sequences indicated that the *Arabidopsis* AGC kinases fall into two groups (Bogre et al. 2003). On the basis of an aliment of their catalytic domain, we conclude however that the *Arabidopsis* AGCVIII kinases fall into the four sub-groups AtAGC1 to 4 (Fig. 3, Galvan and Offringa 2007). It is interesting to note that in this new comparison the phototropins PHOT1 and PHOT2 (AtAGC4) are most closely related to PID and the PID-related kinases (AtAGC3), and that, like PID, the PID-related kinases WAG1 and WAG2 (AtAGC3) and the phototropins are all involved in auxin-mediated plant development. Below we will shortly discuss the roles of these kinases in root waving and blue light-induced responses.

3.5.1 Phototropins Translate Unidirectional Blue Light into Phototropic Growth

The phototropins PHOT1 and 2 are blue light receptors that control blue light-induced processes such as phototropism, chloroplast relocation and stomatal opening (Briggs and Huala 1999; Jarillo et al. 1998). Both proteins have been shown to act redundantly, with PHOT1 being active under all light fluence rate conditions, and PHOT2 functioning specifically under high fluence rate conditions (Sakai et al. 2001; Matsuoka and Tokutomi 2005). PHOT1 and 2 consist of a light perception domain, comprising two light, oxygen or voltage (LOV) motifs that non-covalently bind the blue light sensitive molecule flavin mononucleotide (Huala et al. 1997), and a protein serine/threonine kinase domain (Reymond et al. 1992; Huala et al. 1997).

The molecular mechanism of phototropin action has been partly elucidated. Phototropins are membrane-associated proteins, and in the dark the kinase domain is repressed through binding of the photoreceptor domain. Light induces a conformational change in the photoreceptor that activates its protein kinase activity and at the same time releases the kinase from the plasma membrane (Sakamoto and Briggs 2002; Matsuoka and Tokutomi 2005). The BTB/POZ domain proteins nonphototropic hypocotyl 3 (NPH3) and root phototropism 2 (RPT2) were shown to be additional components in phototropism signaling (Motchoulski and Liscum 1999; Sakai et al. 2000; Inada et al. 2004; Haga et al. 2005), and due to their domain structure they are believed to mediate protein–protein interactions. PHOT1, NPH3 and RPT2 are all associated with the plasma membrane and can bind to each other (Motchoulski and Liscum 1999; Inada et al. 2004; Sakamoto and Briggs 2002). A rice ortholog of NPH3 called coleoptile phototropism 1 was shown to act upstream of the redistribution of auxin induced by unilateral illumination of the seedling (Haga et al. 2005). Interestingly, NPH3 appears to be a phosphoprotein that is in the phosphorylated state in the dark and becomes rapidly de-phosphorylated in the light. The preliminary finding that NPH3 interacts with CULLIN3 suggests that this protein is involved in targeting proteins for degradation by the proteasome (Pedmale and Liscum 2007). Although these observations place these proteins in early stages of this blue light-triggered signaling pathway, the molecular mechanism initiated by light and ultimately leading to PIN-mediated differential auxin transport and auxin signaling is still far from clear (Friml et al. 2002; Esmon et al. 2006).

3.5.2 WAGs in Root Waving

The search for factors involved in photoregulated development of young seedlings of pea led to the finding of a kinase negatively modulated by light named *Pisum sativum* Protein Kinase 3 (PsPK3) (Khanna et al. 1999). The completion of the *Arabidopsis* genome allowed for the identification of PsPK3 orthologs in this organism, named initially AtPK3 and AGC1-11 (Bogre et al. 2003). The function of AtPK3 and AGC1-11 in *Arabidopsis* was later partly elucidated with the observation that *atpk3 agc1-11* double mutants presented enhanced root waving, a finding that resulted in the re-naming of the kinases to WAG1 and WAG2, respectively (Santner and Watson 2006). Santner and Watson further demonstrated that *wag1 wag2* plants display increased resistance to the root curling-inhibition mediated by NPA, suggesting that the double mutants are defective in auxin-related processes such as PAT. The conclusion that WAG1 and WAG2 participate in the control of PAT is in part corroborated by their amino acid sequence, as they classify in the same AtAGC3 group as PID (Fig. 3). The actual role of the WAGs in PAT and their molecular mechanism of action, however, remains to be elucidated.

3.6 The Regulation of AGC VIII Kinases by PDK1

It is known from animal and yeast systems that kinases of the AGC family are targets of the 3-phosphoinositide-dependent-protein kinase PDK1. PDK1 contains a lipid binding pleckstrin homology (PH) domain and binds to PtdIns(3,4,5)P_3 and PtdIns(3,4)P_2 in animals (Bogre et al. 2003), and phosphatidic acid, PtdIns3P and PtdIns(3,4)P_2, among others, in *Arabidopsis* (Deak et al. 1999). These lipid molecules are generated in response to signals from phosphorylation events or by cleavage of their inositol groups by phosphoinositide kinases, phosphoinositide phosphatases or phospholipases (Mueller-Roeber and Pical 2002; Meijer and Munnik 2003) and participate in many signal transduction pathways and cellular processes in various organisms (Vanhaesebroeck et al. 2001).

In animals, the acknowledged targets of PDK1 are members of the AGC family, for example PKA, PKB, PKC, p70 ribosomal S6 kinase (S6K), p90 S6 kinases (RSKs) and the serum and glucocorticoid-inducible kinase (SGK) (Storz and Toker 2002; Bogre et al. 2003). Being the predicted orthologs of animal PKAs and PKCs, the plant-specific AGCVIII kinases like PID, and the WAG kinases are also predicted targets of PDK1, and indeed recent data have partly confirmed this. For example, Anthony and co-workers (Anthony et al. 2004) showed that the kinase AGC2-1, which probably participates in cell growth and division induced by auxin and cytokinin, interacts with PDK1 and is activated by it in a process mediated by phosphatidic acid. In another study, it was demonstrated that PID interacts with and is activated by PDK1, both through in vitro and in semi in vivo experiments using protein extracts from flowers or seedlings shoots, organs where PID is naturally more active (Zegzouti et al. 2006a). Finally, it has been shown that several other members of the AGCVIII subfamily are also direct targets of PDK1 (Zegzouti et al. 2006b). The *in planta* relevance of PDK1-mediated phosphorylation of the AGCVIII kinases, however, remains to be addressed.

4 Conclusion

The mechanism of action of the phytohormone auxin is unique. The central auxin signaling pathway is triggered by direct binding of auxin to intracellular receptors, which activate the auxin-related proteolysis machinery consequently leading to auxin-responsive gene expression. There is only limited data indicating the direct involvement of kinases in this process, although there are suggestions of indirect phosphorylation events, or that ABP1 might be the apoplastic auxin receptor initiating a classical signaling pathway. However, most of the findings tend to suggest that auxin signaling is unique in its mechanism, strongly contrasting with classical routes occurring via membrane-bound hormone-receptors followed by a signaling cascade leading to changes in gene expression. As the tissue/cell specificity of the auxin signal is probably not provided by typical receptors, an alternate and very ingenious system evolved to likely solve this problem: the signal is delivered through the generation of tissue-specific concentration maxima and gradients of auxin, and this is sufficient to promote proper plant development.

The mechanisms behind the generation of such gradients are significantly complex. It involves transporter proteins such as the auxin influx (AUX1) and efflux (PINs) carriers, whose activities depend on their proper asymmetric subcellular localization, and are regulated by a plethora of components. The complex regulatory network of the polar auxin transport system evidently relies on the activity of kinases. Kinases of the AGCVIII subfamily seem to be particularly involved, with PID modulating the direction of PAT, the WAGs being involved in root waving, and the PHOTs controlling phototropism. On the other hand, while many other proteins play a role in the regulation of the activity of these kinases, one kinase, PDK1, seems to be the upstream regulator that links phospholipid signaling with auxin transport.

From the network regulating auxin distribution, only a few links have yet been demonstrated, such as the relationship between PDK1 and PID, PID/PP2A and PINs, and PINs and GNOM. Most connections in this network are, however, still unclear or missing. For example, although we know that the PIN proteins that have been shown to be responsive to PID activity play a role in several tissues, PID expression seems to be restricted to much fewer organs. Considering that PID-related kinases such as WAG1 and WAG2 act in different tissues than PID, it is tempting to speculate that they perform a PID-like function towards the PINs in tissues where PID itself is not present. Furthermore, although the activity of PDK1 towards most of the AGCVIII subfamily kinases has been demonstrated, there is still a gap in our knowledge on the relationship between PDK1 and the phototropins. Our future goal will be to obtain a much more comprehensive understanding of the PAT signaling network, and we expect to answer at least some of the above questions in the near future.

References

Abel S, Oeller PW, Theologis A (1994) Early auxin-induced genes encode short-lived nuclear proteins. Proc Natl Acad Sci USA 91:326–330

Abel S, Theologis A (1996) Early genes and auxin action. Plant Physiol 111:9–17

Anthony RG, Henriques R, Helfer A, Meszaros T, Rios G, Testerink C, Munnik T, Deak M, Koncz C, Bogre L (2004) A protein kinase target of a PDK1 signalling pathway is involved in root hair growth in *Arabidopsis*. EMBO J 23:572–581

Baker DA (2000) Long-distance vascular transport of endogenous hormones in plants and their role in source: sink regulation. Israel J Plant Sci 48:199–203

Benjamins R (2004) Functional analysis of the PINOID protein kinase in *Arabidopsis thaliana*. Leiden University, Thesis

Benjamins R, Ampudia CS, Hooykaas PJ, Offringa R (2003) PINOID-Mediated Signaling Involves Calcium-Binding Proteins. Plant Physiol 132:1623–1630

Benjamins R, Quint A, Weijers D, Hooykaas P, Offringa R (2001) The PINOID protein kinase regulates organ development in *Arabidopsis* by enhancing polar auxin transport. Development 128:4057–4067

Benkova E, Michniewicz M, Sauer M, Teichmann T, Seifertova D, Jurgens G, Friml J (2003) Local, efflux-dependent auxin gradients as a common module for plant organ formation. Cell 115:591–602

Bennett M, Marchant A, Green H, May S, Ward S, Millner P, Walker A, Schulz B, Feldmann K (1996) *Arabidopsis* AUX1 gene: a permease-like regulator of root gravitropism. Science 273:948–950

Bennett S, Alvarez J, Bossinger G, Smyth D (1995) Morphogenesis in *pinoid* mutants of *Arabidopsis thaliana.* Plant J 8:505

Benschop JJ, Mohammed S, O'Flaherty M, Heck AJR, Slijper M, Menke FLH (2007) Quantitative phospho-proteomics of early elicitor signalling in *Arabidopsis.* Mol Cell Proteomics M600429-MMCP200

Bilang J, Sturm A (1995) Cloning and characterization of a glutathione S-transferase that can be photolabeled with 5-azido-indole-3-acetic acid. Plant Physiol 109:253–260

Bogre L, Okresz L, Henriques R, Anthony RG (2003) Growth signalling pathways in *Arabidopsis* and the AGC protein kinases. Trends Plant Sci 8:424–431

Bostick M, Lochhead SR, Honda A, Palmer S, Callis J (2004) Related to ubiquitin 1 and 2 are redundant and essential and regulate vegetative growth, auxin signaling, and ethylene production in *Arabidopsis.* Plant Cell 16:2418–2432

Boutte Y, Crosnier MT, Carraro N, Traas J, Satiat-Jeunemaitre B (2006) The plasma membrane recycling pathway and cell polarity in plants: studies on PIN proteins. J Cell Sci 119:1255–1265

Briggs WR, Huala E (1999) Blue-light photoreceptors in higher plants. Annu Rev Cell Dev Biol 15:33–62

Cambridge AP, Morris DA (1996) Transfer of exogenous auxin from the phloem to the polar auxin transport pathway in pea (*Pisum sativum* L.). Planta 199:583–588

Chamovitz DA, Wei N, Osterlund MT, von Arnim AG, Staub JM, Matsui M, Deng XW (1996) The COP9 complex, a novel multisubunit nuclear regulator involved in light control of a plant developmental switch. Cell 86:115–121

Champion A, Kreis M, Mockaitis K, Picaud A, Henry Y (2004) *Arabidopsis* kinome: after the casting. Funct Integr Genomics 4:163–187

Cholodny N (1927) Wuchshormone und Tropismen bei den Pflanzen. Biol Zent 47:604–626

Christensen SK, Dagenais N, Chory J, Weigel D (2000) Regulation of auxin response by the protein kinase PINOID. Cell 100:469–478

Colon-Carmona A, Chen DL, Yeh KC, Abel S (2000) Aux/IAA proteins are phosphorylated by phytochrome in vitro. Plant Physiol 124:1728–1738

Cope GA, Deshaies RJ (2003) COP9 signalosome: a multifunctional regulator of SCF and other cullin-based ubiquitin ligases. Cell 114:663–671

Dai Y, Wang H, Li B, Huang J, Liu X, Zhou Y, Mou Z, Li J (2006) Increased expression of MAP KINASE KINASE7 causes deficiency in polar auxin transport and leads to plant architectural abnormality in *Arabidopsis.* Plant Cell 18:308–320

Darwin C, Darwin F (1881) The Power of Movement in Plants. Murray, London

Deak M, Casamayor A, Currie RA, Downes CP, Alessi DR (1999) Characterisation of a plant 3-phosphoinositide-dependent protein kinase-1 homologue which contains a pleckstrin homology domain. FEBS Lett 451:220–226

del Pozo JC, Estelle M (1999) The *Arabidopsis* cullin AtCUL1 is modified by the ubiquitin-related protein RUB1. Proc Natl Acad Sci USA 96:15342–15347

Dharmasiri N, Dharmasiri S, Estelle M (2005a) The F-box protein TIR1 is an auxin receptor. Nature 435:441–445

Dharmasiri N, Dharmasiri S, Jones AM, Estelle M (2003a) Auxin action in a cell-free system. Curr Biol 13:1418–1422

Dharmasiri N, Dharmasiri S, Weijers D, Lechner E, Yamada M, Hobbie L, Ehrismann JS, Jurgens G, Estelle M (2005b) Plant development is regulated by a family of auxin receptor F-box proteins. Dev Cell 9:109–119

Dharmasiri S, Dharmasiri N, Hellmann H, Estelle M (2003b) The RUB/Nedd8 conjugation pathway is required for early development in *Arabidopsis*. EMBO J 22:1762–1770

Dharmasiri S, Swarup R, Mockaitis K, Dharmasiri N, Singh SK, Kowalchyk M, Marchant A, Mills S, Sandberg G, Bennett MJ, Estelle M (2006) AXR4 is required for localization of the auxin influx facilitator AUX1. Science 312:1218–1220

Droog F, Hooykaas P, van der Zaal BJ (1995) 2,4-Dichlorophenoxyacetic acid and related chlorinated compounds inhibit two auxin-regulated type-III tobacco glutathione *S*-transferases. Plant Physiol 107:1139–1146

Du L, Poovaiah BW (2004) A novel family of Ca2+/calmodulin-binding proteins involved in transcriptional regulation: interaction with fsh/Ring3 class transcription activators. Plant Mol Biol 54:549–569

Esmon CA, Tinsley AG, Ljung K, Sandberg G, Hearne LB, Liscum E (2006) A gradient of auxin and auxin-dependent transcription precedes tropic growth responses. Proc Natl Acad Sci USA 103:236–241

Friml J, Vieten A, Sauer M, Weijers D, Schwarz H, Hamann T, Offringa R, Jurgens G (2003) Efflux-dependent auxin gradients establish the apical-basal axis of *Arabidopsis*. Nature 426:147–153

Friml J, Wisniewska J, Benkova E, Mendgen K, Palme K (2002) Lateral relocation of auxin efflux regulator PIN3 mediates tropism in *Arabidopsis*. Nature 415:806–809

Friml J, Yang X, Michniewicz M, Weijers D, Quint A, Tietz O, Benjamins R, Ouwerkerk PB, Ljung K, Sandberg G, Hooykaas PJ, Palme K, Offringa R (2004) A PINOID-dependent binary switch in apical-basal PIN polar targeting directs auxin efflux. Science 306:862–865

Fu H, Reis N, Lee Y, Glickman MH, Vierstra RD (2001) Subunit interaction maps for the regulatory particle of the 26S proteasome and the COP9 signalosome. EMBO J 20:7096–7107

Garbers C, DeLong A, Deruere J, Bernasconi P, Soll D (1996) A mutation in protein phosphatase 2A regulatory subunit A affects auxin transport in *Arabidopsis*. EMBO J 15:2115–2124

Gehring CA, Irving HR, Parish RW (1990) Effects of auxin and abscisic acid on cytosolic calcium and pH in plant cells. Proc Natl Acad Sci USA 87:9645–9649

Geisler M, Murphy AS (2006) The ABC of auxin transport: The role of *p*-glycoproteins in plant development. FEBS Lett 580:1094–1102

Geldner N, Anders N, Wolters H, Keicher J, Kornberger W, Muller P, Delbarre A, Ueda T, Nakano A, Jurgens G (2003) The *Arabidopsis* GNOM ARF-GEF mediates endosomal recycling, auxin transport, and auxin-dependent plant growth. Cell 112:219–230

Geldner N, Friml J, Stierhof YD, Jurgens G, Palme K (2001) Auxin transport inhibitors block PIN1 cycling and vesicle trafficking. Nature 413:425–428

Geldner N, Richter S, Vieten A, Marquardt S, Torres-Ruiz RA, Mayer U, Jurgens G (2004) Partial loss-of-function alleles reveal a role for GNOM in auxin transport-related, post-embryonic development of *Arabidopsis*. Development 131:389–400

Goetz M, Vivian-Smith A, Johnson SD, Koltunow AM (2006) AUXIN RESPONSE FACTOR8 is a negative regulator of fruit initiation in *Arabidopsis*. Plant Cell 18:1873–1886

Gray WM, del Pozo JC, Walker L, Hobbie L, Risseeuw E, Banks T, Crosby WL, Yang M, Ma H, Estelle M (1999) Identification of an SCF ubiquitin-ligase complex required for auxin response in *Arabidopsis thaliana*. Genes Dev 13:1678–1691

Gray WM, Hellmann H, Dharmasiri S, Estelle M (2002) Role of the *Arabidopsis* RING-H2 protein RBX1 in RUB modification and SCF function. Plant Cell 14:2137–2144

Gray WM, Kepinski S, Rouse D, Leyser O, Estelle M (2001) Auxin regulates SCF(TIR1)-dependent degradation of AUX/IAA proteins. Nature 414:271–276

Grebe M (2005) Plant biology. Enhanced: growth by auxin: when a weed needs acid. Science 310:60–61

Guilfoyle T (1986) Auxin-regulated gene expression in higher plants. CRC Crit Rev Plant Sci 4:247–276

Guilfoyle T, Hagen G, Ulmasov T, Murfett J (1998a) How does auxin turn on genes? Plant Physiol 118:341–347

Guilfoyle TJ, Ulmasov T, Hagen G (1998b) The ARF family of transcription factors and their role in plant hormone-responsive transcription. Cell Mol Life Sci 54:619–627

Haga K, Takano M, Neumann R, Iino M (2005) The Rice COLEOPTILE PHOTOTROPISM1 gene encoding an ortholog of *Arabidopsis* NPH3 is required for phototropism of coleoptiles and lateral translocation of auxin. Plant Cell 17:103–115

Hagen G, Guilfoyle TJ (1985) Rapid induction of selective transcription by auxins. Mol Cell Biol 5:1197–1203

Hanks SK, Hunter T (1995) Protein kinases 6. The eukaryotic protein kinase superfamily: kinase (catalytic) domain structure and classification. FASEB J 9:576–596

Hardtke CS, Berleth T (1998) The *Arabidopsis* gene MONOPTEROS encodes a transcription factor mediating embryo axis formation and vascular development. EMBO J 17:1405–1411

Hardtke CS, Ckurshumova W, Vidaurre DP, Singh SA, Stamatiou G, Tiwari SB, Hagen G, Guilfoyle TJ, Berleth T (2004) Overlapping and non-redundant functions of the *Arabidopsis* auxin response factors MONOPTEROS and NONPHOTOTROPIC HYPOCOTYL 4. Development 131:1089–1100

Hellmann H, Hobbie L, Chapman A, Dharmasiri S, Dharmasiri N, del Pozo C, Reinhardt D, Estelle M (2003) *Arabidopsis* AXR6 encodes CUL1 implicating SCF E3 ligases in auxin regulation of embryogenesis. EMBO J 22:3314–3325

Huala E, Oeller PW, Liscum E, Han IS, Larsen E, Briggs WR (1997) *Arabidopsis* NPH1: a protein kinase with a putative redox-sensing domain. Science 278:2120–2123

Inada S, Ohgishi M, Mayama T, Okada K, Sakai T (2004) RPT2 is a signal transducer involved in phototropic response and stomatal opening by association with phototropin 1 in *Arabidopsis thaliana*. Plant Cell 16:887–896

Jarillo JA, Ahmad M, Cashmore R (1998) NPL1: a second member of the NPH1 serine/threonine kinase family of *Arabidopsis*. Plant Physiol 117:719

Karcz W, Burdach Z (2002) A comparison of the effects of IAA and 4-Cl-IAA on growth, proton secretion and membrane potential in maize coleoptile segments. J Exp Bot 53:1089–1098

Kepinski S, Leyser O (2005) The *Arabidopsis* F-box protein TIR1 is an auxin receptor. Nature 435:446–451

Key JL, Kroner P, Walker J, Hong JC, Ulrich TH, Ainley WM, Gantt JS, Nagao RT (1986) Auxin-regulated gene expression. Philos Trans R Soc Lond B Biol Sci 314:427–440

Khanna R, Lin X, Watson JC (1999) Photoregulated expression of the PsPK3 and PsPK5 genes in pea seedlings. Plant Mol Biol 39:231–242

Kim J, Harter K, Theologis A (1997) Protein–protein interactions among the Aux/IAA proteins. Proc Natl Acad Sci USA 94:11786–11791

Kogl F, Haagen-Smit A (1931) Über die Chemie des Wuchsstoffs K. Akad Wetenschap Amsterdam. Proc Sect Sci 34:1411–1416

Kovtun Y, Chiu WL, Zeng W, Sheen J (1998) Suppression of auxin signal transduction by a MAPK cascade in higher plants. Nature 395:716–720

Larkin PJ, Gibson JM, Mathesius U, Weinman JJ, Gartner E, Hall E, Tanner GJ, Rolfe BG, Djordjevic MA (1996) Transgenic white clover. Studies with the auxin-responsive promoter, GH3, in root gravitropism and lateral root development. Transgenic Res 5:325–335

Larsen PB, Cancel JD (2004) A recessive mutation in the RUB1-conjugating enzyme, RCE1, reveals a requirement for RUB modification for control of ethylene biosynthesis and proper induction of basic chitinase and PDF1.2 in *Arabidopsis*. Plant J 38:626–638

Leonard CJ, Aravind L, Koonin EV (1998) Novel families of putative protein kinases in bacteria and archaea: evolution of the eukaryotic protein kinase superfamily. Genome Res 8:1038–1047

Li Y, Hagen G, Guilfoyle TJ (1991) An auxin-responsive promoter is differentially induced by auxin gradients during tropisms. Plant Cell 3:1167–1175

Lincoln C, Britton JH, Estelle M (1990) Growth and development of the axr1 mutants of *Arabidopsis*. Plant Cell 2:1071–1080

Liscum E, Briggs WR (1995) Mutations in the NPH1 locus of *Arabidopsis* disrupt the perception of phototropic stimuli. Plant Cell 7:473–485

Liscum E, Reed JW (2002) Genetics of Aux/IAA and ARF action in plant growth and development. Plant Mol Biol 49:387–400

Luschnig C, Gaxiola RA, Grisafi P, Fink GR (1998) EIR1, a root-specific protein involved in auxin transport, is required for gravitropism in *Arabidopsis thaliana*. Genes Dev 12:2175–2187

Manning G, Whyte DB, Martinez R, Hunter T, Sudarsanam S (2002) The protein kinase complement of the human genome. Science 298:1912–1934

Matsuoka D, Tokutomi S (2005) Blue light-regulated molecular switch of Ser/Thr kinase in phototropin. Proc Natl Acad Sci USA 102:13337–13342

Meijer G, Munnik T (2003) Phospholipid-based signalling in plants. Annu Rev Plant Biol 54:265–306

Michniewicz M, Zago MK, Abas L, Weijers D, Schweighofer A, Meskiene I, Heisler MG, Ohno C, Huang F, Weigel D, Meyerowitz EM, Luschnig C, Offringa R, Friml J (2007) Phosphatase 2A and PID kinase activities antagonistically mediate PIN phosphorylation and apical/basal targeting in *Arabidopsis*. Cell 130:1044–1056

Milanesi L, Petrillo M, Sepe L, Boccia A, D'Agostino N, Passamano M, Di Nardo S, Tasco G, Casadio R, Paolella G (2005) Systematic analysis of human kinase genes: a large number of genes and alternative splicing events result in functional and structural diversity. BMC Bioinformatics 6(4):S20

Mockaitis K, Howell SH (2000) Auxin induces mitogenic activated protein kinase (MAPK) activation in roots of *Arabidopsis* seedlings. Plant J 24:785–796

Moon J, Parry G, Estelle M (2004) The ubiquitin-proteasome pathway and plant development. Plant Cell 16:3181–3195

Morre DJ, Morre DM, Ternes P (2003) Auxin-activated NADH oxidase activity of soybean plasma membranes is distinct from the constitutive plasma membrane NADH oxidase and exhibits prion-like properties. In Vitro Cell Dev Biol Plant 39:368–376

Morris DA, Friml J, Zazimalova E (2004) The functioning of hormones in plant growth and development; The transport of auxins. In: Davies PJ (ed) Plant hormones: Biosynthesis, signal trasnduction, action! Kluwer, Dordrecht, pp 437–470

Morris DA, Thomas (1978) A microautoradiographic study of auxin transport in the stem of intact pea seedlings (*Pisum sativum* L.). J Exp Bot 29:147–157

Motchoulski A, Liscum E (1999) *Arabidopsis* NPH3: A NPH1 photoreceptor-interacting protein essential for phototropism. Science 286:961–964

Mueller-Roeber B, Pical C (2002) Inositol phospholipid metabolism in *Arabidopsis*. Characterized and putative isoforms of inositol phospholipid kinase and phosphoinositide-specific phospholipase C. Plant Physiol 130:22–46

Napier RM, David KM, Perrot-Rechenmann C (2002) A short history of auxin-binding proteins. Plant Mol Biol 49:339–348

Nowacki J, Bandurski RS (1980) Myo-inositol esters of indole-3-acetic acid as seed auxin precursors of *Zea mays* L. Plant Physiol 65:422–427

Nuhse TS, Stensballe A, Jensen ON, Peck SC (2004) Phosphoproteomics of the *Arabidopsis* plasma membrane and a new phosphorylation site database. Plant Cell 16:2394–2405

Osterlund MT, Hardtke CS, Wei N, Deng XW (2000a) Targeted destabilization of HY5 during light-regulated development of *Arabidopsis*. Nature 405:462–466

Osterlund MT, Wei N, Deng XW (2000b) The roles of photoreceptor systems and the COP1-targeted destabilization of HY5 in light control of *Arabidopsis* seedling development. Plant Physiol 124:1520–1524

Ottenschlager I, Wolff P, Wolverton C, Bhalerao RP, Sandberg G, Ishikawa H, Evans M, Palme K (2003) Gravity-regulated differential auxin transport from columella to lateral root cap cells. PNAS 100:2987–2991

Paponov IA, Teale WD, Trebar M, Blilou I, Palme K (2005) The PIN auxin efflux facilitators: evolutionary and functional perspectives. Trends Plant Sci 10:170–177

Parry G, Marchant A, May S, Swarup R, Swarup K, James N, Graham N, Allen T, Martucci T, Yemm A, Napier R, Manning K, King G, Bennett M (2001) Quick on the Uptake: Characterization of a Family of Plant Auxin Influx Carriers. J Plant Growth Regul 20:217–225

Pedmale UV, Liscum E (2007) Regulation of phototropic signaling in *Arabidopsis* via phosphorylation state changes in the phototropin 1-interacting protein NPH3. J Biol Chem M702551200

Petrasek J, Mravec J, Bouchard R, Blakeslee JJ, Abas M, Seifertova D, Wisniewska J, Tadele Z, Kubes M, Covanova M, Dhonukshe P, Skupa P, Benkova E, Perry L, Krecek P, Lee OR, Fink GR, Geisler M, Murphy AS, Luschnig C, Zazimalova E, Friml J (2006) PIN proteins perform a rate-limiting function in cellular auxin efflux. Science 312:914–918

Ramos JA, Zenser N, Leyser O, Callis J (2001) Rapid degradation of auxin/indoleacetic acid proteins requires conserved amino acids of domain II and is proteasome dependent. Plant Cell 13:2349–2360

Rashotte AM, DeLong A, Muday GK (2001) Genetic and chemical reductions in protein phosphatase activity alter auxin transport, gravity response, and lateral root growth. Plant Cell 13:1683–1697

Raven J (1975) Transport of indole-3-acetic acid in plant cells in relation to pH and electrical potential gradients, and its significance for polar IAA transport. New Phytologist 74:163–172

Reinhardt D, Pesce ER, Stieger P, Mandel T, Baltensperger K, Bennett M, Traas J, Friml J, Kuhlemeier C (2003) Regulation of phyllotaxis by polar auxin transport. Nature 426:255–260

Reymond P, Short TW, Briggs WR (1992) Blue light activates a specific protein kinase in higher plants. Plant Physiol 100:655–661

Rubery PH, Sheldrake AR (1974) Carrier-mediated auxin transport. Planta 118:101–121

Ruegger M, Dewey E, Gray WM, Hobbie L, Turner J, Estelle M (1998) The TIR1 protein of *Arabidopsis* functions in auxin response and is related to human SKP2 and yeast grr1p. Genes Dev 12:198–207

Sabatini S, Beis D, Wolkenfelt H, Murfett J, Guilfoyle T, Malamy J, Benfey P, Leyser O, Bechtold N, Weisbeek P, Scheres B (1999) An auxin-dependent distal organizer of pattern and polarity in the *Arabidopsis* root. Cell 99:463–472

Sachs T (1991) Pattern formation in plant tissues. Cambridge University Press, Cambridge

Sakai T, Kagawa T, Kasahara M, Swartz TE, Christie JM, Briggs WR, Wada M, Okada K (2001) *Arabidopsis* nph1 and npl1: blue light receptors that mediate both phototropism and chloroplast relocation. Proc Natl Acad Sci USA 98:6969–6974

Sakai T, Wada T, Ishiguro S, Okada K (2000) RPT2. A signal transducer of the phototropic response in *Arabidopsis*. Plant Cell 12:225–236

Sakamoto K, Briggs WR (2002) Cellular and subcellular localization of phototropin 1. Plant Cell 14:1723–1735

Santner AA, Watson JC (2006) The WAG1 and WAG2 protein kinases negatively regulate root waving in *Arabidopsis*. Plant J 45:752–764

Schwechheimer C (2004) The COP9 signalosome (CSN): an evolutionary conserved proteolysis regulator in eukaryotic development. Biochim Biophys Acta 1695:45–54

Schwechheimer C, Deng XW (2000) The COP/DET/FUS proteins-regulators of eukaryotic growth and development. Semin Cell Dev Biol 11:495–503

Schwechheimer C, Serino G, Callis J, Crosby WL, Lyapina S, Deshaies RJ, Gray WM, Estelle M, Deng XW (2001) Interactions of the COP9 signalosome with the E3 ubiquitin ligase SCFTIR1 in mediating auxin response. Science 292:1379–1382

Seo HS, Yang JY, Ishikawa M, Bolle C, Ballesteros ML, Chua NH (2003) LAF1 ubiquitination by COP1 controls photomorphogenesis and is stimulated by SPA1. Nature 423:995–999

Sessions A, Nemhauser JL, McColl A, Roe JL, Feldmann KA, Zambryski PC (1997) ETTIN patterns the *Arabidopsis* floral meristem and reproductive organs. Development 124:4481–4491

Sieburth LE, Muday GK, King EJ, Benton G, Kim S, Metcalf KE, Meyers L, Seamen E, Van Norman JM (2006) SCARFACE encodes an ARF-GAP that is required for normal auxin efflux and vein patterning in *Arabidopsis*. Plant Cell 18:1396–1411

Steinmann T, Geldner N, Grebe M, Mangold S, Jackson CL, Paris S, Galweiler L, Palme K, Jurgens G (1999) Coordinated polar localization of auxin efflux carrier PIN1 by GNOM ARF GEF. Science 286:316–318

Storz P, Toker A (2002) 3′-phosphoinositide-dependent kinase-1 (PDK-1) in PI 3-kinase signaling. Front Biosci 7:d886-d902

Tanaka H, Dhonukshe P, Brewer PB, Friml J (2006) Spatiotemporal asymmetric auxin distribution: a means to coordinate plant development. Cell Mol Life Sci 63:2738–2754

Theologis A (1986) Rapid Gene Regulation by Auxin. Ann Rev Plant Physiol 37:407–438

Theologis A, Huynh TV, Davis RW (1985) Rapid induction of specific mRNAs by auxin in pea epicotyl tissue. J Mol Biol 183:53–68

Tiwari SB, Hagen G, Guilfoyle TJ (2004) Aux/IAA proteins contain a potent transcriptional repression domain. Plant Cell 16:533–543

Trewavas AJ, Rodrigues C, Rato C, Malho R (2002) Cyclic nucleotides: the current dilemma! Curr Opin Plant Biol 5:425–429

Ulmasov T, Hagen G, Guilfoyle TJ (1999a) Activation and repression of transcription by auxin-response factors. Proc Natl Acad Sci USA 96:5844–5849

Ulmasov T, Hagen G, Guilfoyle TJ (1999b) Dimerization and DNA binding of auxin response factors. Plant J 19:309–319

Ulmasov T, Murfett J, Hagen G, Guilfoyle TJ (1997) Aux/IAA proteins repress expression of reporter genes containing natural and highly active synthetic auxin response elements. Plant Cell 9:1963–1971

van der Zaal E, Memelink J, Mennes A, Quint A, Libbenga K (1987) Auxin-induced mRNA species in tobacco cell cultures. Plant Mol Biol 10:145–157

Vanhaesebroeck B, Leevers SJ, Ahmadi K, Timms J, Katso R, Driscoll PC, Woscholski R, Parker PJ, Waterfield MD (2001) Synthesis and function of 3-phosphorylated inositol lipids. Annu Rev Biochem 70:535–602

von Arnim AG, Osterlund MT, Kwok SF, Deng XW (1997) Genetic and developmental control of nuclear accumulation of COP1, a repressor of photomorphogenesis in *Arabidopsis*. Plant Physiol 114:779–788

Weijers D, Benkova E, Jager KE, Schlereth A, Hamann T, Kientz M, Wilmoth JC, Reed JW, Jurgens G (2005a) Developmental specificity of auxin response by pairs of ARF and Aux/IAA transcriptional regulators. EMBO J 24:1874–1885

Weijers D, Jurgens G (2004) Funneling auxin action: specificity in signal transduction. Curr Opin Plant Biol 7:687–693

Weijers D, Sauer M, Meurette O, Friml J, Ljung K, Sandberg G, Hooykaas P, Offringa R (2005b) Maintenance of embryonic auxin distribution for apical-basal patterning by PIN-FORMED-dependent auxin transport in *Arabidopsis*. Plant Cell 17:2517–2526

Went F, Thimann K (1937) Phytohormones. MacMillan, New York

Went FW (1926) On growth accelerating substances in the coleoptile of *Avena sativa*. Proc K Akad Wet Amsterdam 30:10–19

Wisniewska J, Xu J, Seifertova D, Brewer PB, Ruzicka K, Blilou I, Roquie D, Benkova E, Scheres B, Friml J (2006) Polar PIN localization directs auxin flow in plants. Science 312:883

Xu J, Scheres B (2005) Cell polarity: ROPing the ends together. Curr Opin Plant Biol 8:613–618

Yang Y, Hammes UZ, Taylor CG, Schachtman DP, Nielsen E (2006) High-affinity auxin transport by the AUX1 influx carrier protein. Curr Biol 16:1123–1127

Zago MK Components and targets of the PINOID signaling complex in *Arabidopsis thaliana*. 2006. Leiden University, Thesis

Zegzouti H, Anthony RG, Jahchan N, Bogre L, Christensen SK (2006a) Phosphorylation and activation of PINOID by the phospholipid signaling kinase 3-phosphoinositide-dependent protein kinase 1 (PDK1) in *Arabidopsis*. Proc Natl Acad Sci USA

Zegzouti H, Li W, Lorenz TC, Xie M, Payne CT, Smith K, Glenny S, Payne GS, Christensen SK (2006b) Structural and functional insights into the regulation of arabidopsis AGC VIIIa kinases. J Biol Chem 103:6404–6409

Plant Cell Monogr (10)
L. Bögre and G. Beemster: Plant Growth Signaling
DOI 10.1007/7089_2007_141/Published online: 25 July 2007

Brassinosteroid Signaling

Steven D. Clouse

Department of Horticultural Science, North Carolina State University,
Raleigh, NC 27695-7609, USA
steve_clouse@ncsu.edu

Abstract Brassinosteroids (BRs) regulate multiple aspects of plant growth and development and require membrane-localized receptor kinases for BR perception and signal transduction. Recent biochemical and genetic analyses have demonstrated that these plant proteins share many features of the general paradigm of animal receptor kinase function, including ligand-dependent heterodimerization and phosphorylation on specific residues of the cytoplasmic kinase domain. These studies have enhanced our understanding of early events in BR signaling as well as general modes of action for plant receptor kinases. Detailed studies of downstream events in BR signal transduction have now clarified nuclear events associated with the terminal end of BR signaling and have provided novel mechanisms of BR-regulated gene expression. This chapter summarizes the current status of BR signaling knowledge, focusing on these recent developments.

1 Introduction

BRs are steroidal plant hormones that are essential regulators of the expansion, division, and differentiation of plant cells. Functional BR signal transduction is required for normal expression of developmental pathways leading to organ elongation, vascular differentiation, leaf expansion, seed germination, and responses to the environment. The structural characterization in 1979 of the most active endogenously occurring BR, brassinolide (BL), demonstrated for the first time that a steroid signal could modulate the growth and development of plants at nanomolar concentrations, analogous to the profound effect of steroid hormones on animal development (Grove et al. 1979). Physiological experiments in the 1980s showed that under specific cultural conditions, exogenous application of BRs could have dramatic effects on overall plant growth, fruit set, grain filling, and adaptation to stress (Ikekawa and Zhao 1991; Kamuro and Taktsuto 1999). The subsequent identification of BR-deficient mutants in Arabidopsis, rice, tomato and pea with dramatic developmental phenotypes, coupled with the demonstration that these mutant phenotypes could be rescued by exogenous BR application, provided genetic confirmation of the absolute requirement of BRs for normal plant development (Bishop et al. 1996; Li et al. 1996; Szekeres et al. 1996; Nomura et al. 1997; Koka et al. 2000; Hong et al. 2002).

Research on BR signal transduction began in the early 1990s with studies of BR-regulated gene expression and the identification of the *brassinosteroid-insensitive 1* mutant (*bri1*), which exhibited an extremely dwarfed phenotype, dark-green curled leaves, reduced male fertility, delayed senescence and reduced apical dominance (Clouse et al. 1993, 1996; Kauschmann et al. 1996). The phenotype of *bri1* suggested a critical role for the *BRI1* gene in BR signaling. Cloning of the gene in Arabidopsis revealed that *BRI1* encoded a leucine-rich repeat receptor-like kinase (LRR RLK), consistent with a possible role for this protein as the BR receptor (Li and Chory 1997). Subsequent work confirmed BRI1 as a plant steroid receptor with obvious structural differences from the widely studied nuclear steroid receptor superfamily found in animals (Wang et al. 2001; Kinoshita et al. 2005). Thus, plants and animals both require steroid signaling molecules for normal development but present strikingly different pathways for steroid signaling.

Work in several laboratories during the past decade has provided finer details of BRI1 action, identified BRI1-interacting proteins, revealed several downstream regulators required for propagation of the BR signal, and cataloged numerous BR-regulated genes. BR signaling has been extensively reviewed recently (Wang and He 2004; Li 2005; Vert et al. 2005; Belkhadir and Chory 2006; Li and Jin 2007) and this chapter will provide an overview of each component of the BR signal transduction pathway, emphasizing studies in the past few years that have expanded our understanding of the molecular mechanisms of BR signaling and its role in plant development.

2 The Role of Receptor Kinases in BR Signaling

Examination of the completed genome sequences of Arabidopsis and rice reveals very large multigene families predicted to encode proteins with an organization of functional domains similar to that of animal receptor kinases, such as the mammalian insulin and epidermal growth factor receptor tyrosine kinases and the transforming growth factor β (TGF-β)receptor serine/threonine kinases (Shiu et al. 2004). The three most prominent structural features of this class of proteins are a putative extracellular ligand-binding domain, a single-pass transmembrane sequence, and an intracellular cytoplasmic domain consisting of a juxtamembrane region, a catalytic kinase domain with 12 conserved subdomains, and a short carboxy terminal region. Such structural features suggest a role for receptor kinases in extracellular signal perception followed by intracellular signal transduction via phosphorylation of specific targets.

More than 220 of these receptor-like kinases in Arabidopsis and nearly 400 in rice have an extracellular domain with multiple leucine-rich repeats and are classified as LRR RLKs (Shiu and Bleecker 2001, 2003; Shiu et al. 2004).

Only a small subset of LRR RLKs have been characterized in any detail by genetic and biochemical analyses, but those that have been so studied are known to play essential roles in regulating plant growth and development, as well as in defense responses to various pathogens (Torii 2004). Examples of Arabidopsis LRR RLKs that have been well characterized include CLAVATA1, controlling meristematic cell fate (Clark et al. 1997; Trotochaud et al. 1999); ERECTA, specifying organ shape (Torii et al. 1996); HAESA, involved in organ abscission (Jinn et al. 2000); FLS2, which binds the flagellin peptide ligand (Gomez-Gomez and Boller 2000); and AtSERK1 (SOMATIC EMBRYOGENESIS RECEPTOR KINASE), associated with early embryogenesis (Hecht et al. 2001). Besides the critical role of BRI1 in BR signaling, another LRR RLK, BRI1-ASSOCIATED RECEPTOR KINASE 1 (BAK1), has been shown to associate with BRI1 in planta and modulate BR signaling outputs (Li et al. 2002; Nam and Li 2002).

To fully characterize LRR RLK function in BR signaling, it is essential to understand the role of ligand-dependent BRI1/BAK1 dimerization and cytoplasmic domain phosphorylation, including identification of specific in vivo phosphorylation sites and their functional significance. Moreover, identification of BRI1 and BAK1 kinase domain substrates and cataloging of structural requirements for substrate recognition and phosphorylation are also necessary for completing the picture of LRR RLK action in BR signal transduction. Several studies published in the last few years have addressed these issues and suggest that plant LRR RLKs such as BRI1 and BAK1 share several conserved features with the mechanism of animal receptor kinase action, including ligand-dependent oligomerization, followed by phosphorylation-dependent activation of the kinase domain, leading ultimately to changes in the expression level of numerous BR-responsive genes (Nam and Li 2004; Ehsan et al. 2005; Wang et al. 2005a,b; Wang and Chory 2006).

2.1 BRI1 is the BR Receptor

The BRI receptor kinase has been shown by mutational analysis in Arabidopsis, rice, tomato and pea to be absolutely required for normal BR perception and plant growth (Clouse et al. 1996; Kauschmann et al. 1996; Li and Chory 1997; Noguchi et al. 1999; Friedrichsen et al. 2000; Koka et al. 2000; Yamamuro et al. 2000; Montoya et al. 2002; Nomura et al. 2003). Arabidopsis BRI (a member of LRR RLK subfamily X) is an 1196 amino acid protein with a 791 amino acid extracellular domain containing more than 20 LRR motifs, followed by a predicted hydrophobic transmembrane domain spanning amino acids 792–814 (Fig. 1). The cytoplasmic portion of BRI1 consists of a juxtamembrane region (amino acids 815–882), followed by a Ser/Thr kinase domain (amino acids 883-1155) with a final short sequence of 41 amino acids comprising the carboxy-terminal domain. The large number of *bri1* mutant alleles available

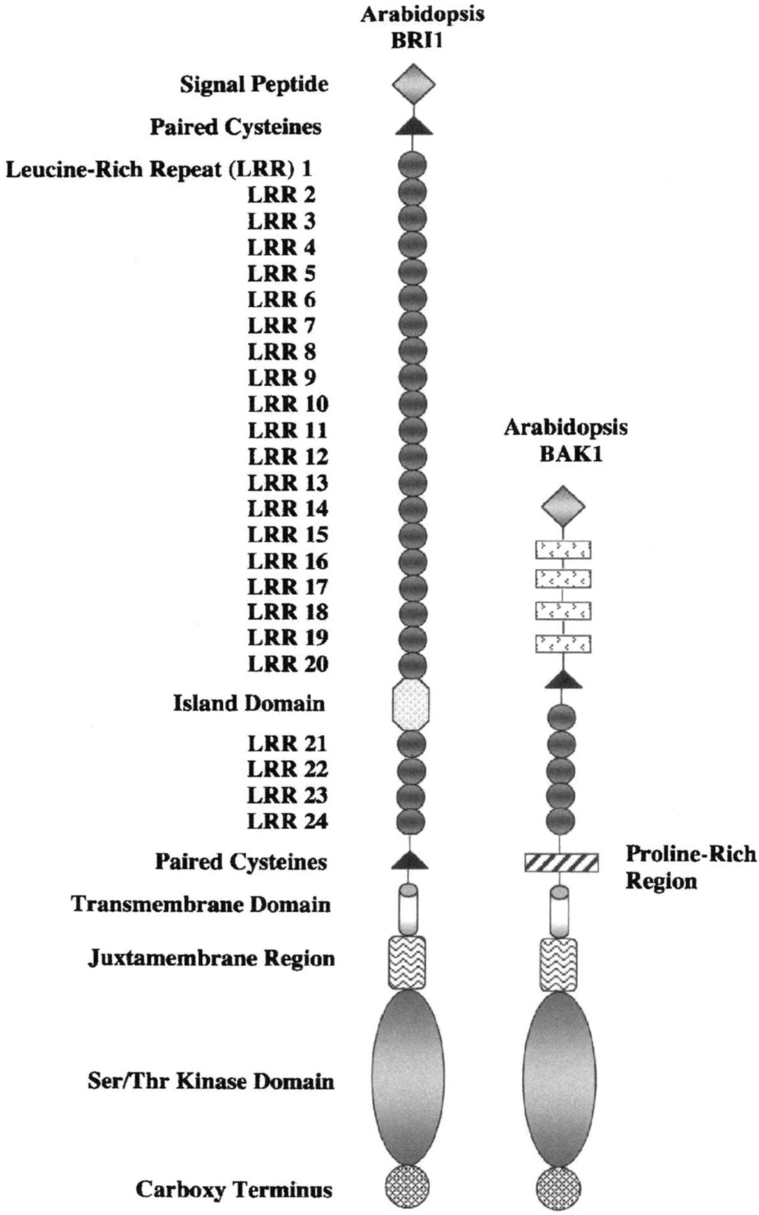

Fig. 1 Structure of two LRR RLKs involved in BR signaling. BRI1 and BAK1 form a ligand-dependent heterodimer in cell membranes and are both involved in BR signaling. All segments of the protein N-terminal to the transmembrane domain are extracellular, while those portions C-terminal of the transmembrane domain lie in the cytoplasm. Annotation of BRI1 is based on Vert et al. 2005; and BAK1 on Nam and Li 2002, and Li et al. 2002

clearly shows the biological significance of both the extracellular and kinase domains. The original annotation of Arabidopsis BRI1 called for 25 tandem LRRs with an average length of 24 amino acids, with a 70 amino acid "island domain" embedded between LRR 21 and 22 (Li and Chory 1997). A recent re-annotation suggests there are only 24 LRRs with the island domain lying between LRR 20 and 21 (Vert et al. 2005). BRI1 has been cloned in several crop species and the number of LRR motifs varies between 22 and 25 with an island domain of 68–70 amino acids inserted four LRRs upstream of the membrane-spanning region in all BRI1 sequences examined (Bishop 2003).

Direct biochemical approaches using radiolabeled BL and transgenic plants overexpressing BRI1-GFP fusions demonstrated the role of BRI1 as at least one component of the BR receptor. The BL-binding activity was precipitated by antibodies to GFP, was competitively inhibited by active, but not by inactive, BRs, and was of an affinity ($K_d = 7.4 \pm 0.9$ nM) consistent with known physiological concentrations of BR required to regulate physiological responses in planta (Wang et al. 2001). Subsequent experiments using a photo-affinity labeled BR analog showed that BR binds directly to BRI1 without an intermediate accessory protein (Kinoshita et al. 2005). LRRs are involved in protein–protein interaction and are generally not known for binding small molecules such as steroids, suggesting that BR binding to BRI1 might be through the island domain embedded in the LRR region. This was experimentally confirmed by using radiolabeled BL and recombinant proteins consisting of the island domain with various combinations of adjacent LRRs. The 70 amino acid island domain in conjunction with the adjacent C-terminal LRR (LRR 21 or 22, depending on annotation) was necessary and sufficient for BL binding and thus defines a novel 94 amino acid steroid-binding motif distinctly different in structure to steroid-binding sequences in animals (Kinoshita et al. 2005). The experiments described above clearly show that BRI1 is a BR receptor. Interestingly, based on biochemical evidence tomato BRI1 has been proposed to have a dual function, binding both BR and the peptide hormone systemin, involved in wound response signaling (Montoya et al. 2002; Szekeres 2003; Boller 2005). Systemin is not found in Arabidopsis, although several other Arabidopsis LRR RLKs are known to have small peptides as ligands (Matsubayashi and Sakagami 2006).

Recent evidence suggests that BL may bind to a preformed ligand-independent BRI1 homodimer in planta. Russinova et al. (2004) found that BRI1-CFP and BRI1-YFP interacted in plant protoplasts using fluorescence lifetime imaging microscopy (FLIM) and Forster resonance energy transfer (FRET) analysis. Moreover, Wang et al. (2005a) found that BRI-CFP and BRI-Flag co-immunoprecipitated in transgenic Arabidopsis plants both in the presence and absence of BR. The addition of BR appeared to stabilize the preformed BRI1 homodimer. Expression of a kinase-inactive BRI1 mutant in either *bri1-5* or wild-type plants leads to a dominant negative phenotype, further suggesting BRI1–BRI1 interaction in vivo.

In animal receptor kinases, activation of the cytoplasmic kinase domain by phosphorylation of specific residues is a consequence of ligand binding to the extracellular domain. The activation of many protein kinases occurs by autophosphorylation of one to three residues within the activation loop of subdomain VII/VIII, which is thought to allow substrate access to the catalytic site in subdomain VIb (Johnson et al. 1996). Further phosphorylation of multiple residues in the juxtamembrane and carboxy-terminal domains generates docking sites for binding downstream kinase substrates in the specific signaling pathway. Immunoprecipitation of BRI-Flag from 11-day old light-grown Arabidopsis plants treated with or without BL, followed by immunoblot analysis with anti-phosphoThr antibody, demonstrated that BRI1 phosphorylation (at least on Thr residues) was BL-dependent in planta (Wang et al. 2005b). The use of a more general phosphorylation stain also suggested that at least one Ser residue was constitutively phosphorylated. Using the same experimental system, immunoprecipitation of BRI1-Flag followed by liquid chromatography tandem mass spectrometry (LC/MS/MS) showed that at least 11 Ser or Thr residues were phosphorylated in vivo in BL-treated Arabidopsis plants, including six sites in the juxtamembrane region, one in the carboxy terminal domain and four in the kinase catalytic domain (Wang et al. 2005b). At least three residues in the subdomain VII/VIII activation loop were phosphorylated in vivo. Most of the BRI1 in vivo phosphorylation sites had been previously identified in vitro using recombinant BRI1 cytoplasmic domain expressed in bacteria (Oh et al. 2000). Therefore, at least in the case of BRI1, the in vitro autophosphorylation sites were highly predictive of in vivo phosphorylation.

The functional significance of each of the identified and predicted phosphorylation sites in Arabidopsis BRI1 was assessed by site-directed mutagenesis of each specific Ser or Thr to Ala followed by biochemical analysis in vitro and testing for the ability of the altered construct to rescue the weak *bri1-5* BR-insensitive mutant in planta. The mutations T-1049-A and S-1044-A in the kinase domain activation loop nearly abolished kinase activity, with respect to both autophosphorylation and peptide substrate phosphorylation. Moreover, these constructs failed to rescue *bri1-5* and resulted in a dominant negative effect (Fig. 2), similar to transformation with a kinase inactive mutant (Wang et al. 2005b). Thus, BRI1 appears to share the activation mechanism of many animal kinases by requiring phosphorylation within the activation loop for kinase function. Autophosphorylation within this region is also likely to occur in other plant RLKs. Sequence alignment of RLKs reveals that the activation loop is highly conserved including several Ser and Thr residues that are routinely present. In fact, the residue corresponding to T-1049 in the BRI1 activation loop is nearly invariant, being either a Ser or Thr in 100 of the most closely related Arabidopsis LRR RLKs. Juxtamembrane substitutions did not result in appreciable differences in autophosphorylation compared with wild-type constructs, but resulted in

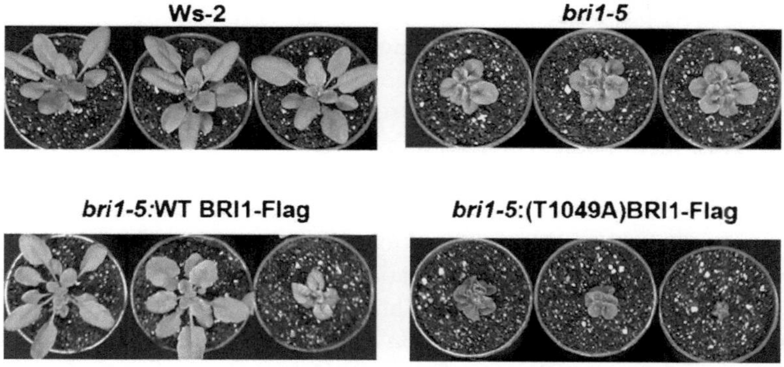

Fig. 2 Effect of mutating BRI1 kinase domain activation loop residue Thr-1049 on rescue of the *bri1-5* mutant. Transgenic constructs contained 1699 bp of 5′ upstream BRI1 sequence (covering the BRI1 promoter and 5′ UTR), the entire coding region, and an in-frame C-terminal epitope tag (WT BRI1-Flag). Three independent transgenic lines for each construct are shown. All lines were grown under the same conditions and are the same age (65 days). Adapted with permission from Fig. 8 of Wang et al. 2005b

74–88% reduction in phosphorylation of a synthetic peptide substrate when compared to the non-mutated kinase control (Wang et al. 2005b). These findings are consistent with the model that autophosphorylation of the activation loop is required for general kinase activity, while autophosphorylation of juxtamembrane and carboxy terminal residues either generates docking sites for specific downstream substrates or affects catalytic activity towards those substrates.

Besides generating docking sites, phosphorylation of juxtamembrane and carboxy terminal domains may also lead to a general activation of the kinase catalytic domain by a variety of mechanisms (Pawson 2002). An inhibitory affect of the C-terminal domain on BRI1 kinase activity was demonstrated both in vitro and in planta (Wang et al. 2005a). Deletion of the C-terminal domain in BRI1-Flag constructs in transgenic plants led to a hypersensitive BR phenotype including elongated hypocotyls, expanded leaves, and elongated petioles. Deletion of the C-terminus also enhanced BRI1 kinase activity in vitro and in planta as did substitution of specific Ser and Thr residues in the C-terminus with Asp, which can mimic constitutive phosphorylation. This data suggests that BL binding to BRI1 results in phosphorylation of the C-terminal region of the cytoplasmic domain, leading to a release of inhibition and subsequent BRI1 kinase activation, most likely via phosphorylation in the activation loop (Wang et al. 2005a). However, another study in which specific Ser and Thr residues in the BRI1 C-terminal domain were substituted with Ala, which prevents phosphorylation at those residues, showed very little effect on general kinase activity or in planta function (Wang et al. 2005b). Thus, the precise mechanism by which the C-terminus regulates BRI1 function requires further examination.

BRI1 activity plays a critical role in plant development and the biochemical mechanisms of its ligand-dependent activation are beginning to be clarified, as described above. Several novel reports on the impact of BRI1 on overall plant development and physiology also recently appeared. Limiting expression of BRI1 to the epidermal layer by transforming a *bri1* null mutant in Arabidopsis with a full-length construct driven by an L1-specific promoter, resulted in nearly complete rescue of several developmental defects including hypocotyl length and leaf and petiole size (Savaldi-Goldstein et al. 2007). However, defects in vascular organization could only be fully rescued by expression of BRI1 in ground layers. Thus, the role of BRI1 in growth can be uncoupled from vascular organization and differentiation. Several new mutant alleles of the rice BRI1 ortholog, OsBRI1, were identified and functional characterization suggested that OsBRI1 is critical for organ development by controlling cell expansion and division, but that it is not essential for organ initiation or pattern development in rice (Nakamura et al. 2006b). A second study in rice revealed potential practical agricultural applications of regulating BRI1 expression. A partial suppression of OsBRI1 expression in transgenic rice gave plants with an erect-leaf phenotype that led to 30% increases in yield compared to control when planted at high densities (Morinaka et al. 2006).

2.2 BAK1 Forms a Heterodimer with BRI1 and Enhances BR Signaling

BAK1 is a member of the LRRII RLK subfamily and was identified independently by activation tagging for suppressors of *bri1-5* (Li et al. 2002) and by a yeast two-hybrid screen for BRI1 interacting proteins using the BRI1 cytoplasmic domain as bait (Nam and Li 2002). The extracellular domain of BAK1 lacks the embedded island domain and has only five LRR motifs, differing markedly from BRI1 (Fig. 1). Therefore, it is highly unlikely that BAK1 binds BL directly and it has also been shown that BRI1 can bind BL in a BAK1 mutant background (Kinoshita et al. 2005). Genetic analysis has demonstrated a role for BAK1 in BR signaling, and overexpression of a kinase-inactive mutant form of *BAK1* in *bri1-5* led to a dominant-negative effect, most likely arising by the disruption of a heterodimeric complex between BRI1 and BAK1 (Li et al. 2002). Direct physical interaction between BRI1 and BAK1 has been demonstrated repeatedly (Li et al. 2002, Nam and Li 2002, Russinova et al. 2004), and co-immunoprecipitation experiments in transgenic plants expressing both BRI1-Flag and BAK1-GFP showed that this association was BR-dependent (Wang et al. 2005b). Using the same experimental system as described above for BRI1, it was also demonstrated that the in vivo phosphorylation of BAK1 on Thr residues was BL-dependent (Fig. 3).

The finding that BRI1 forms a heterodimer with BAK1 in a ligand-dependent manner supports the hypothesis that BR signaling shares some

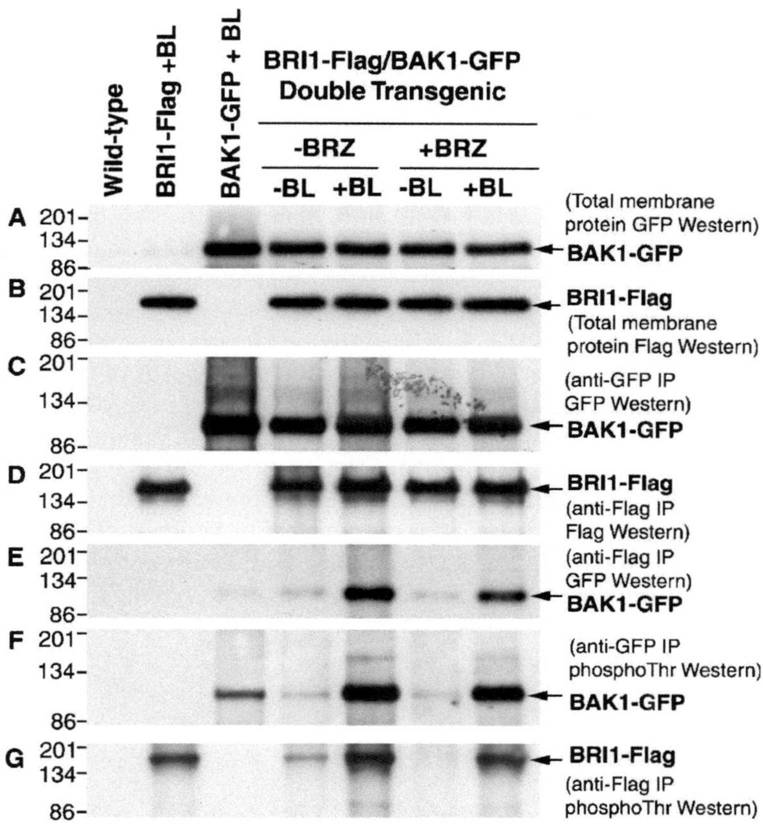

Fig. 3 BL-dependence of early events in BR signaling. Transgenic Arabidopsis plants expressing both BRI1-Flag and BAK1-GFP were grown in shaking liquid culture in the light for 6 days. Half of the flasks were then treated with the BR biosynthesis inhibitor brassinazole (+BRZ), for 5 additional days to reduce endogenous BR levels. Plants were then treated with 100 nM brassinolide (+BL) or solvent (–BL) for 90 min. Total membrane protein was purified from each sample and BRI-Flag or BAK1-GFP were immunoprecipitated from the solubilized membranes. Equal amounts of protein were separated by SDS-PAGE and Western blot analysis was performed as indicated. *A–D* Approximately equal amounts of protein were present in each treatment. *E* Association of BRI1 and BAK1 in vivo, as determined by co-immunoprecipitation is BL-dependent. (*F*, *G*) Phosphorylation of Thr residues in BRI1-FLAG and BAK1-GFP is BL-dependent. Adapted with permission from Fig. 1 of Wang et al. 2005b

mechanistic similarities to mammalian receptor tyrosine kinase and TGF-β receptor kinase signaling. In mammals, the TGF-β family of polypeptides regulate multiple aspects of development and are perceived at the cell surface by a complex of Type I (RI) and Type II (RII) TGF-β receptor serine/threonine kinases. TGF-β RII homodimerizes in a ligand-independent manner and exhibits constitutive kinase activity. TGF-β binding by RII induces formation of the heterotetramer with RI and results in phosphorylation of RI by RII on specific Thr and Ser resides. Phosphorylated RI then propagates the signal by phosphorylating substrates, termed Smads, which translocate to the nucleus where they associate with transcription factors to regulate the expression of TGF-β-responsive genes (Massague 1998). The RII receptor also phosphorylates specific cytoplasmic substrates, including TGF-β receptor interacting protein 1 (TRIP-1), a WD-40 domain protein that has dual functions as a modulator of TGF-β regulated gene expression and as an essential subunit of the eukaryotic translation initiation factor, eIF3 (Chen et al. 1995; Choy and Derynck 1998). The recent demonstration of several BR signaling features, including BRI1 homodimerization (Russinova et al. 2004; Wang et al. 2005a), the binding of BL directly to BRI1 but not BAK1, the ligand-dependent in vivo association of BRI1 and BAK1 (Wang et al. 2005b), and, as discussed below, the fact that the Arabidopsis ortholog of mammalian TRIP-1 is a putative in vivo substrate of BRI1 (Ehsan et al. 2005), suggest intriguing parallels between BR and TGF-β signaling. However, there are numerous features that are not consistent, including phosphorylation of BRI1 by BAK1, lack of Smad orthologs as BAK1 substrates, and significantly different ligand and extracellular domain structures. Thus, BR signaling shares some of the signaling logic of the TGF-β pathway without a direct evolutionary relationship.

2.3
SERK1 is Part of the BRI1 Signaling Complex

SERK1, like BAK1, is a member of the LRRII subfamily and shares 80% amino acid sequence identity with BAK1. Immunoprecipitation of cyan fluorescent protein-tagged SERK1 from transgenic Arabidopsis plants, followed by LC/MS analysis of the immunoprecipitate, revealed that both BRI1 and BAK1 co-immunoprecipitated with SERK1 (Karlova et al. 2006). The direct in vivo physical association of these proteins was further confirmed by FLIM and FRET analysis in protoplasts. Genetic evidence for a role of SERK1 in BR signaling was demonstrated by crossing the *serk1-1* mutant allele that results in a kinase inactive SERK1 protein, with a weak allele of *BRI1*, *bri1-119*. The *serk1-1/bri1-119* double mutant had shorter petioles, reduced rosette size, and shorter inflorescences compared to *bri1-119*. This suggests a role for SERK1 in BR signaling by modulating BRI1 action in a manner similar to, although not quite as strong, as BAK1. To determine if SERK1, BAK1, and BRI1 all associated together at the same time, as opposed to multiple forms of SERK1 complexed to either BRI1 or BAK1 in the immunoprecipitate, blue native electrophoresis was employed to demonstrate a core complex of SERK1, BAK1, and BRI1 of approximately 350 kDa (Karlova et al. 2006). Three other LRRII subfamily members share high sequence similarity to BAK1 and SERK1, and it would be of interest to determine whether any of these can also interact with BRI1. Such variable multimeric complexes could work to fine-tune BRI1 function in different growth conditions.

3 BRI1 Substrates and Interacting Proteins

Studies addressing the number and nature of cytoplasmic binding partners of the BRI1 and BAK1 kinase domains that propagate the signal downstream are essential for a complete understanding of BR action. Yeast two-hybrid screens were used to identify two probable in vivo substrates of BRI1 in Arabidopsis. TRANSTHYRETIN-LIKE protein (TTL) has substantial sequence identity to the vertebrate thyroid-binding protein transthyretin and interacts with BRI1 in yeast cells in a kinase-dependent manner (Nam and Li 2004). TTL is phosphorylated by recombinant BRI1 kinase domain in vitro and overexpression of the TTL gene results in a semi-dwarf phenotype similar to weak *bri1* and null *bak1* mutants, while null mutants of TTL enhance BR sensitivity and promote plant growth. Thus, genetic evidence suggests that TTL is a negative regulator of BR signaling, while in vitro evidence supports a role for TTL as a putative BRI1 substrate, although a direct interaction of the two proteins in planta has not been demonstrated (Nam and Li 2004).

A second interactor from the yeast two-hybrid screen, BKI1 (BRI1 KINASE INHIBITOR 1), is also a negative regulator of BR signaling. It is membrane-associated in the absence of BR and binds to BRI1, presumably inactivating some aspect of its function (Wang and Chory 2006). BR treatment causes dissociation of BKI1 from the membrane, which releases repression of the BR signaling pathway. Tethering BKI1 to the membrane by addition of a myristoylation site results in an enhanced BR dwarf phenotype. BRI1 and BKI1 interact directly in vitro and in planta and BKI1 is phosphorylated by recombinant BRI1 kinase domain in vitro and is a phosphoprotein in vivo. The favored model for BKI1 action suggests that membrane-associated BKI1 binds directly to the kinase domain of BRI1, preventing its association with BAK1. Binding of BL to the extracellular domain of BRI1 activates the kinase domain, leading to dissociation of BKI1 from the membrane and allowing BRI1 and BAK1 to heterodimerize to initiate BR signaling (Wang and Chory 2006).

The third putative cytoplasmic substrate of BRI1 is the plant ortholog of mammalian TRIP-1. LC/MS/MS approaches show that recombinant Arabidopsis TRIP-1 is strongly phosphorylated by the BRI kinase domain in vitro, predominately on three specific residues (Thr-14, Thr-89, and either Thr-197 or Ser-198). Moreover, BRI1 and TRIP-1 co-immunoprecipitate from Arabidopsis plant extracts using native antibodies or various combinations of tagged proteins, suggesting an in vivo interaction between the two proteins (Ehsan et al. 2005). Some of the morphological characteristics of transgenic lines expressing antisense TRIP-1 RNA are also consistent with a possible role for TRIP-1 in BR signaling, although this in itself is not definitive proof (Jiang and Clouse 2001). Thus, TRIP-1 is a likely candidate for a BRI1 cytoplasmic substrate, but its functional role in BR signaling remains unclear. TRIP-1 (also known as eIF3i) is a dual function protein that has also been shown to be an essential subunit of the eIF3 translation initiation complex in mammals, yeast, and plants (Asano et al. 1997; Burks et al. 2001). This finding raises the intriguing possibility that BR-dependent phosphorylation of TRIP-1 by BRI1 may affect eIF3 activity and/or assembly and thus impact the global cellular phenomenon of protein translation, providing a novel mechanism for BR regulation of plant growth.

4
Downstream Components of BR Signaling

Four components of the downstream segment of BR signaling have been characterized in some detail, including a kinase, a phosphatase and two novel transcription factors (Vert et al. 2005; Li and Jin 2007). The *brassinosteroid insensitive 2* (*bin2*) mutant is a semi-dominant gain-of-function allele that in the homozygous state closely resembles the dwarf *bri1* phenotype (Li and Nam 2002). *BIN2* encodes a Ser/Thr kinase with sequence similarity to the Drosophila shaggy kinase and mammalian glycogen synthase kinase 3 (GSK-3), which often function as negative regulators of signaling pathways controlling metabolism, cell fate determination, and pattern formation (Kim and Kimmel 2000). Loss-of-function alleles of *bin2* show no phenotype, but when combined as a triple mutant with null alleles of the two most closely related GSKs in Arabidopsis, a constitutive BR phenotype is observed similar to the overexpression of BRI1 (Vert and Chory 2006). Thus, BIN2 is a negative regulator of BR signaling and considerable evidence suggests that it functions by phosphorylating two transcription factors that are required for the expression of BR-regulated genes.

The semidominant or dominant mutants, *bri1-ems-suppressor1* (*bes1-D*) and *brassinazole-resistant1* (*bzr1-D*), show constitutive brassinosteroid responses and the BES1 and BZR1 proteins share 88% amino acid identity (He et al. 2002; Wang et al. 2002; Yin et al. 2002). Recombinant BIN2 phosphorylates BES1 and BZR1 in vitro, and the in planta expression level of BIN2 is correlated with BES1 and BZR1 phosphorylation levels in vivo. BR treatment leads to rapid dephosphorylation of BES1 and BZR1, most likely by inactivating BIN2, allowing the BRI1 SUPPRESSOR 1 (BSU1) phosphatase (Mora-Garcia et al. 2004) to increase the levels of hypophosphorylated BES1 and BZR1 in the nucleus. Recent studies have shown that BES1 and BZR1 are novel transcription factors that bind to specific BR response elements in the promoters of BR-regulated genes, either as homodimers or heterordimers with other transcription factors (He et al. 2005; Yin et al. 2005). Recent microarray analyses have cataloged hundreds of BR-regulated genes in functional categories ranging from wall-modifying proteins to transcription factors (Goda et al. 2002, 2004; Mussig et al. 2002; Nemhauser et al. 2004, 2006; Vert et al. 2005, Nakamura et al. 2006a).

GSK-3/shaggy kinases play an important negative regulatory role in the animal Wingless/wnt signaling pathways by phosphorylating β-catenin, promoting its proteasome-dependent degradation. Ligand binding to a transmembrane receptor leads to GSK-3/shaggy kinase inactivation resulting in an accumulation of unphosphorylated β-catenin, which escapes degradation by the proteasome and translocates to the nucleus. There, it interacts with transcription factors to regulate the expression of genes essential for developmental pattern formation (Kim and Kimmel 2000). There is evidence that the phosphorylated forms of BES1/BZR1 are subject to proteasome-mediated degradation. A parallel mechanism for BR signaling was proposed in which BR binding to BRI1 initiated a signaling cascade that inactivated BIN2 kinase by an unknown mechanism (Wang et al. 2002; Yin et al. 2002; He et al. 2005). This allowed the accumulation of hypophosphorylated BES1/BZR1, which then translocated to the nucleus to activate BR-regulated gene transcription.

Recent studies, however, have suggested an alternative mechanism. In this model, BES1 is constitutively nuclear-localized and is phosphorylated by BIN2 and dephosphorylated by BSU1 in the nucleus. Proteasome-mediated degradation of the hyperphosphorylated form is proposed to be important for protein turnover but not for BR signaling. Instead, the hyperphosphorylated form of BES1 is inactive because of its inability to bind to BR response elements in the promoters of BR-regulated genes. The rapid loss of phosphorylation on BES1/BZR1 after BR binding to BRI1, allows DNA binding and the activation of BR-regulated gene expression (Vert and Chory 2006).

5 Concluding Remarks

The pace of recent discoveries in BR signaling has been rapid and informative, as summarized in Fig. 4. Only 3 years ago several unanswered but critical questions regarding BR signaling could be posed, such as:

1. Do BRI and BAK1 form true co-receptors in vivo?
2. Are accessory steroid-binding proteins required for BR binding to BRI1 and/or BAK1?
3. What are the in vivo autophosphorylation sites of BRI1 and BAK1, and what are their BR dependencies?
4. What are the number and nature of cytoplasmic binding partners of the BRI1 and BAK1 kinase domains?

5. What is the mechanism of BR-dependent inactivation of BIN2 kinase activity?
6. How do BES and BZR1 participate in the regulation of specific genes?

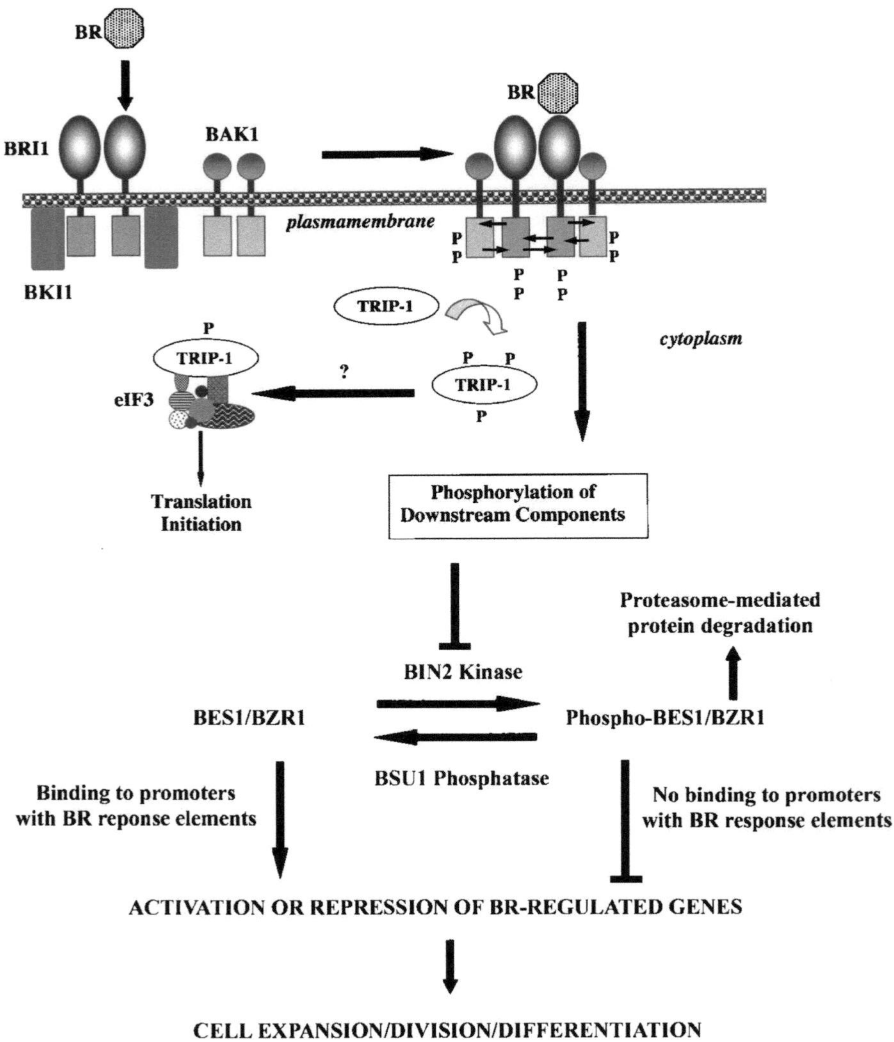

Fig. 4 Summary of components and interactions in BR signaling. See text for discussion of each step in the pathway. The nucleus is not shown due to some uncertainty in the localization of some downstream components. BR-induced transport of hypophosphorylated BES1/BZR1 from the cytoplasm to the nucleus has been proposed as has, more recently, constitutive localization of BES1/BZR1 in the nucleus where phosphorylation and dephosphorylation occur through the action of BIN2 kinase and BSU1 phosphatase

Biochemical approaches, such as immunoprecipitation, LC/MS/MS analysis, FLIM and FRET localization studies, and photo-affinity labeling have clearly shown that BL binds directly to the extracellular domain of BRI1, that BRI1 and BAK1 interact in a ligand-dependent manner and are phosphorylated on specific Ser and Thr residues in response to BR. Detailed analysis of BES and BZR1 action has shown that the hypophosphorylated forms of these novel transcription factors bind to specific BR responses elements. Extensive microarray studies have revealed hundreds of BR-regulated genes for further study. Both genetic screens and biochemical approaches have been useful in characterizing three initial substrates of the BRI1 kinase domain. However, a major gap in our understanding of BR signal transduction lies in the events that follow BRI1/BAK1 heterodimerization and phosphorylation and precede inactivation of the BIN2 kinase. Targets of BRI1/BAK1 phosphorylation that propagate the signal downstream need to be identified and approaches employing proteomic techniques, including LC/MS/MS analysis coupled with isotope coded affinity tagging procedures, may be the most global approach to identifying the full spectrum of BRI1 and BAK1 cytoplasmic substrates.

Acknowledgements Work in the author's laboratory is supported by the US National Science Foundation (MCB-0419819), the US Department of Agriculture (NRI 2004-35304-14930), and the North Carolina Agricultural Research Service.

References

Asano K, Kinzy TG, Merrick WC, Hershey JW (1997) Conservation and diversity of eukaryotic translation initiation factor eIF3. J Biol Chem 272:1101–1109

Belkhadir Y, Chory J (2006) Brassinosteroid signaling: a paradigm for steroid hormone signaling from the cell surface. Science 314:1410–1411

Bishop G, Harrison K, Jones J (1996) The tomato *Dwarf* gene isolated by heterologous transposon tagging encodes the first member of a new cytochrome P450 family. Plant Cell 8:959–969

Bishop GJ (2003) Brassinosteroid mutants of crops. J Plant Growth Regul 22:325–335

Boller T (2005) Peptide signalling in plant development and self/non-self perception. Curr Opin Cell Biol 17:116–122

Burks EA, Bezerra PP, Le H, Gallie DR, Browning KS (2001) Plant initiation factor 3 subunit composition resembles mammalian initiation factor 3 and has a novel subunit. J Biol Chem 276:2122–2131

Chen RH, Miettinen PJ, Maruoka EM, Choy L, Derynck R (1995) A WD-domain protein that is associated with and phosphorylated by the type II TGF-beta receptor. Nature 377:548–552

Choy L, Derynck R (1998) The type II transforming growth factor (TGF)-beta receptor-interacting protein TRIP-1 acts as a modulator of the TGF-beta response. J Biol Chem 273:31455–31462

Clark S, Williams R, Meyerowitz E (1997) The *CLAVATA1* gene encodes a putative receptor kinase that controls shoot and floral meristem size in *Arabidopsis*. Cell 89:575–585

Clouse SD, Langford M, McMorris TC (1996) A brassinosteroid-insensitive mutant in *Arabidopsis thaliana* exhibits multiple defects in growth and development. Plant Physiol 111:671–678

Clouse SD, Hall AF, Langford M, McMorris TC, Baker ME (1993) Physiological and molecular effects of brassinosteroids on *Arabidopsis thaliana*. J Plant Growth Regul 12:61–66

Ehsan H, Ray WK, Phinney B, Wang X, Huber SC, Clouse SD (2005) Interaction of Arabidopsis BRASSINOSTEROID-INSENSITIVE 1 receptor kinase with a homolog of mammalian TGF-beta receptor interacting protein. Plant J 43:251–261

Friedrichsen DM, Joazeiro CA, Li J, Hunter T, Chory J (2000) Brassinosteroid-insensitive-1 is a ubiquitously expressed leucine-rich repeat receptor serine/threonine kinase. Plant Physiol 123:1247–1256

Goda H, Shimada Y, Asami T, Fujioka S, Yoshida S (2002) Microarray analysis of brassinosteroid-regulated genes in Arabidopsis. Plant Physiol 130:1319–1334

Goda H, Sawa S, Asami T, Fujioka S, Shimada Y, Yoshida S (2004) Comprehensive comparison of auxin-regulated and brassinosteroid-regulated genes in Arabidopsis. Plant Physiol 134:1555–1573

Gomez-Gomez L, Boller T (2000) FLS2: an LRR receptor-like kinase involved in the perception of the bacterial elicitor flagellin in Arabidopsis. Mol Cell 5:1003–1011

Grove MD, Spencer GF, Rohwedder WK, Mandava NB, Worley JF, Warthen JD, Steffens GL, Flippen-Anderson JL, Cook JC (1979) Brassinolide, a plant growth-promoting steroid isolated from *Brassica napus* pollen. Nature 281:216–217

He JX, Gendron JM, Yang Y, Li J, Wang ZY (2002) The GSK3-like kinase BIN2 phosphorylates and destabilizes BZR1, a positive regulator of the brassinosteroid signaling pathway in Arabidopsis. Proc Natl Acad Sci USA 99:10185–10190

He JX, Gendron JM, Sun Y, Gampala SS, Gendron N, Sun CQ, Wang ZY (2005) BZR1 is a transcriptional repressor with dual roles in brassinosteroid homeostasis and growth response. Science 307:1634–1638

Hecht V, Vielle-Calzada JP, Hartog MV, Schmidt ED, Boutilier K, Grossniklaus U, de Vries SC (2001) The Arabidopsis SOMATIC EMBRYOGENESIS RECEPTOR KINASE 1 gene is expressed in developing ovules and embryos and enhances embryogenic competence in culture. Plant Physiol 127:803–816

Hong Z, Ueguchi-Tanaka M, Shimizu-Sato S, Inukai Y, Fujioka S, Shimada Y, Takatsuto S, Agetsuma M, Yoshida S, Watanabe Y, Uozu S, Kitano H, Ashikari M, Matsuoka M (2002) Loss-of-function of a rice brassinosteroid biosynthetic enzyme, C-6 oxidase, prevents the organized arrangement and polar elongation of cells in the leaves and stem. Plant J 32:495–508

Ikekawa N, Zhao Y-J (1991) Application of 24-epibrassinolide in agriculture. In: Cutler H, Yokota T, Adam G (eds) Brassinosteroids chemistry, bioactivity & applications. ACS Symposium Series 474. American Chemical Society, Washington, DC, pp 280–291

Jiang J, Clouse SD (2001) Expression of a plant gene with sequence similarity to animal TGF-beta receptor interacting protein is regulated by brassinosteroids and required for normal plant development. Plant J 26:35–45

Jinn TL, Stone JM, Walker JC (2000) HAESA, an Arabidopsis leucine-rich repeat receptor kinase, controls floral organ abscission. Genes Dev 14:108–117

Johnson LN, Noble ME, Owen DJ (1996) Active and inactive protein kinases: structural basis for regulation. Cell 85:149–158

Kamuro Y, Taktsuto S (1999) Practical application of brassinosteroids in agricultural fields. In: Sakurai A, Yokota T, Clouse S (eds) Brassinosteroids: steroidal plant hormones. Springer, Tokyo, pp 223–241

Karlova R, Boeren S, Russinova E, Aker J, Vervoort J, de Vries S (2006) The Arabidopsis SOMATIC EMBRYOGENESIS RECEPTOR-LIKE KINASE1 protein complex includes BRASSINOSTEROID-INSENSITIVE1. Plant Cell 18:626–638

Kauschmann A, Jessop A, Koncz C, Szekeres M, Willmitzer L, Altmann T (1996) Genetic evidence for an essential role of brassinosteroids in plant development. Plant J 9:701–713

Kim L, Kimmel AR (2000) GSK3, a master switch regulating cell-fate specification and tumorigenesis. Curr Opin Genet Dev 10:508–514

Kinoshita T, Cano-Delgado A, Seto H, Hiranuma S, Fujioka S, Yoshida S, Chory J (2005) Binding of brassinosteroids to the extracellular domain of plant receptor kinase BRI1. Nature 433:167–171

Koka C, Cerny R, Gardner R, Noguchi T, Fujioka S, Takatsuto S, Yoshida S, Clouse S (2000) A putative role for the tomato genes *DUMPY* and *CURL-3* in brassinosteroid biosynthesis and response. Plant Physiol 122:85–98

Li J (2005) Brassinosteroid signaling: from receptor kinases to transcription factors. Curr Opin Plant Biol 8:526–531

Li J, Chory J (1997) A putative leucine-rich repeat receptor kinase involved in brassinsteroid signal transduction. Cell 90:929–938

Li J, Nam KH (2002) Regulation of brassinosteroid signaling by a GSK3/SHAGGY-like kinase. Science 295:1299–1301

Li J, Jin H (2007) Regulation of brassinosteroid signaling. Trends Plant Sci 12:37–41

Li J, Nagpal P, Vitart V, McMorris TC, Chory J (1996) A role for brassinosteroids in light-dependent development of *Arabidopsis*. Science 272:398–401

Li J, Wen J, Lease KA, Doke JT, Tax FE, Walker JC (2002) BAK1, an Arabidopsis LRR receptor-like protein kinase, interacts with BRI1 and modulates brassinosteroid signaling. Cell 110:213–222

Massague J (1998) TGF-beta signal transduction. Annu Rev Biochem 67:753–791

Matsubayashi Y, Sakagami Y (2006) Peptide hormones in plants. Annu Rev Plant Biol 57:649–674

Montoya T, Nomura T, Farrar K, Kaneta T, Yokota T, Bishop GJ (2002) Cloning the tomato curl3 gene highlights the putative dual role of the leucine-rich repeat receptor kinase tBRI1/SR160 in plant steroid hormone and peptide hormone signaling. Plant Cell 14:3163–3176

Mora-Garcia S, Vert G, Yin Y, Cano-Delgado A, Cheong H, Chory J (2004) Nuclear protein phosphatases with Kelch-repeat domains modulate the response to brassinosteroids in Arabidopsis. Genes Dev 18:448–460

Morinaka Y, Sakamoto T, Inukai Y, Agetsuma M, Kitano H, Ashikari M, Matsuoka M (2006) Morphological alteration caused by brassinosteroid insensitivity increases the biomass and grain production of rice. Plant Physiol 141:924–931

Mussig C, Fischer S, Altmann T (2002) Brassinosteroid-regulated gene expression. Plant Physiol 129:1241–1251

Nakamura A, Nakajima N, Goda H, Shimada Y, Hayashi K, Nozaki H, Asami T, Yoshida S, Fujioka S (2006a) Arabidopsis Aux/IAA genes are involved in brassinosteroid-mediated growth responses in a manner dependent on organ type. Plant J 45:193–205

Nakamura A, Fujioka S, Sunohara H, Kamiya N, Hong Z, Inukai Y, Miura K, Takatsuto S, Yoshida S, Ueguchi-Tanaka M, Hasegawa Y, Kitano H, Matsuoka M (2006b) The role of OsBRI1 and its homologous genes, OsBRL1 and OsBRL3, in rice. Plant Physiol 140:580–590

Nam KH, Li J (2002) BRI1/BAK1, a receptor kinase pair mediating brassinosteroid signaling. Cell 110:203–212

Nam KH, Li J (2004) The Arabidopsis transthyretin-like protein is a potential substrate of BRASSINOSTEROID-INSENSITIVE 1. Plant Cell 16:2406–2417

Nemhauser JL, Mockler TC, Chory J (2004) Interdependency of brassinosteroid and auxin signaling in Arabidopsis. PLoS Biology 2:e258

Nemhauser JL, Hong F, Chory J (2006) Different plant hormones regulate similar processes through largely nonoverlapping transcriptional responses. Cell 126:467–475

Noguchi T, Fujioka S, Choe S, Takatsuto S, Yoshida S, Yuan H, Feldmann K, Tax F (1999) Brassinosteroid-insensitive dwarf mutants of Arabidopsis accumulate brassinosteroids. Plant Physiol 121:743–752

Nomura T, Nakayama M, Reid JB, Takeuchi Y, Yokota T (1997) Blockage of brassinosteroid synthesis and sensitivity causes dwarfism in *Pisum sativum*. Plant Physiol 113:31–37

Nomura T, Bishop GJ, Kaneta T, Reid JB, Chory J, Yokota T (2003) The LKA gene is a BRASSINOSTEROID INSENSITIVE 1 homolog of pea. Plant J 36:291–300

Oh MH, Ray WK, Huber SC, Asara JM, Gage DA, Clouse SD (2000) Recombinant brassinosteroid insensitive 1 receptor-like kinase autophosphorylates on serine and threonine residues and phosphorylates a conserved peptide motif in vitro. Plant Physiol 124:751–766

Pawson T (2002) Regulation and targets of receptor tyrosine kinases. Eur J Cancer 38(Suppl 5):S3–10

Russinova E, Borst JW, Kwaaitaal M, Cano-Delgado A, Yin Y, Chory J, de Vries SC (2004) Heterodimerization and endocytosis of Arabidopsis brassinosteroid receptors BRI1 and AtSERK3 (BAK1). Plant Cell 16:3216–3229

Savaldi-Goldstein S, Peto C, Chory J (2007) The epidermis both drives and restricts plant shoot growth. Nature 446:199–202

Shiu S-H, Bleecker AB (2003) Expansion of the receptor-like kinase/pelle gene family and receptor-like proteins in Arabidopsis. Plant Physiol 132:530–543

Shiu SH, Bleecker AB (2001) Receptor-like kinases from Arabidopsis form a monophyletic gene family related to animal receptor kinases. Proc Natl Acad Sci USA 98:10763–10768

Shiu SH, Karlowski WM, Pan R, Tzeng YH, Mayer KF, Li WH (2004) Comparative analysis of the receptor-like kinase family in Arabidopsis and rice. Plant Cell 16:1220–1234

Szekeres M, Nemeth K, Koncz-Kalman Z, Mathur J, Kauschmann A, Altmann T, Redei GP, Nagy F, Schell J, Koncz C (1996) Brassinosteroids rescue the deficiency of CYP90, a cytochrome P450, controlling cell elongation and de-etiolation in Arabidopsis. Cell 85:171–182

Szekeres M (2003) Brassinosteroid and systemin: two hormones perceived by the same receptor. Trends Plant Sci 8:102–104

Torii K, Mitsukawa N, Oosumi T, Matsuura Y, Yokoyama R, Whittier R, Komeda Y (1996) The *Arabidopsis ERECTA* gene encodes a putative receptor protein kinase with extracellular leucine-rich repeats. Plant Cell 8:735–746

Torii KU (2004) Leucine-rich repeat receptor kinases in plants: structure, function, and signal transduction pathways. Int Rev Cytol 234:1–46

Trotochaud A, Hao T, Wu G, Yang Z, Clark S (1999) The CLAVATA1 receptor-like kinase requires CLAVATA3 for its assembly into a signalling complex that includes KAPP and a Rho-related protein. Plant Cell 11:393–405

Vert G, Chory J (2006) Downstream nuclear events in brassinosteroid signalling. Nature 441:96–100

Vert G, Nemhauser JL, Geldner N, Hong F, Chory J (2005) Molecular mechanisms of steroid hormone signaling in plants. Ann Rev Cell Dev Biol 21:177–201

Wang X, Chory J (2006) Brassinosteroids regulate dissociation of BKI1, a negative regulator of BRI1 signaling, from the plasma membrane. Science 313:1118–1122

Wang X, Li X, Meisenhelder J, Hunter T, Yoshida S, Asami T, Chory J (2005a) Autoregulation and homodimerization are involved in the activation of the plant steroid receptor BRI1. Dev Cell 8:855–865

Wang X, Goshe MB, Soderblom EJ, Phinney BS, Kuchar JA, Li J, Asami T, Yoshida S, Huber SC, Clouse SD (2005b) Identification and functional analysis of in vivo phosphorylation sites of the Arabidopsis BRASSINOSTEROID-INSENSITIVE1 receptor kinase. Plant Cell 17:1685–1703

Wang ZY, He JX (2004) Brassinosteroid signal transduction–choices of signals and receptors. Trends Plant Sci 9:91–96

Wang ZY, Seto H, Fujioka S, Yoshida S, Chory J (2001) BRI1 is a critical component of a plasma-membrane receptor for plant steroids. Nature 410:380–383

Wang ZY, Nakano T, Gendron J, He J, Chen M, Vafeados D, Yang Y, Fujioka S, Yoshida S, Asami T, Chory J (2002) Nuclear-localized BZR1 mediates brassinosteroid-induced growth and feedback suppression of brassinosteroid biosynthesis. Dev Cell 2:505–513

Yamamuro C, Ihara Y, Wu X, Noguchi T, Fujioka S, Takatsuto S, Ashikari M, Kitano H, Matsuoka M (2000) Loss of function of a rice brassinosteroid insensitive1 homolog prevents internode elongation and bending of the lamina joint. Plant Cell 12:1591–1606

Yin Y, Vafeados D, Tao Y, Yoshida S, Asami T, Chory J (2005) A new class of transcription factors mediates brassinosteroid-regulated gene expression in Arabidopsis. Cell 120:249–259

Yin Y, Wang ZY, Mora-Garcia S, Li J, Yoshida S, Asami T, Chory J (2002) BES1 accumulates in the nucleus in response to brassinosteroids to regulate gene expression and promote stem elongation. Cell 109:181–191

Plant Cell Monogr (10)
L. Bögre and G. Beemster: Plant Growth Signaling
DOI 10.1007/7089_2007_151/Published online: 26 January 2008

Ethylene: Inhibitor and Stimulator of Plant Growth

Jasper Dugardeyn · Dominique Van Der Straeten (✉)

Unit of Plant Hormone Signaling and Bioimaging, Department of Molecular Genetics, Ghent University, Ledeganckstraat 35, 9000 Ghent, Belgium
dominique.vanderstraeten@ugent.be

Abstract Ethylene is a gaseous hormone which plays an essential role in a myriad of plant developmental processes. It promotes root hair formation, flowering in a number of species, fruit ripening and abscission and leaf and petal abscission. Ethylene can stimulate growth in hypocotyls of light-grown plants, and shoot growth in shaded conditions. On the other hand, it inhibits root growth, and hypocotyl elongation in the dark. In recent years, compelling molecular evidence has been gathered to support intricate connections between ethylene and other hormonal pathways that yield its well-known effects on plant growth. In this chapter, we will discuss the role of ethylene in both growth-stimulating and growth-inhibiting processes.

1 Ethylene Synthesis

Plant hormones, just like animal hormones, function in a dose-dependent manner (Taiz and Zeiger 2006). The most direct way to regulate endogenous ethylene concentrations is to change the rate at which it is synthesized. In this paragraph we will briefly summarize the ethylene biosynthesis pathway and discuss the different mechanisms that influence the rate of ethylene synthesis.

1.1 Biosynthesis Pathway

The precursor for ethylene synthesis is methionine. This amino acid is converted to *S*-adenosyl methionine (SAM) from ATP and methionine. The reaction is catalyzed by SAM synthetase (Ravanel et al. 1998). *Arabidopsis* has two genes encoding this enzyme (Peleman et al. 1989). Analysis for subcellular localization signals using Target P (Emanuelsson et al. 2000) did not reveal potential subcellular targeting, suggesting cytosolic localization of the enzyme.

The next step is the conversion of SAM to 1-aminocyclopropane-1-carboxylic acid (ACC). This step is catalyzed by ACC synthase (ACS). This is a cytosolic enzyme that requires PLP (pyridoxal-5′-phosphate) as a cofactor (Adams and Yang 1979; Yang and Hoffman 1984). ACS isozymes function as homodimers (Capitani et al. 1999; Yamagami et al. 2003). In *Arabidopsis*,

ACS is encoded by a gene family containing 12 members (*ACS1* and *ACS2* of Van Der Straeten et al. 1992 are named *ACS2* and *ACS4* respectively by Yamagami et al. 2003). Ten of the 12 family members encode ACS isozymes; of these ACS1 and ACS3 are not biologically active, and ACS10 and ACS12 function as aminotransferases. It has been proven that ACS is encoded by a multigene family in other plants too (for a review, see Vandenbussche et al. 2006). The catalytic activity of ACS results in not only ACC but also 5′-methylthioadenosine (MTA). MTA is recycled to methionine in the Yang cycle (Miyazaki and Yang 1987). ACS is the main enzyme that controls the synthesis of ethylene and is, in turn, controlled by multiple signals. We will discuss these signal interactions in the next paragraphs.

ACO (ACC oxidase) catalyzes the conversion of ACC to ethylene. During this reaction ACC is oxidized and forms ethylene, CO_2 and HCN (Yang and Hoffman 1984). In *Arabidopsis ACO* is part of a multigene family, as is *ACS* (Gomez-Lim et al. 1993). It was proposed that under particular conditions, such as upon wounding or during ripening and senescence, ACO also plays a role in regulating ethylene levels in plants (Kende 1993).

1.2 Regulation of Synthesis

ACS is the major factor regulating the rate of ethylene synthesis. This regulation is in part dependent on the level of ACS. One of the mechanisms used by plants to control the concentration of ACS is the transcriptional regulation of *ACS* genes.

Using promoter-GUS fusions, Rodrigues-Pousada et al. (1993, 1999) and Tsuchisaka and Theologis (2004b) showed that the different functional *ACS* genes in *Arabidopsis* each have a unique pattern of expression. Although the patterns are specific to each gene, they show overlapping regions of expression. The GUS expression also differs according to changes in environmental conditions. For instance, the expression of *ACS1/2*, *2/4* (see remark concerning *ACS* gene numbering above), *6*, *7*, *8* and *11* in five-day-old etiolated seedlings is confined to the elongation zone of the hypocotyl, the embryonic root region, the cotyledons and the root vascular tissue. In the light, however, these genes are expressed in the cotyledons, the embryonic root, the roots, and in primary leaves, while *ACS1/2*, *5*, *8* and *11* are active in the shoot apex.

In addition, different stress-promoting factors (cold, wounding, heat) have been shown to alter the transcriptions of *ACS* genes, with each factor altering the transcription of each individual *ACS* gene in a specific manner (Tsuchisaka and Theologis 2004b). Wounding (by cutting) the hypocotyls of five-day-old light-grown seedlings inhibits the expression of the genes that are constitutively expressed in the intact tissue, like *ACS1* and *ACS5*, and induces the expression of *ACS1/2*, *2/4*, *6*, *7* and *8*, which were not expressed before wounding. Cold treatment inhibits the expression of *ACS5* and *ACS11* and alters the pattern of *ACS8* expression, whereas heat enhances the expression of *ACS4* and alters the pattern of *ACS8* and *ACS11*. Moreover, it was shown by RT-PCR that *ACS1/2* and *ACS7* are salt-inducible (Achard et al. 2006), and that *ACS 6, 8, 9,* and *11* are induced in low-light conditions (Vandenbussche et al. 2003b). Finally, *ACS8* appears to be subject to strong circadian control (Thain et al. 2004).

Hormonal induction of *ACS* genes has also been documented extensively. First, it was demonstrated that auxins are inducers of ethylene production (Yang and Hoffman 1984; Abeles et al. 1992). Abel et al. (1995) and Tsuchisaka and Theologis (2004b) provided proof of auxin-enhanced transcription and differential expression patterns of *ACS 1/2, 2/4, 5, 6, 7, 8* and *11* in the root. Furthermore, *ACS1/2* also was proven to be up-regulated by cytokinins in the root (Rodrigues-Pousada et al. 1999). In light-grown seedlings, *ACS4, 5,* and *7* are responsive to ABA (Wang et al. 2005), while *ACS7* is also responsive to GA (gibberellins). Finally, brassinosteroids can stimulate *ACS2/4* expression in dark-grown *Arabidopsis* seedlings (Joo et al. 2006).

As mentioned before, it was shown that ACS functions as a homodimer. This implies the possibility that functional heterodimers can also form. Tsuchisaka and Theologis (2004a) have been able to prove that 17 functional heterodimers can form in *E. coli*. As yet this finding has not been confirmed in plants, but this may indicate why ACS is encoded by a multigene family. The unique expression pattern of these genes (which often overlaps) could therefore offer a combination of different isozymes, leading to a high number of heterodimers, which in turn can have different biochemical properties. It was hypothesized that ACC can be synthesized in different tissues and under different conditions due to these different biochemical properties (Tsuchisaka and Theologis 2004b).

Plants also control ACS levels post-transcriptionally. In *Arabidopsis* this mechanism was revealed by the analysis of *eto* (ethylene overproduction) mutants. There are three *ETO* genes in *Arabidopsis*, and mutants have been characterized for each. *eto2* is an allele of *ACS5*. A dominant mutation (insertion in the C-terminus) in this gene causes an increase of ACS5 protein accumulation (Vogel et al. 1998). Likewise, a recessive mutation in *ETO1* causes an increase of ACS5. ETO1 is a BTB (broad complex – tramtrack – brick-à-brack)-domain-containing protein that also possesses a TPR (tetratricopeptide repeat) motif (known to interact with other proteins) and could function as an adaptor connecting ACS to a CUL3 ubiquitin E3 ligase, thus promoting ACS degradation (Wang et al. 2004). ETO1 exercises its function through interaction with the C-terminus of ACS5, which also explains the higher stability of ACS5 in the *eto2* mutant (Wang et al. 2004). Much like *eto2*, *eto3* is modified in the C-terminal region of ACS9, an ACS closely related to ACS5, rendering it more stable (Chae et al. 2003).

The levels of ethylene are also regulated by ACO. In *Arabidopsis* it was proven that *ACO* is regulated by ethylene at the transcriptional level. This regulation is different for different isozymes (van Zhong and Burns 2003; De Paepe et al. 2004b). *ACO* genes are also differentially regulated by light intensity (Vandenbussche et al. 2003b).

2 Ethylene Signaling

In *Arabidopsis*, ethylene is perceived by five receptors, which are partially redundant in function. They show homology to bacterial two-component histidine kinases and are localized on the endoplasmatic reticulum membrane (Chen et al. 2002; Gao et al. 2003). Ethylene binds to the aminoterminal end of these proteins (Schaller and Bleecker 1995). To be functional, the receptors need copper as a cofactor, which is delivered by RAN1 (responsive to antagonist) (Hirayama et al. 1999; Woeste and Kieber 2000). The ethylene receptors can be divided into two groups. Subfamily 1 contains ETR1 (ethylene resistant) and ERS1 (ethylene response sensor). They have four conserved hydrophobic regions in the amino terminus and their C-terminus strongly resembles the response regulator part of bacterial two-component systems. The second subfamily comprises ETR2, ERS2, and EIN4 (ethylene insensitive). These receptors have only three hydrophobic regions and a C-terminus that may function as a Ser/Thr domain (Hua et al. 1998; Hall and Bleecker 2003). The receptors act as negative regulators of ethylene signaling by activating CTR1 (constitutive triple response) in the absence of the hormone (Hua and Meyerowitz 1998; reviewed by De Paepe and Van Der Straeten (2005)).

CTR1 belongs to the Raf family of Ser/Thr protein kinases, but it is unclear whether it functions as a MAPKKK (mitogen-activated protein kinase kinase kinase). It is supposed to be part of, and a negative regulator of, a MAPK cascade (Kieber et al. 1993). The MAPK cascade acts as a positive regulator of ethylene signaling (Chang 2003; Ouaked et al. 2003), however a direct biochemical connection between CTR1 and SIMKK (the next step in the MAPK cascade) remains to be confirmed. Furthermore, a CTR1-independent pathway is supposed to exist as well, since quadruple loss-of-function receptor mutants show a more severe phenotype than *ctr1*-null mutants (Hua and Meyerowitz 1998) and *ctr1*-null mutants still respond to ethylene (Larsen and Chang 2001).

EIN2 is an essential positive regulator in the ethylene signaling pathway. It functions downstream of CTR1 and upstream of EIN3. A loss of function mutation in the gene encoding this protein causes a very severe ethylene insensitivity (Hall and Bleecker 2003). EIN2 contains 12 trans-membrane domains in its N-terminal part, which shows substantial similarity to Nramp (natural resistance-associated macrophage protein) proteins. This region is necessary for the regulation of the ethylene signal, possibly through the regulation of the activity of the C-terminus of EIN2, since constitutive expression of this C-terminal part confers a phenotype similar to *ctr1* (Alonso et al. 1999).

The next components in the ethylene signaling pathway are EIN3 and the EILs (EIN3-like proteins). *eil1ein3* double mutants are almost completely ethylene-insensitive and almost indistinguishable from *ein2-5* (Alonso et al. 2003). Over-expression of these genes confers a *ctr1*-like phenotype. This raises questions about the function of the other EILs. The latter proteins were suggested to be important for more specific ethylene responses at particular stages of development (Chao et al. 1997). EIN3 is constitutively synthesized and degraded in a ubiquitin-dependent way. EBF1 (EIN3-binding F-box protein) and EBF2 mediate this breakdown (Guo and Ecker 2003; Potuschak et al. 2003; Gagne et al. 2004). The stability of EIN3 is raised in the presence of ethylene, while other signals (e.g. glucose) destabilize EIN3 (Yanagisawa et al. 2003). This might be caused by EIN5, an exoribonuclease. *EIN5* expression is up-regulated by ethylene and the corresponding protein is supposed to act in ethylene signaling by breaking down EBF1/2 mRNA (Olmedo et al. 2006; Potuschak et al. 2006). EIN3 dimers are able to bind the PERE (primary ethylene response element) in the promoter of *ERF1* (ethylene response factor; Solano et al. 1998). ERF1 is part of the family of EREBP (ethylene response element binding proteins). These proteins are able to bind the SERE (secondary ethylene-response element), a GCC box in the promoter region of ethylene-regulated genes (Fujimoto et al. 2000).

3 Growth Inhibition

The phenotype used to discover most ethylene signaling mutants is the triple response. When dark-grown plants are treated with ethylene, the hypocotyl becomes shorter and thicker, root growth is inhibited, and an exaggerated apical hook is formed (Guzman and Ecker 1990). This phenotype indicates a growth-inhibiting function of ethylene. We will discuss the mechanism behind this growth inhibition and illustrate that growth inhibition by ethylene is not limited to the triple response.

3.1 Ethylene Inhibits Root Growth

Exogenous ethylene inhibits root elongation in different plant species. This was shown for *Arabidopsis* (reviewed by Bleecker et al. 1988; Smalle et al. 1997), cucumber (Pierik et al. 1999), and *Rumex* (Visser et al. 1997). This inhibition of elongation is at least partially dependent on an ethylene-induced inhibition of cell elongation. In *Arabidopsis*, wild-type epidermal cells with a visible root hair bulge (the first sign of root hair outgrowth) are shorter when treated with ACC or ethylene (Le et al. 2001). *eto2* roots show this phenotype too (De Cnodder et al. 2005). Wild-type epidermal cells from plants treated with AVG (aminoethoxyvinylglycine, ACC synthase inhibitor) are larger than those of untreated seedlings (Le et al. 2001). This inhibition of elongation can be partly explained by a callose deposition in epidermis and cortex cells in the elongation and differentiation zone under the influence of ACC (De Cnodder et al. 2005). Since callose is a structural component of plasmodesmata (Roberts and Oparka 2003), it is possible that cell-to-cell transport is also part of the control mechanism for cell elongation. ACC also raises the hydrogen peroxide concentration (by NADPH oxidases), leading to cross-linking of hydroxyproline-rich glycoproteins (HRPG) in the cell wall by the oxidation of a tyrosine residue (De Cnodder et al. 2005), which presumably restricts cell wall extensibility.

To achieve root-growth inhibition, both ethylene and auxin are required. This is illustrated by the phenotype of *wei*-mutants (weak ethylene insensitive). Root elongation of dark-grown *wei2* (Alonso et al. 2003) and *wei7* (Stepanova et al. 2005) seedlings is not inhibited upon the application of ethylene. Nevertheless these mutants display a short hypocotyl and an exaggerated hook upon ethylene treatment, indicating uncoupling of different triple response phenotypes. *Wei2* is an allele of *ASA1* (anthranilate synthase α), necessary for Trp synthesis. A previously identified mutant in this gene, *tir7* (transport inhibitor response), is resistant to auxin transport inhibitors (Ljung et al. 2005). The effect of *wei7* is caused by a mutation in *ASB1* (anthranilate synthase β). Both genes are expressed in the columella of roots and their expression is enhanced when ethylene is present. Applying ethylene to wild-type plants increases the expression of DR5-GUS (synthetic auxin reporter) in the root cap (Stepanova et al. 2005). DR5-GUS activity has been shown to correlate well with endogenous auxin levels in roots (Casimiro et al. 2001; Benkova et al. 2003). An ethylene-regulated increase in DR5-GUS expression is absent in *ein2*-5, *wei2* and *wei7*, which is indicative of a lower IAA (indole-3-acetic acid) content, resulting in a longer root (Stepanova et al. 2005). Further proof of the need for both auxin and ethylene in the inhibition of root growth is delivered by *wei1*. This mutation is an allele of *TIR1*, lacking 21 carboxy terminal amino acids (Alonso et al. 2003). TIR1 normally interacts with the AUX/IAA proteins AXR2/IAA7 (auxin resistant) and AXR3/IAA17, labeling them for ubiquitin-dependent degradation and subsequent regulation of auxin-regulated genes (Gray et al. 2001). In *wei1*, ubiquitin labeling of AUX/IAA is abolished, blocking auxin response and hence conferring a longer root than that of the wild type when treated with ethylene or auxin.

Furthermore, upon treatment with ethylene, *ARF19* (auxin response factor) expression increases (Li et al. 2006). Loss-of-function mutants in this gene have ethylene-insensitive roots. It is not yet clear whether this is a direct effect of ethylene on the expression of *ARF19* or a consequence of the ethylene-dependent rise in IAA concentration.

It can be concluded that the addition of ethylene results in the inhibition of root elongation, an effect that is largely mediated by a change in auxin concentration and/or auxin signaling.

3.2 Ethylene Inhibits Hypocotyl Growth in the Dark

Another trait of the triple response is the shortening and thickening of the hypocotyl (Guzman and Ecker 1990; Fig. 1). The hypocotyl of *Arabidopsis* mainly grows by longitudinal cell expansion (Saibo et al. 2003). This elongation is not the same in all cells along the hypocotyl and differs in the light and the dark (Gendreau et al. 1997). Applying ethylene to dark-grown *Arabidopsis* seedlings inhibits hypocotyl cell elongation (Le et al. 2005). The decrease in cell elongation is correlated with a change in microtubule orientation. In the dark, the strongest elongation is seen in the top two-thirds of the hypocotyl, from cell 8 upwards. This correlates with a transverse microtubule orientation in the apical cells, whereas in the basal cells microtubules are longitudinally oriented (Le et al. 2005). Under the influence of ACC, dark-grown seedlings had a smaller zone at the top of the hypocotyl showing transverse microtubule orientation (Le et al. 2005). The decrease in the hypocotyl growth rate is already visible 15 minutes after ethylene addition and is reversible (Binder et al. 2004). A partial ethylene oversensitive mutant, *eer1* (enhanced ethylene response), shows a thicker and shorter basal part of the hypocotyl as compared to the wild type when grown in the dark over a broad range of ACC concentrations (Larsen and Chang 2001). Since cell elongation in etiolated seedlings requires GA (Cowling and Harberd 1999), crosstalk between ethylene and GA was suggested. The GA signaling pathway is negatively regulated by DELLA proteins. These proteins are destabilized by GA in a 26S proteasome-dependent manner (Peng et al. 1997; Fu and Harberd 2003; Itoh et al. 2003). In contrast, ethylene stabilizes DELLA proteins (Achard et al. 2003; Vriezen et al. 2004), repressing cell elongation and hence resulting in shorter hypocotyls.

3.3 Ethylene Inhibits Stem Growth

Ethylene also causes an inhibition of stem elongation. Treatment of light-grown *Arabidopsis* with ethylene results in a stunted and thick inflorescence stem. The same can be seen in untreated *ctr1* mutants (Kieber et al. 1993). Furthermore, the internodes are shorter in an ethylene-overproducing transgenic tobacco line (Knoester et al. 1997).

3.4 Ethylene Inhibits Leaf Expansion

The constitutive ethylene mutant *ctr1* has a dwarfed phenotype in the light with very limited leaf expansion. This is mainly caused by a reduction in cell size (Kieber et al. 1993; Rodrigues-Pousada et al. 1993). The opposite is seen in ethylene-insensitive mutants, with a slightly larger leaf surface than the wild type. Although the extra leaf area is thought to result from larger cells (Bleecker et al. 1988; Hua et al. 1995), no detailed measurements of cell-elongation rates in leaves of ethylene-insensitive plants are available. Therefore, it is possible that the greater leaf surface can be explained either by an extended expansion phase or by a higher expansion rate (Pierik et al. 2006). Smalle et al. (1999) showed that in wild-type plants *ACS1* mRNA levels are lower during leaf surface expansion than during leaf emergence or senescence. This is also reflected in a lower ethylene production. Although it cannot be excluded that this pattern is an effect of leaf development rather than being the causal factor, it does implement an inhibiting role of ethylene in leaf cell expansion (Smalle et al. 1999). Lower ethylene concentrations result in a relief of repression and consequent leaf expansion.

3.5 Ethylene Stimulates Exaggeration of the Apical Hook in Etiolated Seedlings

In darkness, *Arabidopsis* seedlings display an apical hook. The hook structure protects the shoot apical meristem against mechanical stress during germination and early seedling growth (Vandenbussche and Van Der Straeten 2004). Adding ethylene to etiolated seedlings causes an exaggeration of the curvature of the hook. This is also one of the traits of the triple response (Guzman and Ecker 1990). This exaggeration of curvature is accompanied by a longer arc that can be explained by extra cell divisions and differential elongation of the concave versus the convex side of the hook (Vriezen et al. 2004).

The formation of the hook results from the differential growth of cells in the apical region. Cells at the outer side of the hook are larger than those at the inner side (Vriezen et al. 2004). The phenotype of *eto1*, *eto2*, *eto3* and *ctr1* mutants proves that the formation of an exaggerated hook is at least partly dependent on ethylene, since they show an exaggerated hook even when grown in the absence of ethylene (Guzman and Ecker 1990; Kieber et al. 1993). Also, ethylene is required for normal hook formation since ethylene-insensitive mutants show a reduction in curvature (Roman et al. 1995). This difference in cell elongation can result from differential ethylene production. However, the literature does not appear fully consistent on this. While *AtACO1* appears to be more strongly expressed at the concave side of the hook in pea (Peck et al. 1998), *AtACO2* is expressed more at the convex side of the hook in *Arabidopsis* (Raz and Ecker 1999). Furthermore, ACC is asymmetrically localized in cells of the apical hook in bean (Schwark and Bopp 1993). However, it was proven that seedlings are only sensitive to exogenous ethylene 60–72 hours after germination (Raz and Ecker 1999).

To exert its effects on apical hook formation, ethylene interacts with other hormones (Fig. 1). There is compelling evidence for the involvement of auxins, gibberellins, and brassinosteroids in this process. Differential auxin distribution is a prerequisite for hook formation in etiolated seedlings, since blocking auxin transport results in a hookless phenotype (Lehman et al. 1996). The auxin efflux transporter PIN3 (pin formed) is required for hook formation, since dark-grown *pin3* mutants show a faster opening of the apical hook (Friml et al. 2002). When ethylene is added to *pin1* and *pin3* an exaggerated hook is not formed, as evidenced by the smaller ratios of the length of cells in the outer to the length of cells at the inner side of the arc in both ACC and untreated seedlings (De Grauwe et al. 2005). Other auxin mutants like *axr1* (Lincoln et al. 1990; del Pozo et al. 1998; auxin resistant, positive regulator of auxin response, as achieved by regulating activity of TIR1 SCF ubiquitin ligase), *hls1* (Lehman et al. 1996; Li et al. 2004; hookless, negative regulator of auxin response, as achieved by degrading ARF2, auxin response factor), *hls3* (King et al. 1995; Gopalraj et al. 1996; Lehman et al. 1996; aminotransferase with higher levels of endogenous IAA, as achieved by stimulating the synthesis thereof), and *yuc* (Zhao et al. 2001; yucca, flavin monooxygenase-like enzyme, with elevated IAA biosynthesis) do not display a normal apical hook.

HLS1 is a downstream target of ethylene-dependent transcriptional regulators. It is equally expressed throughout the apical hook and its expression is enhanced under the influence of ethylene. *HLS1* over-expression results in an exaggerated hook. In *hls1* the cells where the apical hook is supposed to arise, are elongated. Cells at the outer and inner sides of the "hook" elongate two- and tenfold, respectively, compared to those in the wild type, resulting in an equal size on both sides in the *hls1* mutant, which explains why no hook is formed (Lehman et al. 1996). HLS1 in turn lowers the concentration of ARF2 (Li et al. 2004). ARF2 (auxin response factor) is a negative regulator of differential growth responses (Li et al. 2004) that binds TGTCTC sequences in the promoter of primary auxin response genes (Ulmasov et al. 1999). When ethylene is applied to wild-type plants, the concentration of HLS1 is enhanced, leading to a lower concentration of ARF2, ultimately resulting in the formation of the apical hook. Adding auxin to the medium does not alter the concentration of ARF2 (Li et al. 2004).

In dark-grown seedlings, DR5::GUS expression is localized at the inner side of the apical hook. Ethylene exposure enhances this expression. The differential localization is completely abolished in the *hls1* mutant and partially restored in the double mutant *hls1arf2* (Li et al. 2004). These results lead to a model where ethylene modifies auxin response genes through the modulation of HLS1 and ARF2. Taking into account the equal expression of *HLS1*

(Lehman et al. 1996), tissue-specific signals and different auxin concentrations (Friml et al. 2002) are supposed to modify the effect of ARF2 on the expression of auxin-dependent genes.

Ethylene also interacts with gibberellin signaling to maintain the apical hook (Fig. 1). Gibberellin biosynthesis and signaling are required to form an apical hook. ACC enhances the stimulating effect of GA on cell division in the formation of an exaggerated apical hook of etiolated seedlings, and stabilizes RGA (a negative GA regulator), leading to smaller cells. There seems to be no significant difference in the stabilization of RGA between the outer and inner sides of the hook. Despite this, a stronger up-regulation of GASA1::GUS (a GA reporter line) was noticed at the outer side of the hook. This can be explained by changes in ethylene sensitivity or gibberellin sensitivity (Vriezen et al. 2004).

Finally, brassinosteroids are also required to form the apical hook. Adding ACC to BR biosynthesis mutants *cbb1/dwf1* (Kauschmann et al. 1996; cabbage1/dwarf1) and *det2* (Chory et al. 1991; de-etiolated2) does not induce an exaggeration of the hook. Moreover, the expression of CPD::GUS (Fujioka and Yokota 2003; CPD is involved in brassinolide synthesis) in the apical hook is up-regulated by the addition of ACC. This expression is stronger in the convex side of the hook than in the concave side. Auxin disrupts the differential expression, whereas NPA (*N*-1-napthylphthalamic acid, polar auxin transport inhibitor) limits the expression to the stele. These results indicate that ethylene causes a stronger CPD promoter activity at the outer side of the hook (De Grauwe et al. 2005).

4 Growth Stimulation

When studying the effect of ethylene on hook formation, it is clear that ethylene cannot be strictly defined as a growth-inhibiting hormone. Although the growth-inhibiting role of ethylene has been known since its discovery as a plant growth regulator (Neljubov 1901), in the past decade evidence has accumulated for a growth-stimulating function of ethylene. In the next few sections we will discuss the physiological conditions at different stages in a plant's life where ethylene exercises a growth-promoting function.

4.1 Ethylene Stimulates Hypocotyl Growth in the Light

When plants are grown on LNM (low-nutrient medium) in the light, hypocotyl elongation (up to twofold) is induced by concentrations of ACC above 1 μM (Smalle et al. 1997). The response is saturated at 20 μM ACC. When grown on MS/2, the addition of ACC results in an up to 30% longer hypocotyl. Adding $AgNO_3$—a known blocker of ethylene action—to the medium inhibits this response. This result correlates with *ctr1* displaying a longer hypocotyl than the wild type when grown in the light in the absence of ACC, whereas *etr1-3* does not elongate when treated with ACC. Adding ethylene to the seedlings induces the same effect as ACC, and the response in ethylene is abolished by combination with 1-MCP—another inhibitor of ethylene perception (De Paepe and Van Der Straeten 2005). The elongation with ethylene/ACC is a result of cell elongation. Several other hormonal signals confer hypocotyl elongation on LNM. Plants grown on LNM supplied with IAA also have longer hypocotyls (Smalle et al. 1997; Saibo et al. 2003). Vandenbussche et al. (2003a) demonstrated that the ethylene-regulated hypocotyl elongation depends on a functional auxin transport system. In addition, GA_3 stimulates hypocotyl elongation (Cowling and Harberd 1999). Mutants that are defective in GA synthesis (Sun et al. 1992) or auxin signaling (Timpte et al. 1992) were shown to have shorter hypocotyls, indicating the need for both hormones for hypocotyl elongation. Saibo et al. (2003) presented a model for a network of interactions between ethylene, GA_3 and auxins regulating hypocotyl growth. Growth primarily occurs within the first three days after germination, but when ACC is applied to the medium, this fast growth phase is prolonged by one day. In contrast, GA_3 does not prolong the growth phase, but enhances the growth rate between day 2 and day 3 after germination. After inhibiting growth during the first two days, IAA prolongs the growth period until the sixth day after germination. Nevertheless, ACC also works independently of GA since the GA biosynthesis inhibitor paclobutrazol (PAC) only reduced the effect of ACC, whereas IAA-treated plants show no elongation when PAC is added to the medium (Saibo et al. 2003).

The ethylene signal responsible for elongation is at least partially separated from other ethylene-dependent effects. The over-expression of the EIN2 carboxyl terminus is sufficient to rescue the ethylene-dependent elongation in *ein2-5*, but it does not rescue the triple response (Alonso et al. 1999). Furthermore, ACC treatment also causes radial expansion of the hypocotyl which is independent of GA (Saibo et al. 2003). This indicates that at least partially different pathways control the effects of ethylene upon elongation and radial expansion.

The three abovementioned hormones achieve their effects by stimulating cell elongation. This is in part accompanied by endoreduplication. GA_3 is required for endoreduplication (Gendreau et al. 1999). The effect of GA_3 was enhanced by adding ACC (Saibo et al. 2003).

Recent studies also implicate a role for brassinosteroids in ethylene-regulated elongation. Wild-type seedlings treated with Brz2001 (brassinosteroid synthesis inhibitor), *cbb1* and *det2* (mutants in BR biosynthesis) do not show an increase in hypocotyl length in the light when treated with ACC (De Grauwe et al. 2005). The hypocotyl of *hls1* is ethylene-insensitive in the light. When both ethylene and EBR (epi-brassinolide, physiological active brassi-

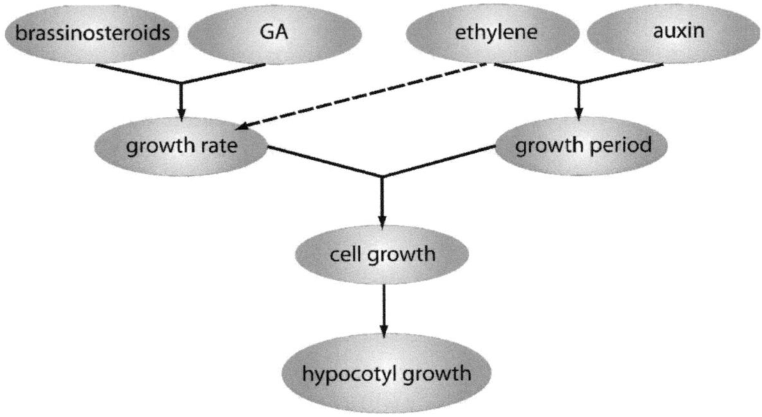

Fig. 1 Hypocotyl elongation in the light. Hypocotyl growth is determined by two major factors: growth rate and duration of the growth period. Ethylene and auxin prolong the duration of elongation growth, whereas GA and possibly brassinosteroids affect the growth rate. If the latter is disturbed, prolonging the growth period will have no effect, supporting the dominant role of GA in this process. The *dotted arrow* indicates the fact that there also is a gibberellin-independent mechanism by which ethylene exerts its role

nosteroid) are applied to *hls1*, they show a synergistic effect. This indicates that brassinosteroids function downstream or independent of the ethylene signal (De Grauwe et al. 2005). The hypocotyl of *sax1* (Ephritikhine et al. 1999; hyperSensitive to Abscisic acid and auXin, necessary for BR biosynthesis) is insensitive to ACC and GA. When EBR is applied, the hypocotyl regains its sensitivity to GA, but not to ACC (Ephritikhine et al. 1999). A schematic overview of these interactions is presented in Fig. 1.

4.2 Ethylene Stimulates Shade Avoidance Responses

When shaded, plants receive lower doses of light, and the light actually transmitted contains proportionally more far red light than light received by unshaded plants (Fig. 2). Plants will try to move out of the shade in order to maximize photosynthesis (Vandenbussche et al. 2005). Not only is ethylene production enhanced under shaded conditions, optimal shade avoidance responses also require ethylene (Vandenbussche et al. 2003b; Pierik et al. 2004a,b).

In tobacco, low R:FR (ratio of red to far red light) ratios induce hyponasty independent of ethylene, but ethylene determines the rate of low R:FR-induced stem and petiole elongation (Pierik et al. 2004a). Applying ethylene to *Arabidopsis* plants causes an upward movement of leaves, although this response varies depending on the conditions (Vandenbussche et al. 2003b; Millenaar et al. 2005). This is caused by differential growth at the ends of

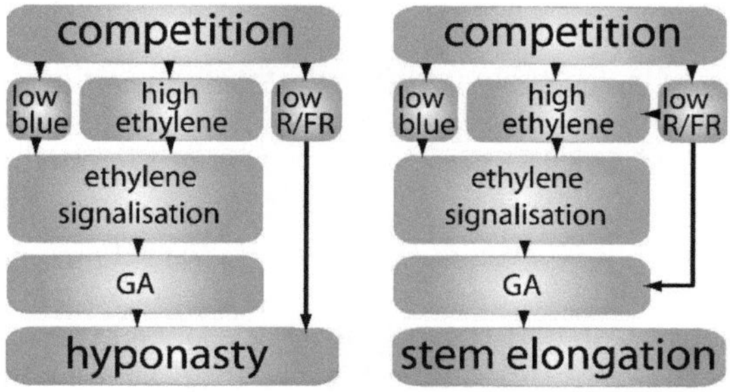

Fig. 2 Shade avoidance. When plants face competition in dense canopies, different environmental signals change. The dose of transmitted light under the canopy decreases and the ratio *red/far red light* decreases. Under low R : FR conditions, ethylene synthesis is stimulated, leading to a higher ethylene concentration. Both a high ethylene concentration and blue light need a fully functioning ethylene and GA signaling pathway in order to affect hyponasty and stem elongation. Low R : FR ratios can stimulate hyponastic movements independently of ethylene and GA. In contrast, stem elongation under low R : FR conditions is dependent on a functioning GA pathway. Stem elongation is possible without ethylene signaling, although it occurs more slowly in this case

both the petiole and the leaf blade. Combining ethylene treatment with low light does not yield additive effects, except for a higher petiole angle after 20 hours of exposure (Millenaar et al. 2005). When the light dose is lowered to a level comparable with that under a canopy, ethylene-resistant tobacco plants showed no stem elongation and no hyponastic response, whereas wild-type plants did show these responses (Pierik et al. 2004a). They are caused by a lower dose of blue light. Changes in leaf angle and stem elongation of competing plants occur faster in wild-type than in ethylene-resistant tobacco plants. This delay causes a competitive disadvantage, resulting in a lower biomass of ethylene-resistant plants, whereas they have the same biomass in a noncompeting setup (Pierik et al. 2004a). When GA synthesis is blocked in tobacco, ethylene-induced hyponastic movements and stem and petiole elongation are prevented, indicating the essential role of GA in these responses. However, when R:FR ratios are lowered in the presence of PAC, stem and petiole elongation are inhibited but hyponastic movement is not. In the wild-type and ethylene-resistant tobacco mutants, petiole elongation seems to be equally sensitive to GA. In contrast, stem elongation is more sensitive to GA in wild-type than in ethylene-resistant tobacco mutants. Both processes are more pronounced for low R : FR ratios (Pierik et al. 2004b). An overview of the process involvrd in shade avoidance is given in Fig. 2.

4.3 Ethylene is Responsible for Submergence-Induced Shoot Elongation

When plants are flooded, the gas diffusion rate is impeded (Jackson 1985). Hence flooding not only leads to lower internal O_2 concentrations (hypoxic conditions) but also higher CO_2 and ethylene concentrations, as reviewed by Vriezen et al. (2003).

In *Rumex palustris*, one of the responses to submergence is a hyponastic movement of the leaves, caused by cell elongation at the abaxial side in the basal region of the petiole (Cox et al. 2004). The effect can be mimicked by adding ethylene to unsubmerged plants. Under submergence, ethylene concentrations in *Rumex* rise due to ethylene entrapment and not due to ethylene synthesis (Voesenek et al. 1993). However, ethylene is not the only signal that causes this response. Submerged plants treated with 1-MCP do show a hyponastic movement, although it is not as pronounced as that seen in untreated plants. Auxins are also necessary for hyponastic movements. Auxin deficiency cannot be rescued by ethylene or GA. GA promotes hyponastic movements, while ABA has inhibitory effects (Cox et al. 2004).

Ethylene biosynthesis of rice seedlings is enhanced by hypoxic conditions (Satler and Kende 1985; Van Der Straeten et al. 1992). As a result of the entrapment of ethylene, a positive feedback mechanism further enhances its synthesis (Chae et al. 2003). Growth upon submergence occurs in the youngest internode. An intercalary meristem is located just above the second node (Kende et al. 1998). Applying ethylene to rice plants that were older than 28 days stimulated the growth of internodes only in a deepwater rice variety (Metraux and Kende 1983). When AVG is present, this growth is inhibited.

As reviewed above (Sect. 1.2 Regulation of Synthesis), ethylene concentration can be enhanced by modulating *ACS* and *ACO* expression. In air-grown plants, *OsACS5* is expressed at low levels in the shoot apex, meristems, leaves, adventitious root primordia, and in vascular tissues of unelongated stems and leaf sheaths. Upon submergence, *OsACS5* expression is enhanced in vascular bundles of young stems and leaf sheaths. Furthermore, under hypoxic conditions, exogenously applied GA up-regulates *OsACS5* expression, whereas ABA has the opposite effect (Van Der Straeten et al. 2001). The transcript levels of *OsACO1* reached a maximum in the intercalary meristem, elongation zone and differentiation zone of the internodes after 15 hours of submergence (Mekhedov and Kende 1996).

In deepwater rice, submergence is accompanied by a decrease in ABA. This effect can be mimicked by applying exogenous ethylene (Hoffmann-Benning and Kende 1992; Azuma et al. 1995). The same effect was noted for a submergence-tolerant *Rumex* species (*Rumex palustris*). When *R. palustris* plants are flooded or treated with ethylene, a fast down-regulation of NCED (neoxanthine cis-epoxycarotenoid dioxygenase) and an increase in ABA breakdown is observed (Benschop et al. 2005). However, when 1-MCP and fluridone (ABA biosynthesis inhibitor) are applied simultaneously, no elongation is observed. This indicates that a reduction in ABA must be accompanied by an increase in the ethylene concentration in order to induce stem elongation. Submergence-intolerant *Rumex acetosa* plants do not show a decrease in ABA levels (Benschop et al. 2005).

The submergence of deepwater rice enhances the concentrations of GA_1 and GA_{20} (Hoffmann-Benning and Kende 1992; Van Der Straeten et al. 2001). Applying ABA to air-grown deepwater rice inhibited cell elongation, an effect that can be reversed by adding GA. It is thus reasonable to hypothesize that ethylene exerts its effect on GA through ABA (Hoffmann-Benning and Kende 1992). Both submergence and ethylene treatment increase the level of bioactive GA_1 in *R. palustris* (Rijnders et al. 1997). Also, in submerged *R. palustris* plants, ABA inhibits the increase in the level of GA seen in submerged plants without exogenously applied ABA (Benschop et al. 2005).

Stem elongation in rice is an effect of both cell division and cell elongation (Vriezen et al. 2003). Cell elongation can be partially caused by increased expression of expansins. In rice, both submergence and GA treatment enhanced expansin A and B expression (Cho and Kende 1997). In *Rumex*, however, out of 13 studied expansins, only *RpEXPA1* is induced in petioles upon submergence or ethylene treatment. The addition of ABA or PAC to submerged plants does not inhibit this induction. Also, when GA is applied to air-grown plants, this effect can not be mimicked (Vreeburg et al. 2005). Submergence-induced acidification of the cell wall of *Rumex* is inhibited when petiole tissue is pretreated with 1-MCP. The addition of ABA to submerged plants had no effect on cell wall acidification (Vreeburg et al. 2005). Thus, in *Rumex* petioles, ethylene may be involved in two separate pathways. It is proposed to cause a decrease in ABA (and a subsequent increase in GA) and to enhance expansin expression and cell-wall acidification through a distinct mechanism (Vreeburg et al. 2005).

5 The Balance Between Growth Stimulation and Growth Inhibition: Concentration Dependent?

Pierik et al. (2006) present a biphasic model to explain the differential responses to ethylene. The authors suggest that internal signals, environmental conditions and species-specific characteristics influence the response to ethylene. The authors illustrate that a dose-dependent response to ethylene can be found for different plants/tissues. These plant/tissue-specific ethylene responses can be placed in-between two extremes. One extreme consists of (tissues of) plants showing growth stimulation over a large range of ethylene concentrations (with a limited growth-inhibiting range); the other extreme would be (tissues of) plants that show growth inhibition at most ethylene levels. However, it should be noted that none of the dose–response curves shown in Pierik et al. (2006) have more than one data point behaving differently (induction instead of inhibition, or vice versa) from the other. Therefore more detailed analyses are essential to support this point of view.

It is noteworthy that several genes respond differentially to low/high ethylene concentrations (De Paepe et al. 2004a,b). Using a cDNA-AFLP experiment, the authors isolated a group of 30 genes which were more strongly up-regulated when treated with 10 μl/l ethylene than when treated with 0.1 μl/l ethylene, whereas seven genes were more up-regulated by the lower concentration.

Although a gene-specific response to ethylene can in part explain the different ethylene responses, further research is needed in order to discover the factors that modulate these plant- and tissue-specific ethylene responses (including ethylene-responsive tissue-specific promoters, interactions with other hormones, and environmental signals). Finally, ethylene response may not be placed in species-dependent two-dimensional graphs as suggested (Pierik et al. 2006), but could be seen as an *n*-dimensional function that is dependent on species, environmental conditions, and tissue- and cell-type-specific factors.

Acknowledgements This work was supported by a PhD fellowship to Jasper Dugardeyn from the Institute for the Promotion of Innovation through Science and Technology in Flanders (IWT Vlaanderen).

References

Abel S, Nguyen MD, Chow W, Theologis A (1995) ACS4, a primary indoleacetic acid-responsive gene encoding 1-aminocyclopropane-1-carboxylate synthase in *Arabidopsis thaliana* (Vol 270:19093, 1995). J Biol Chem 270:26020–26020

Abeles FB, Morgan PW, Saltveit ME (1992) Ethylene in plant biology, 2nd edn. Academic, San Diego, CA

Achard P, Vriezen WH, Van Der Straeten D, Harberd NP (2003) Ethylene regulates *Arabidopsis* development via the modulation of DELLA protein growth repressor function. Plant Cell 15:2816–2825

Achard P, Cheng H, De Grauwe L, Decat J, Schoutteten H, Moritz T, Van Der Straeten D, Peng JR, Harberd NP (2006) Integration of plant responses to environmentally activated phytohormonal signals. Science 311:91–94

Adams DO, Yang SF (1979) Ethylene biosynthesis—identification of 1-aminocyclopropane-1-carboxylic acid as an intermediate in the conversion of methionine to ethylene. Proc Natl Acad Sci USA 76:170–174

Alonso JM, Hirayama T, Roman G, Nourizadeh S, Ecker JR (1999) EIN2, a bifunctional transducer of ethylene and stress responses in *Arabidopsis*. Science 284:2148–2152

Alonso JM, Stepanova AN, Solano R, Wisman E, Ferrari S, Ausubel FM, Ecker JR (2003) Five components of the ethylene-response pathway identified in a screen for weak ethylene-insensitive mutants in *Arabidopsis*. Proc Natl Acad Sci USA 100:2992–2997

Azuma T, Hirano T, Deki Y, Uchida N, Yasuda T, Yamaguchi T (1995) Involvement of the decrease in levels of abscisic-acid in the internodal elongation of submerged floating rice. J Plant Physiol 146:323–328

Benkova E, Michniewicz M, Sauer M, Teichmann T, Seifertova D, Jurgens G, Friml J (2003) Local, efflux-dependent auxin gradients as a common module for plant organ formation. Cell 115:591–602

Benschop JJ, Jackson MB, Guhl K, Vreeburg RAM, Croker SJ, Peeters AJM, Voesenek L (2005) Contrasting interactions between ethylene and abscisic acid in *Rumex* species differing in submergence tolerance. Plant J 44:756–768

Binder BM, O'Malley RC, Wang WY, Moore JM, Parks BM, Spalding EP, Bleecker AB (2004) *Arabidopsis* seedling growth response and recovery to ethylene. A kinetic analysis. Plant Physiol 136:2913–2920

Bleecker AB, Estelle MA, Somerville C, Kende H (1988) Insensitivity to ethylene conferred by a dominant mutation in *Arabidopsis thaliana.* Science 241:1086–1089

Capitani G, Hohenester E, Feng L, Storici P, Kirsch JF, Jansonius JN (1999) Structure of 1-aminocyclopropane-1-carboxylate synthase, a key enzyme in the biosynthesis of the plant hormone ethylene. J Mol Biol 294:745–756

Casimiro I, Marchant A, Bhalerao RP, Beeckman T, Dhooge S, Swarup R, Graham N, Inze D, Sandberg G, Casero PJ, Bennett M (2001) Auxin transport promotes *Arabidopsis* lateral root initiation. Plant Cell 13:843–852

Chae HS, Faure F, Kieber JJ (2003) The eto1, eto2, and eto3 mutations and cytokinin treatment increase ethylene biosynthesis in *Arabidopsis* by increasing the stability of ACS protein. Plant Cell 15:545–559

Chang CR (2003) Ethylene signaling: the MAPK module has finally landed. Trend Plant Sci 8:365–368

Chao QM, Rothenberg M, Solano R, Roman G, Terzaghi W, Ecker JR (1997) Activation of the ethylene gas response pathway in *Arabidopsis* by the nuclear protein ETHYLENE-INSENSITIVE3 and related proteins. Cell 89:1133–1144

Chen YF, Randlett MD, Findell JL, Schaller GE (2002) Localization of the ethylene receptor ETR1 to the endoplasmic reticulum of *Arabidopsis.* J Biol Chem 277:19861–19866

Cho HT, Kende H (1997) Expression of expansin genes is correlated with growth in deepwater rice. Plant Cell 9:1661–1671

Chory J, Nagpal P, Peto CA (1991) Phenotypic and genetic-analysis of Det2, a new mutant that affects light-regulated seedling development in *Arabidopsis.* Plant Cell 3:445–459

Cowling RJ, Harberd NP (1999) Gibberellins control *Arabidopsis* hypocotyl growth via regulation of cellular elongation. J Exp Bot 50:1351–1357

Cox MCH, Benschop JJ, Vreeburg RAM, Wagemakers CAM, Moritz T, Peeters AJM, Voesenek L (2004) The roles of ethylene, auxin, abscisic acid, and gibberellin in the hyponastic growth of submerged *Rumex palustris* petioles. Plant Physiol 136:2948–2960

De Cnodder T, Vissenberg K, Van Der Straeten D, Verbelen JP (2005) Regulation of cell length in the *Arabidopsis thaliana* root by the ethylene precursor 1-aminocyclopropane-1-carboxylic acid: a matter of apoplastic reactions. New Phytologist 168:541–550

De Grauwe L, Vandenbussche F, Tietz O, Palme K, Van Der Straeten D (2005) Auxin, ethylene and brassinosteroids: Tripartite control of growth in the *Arabidopsis* hypocotyl. Plant Cell Physiol 46:827–836

De Paepe A (2004a) Study of ethylene signalling in *Arabidopsis thaliana* by transcript profiling and mutant analysis. PhD thesis, Ghent University, Ghent, Belgium

De Paepe A, Vuylsteke M, Van Hummelen P, Zabeau M, Van Der Straeten D (2004b) Transcriptional profiling by cDNA-AFLP and microarray analysis reveals novel insights into the early response to ethylene in *Arabidopsis.* Plant J 39:537–559

De Paepe A, Van Der Straeten D (2005) Ethylene biosynthesis and signaling: an overview. Vitam Horm 72:399–430

del Pozo JC, Timpte C, Tan S, Callis J, Estelle M (1998) The ubiquitin-related protein RUB1 and auxin response in *Arabidopsis*. Science 280:1760–1763

Emanuelsson O, Nielsen H, Brunak S and von Heijne G (2000) Predicting subcellular localization of proteins based on their N-terminal amino acid sequence. J Mol Biol 300:1005–1016

Ephritikhine G, Fellner M, Vannini C, Lapous D, Barbier-Brygoo H (1999) The sax1 dwarf mutant of *Arabidopsis thaliana* shows altered sensitivity of growth responses to abscisic acid, auxin, gibberellins and ethylene and is partially rescued by exogenous brassinosteroid. Plant J 18:303–314

Friml J, Wisniewska J, Benkova E, Mendgen K, Palme K (2002) Lateral relocation of auxin efflux regulator PIN3 mediates tropism in *Arabidopsis*. Nature 415:806–809

Fu XD, Harberd NP (2003) Auxin promotes *Arabidopsis* root growth by modulating gibberellin response. Nature 421:740–743

Fujimoto SY, Ohta M, Usui A, Shinshi H, Ohme-Takagi M (2000) *Arabidopsis* ethylene-responsive element binding factors act as transcriptional activators or repressors of GCC box-mediated gene expression. Plant Cell 12:393–404

Fujioka S, Yokota T (2003) Biosynthesis and metabolism of brassinosteroids. Annu Rev Plant Biol 54:137–164

Gagne JM, Smalle J, Gingerich DJ, Walker JM, Yoo SD, Yanagisawa S, Vierstra RD (2004) *Arabidopsis* EIN3-binding F-box 1 and 2 form ubiquitin-protein ligases that repress ethylene action and promote growth by directing EIN3 degradation. Proc Natl Acad Sci USA 101:6803–6808

Gao ZY, Chen YF, Randlett MD, Zhao XC, Findell JL, Kieber JJ, Schaller GE (2003) Localization of the Raf-like kinase CTR1 to the endoplasmic reticulum of *Arabidopsis* through participation in ethylene receptor signaling complexes. J Biol Chem 278:34725–34732

Gendreau E, Traas J, Desnos T, Grandjean O, Caboche M, Hofte H (1997) Cellular basis of hypocotyl growth in *Arabidopsis thaliana*. Plant Physiol 114:295–305

Gendreau E, Orbovic V, Hofte H, Traas J (1999) Gibberellin and ethylene control endoreduplication levels in the *Arabidopsis thaliana* hypocotyl. Planta 209:513–516

Gomez-Lim MA, Valdeslopez V, Cruzhernandez A, Saucedoarias LJ (1993) Isolation and characterization of a gene involved in ethylene biosynthesis from *Arabidopsis thaliana*. Gene 134:217–221

Gopalraj M, Tseng TS, Olszewski N (1996) The rooty gene of *Arabidopsis* encodes a protein with highest similarity to aminotransferases. Plant Physiol 111:469–469

Gray WM, Kepinski S, Rouse D, Leyser O, Estelle M (2001) Auxin regulates SCFTIR1-dependent degradation of AUX/IAA proteins. Nature 414:271–276

Guo HW, Ecker JR (2003) Plant responses to ethylene gas are mediated by SCF (EBF1/EBF2)-dependent proteolysis of EIN3 transcription factor. Cell 115:667–677

Guzman P, Ecker JR (1990) Exploiting the triple response of *Arabidopsis* to identify ethylene-related mutants. Plant Cell 2:513–523

Hall AE, Bleecker AB (2003) Analysis of combinatorial loss-of-function mutants in the *Arabidopsis* ethylene receptors reveals that the ers1 etr1 double mutant has severe developmental defects that are EIN2 dependent. Plant Cell 15:2032–2041

Hirayama T, Kieber JJ, Hirayama N, Kogan M, Guzman P, Nourizadeh S, Alonso JM, Dailey WP, Dancis A, Ecker JR (1999) Responsive-to-antagonist1, a Menkes/Wilson disease-related copper transporter, is required for ethylene signaling in *Arabidopsis*. Cell 97:383–393

Hoffmann-Benning S, Kende H (1992) On the role of abscisic-acid and gibberellin in the regulation of growth in rice. Plant Physiol 99:1156–1161

Hua J, Chang C, Sun Q, Meyerowitz EM (1995) Ethylene insensitivity conferred by *Arabidopsis* ers gene. Science 269:1712–1714

Hua J, Sakai H, Nourizadeh S, Chen QHG, Bleecker AB, Ecker JR, Meyerowitz EM (1998) EIN4 and ERS2 are members of the putative ethylene receptor gene family in *Arabidopsis.* Plant Cell 10:1321–1332

Hua J, Meyerowitz EM (1998) Ethylene responses are negatively regulated by a receptor gene family in *Arabidopsis thaliana.* Cell 94:261–271

Itoh H, Matsuoka M, Steber CM (2003) A role for the ubiquitin-26S-proteasome pathway in gibberellin signaling. Trend Plant Sci 8:492–497

Jackson MB (1985) Ethylene, responses of plants to soil waterlogging and submergence. Annu Rev Plant Physiol Plant Mol Biol 36:145–174

Joo S, Seo YS, Kim SM, Hong DK, Park KY, Kim WT (2006) Brassinosteroid induction of AtACS4 encoding an auxin-responsive 1-aminocyclopropane-1-carboxylate synthase 4 in *Arabidopsis* seedlings. Physiol Plant 126:592–604

Kauschmann A, Jessop A, Koncz C, Szekeres M, Willmitzer L, Altmann T (1996) Genetic evidence for an essential role of brassinosteroids in plant development. Plant J 9:701–713

Kende H (1993) Ethylene biosynthesis. Annu Rev Plant Physiol Plant Mol Biol 44:283–307

Kende H, van der Knaap E, Cho HT (1998) Deepwater rice: A model plant to study stem elongation. Plant Physiol 118:1105–1110

Kieber JJ, Rothenberg M, Roman G, Feldmann KA, Ecker JR (1993) Ctr1, a negative regulator of the ethylene response pathway in *Arabidopsis*, encodes a member of the Raf family of protein-kinases. Cell 72:427–441

King JJ, Stimart DP, Fisher RH, Bleecker AB (1995) A mutation altering auxin homeostasis and plant morphology in *Arabidopsis.* Plant Cell 7:2023–2037

Knoester M, Linthorst HJM, Bol JF, van Loon LC (1997) Modulation of stress-inducible ethylene biosynthesis by sense and antisense gene expression in tobacco. Plant Sci 126:173–183

Larsen PB, Chang C (2001) The *Arabidopsis* eer1 mutant has enhanced ethylene responses in the hypocotyl and stem. Plant Physiol 125:1061–1073

Le J, Vandenbussche F, Van Der Straeten D, Verbelen JP (2001) In the early response of *Arabidopsis* roots to ethylene, cell elongation is up- and down-regulated and uncoupled from differentiation. Plant Physiol 125:519–522

Le J, Vandenbussche F, De Cnodder T, Van Der Straeten D, Verbelen JP (2005) Cell elongation and microtubule behavior in the *Arabidopsis* hypocotyl: Responses to ethylene and auxin. J Plant Growth Regul 24:166–178

Lehman A, Black R, Ecker JR (1996) HOOKLESS1, an ethylene response gene, is required for differential cell elongation in the *Arabidopsis* hypocotyl. Cell 85:183–194

Li H, Johnson P, Stepanova A, Alonso JM, Ecker JR (2004) Convergence of signaling of differential cell growth pathways in the control in *Arabidopsis.* Dev Cell 7:193–204

Li JS, Dai XH, Zhao YD (2006) A role for auxin response factor 19 in auxin and ethylene signaling in *Arabidopsis.* Plant Physiol 140:899–908

Lincoln C, Britton JH, Estelle M (1990) Growth and development of the Axr1 mutants of *Arabidopsis.* Plant Cell 2:1071–1080

Ljung K, Hull AK, Celenza J, Yamada M, Estelle M, Nonmanly J, Sandberg G (2005) Sites and regulation of auxin biosynthesis in *Arabidopsis* roots. Plant Cell 17:1090–1104

Mekhedov SL, Kende H (1996) Submergence enhances expression of a gene encoding 1-aminocyclopropane-1-carboxylate oxidase in deepwater rice. Plant Cell Physiol 37:531–537

Metraux JP, Kende H (1983) The role of ethylene in the growth-response of submerged deep-water rice. Plant Physiol 72:441–446

Millenaar FF, Cox MCH, van Berkel Y, Welschen RAM, Pierik R, Voesenek L, Peeters AJM (2005) Ethylene-induced differential growth of petioles in *Arabidopsis*. Analyzing natural variation, response kinetics, and regulation. Plant Physiol 137:998–1008

Miyazaki JH, Yang SF (1987) Inhibition of the methionine cycle enzymes. Phytochem 26:2655–2660

Neljubov D (1901) Über die horizontale Nutation der Stengel von Pisum sativum und einiger Anderer. Pflanzen Beih Bot Zentralb 10:128–139

Olmedo G, Guo HW, Gregory BD, Nourizadeh SD, Aguilar-Henonin L, Li HJ, An FY, Guzman P, Ecker JR (2006) ETHYLENE-INSENSITIVE5 encodes a $5' \rightarrow 3'$ exoribonuclease required for regulation of the EIN3-targeting F-box proteins EBF1/2. Proc Natl Acad Sci USA 103:13286–13293

Ouaked F, Rozhon W, Lecourieux D, Hirt H (2003) A MAPK pathway mediates ethylene signaling in plants. EMBO J 22:1282–1288

Peck SC, Pawlowski K, Kende H (1998) Asymmetric responsiveness to ethylene mediates cell elongation in the apical hook of peas. Plant Cell 10:713–719

Peleman J, Boerjan W, Engler G, Seurinck J, Botterman J, Alliotte T, Van Montagu M, Inze D (1989) Strong cellular preference in the expression of a housekeeping gene of *Arabidopsis thaliana* encoding *S*-adenosylmethionine synthetase. Plant Cell 1:81–93

Peng JR, Carol P, Richards DE, King KE, Cowling RJ, Murphy GP, Harberd NP (1997) The *Arabidopsis* GAI gene defines a signaling pathway that negatively regulates gibberellin responses. Genes Dev 11:3194–3205

Pierik R, Verkerke W, Voesenek R, Blom K, Visser EJW (1999) Thick root syndrome in cucumber (*Cucumis sativus* L.): A description of the phenomenon and an investigation of the role of ethylene. Ann Bot 84:755–762

Pierik R, Whitelam GC, Voesenek L, de Kroon H, Visser EJW (2004a) Canopy studies on ethylene-insensitive tobacco identify ethylene as a novel element in blue light and plant-plant signalling. Plant J 38:310–319

Pierik R, Cuppens MLC, Voesenek L, Visser EJW (2004b) Interactions between ethylene and gibberellins in phytochrome-mediated shade avoidance responses in tobacco. Plant Physiol 136:2928–2936

Pierik R, Tholen D, Poorter H, Visser EJW, Voesenek L (2006) The Janus face of ethylene: growth inhibition and stimulation. Trend Plant Sci 11:176–183

Potuschak T, Lechner E, Parmentier Y, Yanagisawa S, Grava S, Koncz C, Genschik P (2003) EIN3-dependent regulation of plant ethylene hormone signaling by two *Arabidopsis* F-box proteins: EBF1 and EBF2. Cell 115:679–689

Potuschak T, Vansiri A, Binder BM, Lechner E, Vierstra RD, Genschik P (2006) The exoribonuclease XRN4 is a component of the ethylene response pathway in *Arabidopsis*. Plant Cell 18:3047–3057

Ravanel S, Gakiere B, Job D, Douce R (1998) The specific features of methionine biosynthesis and metabolism in plants. Proc Natl Acad Sci USA 95:7805–7812

Raz V, Ecker JR (1999) Regulation of differential growth in the apical hook of *Arabidopsis*. Development 126:3661–3668

Rijnders J, Yang YY, Kamiya Y, Takahashi N, Barendse GWM, Blom C, Voesenek L (1997) Ethylene enhances gibberellin levels and petiole sensitivity in flooding-tolerant *Rumex palustris* but not in flooding-intolerant *R-acetosa*. Planta 203:20–25

Roberts AG, Oparka KJ (2003) Plasmodesmata and the control of symplastic transport. Plant Cell Environ 26:103–124

Rodrigues-Pousada R, Derycke R, Dedonder A, Van Caeneghem W, Engler G, Van Montagu M, Van Der Straeten D (1993) The *Arabidopsis* 1-aminocyclopropane-1-carboxylate synthase gene-1 is expressed during early development. Plant Cell 5:897–911

Rodrigues-Pousada R, Van Caeneghem W, Chauvaux N, Van Onekelen H, Van Montagu M, Van Der Straeten D (1999) Hormonal cross-talk regulates the *Arabidopsis thaliana* 1-aminocyclopropane-1-carboxylate synthase gene 1 in a developmental and tissue-dependent manner. Physiol Plant 105:312–320

Roman G, Lubarsky B, Kieber JJ, Rothenberg M, Ecker JR (1995) Genetic-analysis of ethylene signal-transduction in *Arabidopsis thaliana*—5 novel mutant loci integrated into a stress-response pathway. Genetics 139:1393–1409

Saibo NJM, Vriezen WH, Beemster GTS, Van Der Straeten D (2003) Growth and stomata development of *Arabidopsis* hypocotyls are controlled by gibberellins and modulated by ethylene and auxins. Plant J 33:989–1000

Satler SO, Kende H (1985) Ethylene and the growth of rice seedlings. Plant Physiol 79:194–198

Schaller GE, Bleecker AB (1995) Ethylene-binding sites generated in yeast expressing the *Arabidopsis* Etr1 gene. Science 270:1809–1811

Schwark A, Bopp M (1993) Interaction of ethylene and auxin in the regulation of hook growth. 2. The role for ethylene in different growing regions of the hypocotyl hook of *Phaseolus vulgaris*. J Plant Physiol 142:585–592

Smalle J, Haegman M, Kurepa J, Van Montagu M, Van Der Straeten D (1997) Ethylene can stimulate *Arabidopsis* hypocotyl elongation in the light. Proc Natl Acad Sci USA 94:2756–2761

Smalle J, Haegman M, Mertens J, Vangronsveld J, Van Montagu M, Van Der Straeten D (1999) The expression pattern of the *Arabidopsis* ACC synthase gene 1 during rosette leaf development. J Exp Bot 50:1561–1566

Solano R, Stepanova A, Chao QM, Ecker JR (1998) Nuclear events in ethylene signaling: a transcriptional cascade mediated by ETHYLENE-INSENSITIVE3 and ETHYLENE-RESPONSE-FACTOR1. Genes Dev 12:3703–3714

Stepanova AN, Hoyt JM, Hamilton AA, Alonso JM (2005) A link between ethylene and auxin uncovered by the characterization of two root-specific ethylene-insensitive mutants in *Arabidopsis*. Plant Cell 17:2230–2242

Sun TP, Goodman HM, Ausubel FM (1992) Cloning the *Arabidopsis* Ga1 locus by genomic subtraction. Plant Cell 4:119–128

Taiz L, Zeiger E (2006) Plant physiology, 4th edn. Sinauer Associates, Sunderland, MA

Thain SC, Vandenbussche F, Laarhoven LJJ, Dowson-Day MJ, Wang ZY, Tobin EM, Harren FJM, Millar AJ, Van Der Straeten D (2004) Circadian rhythms of ethylene emission in *Arabidopsis*. Plant Physiol 136:3751–3761

Timpte CS, Wilson AK, Estelle M (1992) Effects of the Axr2 mutation of *Arabidopsis* on cell-shape in hypocotyl and inflorescence. Planta 188:271–278

Tsuchisaka A, Theologis A (2004a) Heterodimeric interactions among the 1-amino-cyclopropane-1-carboxylate synthase polypeptides encoded by the *Arabidopsis* gene family. Proc Natl Acad Sci USA 101:2275–2280

Tsuchisaka A, Theologis A (2004b) Unique and overlapping expression patterns among the *Arabidopsis* 1-amino-cyclopropane-1-carboxylate synthase gene family members. Plant Physiol 136:2982–3000

Ulmasov T, Hagen G, Guilfoyle TJ (1999) Dimerization and DNA binding of auxin response factors. Plant J 19:309–319

Vandenbussche F, Smalle J, Le J, Saibo NJM, De Paepe A, Chaerle L, Tietz O, Smets R, Laarhoven LJJ, Harren FJM, Van Onckelen H, Palme K, Verbelen JP, Van Der Straeten D (2003a) The *Arabidopsis* mutant alh1 illustrates a cross talk between ethylene and auxin. Plant Physiol 131:1228–1238

Vandenbussche F, Vriezen WH, Smalle J, Laarhoven LJJ, Harren FJM, Van Der Straeten D (2003b) Ethylene and auxin control decreased light intensity. Plant Physiol 133:517–527

Vandenbussche F, Van Der Straeten D (2004) Shaping the shoot: a circuitry that integrates multiple signals. Trend Plant Sci 9:499–506

Vandenbussche F, Pierik R, Millenaar FF, Voesenek LA, Van Der Straeten D (2005) Reaching out of the shade. Curr Opin Plant Biol 8:462–468

Vandenbussche F, Vriezen W, Van Der Straeten D (2006) Ethylene biosynthesis and signaling: a puzzle yet to be completed. In: Hedden P, Thomas S (eds) Annual plant reviews, vol 24: Plant hormone signaling. Blackwell, Oxford, pp 125–145 [ISBN 1-4051-3887-4]

Van Der Straeten D, Rodriguespousada RA, Villarroel R, Hanley S, Goodman HM, Van Montagu M (1992) Cloning, genetic-mapping, and expression analysis of an *Arabidopsis thaliana* gene that encodes 1-aminocyclopropane-1-carboxylate synthase. Proc Natl Acad Sci USA 89:9969–9973

Van Der Straeten D, Zhou ZY, Prinsen E, Van Onckelen HA, Van Montagu MC (2001) A comparative molecular-physiological study of submergence response in lowland and deepwater rice. Plant Physiol 125:955–968

Van Zhong G, Burns JK (2003) Profiling ethylene-regulated gene expression in *Arabidopsis thaliana* by microarray analysis. Plant Mol Biol 53:117–131

Visser EJW, Nabben RHM, Blom C, Voesenek L (1997) Elongation by primary lateral roots and adventitious roots during conditions of hypoxia and high ethylene concentrations. Plant Cell Environ 20:647–653

Voesenek L, Banga M, Thier RH, Mudde CM, Harren FJM, Barendse GWM, Blom CWPM (1993) Submergence-induced ethylene synthesis, entrapment, and growth in 2 plant-species with contrasting flooding resistances. Plant Physiol 103:783–791

Vogel JP, Woeste KE, Theologis A, Kieber JJ (1998) Recessive and dominant mutations in the ethylene biosynthetic gene ACS5 of *Arabidopsis* confer cytokinin insensitivity and ethylene overproduction, respectively. Proc Natl Acad Sci USA 95:4766–4771

Vreeburg RAM, Benschop JJ, Peeters AJM, Colmer TD, Ammerlaan AHM, Staal M, Elzenga TM, Staals RHJ, Darley CP, McQueen-Mason SJ, Voesenek L (2005) Ethylene regulates fast apoplastic acidification and expansin A transcription during submergence-induced petiole elongation in *Rumex palustris*. Plant J 43:597–610

Vriezen WH, Zhou ZY, Van Der Straeten D (2003) Regulation of submergence-induced enhanced shoot elongation in *Oryza sativa* L. Ann Bot 91:263–270

Vriezen WH, Achard P, Harberd NP, Van Der Straeten D (2004) Ethylene-mediated enhancement of apical hook formation in etiolated *Arabidopsis thaliana* seedlings is gibberellin dependent. Plant J 37:505–516

Wang KLC, Yoshida H, Lurin C, Ecker JR (2004) Regulation of ethylene gas biosynthesis by the *Arabidopsis* ETO1 protein. Nature 428:945–950

Wang NN, Shih MC, Li N (2005) The GUS reporter-aided analysis of the promotor activities of Arabidopsis ACC synthase genes AtACS4, AtACS5, AtACS7 induced by hormones and stresses. J Exp Bot 56:909–920

Woeste KE, Kieber JJ (2000) A strong loss-of-function mutation in RAN1 results in constitutive activation of the ethylene response pathway as well as a rosette-lethal phenotype. Plant Cell 12:443–455

Yamagami T, Tsuchisaka A, Yamada K, Haddon WF, Harden LA, Theologis A (2003) Biochemical diversity among the 1-amino-cyclopropane-1-carboxylate synthase isozymes encoded by the *Arabidopsis* gene family. J Biol Chem 278:49102–49112

Yanagisawa S, Yoo SD, Sheen J (2003) Differential regulation of EIN3 stability by glucose and ethylene signalling in plants. Nature 425:521–525

Yang SF, Hoffman NE (1984) Ethylene biosynthesis and its regulation in higher-plants. Annu Rev Plant Physiol Plant Mol Biol 35:155–189

Zhao YD, Christensen SK, Fankhauser C, Cashman JR, Cohen JD, Weigel D, Chory J (2001) A role for flavin monooxygenase-like enzymes in auxin biosynthesis. Science 291:306–309

Plant Cell Monogr (10)
L. Bögre and G. Beemster: Plant Growth Signaling
DOI 10.1007/7089_2007_162/Published online: 7 February 2008

Light and the Control of Plant Growth

Enrique López-Juez (✉) · Paul F. Devlin

School of Biological Sciences, Royal Holloway, University of London, Egham TW20 0EX, UK
e.lopez@rhul.ac.uk

Abstract Light is the source of energy for plants, and as a result is a key environmental cue controlling their growth. The need to reach the appropriate light environment, and then to maximise its capture, leads to different organs showing contrasting growth regulation by light: generally negative in aerial elongating organs (hypocotyl, internodes, petioles) and positive in photosynthetic ones (leaf blades), as well as in roots. The basis of the growth repression of elongating organs by photoreceptors appears to involve the suppression of hormone signalling pathways, particularly those of auxin and gibberellin, although the mechanisms of control prior or subsequent to light exposure differ. Less is known about growth promotion by light in cotyledons, shoot meristem and leaves, but this promotion is closely associated both with the repression of elongation growth in hypocotyls and with the differentiation of photosynthetic cells: molecular interference with one process generally affects the others. The possible nature of these links is discussed. Lastly, light fluence rate controls the internal anatomy of leaves, and this highlights a role for chloroplasts themselves as sources of growth-regulatory signals.

1 Introduction

The responses of higher plants to light have an impact on almost all aspects of growth. This is inevitable given the nature of plants as light-harvesting devices, which satisfy their energy needs, and ultimately those of most of the biosphere, from solar radiation. From germination, through seedling establishment, leaf development, elongation and flowering to seed maturation, information on the light environment steers plant growth. That steering is a combination of positive and negative growth stimuli. Induction of germination is a positive stimulus promoting growth of the embryo, causing first the radicle, then the cotyledons, to break out of the seed coat. Seeds whose germination is light promoted are generally small with few reserves. Following germination, if the seed is under soil the hypocotyl, epicotyl or mesocotyl (depending on the architecture of the seedling) elongates rapidly pushing the shoot apical meristem into the light. However, upon emergence, light simultaneously acts negatively to inhibit this elongation and positively to promote expansion of the cotyledons and to stimulate the apical meristem to produce new leaves. Once the plant is established, light promotes expansion of the new leaves, but also inhibits elongation of the leaf petiole and of internodes formed between leaves. Overall this combination of responses serves a fitness purpose: to promote photosynthetic development under the best possible environment, or to otherwise invest in elongation growth to reach such an environment (Whitelam and Halliday 2007) (Fig. 1).

The photosynthetic environment can, nevertheless, be a highly varied one, with maximum light intensity (fluence rate or irradiance) levels that individual leaves find themselves under varying naturally over at least two orders of magnitude, up to over 2000 μmol m^{-2} s^{-1} at noon under full sunlight in temperate latitudes. As a result growth-related developmental decisions will modulate the rate of leaf production, the surface area occupied by the available leaf biomass, and the internal anatomy and leaf absorption characteristics of those leaves (Walters 2005).

One other aspect of growth is also affected by light: the promotion of flowering is strongly influenced in many species. The transition to flowering involves a dramatic promotion of growth involving the production and expansion of the floral organs, often accompanied by the rapid elongation of a flowering stem. Depending on the species, light can serve to monitor the growth environment or the season, and either act as a promoter or an inhibitor of flowering (Whitelam and Halliday 2007). However, an analysis of this light control of flowering responses is beyond the scope of this chapter.

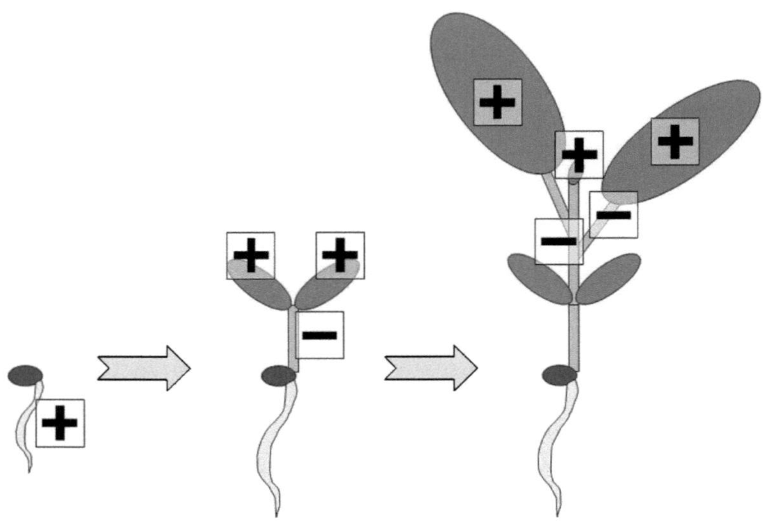

Fig. 1 Contrasting growth responses of different seedling organs to light. Light promotes (+) root extrusion during seed germination, root elongation (at least in response to high red to far-red ratio light) and leaf initiation and expansion during seedling growth, and leaf palisade division and anticlinal elongation during acclimation to high fluence rate. Light inhibits (−) hypocotyl and internode elongation during seedling growth and during plant acclimation to sun versus shade

Overall a consideration of photobiology is an indispensable component of a dissection of plant growth responses. Photobiology can also serve as a tool in this respect: light is a relatively easy-to-modulate environmental cue that can place such developmental responses under the tight control of an environmental switch. Although many years and a bewildering number of studies have been devoted to this area of research, interest in it is bound to continue.

2 Biology of the Photoreceptors

Plants possess three main classes of photomorphogenic photoreceptors: phytochromes, cryptochromes and phototropins (see Jiao et al. 2007). Those first identified, the phytochromes, act "photochromically" in the red and far-red region of the spectrum, being reversibly activated by red wavelengths and inactivated by far-red. This "ancestral" type of phytochrome is also termed type II or light stable, and it is encoded by four genes in *Arabidopsis*, *PHYB–PHYE* (Clack et al. 1994), or tomato, *PHYB1*, *PHYB2*, *PHYE*, *PHYF* (Hauser et al. 1997), of which *PHYB* is most highly expressed. In angiosperms, however, a further type of phytochrome, type I or light labile, is encoded by *PHYA*, accumulates in the dark, absorbs a broad range of wavelengths, notably far-red, is exquisitely light sensitive and is not photoreversible (Shinomura et al. 1996). A second class of photoreceptors are the blue light-absorbing cryptochromes, in *Arabidopsis* the products of *CRY1* and *CRY2*. The phototropins, in *Arabidopsis* the products of *PHOT1* and *PHOT2*, also absorb blue light. The phototropins harbour a flavin chromophore in a LOV domain. Plant genomes also encode other LOV domain-containing proteins, at least some of which have recently been found to be capable of photoperception (Kim et al. 2007).

During the initial transition from darkness to light, a dramatic modification of plant growth programme takes place, from skoto-morphogenesis (dark form) to photo-morphogenesis (light form), and this is accompanied by a massive transcriptional reprogramming (Jiao et al. 2007). The vast majority of such gene expression changes are controlled by the phytochromes and the cryptochromes (Ma et al. 2001). Phototropins, associated with plasma membrane, primarily control short-term responses to light, often directional light, for example tropisms, stomatal opening and chloroplast relocation. An exception is leaf lamina expansion, seemingly a developmental response, but strongly dependent on blade tropism towards the light (Takemiya et al. 2005). At least in the first few hours after light exposure, cryptochromes mediate the majority of gene expression changes initiated by blue light (Ohgishi et al. 2004), while phototropins mediate their actions through cytoskeletal or membrane-based responses.

The transduction of phytochrome and cryptochrome signals, which often interact, occurs through multiple pathways, many of them based in the nucleus. A key theme is the proteolysis-mediated removal of repressors for photomorphogenesis in the light that are otherwise active in the dark. Those repressors include COP1, DET1 and its associated complex, and the COP9 signalosome (CSN). DET1 and its partner DDB1 can modify chromatin structure and also enhance ubiquitin conjugating enzymes (Yanagawa et al. 2004). COP1 has ubiquitin ligase activity, and the CSN regulates the activity of another type of ubiquitin ligases (SCF) by post-translational modification (de-rubylation, Wei and Deng 2003). Among the positive regulators of photomorphogenesis targeted by these repressors, transcription factors of the bZIP (HY5 and HYH) and bHLH (PIF3 and PIL1) classes are the best known.

3 Light Promotes Radicle Growth and Germination

A classic example of the light control of seed germination is seen in the Grand Rapids cultivar of lettuce. Such control, and the demonstration of its photoreversibility, was a key milestone in the identification of phytochromes (Borthwick et al. 1952).

Germination has traditionally been considered the result of imbibition and the initiation of cell expansion in the embryo, causing it to break through the seed coat. A recent, elegant analysis (Masubelele et al. 2005), however, has unequivocally demonstrated that the activity of the root apical meristem (RAM) is at the heart of the process. Radicle protrusion is preceded by cell cycle entry in the RAM. Global transcriptome analysis revealed that the expression of 2000 genes changed prior to visible radicle protrusion, with regulators of cell cycle re-entry (six D-type and two A-type cyclins) appearing among them; up-regulation of the levels of one of the D-type cyclins successfully accelerated germination.

Light and the plant hormones gibberellins (GAs) are very well documented triggers of germination. *Arabidopsis thaliana* seed germination, like that of lettuce, is promoted by light. The process is exquisitely light sensitive, being initiated by fluences of light present in seconds of moonlight, and this is mediated by the phytochrome A (phyA) photoreceptor (Shinomura et al. 1996). The action of this and other phytochromes is mediated by GA, as shown by the inability of GA-deficient plants to germinate in the dark or light. Indeed, phytochrome rapidly up-regulates the transcription of *GA3ox1*, a gene that converts the last inactive GA into GA_4, the active GA in *Arabidopsis*. This up-regulation is defective in the *phyB* mutant (Yamaguchi et al. 1998).

The control of GA biosynthesis takes place through a short signalling cascade, using established phytochrome-interacting transcriptional regulators. PIF1/PIL5, one such protein, is directly destabilised upon interaction with active phytochrome, and is itself a negative regulator of the GA biosynthetic genes (Oh et al. 2006). In turn, the action of GAs, which directly or indirectly leads to the cell cycle re-entry in the RAM, occurs through the destabilisation of the growth-repressive DELLA proteins. This is shown by the light-independent, constitutive germination of DELLA multiple mutants (Cao et al. 2005).

4
Light Inhibits Hypocotyl, Internode and Leaf Petiole Extension Growth

4.1
Photobiology of Extension Growth

Plants show a tremendous plasticity in their response to light. Negative responses essentially serve the purpose of preventing unnecessary investment of resources, but can also be conceptualised as positive growth responses to the absence of light, or to light conditions sub-optimal for photosynthesis. The default state in a germinated seedling in the absence of any light stimulus is elongation growth. The need for photosynthesis in order to obtain energy requires that all available energy is invested in reaching the light. Once in the light such investment is no longer advantageous and resources can be allocated more efficiently in promotion of photosynthetic development.

The inhibition of hypocotyl elongation was the basis of the first screen for mutants deficient in aspects of photoperception. Koornneef and collaborators (1980) grew mutagenised seed of *Arabidopsis* in white light and identified a number of mutants showing a long hypocotyl. In doing so they identified mutants in the two major photoreceptors responsible for this light-induced inhibition of elongation growth: the red/far-red light photoreceptor, phyB, and the blue light photoreceptor, cry1.

In fact, all of the phytochromes and cryptochromes act in the inhibition of elongation growth. The combined action of this battery of photoreceptors gives plants the ability to respond appropriately to a wide range of conditions. Although the majority of these photoreceptors have little effect on a wild-type seedling in white light, their contribution can be observed under specific conditions. In etiolated seedlings high levels of phyA accumulate making the seedlings very sensitive to small amounts of light. Any wavelength of light is capable of producing the small amount of phyA active phytochrome (P_{fr}) required to initiate a very low fluence response (VLFR). The VLFR triggers an early inhibition of hypocotyl elongation, but this response accounts for only a minor effect and is quickly saturated. In red or white light phyA is then rapidly degraded as the majority of the phyA pool will be converted to the labile P_{fr} form. PhyA appears to play an "antenna" role in de-etiolation under these conditions. The light-stable phytochromes play a major role in inhibiting hypocotyl elongation in red or white light, with phyB the major player and lesser roles for phyC, D and E being revealed in the absence of phyB. By contrast, in far-red light phyA plays a major role. The maintenance of a large pool of phyA in the stable P_r form prevents degradation and allows a phenomenon known as the FR high irradiance response (HIR), a response to prolonged irradiation that causes a dramatic inhibition of elongation growth. PhyA is the only phytochrome capable of inhibiting elongation growth in far-red light.

The needs of the established plant remain the same as those seen during de-etiolation, the priority being to gain maximum advantage from the light environment. The shade avoidance response is a response to competition from neighbouring plants. Plants are able to detect light that has been reflected from a neighbouring plant by perceiving the change in light quality. Plants absorb strongly in the red and blue wavelengths due to absorption by chlorophyll, but they reflect far-red wavelengths. Plants are able to monitor the red/far-red (R : FR) ratio of incident light and can interpret a reduction in the ratio as evidence of a neighbour growing closely alongside that may overtop it in future. A low R : FR ratio triggers a pronounced elongation growth response, reduction in branching and reduction in leaf area. Prolonged shade eventually triggers a flowering response as the plant ensures production of offspring as a last resort. This system of using far-red as a reference wavelength against which to compare a reduction in red wavelengths is vastly superior to a simple response triggered by a reduction in photosynthetically active radiation (PAR). An inanimate object alongside a plant may slightly reduce the intensity of incident PAR to the same extent as a neighbouring plant, but would not pose a future threat of severe shading.

The reversible, photochromic nature of the phytochromes makes them ideal to detect this change in R : FR ratio. As in the control of de-etiolation via the LFR, the P_r conformation is inactive and the P_{fr} conformation is active, suppressing elongation growth and flowering. A low R : FR ratio results in a loss of P_{fr} triggering the shade-avoiding phenotype. PhyB, the major phytochrome responsible for the LFR in de-etiolation, is the main player in shade avoidance affecting all aspects of the response. *phyB* mutants in a range of species display a constitutively shade-avoiding phenotype, due to the constitutive absence of phyB P_{fr} and a greatly reduced response to shade. The majority of work characterising the phytochromes involved in the shade-avoidance response has been carried out in the model plant, *Arabidopsis*. Here phyD and phyE also play more minor roles which are most apparent when looked for in the absence of phyB. PhyD is involved in the control of petiole elongation and flowering, while phyE is involved in the regulation of internode elongation and flowering. PhyA also plays a key role, but here it acts as a moderator of shade avoidance. Remarkably, in a multiple phytochrome *phyA phyB phyE* mutant of the normally rosette-forming plant *Arabidopsis*, internodes are constitutively formed, implying that even the rosette growth habit, resulting from constitutive inhibition of internode elongation, can be reversed by environmental sensory pathways.

4.2 Cellular Basis and Signalling Processes in the Control of Extension Growth

How does the repression of elongation growth by light take place? It is important to note that although the photoreceptors acting in shade avoidance are the same as those acting in de-etiolation, recent work has also shown that signalling elements, and probably elementary processes, unique to each process also exist (Roig-Villanova et al. 2007). An elegant study (Gendreau et al. 1998) analysed the cellular basis of the differential growth response of the hypocotyl in darkness and light, and identified a number of key differences. The dramatically accelerated hypocotyl elongation in the dark was due to increased cellular expansion. However, the extra growth took place almost exclusively at the top of the hypocotyl, i.e. a clear gradient of extensibility was seen longitudinally in the dark. Furthermore, although no new cells were formed, cells endoreduplicated more frequently under those conditions, with 16C (16 times the haploid genome content) nuclei being detected only in the dark. A general correlation between cell size and degree of endoreduplication is well known (Melaragno et al. 1993; Sugimoto-Shirasu et al. 2002). Recently a protein, IPD1, which is actively involved in promoting endoreduplication in hypocotyls in the dark (exclusively) and whose expression is suppressed by light, has been identified (Tsumoto et al. 2006). The study of Gendreau and collaborators also observed differentiation of epidermal hypocotyl cells in the light: endoreduplication became less prominent, with nuclei never exceeding 8C, and all epidermal cells, sampled longitudinally, elongated equally. In other words, a coordinated change of elongation programme along the complete organ took place in the light.

Such an organ-wide change in programme again suggests the possibility of involvement of hormonal growth regulators. Auxins and GAs are the best known hormones in the control of elongation growth. Intercellular transport is central to auxin biology. Interestingly, auxin transport inhibitors have no effect on the ability of etiolated seedlings to elongate, while they reduce the height of hypocotyls in the light (Jensen et al. 1998; Shinkle et al. 1998). This indicates that dark growth has little sensitivity to changes in auxin concentration, either because it is auxin-independent, or because it is supersaturated for auxin levels, while upon transition to light, auxin transport becomes central to growth control. Classical auxin transporters belong to the PIN family, but members of the ATP-binding cassette (ABC) family of transmembrane proteins are also involved in auxin transport (Lin and Wang 2005). Loss of function mutations in the AtPGP1 and AtMDR1 ABC transporters results in enhanced growth inhibition in the light, or what could be described as an enhanced light-sensitivity phenotype (Lin and Wang 2005). This suggests that de-etiolation may be accompanied by a broad loss of auxin flow, presumably from the site of highest auxin concentration, around the seedling apex. Possible support for this notion is the recent observation that shoot phytochrome can control the growth of lateral roots, and that the DR5:GUS reporter, a well-established auxin biosensor, is expressed in phytochrome-deficient *phyB* seedlings closer to the base of the hypocotyl than it is in the wild type (Salisbury et al. 2007). Shade (FR-rich) light, i.e. loss of active phytochrome, promotes a rapid and large auxin response, as revealed in a global transcriptome analysis of whole seedlings by the dramatic elevation of *HAT2* and *HAT4* (Devlin et al. 2003). *HAT2* and *HAT4* are homeodomain-Leu zipper transcription factors which are strongly induced by auxin. The site of this elevation in seedlings is not known, but is likely to be the hypocotyl, given the later expression of these genes in shoot internodes (Carabelli et al. 1993). Shade also rapidly stimulates the expression of two atypical beta helix–loop–helix transcriptional regulators, PAR1 and PAR2, in a negative feedback loop in which they repress specific auxin responsive genes (Roig-Villanova et al. 2007).

Further evidence for a negative role of light in the control of auxin responses is provided by SHY2, also known as IAA3. This protein acts as a repressor of auxin responses. A mutant with constitutively active SHY2 was identified as a suppressor of the *phyB* mutant, causing suppression of the long etiolated hypocotyl of *phyB* (Tian et al. 2002). Auxin responses are mediated by the targeted proteolysis of AUX/IAA proteins, including SHY2/IAA3, and phytochrome interacts with SHY2 and the ubiquitin ligase that specifically targets it for degradation (although no light control of the degradation was observed, Tian et al. 2002).

The sensitivity of tissues to auxin is regulated in other ways. One notable, potentially important observation is that two related transcription factors, HY5 and HYH, desensitise auxin responses (Sibout et al. 2006). HY5-deficient plants appear partially etiolated in the light, and the level of HY5 protein closely correlates with the extent of light response, specifically in the hypocotyl, HY5 levels being minimal in the dark and maximal under the highest fluence rate of light tested (Osterlund et al. 2000). The role of HY5 in auxin responses provides an explanation for the surprising fact that its mutant was re-identified, and the mutated gene cloned, due to its altered root architecture (Oyama et al. 1997).

GA is a second hormone whose role in the suppression of elongation in the light has been extensively tested. Importantly, a study that examined the phenomenon in both *Arabidopsis* and pea observed that chemical or genetic removal of active GA was sufficient to cause a de-etiolated-like phenotype (Alabadi et al. 2004). Could GAs fully account for such a process? This is unlikely to be the case, GA action being more likely one of several growth-triggering pathways under light control. This was serendipitously observed in a study (Lopez-Juez et al. 1995) using the first plant identified to be defective in the phyB protein, the cucumber *long hypocotyl* (*lh*) mutant. Grown under a moderately low fluence rate light, mutant hypocotyls grew to just over twice the length of wild-type ones, with a very small increase in DNA content (quantified from all hypocotyl cells) and with no increase (in fact with a reduction) in active GA levels. Clearly the increased elongation in the phyB-deficient mutant was not mediated by increased GA levels. However, under a higher light fluence a second growth component emerged: elongation was sustained over a longer period, epidermal cells increased in number and total hypocotyl DNA content increased in the mutant. Under those conditions active GA levels were two- to threefold higher in the mutant than in the wild type. This suggested the possibility of GA acting as a mitogen or as a cell proliferation signal, and as an "optional" component of the response to reduced phytochrome activity (Lopez-Juez et al. 1995).

Figure 2 shows the same phenomena in internodes of wild-type and phyB-deficient cucumber. At this time, active GA levels are fourfold greater in apices (where the second internode is developing). Meanwhile the preceding, first internode is already threefold longer, but actually shows smaller epidermal cells in which new transverse cell walls, the result of recent mitosis, are easy to observe (Fig. 2a). As a consequence of this greater cell pool, this internode is capable of much greater additional growth when entering the phase of cell expansion (Fig. 2b; López-Juez, Kobayashi and Kamiya, unpublished observations). This phenomenon strongly resembles the role of active GA in the response to flooding of internodes of deep water rice. In such internodes, when they undergo extraordinarily rapid elongation, GA promotes both cell cycle activation and cell expansion, activating the expression of genes like *Replication Protein A1*—for DNA synthesis—or expansins—for cell wall expansion (van der Knaap et al. 1997). Growth responsive factor 1, or GRF1, is the prototype of a novel class of plant transcription factors and was identified in rice through its role in this response (van der Knaap et al. 2000). Members of the GRF family in *Arabidopsis* also control growth, and distinct GRFs play roles either during cell proliferation, involving both cell growth and division (Horiguchi et al. 2005), or during cell wall expansion (Kim et al. 2003; Doerner, this volume. Indeed, at least two GRF genes appear to be light-regulated in *Arabidopsis* (Dillon, Bögre and López-Juez, unpublished results).

GA activity could also be modulated by regulation of GA signalling. Phytochrome action has been shown to limit the response to externally applied GAs (Lopez-Juez et al. 1995; Reed et al. 1996). GAs target growth-repressive DELLA proteins for degradation (see Sect. 3). DELLAs have recently been shown to constrain growth in shade, and their degradation, shown to take place in response to far-red light supplementation, is necessary but not sufficient to account for the accelerated growth response (Djakovic-Petrovic et al. 2007).

Unravelling the roles of plant hormones is complicated by their mutual interdependence. This brings about a bewildering array of light effects on hormonal growth responses, including brassinosteroids and ethylene, some

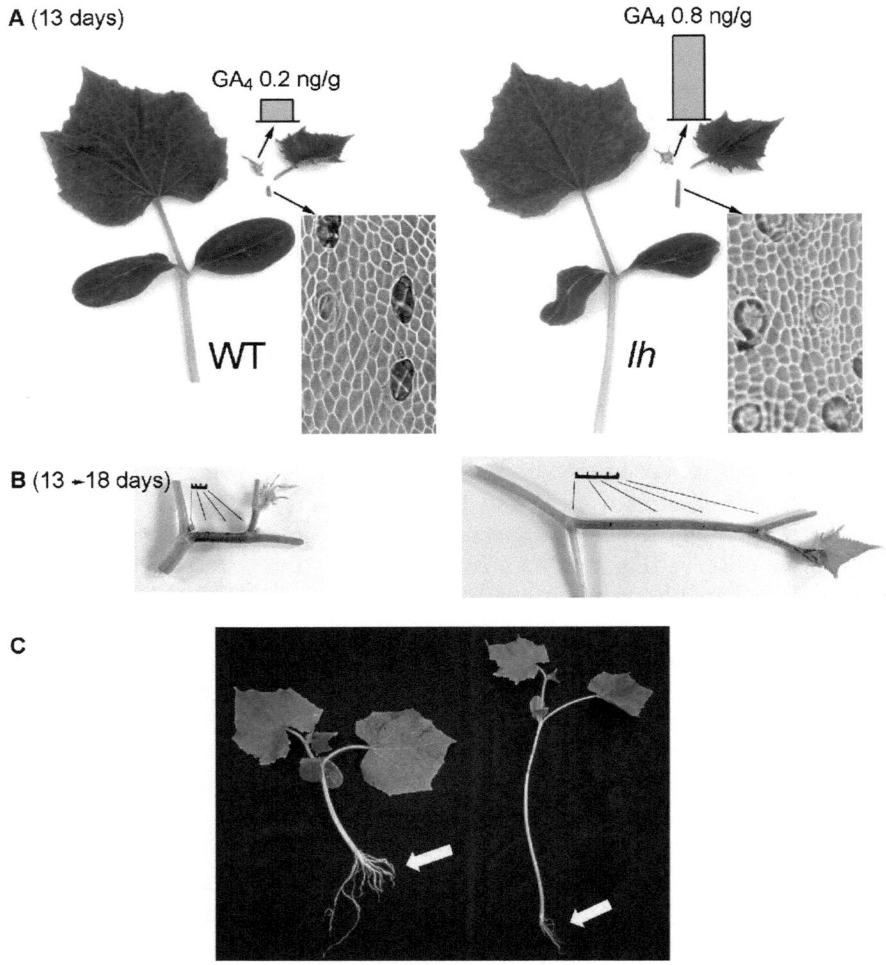

Fig. 2 Elevated active gibberellin (GA) levels precede an increased cell proliferation in extremely elongated internodes of a phytochrome-deficient mutant. **A** Seedlings of wild type (*left*) or phytochrome B-deficient *long hypocotyl* (*lh*) cucumber mutant (*right*) are shown, after dissecting the first internode, the second leaf and the remainder of the shoot apex (including the third leaf, the second internode and the shoot meristem). The first internode of *lh* is already at least threefold longer that that of the wild type, yet its epidermal cells are equal or smaller in size, and in far greater numbers, this pointing to increased cell proliferation. This is preceded, in the shoot apex, by enhanced levels of active GA, in this species GA_4. **B** The increased cell proliferation in the *lh* internode allows for much greater final size after the phase of cell expansion. Tracking of marks placed before this phase started indicates that the entire organ is capable of growth. **C** The *lh* mutant shows increased biomass allocation into hypocotyl and petioles, and reduced into leaf blades and roots (*arrows*)

of which may be indirect (Halliday and Fankhauser 2003; Nemhauser and Chory 2002). A detailed discussion of these is beyond the scope of this chapter. The hormone ethylene may also be one primary mediator of light-dependent growth responses, as shown by the fact that phytochrome defects cause ethylene overproduction in pea, and that restricting ethylene biosynthesis rescues many of the phytochrome deficiency phenotypes and mimics a full light response (Foo et al. 2006). In this case ethylene appears to act, at least in part, by suppressing GA production. Relationships between light and ethylene, and its interactions with GAs and auxin, are discussed in detail by Dugardeyn and Van Der Straeten elsewhere in this volume.

5
Light Promotes Leaf Initiation and Cotyledon and Leaf Blade Expansion

One of the most remarkable aspects of the growth control by light is that it can simultaneously operate in opposite directions in adjacent organs. The transition from skotomorphogenesis to photomorphogenesis, in particular, brings about the expansion of the embryonic cotyledons and the reactivation of the shoot apical meristem (SAM). For example, a combined deficiency in the two major phytochromes, phyA and phyB, and both cryptochromes leads to extraordinarily long hypocotyls in the light, but also to a dramatic delay in leaf initiation (Mazzella et al. 2001). In spite of their importance, among the many light responses, we know little about how photoreceptors control cotyledon expansion and particularly the activity of the SAM.

From the little data available, we do know that the light signal involves, besides the COP/DET group of repressors, the HY5 transcription factor, since its mutant phenotype affects both hypocotyls and cotyledons in same the way that light does. In contrast, the majority of other genes identified as potential primary light targets, through their rapidly regulated expression, appear to have unidirectional effects, i.e. their deletion either causes no phenotype or leads to increased or decreased size of both hypocotyls and cotyledons (Khanna et al. 2006).

An unexpected, interesting observation is the fact that making sucrose directly available to the shoot apex, either by growth in liquid medium or by contact on vertical tissue-culture plates, causes reactivation of the SAM (Araki and Komeda 1993; Roldan et al. 1999). The meristem then gives rise to fully formed leaves, albeit with a very exaggerated petiole and extremely reduced leaf blade. How this sugar response relates to the "natural" light response is unknown, but it is known that loss of a sugar sensor, a specific hexokinase, leads to diminished, rather than enhanced, leaf blade expansion at increased light intensities (Moore et al. 2003). Clearly the potential role of sugar sensing in controlling light responses deserves further attention.

Other information on how light promotes leaf initiation is extremely limited. GAs are potential players in the response, since both GAs and GRFs play roles in leaf blade expansion and in conferring determinacy to the expanding tissue (Hay et al. 2002; Kim et al. 2003). Genetic interference with auxin and brassinosteroid responses also causes some (Tian et al. 2002) or much (Li et al. 1996) reactivation of the shoot apex, but a clear picture is far from emerging. In contrast, downstream agents of the growth response are starting to emerge from transcriptomics data. Earlier gene expression studies analysing responses to light used whole seedlings, this complicating the interpretation of any result from the growth perspective (see Jiao et al. 2007). One study (Ma et al. 2005) for the first time used a spotted oligo microarray to establish the responses to light of dissected cotyledons, hypocotyls and roots. While no key regulators emerged immediately, it was evident that a number of cell wall expansion proteins, including members of the expansin and the xyloglucan endotransglucosylase families, were differentially, sometimes oppositely, regulated by light in cotyledons and hypocotyls. We have recently embarked on a similar analysis of light responses in specific seedling organs, concentrating on the shoot apex and the cotyledons (López-Juez, Dillon, Bögre and Shanahan, unpublished data). Our observations suggest that, while many previously known light-responsive genes follow similar regulation in both organs, changes specific to the shoot apex can also be observed, including rapid down-regulation of RING-finger ubiquitin ligases, and a rapid, transient loss of auxin and ethylene and a gain of cytokinin and GA responses in shoot meristems, preceding or accompanying the time of leaf primordia initiation. Our data also show a large, coordinated promotion of cellular cytoplasmic growth and of cell cycle genes under the control of this environmental cue.

6
A Conflict Between Positive and Negative Responses?

Understanding the opposite growth outcome of photoreceptor action in different organs poses a real challenge, but its ultimate explanation may not be as complex as it seems at first. There are precedents in other light-related responses. For example, the time of flowering is under strict photo-period control in many species, with long days being promotive in some and repressive in others. Within a single species, tobacco, it is indeed possible to find cultivars which are long day, short day and day neutral, implying that a small number of genetic differences can account for the disparity. Although we do not know about the molecular basis of such differences among tobacco cultivars, comparisons between *Arabidopsis* (a long day plant) and rice (a short day one) have been informative (Yanovsky and Kay 2003). In both species the day length is detected through the extent of overlap between the physical presence of light, determined by photoreceptor action, and the timing of the subjective night, during which the CO protein is synthesised. In both the flowering trigger is encoded by the *FT* gene. The difference lies in the regulation of *FT* by CO: in *Arabidopsis* coincident light and CO promote *FT* expression, while in rice they inhibit it (Yanovsky and Kay 2003).

This example could help conceptualise some growth paradoxes. For example, for both radicle and hypocotyl extension GAs act as growth factors. However, during germination phytochrome activity leads to active GA production, through the transcriptional regulation of the corresponding biosynthetic genes, while for hypocotyl extension in the light it is the inactivation of phytochrome (brought about by shade light or by mutations) that brings about the production of the hormone.

Explaining the contrasting, but simultaneous, growth response of different organs might require more complex explanations, and indeed may require organ-specific signal transduction changes. In a few cases, however, the contrasting responses may be explained by a combination of resource allocation and flow of growth regulators between organs. One good example is the reduction in root growth when shade or loss-of-phyB promote longer hypocotyls or shoots (Salisbury et al. 2007). Altered auxin flow from shoot to root has been shown to be associated with this phenomenon. The extent to which loss of phyB in cucumber leads to biomass redistribution, away from leaf blades and roots and into elongating hypocotyl and leaf petioles, is illustrated in Fig. 2c.

7 Is There a Central, Light-Dependent Control of Differentiation?

One fascinating observation is the coordination that is often observed between the responses of different organs. This can lead to unexpected outcomes in one aspect of photomorphogenesis when a separate aspect is altered. Specifically, signals that suppress the elongation of the etiolated hypocotyl often also cause the unfolding of cotyledons, a degree of leaf development and even expression of photosynthetic genes. For example, the loss of brassinosteroids, hormones which also primarily affect cell expansion responses, in *Arabidopsis det2* causes a short hypocotyl and unfolded cotyledons in the dark, but is also accompanied by a degree of leaf development, including the expression of photosynthesis-related genes (Li et al. 1996). The same is true following the chemical or genetic suppression of GA production in pea and *Arabidopsis* (Alabadi et al. 2004), and, to some extent, also following the inhibition of auxin signalling through SHY2 (Tian et al. 2002). A direct screen for mutants, which showed de-repression of photosynthesis-related genes in the dark, yielded the *Dark overexpressor of Cab 1* (*doc1*) mutant. The *doc1* mutation, showing minimal morphological phenotypes at the seedling stage, in darkness or light, nevertheless turned out to be caused by a defect in the *BIG* gene, encoding a callosin-like protein with a role in auxin transport (Gil et al. 2001). Again this indicates a link between a hormone involved in the control of (hypocotyl) organ growth and a response that is part of the differentiation programme of cotyledon photosynthetic cells. The nature of this link is one of the most central, standing questions in the understanding of the control of growth responses by light.

At present it is only possible to speculate about the link above. However, it may be useful to conceptualise the phenomenon as part of the fundamental decision that takes place near meristems, a decision between cellular proliferation (and possibly expansion) on the one hand, and differentiation of mature cells on the other. Cell cycle progression is ultimately driven by genes under the control of transcription factors of the E2F family. E2Fs are post-translationally regulated by the repressive binding of the retinoblastoma-related (RBR1) protein, with the activity of RBR1 being, in turn, controlled by phosphorylation by cyclin-dependent kinases, the primary cell cycle regulators (De Veylder et al. 2007). From deregulated expression studies a picture is gradually emerging in which different E2F factors can have different, even opposite, functions, with E2FA being associated with entry into DNA synthesis (S) phase and E2FB being associated with S phase and with entry into mitosis, while E2FC represses cell cycle activity and initiates differentiation responses (De Veylder et al. 2007). In fact these transcription factors could, themselves, be direct targets of light signalling pathways. For example, expression of E2FC can be observed in dark-grown hypocotyls, and is destabilised in the light (del Pozo et al. 2002). We have directly observed the control of E2FC and E2FB protein levels by light and by photomorphogenesis regulators, COP1, DET1 and CSN subunit 5 (Magyar, López-Juez and Bögre, unpublished observations). This is not completely surprising, given the fact that de-etiolation involves the removal of the repression of the normal growth programme, and that the ancestral function of COP1 and the CSN in other organisms is related to the control of cell cycle and differentiation (Doronkin et al. 2003; Wei and Deng 2003). It is also worth noting that the ectopic expression of RBR in shoot meristems has been shown to result in cellular vacuolation, a phenomenon normally associated with differentiation, and with the increased expression of photosynthesis-related genes (Wyrzykowska et al. 2006). Therefore, photomorphogenic regulators could be at the heart of the decisions involving basic cell growth and differentiation in the proximity of the meristem.

Regulation of cell proliferation or differentiation through RBR1 and E2Fs adds another layer, possibly a basal one, to our understanding of photomorphogenesis. RB, specifically, affects gene regulation by recruiting chromatin remodelling enzymes in animal cells (Du and Pogoriler 2006). Although it is at present unclear whether this is related to RB-dependent gene regulation, it is known that modification of chromatin plays a role in de-etiolation growth responses. For example, DET1 is capable of association with histones and of causing their post-translational modification (Benvenuto et al. 2002), while loss of HAF2, a transcription cofactor, causes reduced acetylation of histone H3 in light-responsive promoters and reduced ability to de-etiolate (Bertrand et al. 2005). Clearly unravelling the growth/differentiation link will be an important milestone in our understanding of the control of growth and of its response to light in plants.

8 Increased Light Quantity Promotes and Adjusts the Growth of Leaves

One final aspect worth considering, the control of the final morphology and internal anatomy of the light-capturing organ, the leaf, by light itself, is being discussed separately (Ferjani et al., this volume). However, we would like to consider here the multiplicity of light sensors in relation to separate aspects of growth control.

Under high fluence rate light, it is well known that the leaf cellular anatomy adapts by forming a thick palisade, made of multiple cell layers, with cells which are anticlinally elongated (perpendicular to the leaf surface). This, therefore, involves both the control of cell divisions and of cell expansion. The ultimate outcome is the generation of a leaf anatomy in which internal shading is high, in which many chloroplasts can position themselves parallel to the direction of the light (minimising its absorption), and in which a number of chloroplasts are exposed to light of a non-photodamaging intensity (Walters 2005). In contrast, under low light, the existing biomass is organised as a thin leaf blade so as to maximise its efficiency in capturing the available irradiance.

These palisade cell responses are under the control of the irradiance of light itself. Interestingly, it has been known for some time that the establishment of a palisade cell morphology has two prerequisites: the presence of blue light (Schuerger et al. 1997, see Fig. 3a) and the presence of functional chloroplasts (Chatterjee et al. 1996, see Fig. 3b). The specificity for blue wavelengths argues for a photomorphogenic light sensor. The nature of that sensor remains, however, elusive, as neither cryptochromes (Weston et al. 2000) nor phototropins alone (Lopez-Juez et al. 2007) appear to be essential. Meanwhile the examination of adjacent green and albino sectors of a variegated mutant has revealed that, while palisade cell elongation was still observable, albeit reduced, in albino sectors, palisade cellular proliferation absolutely requires the presence of green chloroplasts (Tan et al. 2007, see Fig. 3b). In other words, plastids themselves are the photoreceptors for the cell proliferation response of the palisade. Interestingly, the signal for such behaviour is systemic (Yano and Terashima 2001). A candidate signal is photosynthate itself, in the form of mobile sugars. Sophisticated sugar sensory mechanisms are be-

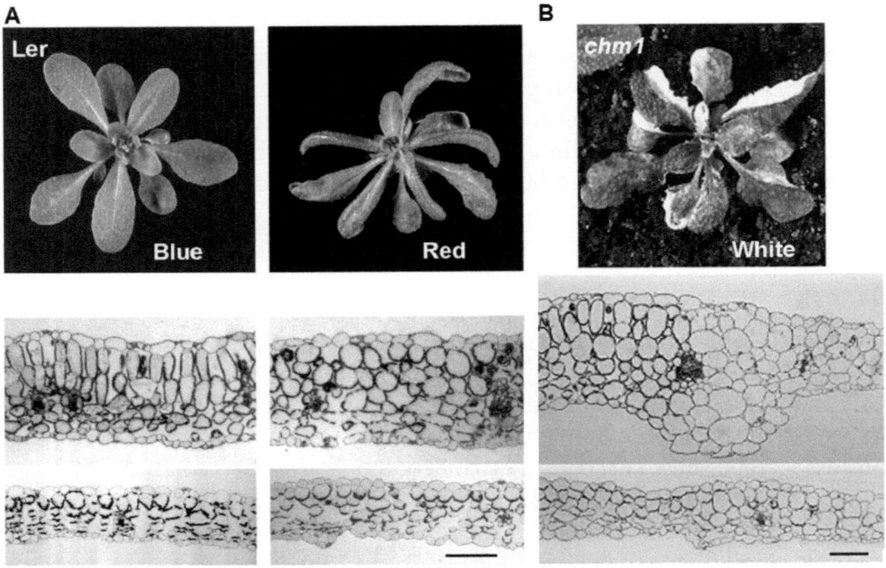

Fig. 3 Multiple light signals control palisade cell division and expansion. **A** *Top row*: *Arabidopsis* wild-type plants grown under elevated irradiance of blue or red light. *Middle row*: sections through leaves from plants corresponding to those shown on the *top row* (at a fluence rate of 200 μmol m^{-2} s^{-1}). *Bottom row*: sections through leaves grown as shown above, but at a reduced fluence rate (15 μmol m^{-2} s^{-1}). Reproduced from López-Juez et al. (2007), with permission. **B** *Top*: *Arabidopsis chm1* variegated mutant plant grown under elevated (600 μmol m^{-2} s^{-1}) fluence rate of white light. *Middle*: section through leaf from a plant identical to the one above. A *green sector* is on the *right* and a *white sector* on the *left*. *Bottom*: section through a *chm1* leaf grown under reduced fluence rate (60 μmol m^{-2} s^{-1}) of white light. *Green sectors* are on the *left*, *white sectors* on the *right*. Note the increased cell number under high light only in the green sector. Reproduced from Tan et al. (2007), with permission. *Scale bar*: 100 μm

ing unravelled (Moore et al. 2003) and, as discussed above (see also Ferjani et al., this volume), they could mediate important aspects of light-regulated growth.

References

Alabadi D, Gil J, Blazquez MA, Garcia-Martinez JL (2004) Gibberellins repress photomorphogenesis in darkness. Plant Physiol 134:1050–1057

Araki T, Komeda Y (1993) Flowering in darkness in *Arabidopsis thaliana*. Plant J 4:801–811

Benvenuto G, Formiggini F, Laflamme P, Malakhov M, Bowler C (2002) The photomorphogenesis regulator DET1 binds the amino-terminal tail of histone H2B in a nucleosome context. Curr Biol 12:1529–1534

Bertrand C, Benhamed M, Li YF, Ayadi M, Lemonnier G, Renou JP, Delarue M, Zhou DX (2005) *Arabidopsis* HAF2 gene encoding TATA-binding protein (TBP)-associated factor TAF1, is required to integrate light signals to regulate gene expression and growth. J Biol Chem 280:1465–1473

Borthwick HA, Hendricks SB, Parker MW, Toole EH, Toole VK (1952) A reversible photoreaction controlling seed germination. Proc Natl Acad Sci USA 38:662–666

Cao D, Hussain A, Cheng H, Peng J (2005) Loss of function of four DELLA genes leads to light- and gibberellin-independent seed germination in *Arabidopsis*. Planta 223:105–113

Carabelli M, Sessa G, Baima S, Morelli G, Ruberti I (1993) The *Arabidopsis Athb-2* and *Athb-4* genes are strongly induced by far-red-rich light. Plant J 4:469–479

Chatterjee M, Sparvoli S, Edmunds C, Garosi P, Findlay K, Martin C (1996) *DAG*, a gene required for chloroplast differentiation and palisade development in *Antirrhinum majus*. EMBO J 15:4194–4207

Clack T, Mathews S, Sharrock RA (1994) The phytochrome apoprotein family in *Arabidopsis* is encoded by five genes: the sequences and expression of *PHYD* and *PHYE*. Plant Mol Biol 25:413–427

De Veylder L, Beeckman T, Inze D (2007) The ins and outs of the plant cell cycle. Nat Rev Mol Cell Biol 8:655–665

del Pozo JC, Boniotti MB, Gutierrez C (2002) *Arabidopsis* E2Fc functions in cell division and is degraded by the ubiquitin-SCFAtSKP2 pathway in response to light. Plant Cell 14:3057–3071

Devlin PF, Yanovsky MJ, Kay SA (2003) A genomic analysis of the shade avoidance response in *Arabidopsis*. Plant Physiol 133:1617–1629

Djakovic-Petrovic T, de Wit M, Voesenek LA, Pierik R (2007) DELLA protein function in growth responses to canopy signals. Plant J 51:117–126

Doerner P (2008) Signals and mechanisms in the control of plant growth (in this volume). Springer, Heidelberg

Doronkin S, Djagaeva I, Beckendorf SK (2003) The COP9 signalosome promotes degradation of cyclin E during early *Drosophila* oogenesis. Dev Cell 4:699–710

Du W, Pogoriler J (2006) Retinoblastoma family genes. Oncogene 25:5190–5200

Dugardeyn J, Van Der Straeten D (2008) Ethylene: inhibitor and stimulator of plant growth (in this volume). Springer, Heidelberg

Ferjani A, Yano S, Horiguchi G, Tsukaya H (2008) Control of leaf morphogenesis by long- and short-distance signalling: differentiation of leaves into sun or shade types and compensated cell enlargement (in this volume). Springer, Heidelberg

Foo E, Ross JJ, Davies NW, Reid JB, Weller JL (2006) A role for ethylene in the phytochrome-mediated control of vegetative development. Plant J 46:911–921

Gendreau E, Hofte H, Grandjean O, Brown S, Traas J (1998) Phytochrome controls the number of endoreduplication cycles in the *Arabidopsis thaliana* hypocotyl. Plant J 13:221–230

Gil P, Dewey E, Friml J, Zhao Y, Snowden KC, Putterill J, Palme K, Estelle M, Chory J (2001) BIG: a calossin-like protein required for polar auxin transport in *Arabidopsis*. Genes Dev 15:1985–1997

Halliday KJ, Fankhauser C (2003) Phytochrome-hormonal signalling networks. New Phytol 157:449–463

Hauser BA, Pratt LH, Cordonnier-Pratt MM (1997) Absolute quantification of five phytochrome transcripts in seedlings and mature plants of tomato (*Solanum lycopersicum* L.). Planta 201:379–387

Hay A, Kaur H, Phillips A, Hedden P, Hake S, Tsiantis M (2002) The gibberellin pathway mediates KNOTTED1-type homeobox function in plants with different body plans. Curr Biol 12:1557–1565

Horiguchi G, Kim GT, Tsukaya H (2005) The transcription factor AtGRF5 and the transcription coactivator AN3 regulate cell proliferation in leaf primordia of *Arabidopsis thaliana*. Plant J 43:68–78

Jensen PJ, Hangarter RP, Estelle M (1998) Auxin transport is required for hypocotyl elongation in light-grown but not dark-grown *Arabidopsis*. Plant Physiol 116:455–462

Jiao Y, Lau OS, Deng XW (2007) Light-regulated transcriptional networks in higher plants. Nat Rev Genet 8:217–230

Khanna R, Shen Y, Toledo-Ortiz G, Kikis EA, Johannesson H, Hwang YS, Quail PH (2006) Functional profiling reveals that only a small number of phytochrome-regulated early response genes in *Arabidopsis* are necessary for optimal deetiolation. Plant Cell 18:2157–2171

Kim JH, Choi D, Kende H (2003) The AtGRF family of putative transcription factors is involved in leaf and cotyledon growth in *Arabidopsis*. Plant J 36:94–104

Kim WY, Fujiwara S, Suh SS, Kim J, Kim Y, Han L, David K, Putterill J, Nam HG, Somers DE (2007) ZEITLUPE is a circadian photoreceptor stabilized by GIGANTEA in blue light. Nature 449:356–360

Koornneef M, Rolff E, Spruit CJP (1980) Genetic control of light-inhibited hypocotyl elongation in *Arabidopsis thaliana* (L) Heynh. Z Pflanzenphysiol 100:147–160

Li J, Nagpal P, Vitart V, McMorris TC, Chory J (1996) A role for brassinosteroids in light-dependent development of *Arabidopsis*. Science 272:398–401

Lin R, Wang H (2005) Two homologous ATP-binding cassette transporter proteins, AtMDR1 and AtPGP1, regulate *Arabidopsis* photomorphogenesis and root development by mediating polar auxin transport. Plant Physiol 138:949–964

Lopez-Juez E, Kobayashi M, Sakurai A, Kamiya Y, Kendrick RE (1995) Phytochrome, gibberellins, and hypocotyl growth—a study using the cucumber (*Cucumis Sativus* L.) *Long Hypocotyl* mutant. Plant Physiol 107:131–140

Lopez-Juez E, Bowyer JR, Sakai T (2007) Distinct leaf developmental and gene expression responses to light quantity depend on blue-photoreceptor or plastid-derived signals, and can occur in the absence of phototropins. Planta 227:113–123

Ma L, Li J, Qu L, Hager J, Chen Z, Zhao H, Deng XW (2001) Light control of *Arabidopsis* development entails coordinated regulation of genome expression and cellular pathways. Plant Cell 13:2589–2607

Ma L, Sun N, Liu X, Jiao Y, Zhao H, Deng XW (2005) Organ-specific expression of *Arabidopsis* genome during development. Plant Physiol 138:80–91

Masubelele NH, Dewitte W, Menges M, Maughan S, Collins C, Huntley R, Nieuwland J, Scofield S, Murray JA (2005) D-type cyclins activate division in the root apex to promote seed germination in *Arabidopsis*. Proc Natl Acad Sci USA 102:15694–15699

Mazzella MA, Cerdan PD, Staneloni RJ, Casal JJ (2001) Hierarchical coupling of phytochromes and cryptochromes reconciles stability and light modulation of *Arabidopsis* development. Development 128:2291–2299

Melaragno JE, Mehrotra B, Coleman AW (1993) Relationship between endopolyploidy and cell size in epidermal tissue of *Arabidopsis*. Plant Cell 5:1661–1668

Moore B, Zhou L, Rolland F, Hall Q, Cheng WH, Liu YX, Hwang I, Jones T, Sheen J (2003) Role of the *Arabidopsis* glucose sensor HXK1 in nutrient, light, and hormonal signaling. Science 300:332–336

Nemhauser JL, Chory J (2002) Photomorphogenesis. In: Somerville CR, Meyerowitz EM (eds) The *Arabidopsis* book. American Society of Plant Biologists, Rockville

Oh E, Yamaguchi S, Kamiya Y, Bae G, Chung WI, Choi G (2006) Light activates the degradation of PIL5 protein to promote seed germination through gibberellin in *Arabidopsis*. Plant J 47:124–139

Ohgishi M, Saji K, Okada K, Sakai T (2004) Functional analysis of each blue light receptor, cry1, cry2, phot1, and phot2, by using combinatorial multiple mutants in *Arabidopsis*. Proc Natl Acad Sci USA 101:2223–2228

Osterlund MT, Hardtke CS, Wei N, Deng XW (2000) Targeted destabilization of HY5 during light-regulated development of *Arabidopsis*. Nature 405:462–466

Oyama T, Shimura Y, Okada K (1997) The *Arabidopsis HY5* gene encodes a bZIP protein that regulates stimulus-induced development of root and hypocotyl. Genes Dev 11:2983–2995

Reed JW, Foster KR, Morgan PW, Chory J (1996) Phytochrome B affects responsiveness to gibberellins in *Arabidopsis*. Plant Physiol 112:337–342

Roig-Villanova I, Bou-Torrent J, Galstyan A, Carretero-Paulet L, Portoles S, Rodriguez-Concepcion M, Martinez-Garcia JF (2007) Interaction of shade avoidance and auxin responses: a role for two novel atypical bHLH proteins. EMBO J 26:4756–4767

Roldan M, Gomez-Mena C, Ruiz-Garcia L, Salinas J, Martinez-Zapater JM (1999) Sucrose availability on the aerial part of the plant promotes morphogenesis and flowering of *Arabidopsis* in the dark. Plant J 20:581–590

Salisbury FJ, Hall A, Grierson CS, Halliday KJ (2007) Phytochrome coordinates *Arabidopsis* shoot and root development. Plant J 50:429–438

Schuerger AC, Brown CS, Stryjewski EC (1997) Anatomical features of pepper plants (*Capsicum annuum* L.) grown under red light-emitting diodes supplemented with blue or far-red light. Ann Bot 79:273–282

Shinkle JR, Kadakia R, Jones AM (1998) Dim-red-light-induced increase in polar auxin transport in cucumber seedlings. I. Development of altered capacity, velocity, and response to inhibitors. Plant Physiol 116:1505–1513

Shinomura T, Nagatani A, Hanzawa H, Kubota M, Watanabe M, Furuya M (1996) Action spectra for phytochrome A- and B-specific photoinduction of seed germination in *Arabidopsis thaliana*. Proc Natl Acad Sci USA 93:8129–8133

Sibout R, Sukumar P, Hettiarachchi C, Holm M, Muday GK, Hardtke CS (2006) Opposite root growth phenotypes of *hy5* versus *hy5 hyh* mutants correlate with increased constitutive auxin signaling. PLoS Genet 2:1898–1911

Sugimoto-Shirasu K, Stacey NJ, Corsar J, Roberts K, McCann MC (2002) DNA topoisomerase VI is essential for endoreduplication in *Arabidopsis*. Curr Biol 12:1782–1786

Takemiya A, Inoue S, Doi M, Kinoshita T, Shimazaki K (2005) Phototropins promote plant growth in response to blue light in low light environments. Plant Cell 17:1120–1127

Tan W, Bögre L, Lopez-Juez E (2007) Light fluence rate and chloroplasts are sources of signals controlling mesophyll cell morphogenesis and division. Cell Biol Int. doi: 10.1016/jcellbi.2007.11.011 (in press)

Tian Q, Uhlir NJ, Reed JW (2002) *Arabidopsis* SHY2/IAA3 inhibits auxin-regulated gene expression. Plant Cell 14:301–319

Tsumoto Y, Yoshizumi T, Kuroda H, Kawashima M, Ichikawa T, Nakazawa M, Yamamoto N, Matsui M (2006) Light-dependent polyploidy control by a CUE protein variant in *Arabidopsis*. Plant Mol Biol 61:817–828

van der Knaap E, Jagoueix S, Kende H (1997) Expression of an ortholog of replication protein A1 (RPA1) is induced by gibberellin in deepwater rice. Proc Natl Acad Sci USA 94:9979–9983

van der Knaap E, Kim JH, Kende H (2000) A novel gibberellin-induced gene from rice and its potential regulatory role in stem growth. Plant Physiol 122:695–704

Walters RG (2005) Towards an understanding of photosynthetic acclimation. J Exp Bot 56:435–447

Wei N, Deng XW (2003) The COP9 signalosome. Annu Rev Cell Dev Biol 19:261–286

Weston E, Thorogood K, Vinti G, Lopez-Juez E (2000) Light quantity controls leaf-cell and chloroplast development in *Arabidopsis thaliana* wild type and blue-light-perception mutants. Planta 211:807–815

Whitelam G, Halliday K (2007) Light and plant development. Blackwell, Oxford

Wyrzykowska J, Schorderet M, Pien S, Gruissem W, Fleming AJ (2006) Induction of differentiation in the shoot apical meristem by transient overexpression of a retinoblastoma-related protein. Plant Physiol 141:1338–1348

Yamaguchi S, Smith MW, Brown RG, Kamiya Y, Sun T (1998) Phytochrome regulation and differential expression of gibberellin 3beta-hydroxylase genes in germinating *Arabidopsis* seeds. Plant Cell 10:2115–2126

Yanagawa Y, Sullivan JA, Komatsu S, Gusmaroli G, Suzuki G, Yin JN, Ishibashi T, Saijo Y, Rubio V, Kimura S, Wang J, Deng XW (2004) *Arabidopsis* COP10 forms a complex with DDB1 and DET1 in vivo and enhances the activity of ubiquitin conjugating enzymes. Genes Dev 18:2172–2181

Yano S, Terashima I (2001) Separate localization of light signal perception for sun or shade type chloroplast and palisade tissue differentiation in *Chenopodium album*. Plant Cell Physiol 42:1303–1310

Yanovsky MJ, Kay SA (2003) Living by the calendar: how plants know when to flower. Nat Rev Mol Cell Biol 4:265–275

Plant Cell Monogr (10)
L. Bögre and G. Beemster: Plant Growth Signaling
DOI 10.1007/7089_2007_139/Published online: 27 November 2007

TOR Signaling in Plants

Garrett H. Anderson

Research Associate, Umen Lab, PBIO, The Salk Institute for Biological Studies,
10010 N. Torrey Pines, La Jolla, CA 92037, USA
ganderson@salk.edu

Abstract Plant development is strikingly plastic. Totipotent cell lines called meristems give rise to all post-embryonic organs—leaves, roots, flowers, for example—throughout the life of the plant. Unlike most multicellular eukaryotes, plants can change their final body plans dramatically in response to environmental and nutrient cues. Understanding how plants respond to their environment may shed light on how plants make developmental decisions that determine their final form. TOR is a conserved eukaryotic cell growth regulator. Recent work suggests that the TOR protein kinase plays a critical role in nutrient-responsive growth in a broad range of eukaryotes. After a brief survey of TOR signaling in yeast and mammals TOR signaling in plants will be surveyed.

1 Introduction—Plants and TOR

Target of rapamycin (TOR) proteins are large, highly conserved protein kinases encoded in almost every eukaryotic genome. All TOR proteins identified, from the basal eukaryote *Giardia lamblia* (Morrison et al. 2002) to yeast (Heitman et al. 1991) and metazoans (Chen et al. 1995), share the same set of structural motifs (Schmelzle and Hall 2000). The first ~1800 residues include a set of ~14 N-terminal HEAT repeats (Kunz et al. 2000), a DNA binding domain, a nuclear export signal, and a nuclear import signal (Li et al. 2006). The final 600 residues contain the rapamycin-binding domain (Chen et al. 1995), the kinase domain, and both halves of the split FAT domain (Bosotti et al. 2000). TOR proteins show sequence similarity to Phoshpatidyl-inositol 3-like kinases (PI3K) but are Ser/Thr protein kinases. HEAT repeats and FAT domains mediate protein–protein interactions, suggesting that TOR has the potential to interact with multiple binding partners.

2 TORC1 and TORC2—the Central Themes of TOR Signaling

2.1 Rapamycin and Genetic Evidence for Distinct TOR Activities

In budding yeast and metazoans, TOR proteins control cell growth. This is accomplished through two distinct activities: temporal growth regulation through upregulation of ribosomal processivity, and spatial growth regulation through modulation of cytoskeletal structure—reviewed in (Arsham and Neufeld 2006; Wullschleger et al. 2006).

The original identification of these distinct functions was facilitated by the antiproliferative drug rapamycin. Rapamycin is a cyclic macrolide isolated from the soil eubacterium *Streptomyces hygroscopicus* collected on the island of Rapa Nui, or Easter Island (Thomas et al. 2004). Treatment of yeast cells with rapamycin blocks cell growth immediately after cell division (Kunz et al. 1993). Mutations suppressing this effect mapped to three loci: null mutations mapped to the FKBP12 locus *FKB*, and missense mutations mapped to similar regions of two distinct TOR loci, *Tor1* and *Tor2*, that were later shown to be rapamycin binding sites (Heitman et al. 1991). Rapamycin in the cytosol binds the immunophilin FPBP12. Rapamycin-FKBP12 complexes disrupt cell growth; in *FKBP12* mutants, rapamycin binds TOR but has no inhibitory effect on TOR signaling (Koltin et al. 1991; Koser et al. 1993). *Tor1* null mutants show only mild rapamycin hypersensitivity, indicating that TOR1 is functionally redundant with TOR2 (Cafferkey et al. 1993). *Tor2* null mutants show lethality after a few rounds of cell division (Schmidt et al. 1996); cells show aberrant bud formation implying a defect in cytoskeletal organization distinct from rapamycin treatment. *Tor1 Tor2* double mutants phenocopy rapamycin treatment: like rapamycin-treated cells, double-mutants divide but fail to grow (Helliwell et al. 1994, 1998). This genetic work led to the conclusion that there are two TOR activities: a rapamycin-sensitive activity promoting cell growth shared by TOR1 and TOR2, and a rapamycin-insensitive activity regulating cytoskeletal organization in cell division unique to TOR2.

2.2 TOR Complexes

These two TOR activities have been linked to two distinct protein complexes, TORC1 and TORC2 (Loewith et al. 2002). Each has a sister complex in mammalian cells (Hara et al. 2002; Kim et al. 2002; Nojima et al. 2003; Sarbassov et al. 2004). TORC1 is comprised of yeast TOR1 or TOR2, LST8, and KOG1; the mammalian orthologues comprising mTORC1 are mTOR, Gβl/mLST8, and Raptor. Raptor/KOG1 functions in TORC1 to recruit substrates for phosphorylation by TOR (Hara et al. 2002; Kim et al. 2002; Nojima et al. 2003). Rapamycin-FKBP12 complexes specifically disrupt TORC1 activity, perhaps by dislodging Raptor from TOR (Oshiro et al. 2004). Nutrient stress may reduce TORC1 activity by increasing the strength of the Raptor-TOR interaction, thus inhibiting Raptor's ability to recruit substrates for phosphorylation by TOR (Hara et al. 2002).

TORC2 is comprised of TOR2, LST8, AVO1, AVO3 (Loewith et al. 2002; Wullschleger et al. 2005), and a few yeast-specific proteins (Reinke et al. 2004); the mammalian orthologues in mTORC2 are mTOR, mLST8, hSin1 and Rictor (Sarbassov et al. 2004; Yang et al. 2006). TORC2 regulates the cytoskeleton in both yeast and mammals (Helliwell et al. 1998; Jacinto et al. 2004; Kamada et al. 2005; Wullschleger et al. 2005).

2.3 Up and Downstream of TOR

There is little conservation among TOR effectors. In budding yeast, TORC1 triggers growth by promoting ribosomal assembly and inhibiting autophagy, largely by mediating the phosphorylation-sensitive subcellular localization of transcription factors—reviewed in Inoki and Guan (2006). Many of these transcription factors govern ribosomal components or autophagy promoting proteins. Active TORC1 simultaneously triggers the nuclear accumulation of transcription factors governing ribosomal components while inhibiting nuclear localization of transcription factors for autophagy-promoting genes. In the absence of nutrients or upon rapamycin treatment, TORC1 is inactive, ribosomal component transcription factors are retained in the cytoplasm, and autophagy-promoting transcription is localized to the nucleus.

Mammalian mTORC1 similarly promotes protein synthesis, but through very different effectors. TORC1 phosphorylates S6K and 4E-BP (Hara et al. 1998). S6K (ribosomal protein rpS6 kinase) phosphorylates the ribosomal regulatory component rpS6; S6K activity requires activating phosphorylation by TORC1. Phosphorylated rpS6 increases ribosomal processivity. 4E-BP (eukaryotic initiation factor 4E binding protein) inhibits mRNA recruitment to the ribosome. TORC1 phosphorylated 4E-BP is inactive, thus increasing mRNA translation.

TORC2 defects can be suppressed by overexpression of the AGC kinase Ypk2 (Kamada et al. 2005), implicating this kinase as a major budding yeast TORC2 effector. mTORC2 is known to phosphorylate the Ser/Thr kinase Akt/PKB (Sarbassov et al. 2005). Akt is also an AGC kinase but not the mammalian homologue of yeast Ypk2.

Like TOR effectors, regulatory elements upstream of TOR are not well conserved. mTORC1 is regulated by amino acid levels, hormonal signals, and cellular AMP levels (Schmelzle and Hall 2000). Budding yeast TORC1 is similarly regulated by amino acid levels and AMP levels, but the specific amino acids sensed are different and the mechanism of sensing is unclear (Crespo and Hall 2002; Wullschleger et al. 2006). AMP concentration, as a measure of cellular energy levels, is sensed by AMPK in yeast and mammals, and perhaps in all eukaryotes (Hardie 2005); however, intermediates that transfer signals from AMPK to TORC1 are not conserved.

2.4 Summary

The central theme of TOR signaling is that there are two TOR activities: (1) a nutrient- and rapamycin-sensitive regulation of protein synthesis located in TORC1, and (2) a nutrient- and rapamycin-insensitive regulation of the cytoskeleton located in TORC2. Although the TORC protein complexes are well known and their global effects on cell growth are quite similar, there is much variation in their effectors and upstream regulators. It should also be pointed out that many effectors of each of these TORCs remain to be identified.

3 TOR Signaling in Plants—Introduction

Plant growth, far more than metazoan growth, is intricately linked to nutrient sensing. In plants, nutrients and stress affect not only the growth rate but also the number of organs produced and the timing of developmental transitions. Given the considerable information about TOR in yeast and mammalian systems, the obvious question for the plant researcher becomes, "how much of this is relevant to plant growth regulation"? Put another way, "how has this ancestral eukaryotic nutrient-sensing mechanism adapted, if at all, to mediate plants' meristem-driven growth"?

To address this question, plant biologists have relied on four tools: bioinformatics, rapamycin, biochemistry, and genetics. For the remainder of this work I will survey the progress made toward understanding TOR signaling in plants using each of these tools.

4 Bioinformatics as a Tool

The complete genomes of a handful of plant and plant-like algal species—*Arabidopsis thaliana* (AGI 2000), *Oryza sativa* (Goff et al. 2002; Yu et al. 2002), *Populus trichocarpa* (Tuskan et al. 2006), *Ostreococcus tauri* and *lucimarinus* (Derelle et al. 2006), and *Cyanidioschyzon merolae* (Matsuzaki et al. 2004)—have been published; the genomes of *Medicago truncatula* (a relative of alfalfa), the moss *Physcomitrella patens*, the club-moss *Selaginella moellendorffii*, the algae *Chlamydomonas reinhardtii*, *Volvox carteri*, *Galdieria sulphuraria* and others are well on their way. Keeping in mind that genes can diverge considerably without losing function and that some conserved proteins have very different functions across species, we can nonetheless ask what TOR signaling components are present in plant genomes. To do this, one need only to blast (Altschul et al. 1997) the Arabidopsis or other plant genome with the protein of interest, and then verify the plant hit by blasting it against the genome of the organism encoding the probe protein; if the strongest hit is to the original probe protein, one can safely conclude that the hit is indeed a homologue.

4.1 TOR Complex Homologues

A clear TOR homologue, AtTOR (At1g50030), has been identified in Arabidopsis (Menand et al. 2002). TOR homologues are easily recognized in all plants with sufficient sequence information. All share the TOR motifs described above.

Arabidopsis encodes two LST8/Gβl homologues (At3g18140, At2g22040). Like LST8 and mammalian Gβl, both Arabidopsis LST8 proteins are comprised of seven WD-40 repeats. WD-40 repeats, like the HEAT repeats and the split FAT domain in TOR, mediate protein–protein interactions.

The TORC1-specific protein Raptor/KOG1 is encoded in two copies in Arabidopsis, AtRaptor1A (At5g01770) and AtRaptor1B (At3g08850)—also called Raptor2 and Raptor1, respectively (Anderson et al. 2005; Deprost et al. 2005). Both *AtRaptor* loci encode proteins comprised of an RNC (Raptor N-terminal Conserved) domain, three HEAT repeats and seven WD-40 repeats, as is common for all Raptor proteins (Shinozaki-Yabana et al. 2000). The RNC domain shows some similarity to a protease domain (Ginalski et al. 2004) but has not been shown to harbor catalytic activity.

The TORC2-specific proteins Rictor/AVO3 and hSin1/AVO1 are not found in any available plant sequence. Indeed, Rictor and hSin1 homologues are found only in the genomes of fungi, metazoans, slime molds, and ciliates.

4.2 Up and Downstream of TOR in Plants

AMPK is a conserved TORC1 regulator in yeast and mammals. AMPK has a readily identifiable homologue in Arabidopsis, At3g01090, and in other plants (Bhalerao et al. 1999; Sugden et al. 1999; Thelander et al. 2004). However, it is unclear how this energy-sensing pathway signals to TORC1.

The best-characterized mammalian TOR effectors are S6K and 4e-BP. There are two clear S6K homologues in Arabidopsis (Turck et al. 1998, 2004), raising the possibility that TOR signaling through this effector is conserved. Biochemical evidence supporting this hypothesis is discussed below. 4E-BP is not identified in the Arabidopsis genome.

Another putative TOR effector in plants is the meiosis regulator Mei2. Though not present in budding yeast or mammals, evidence from fission yeast suggests that Mei2 is intimately connected to TOR signaling.

Mei2 triggers premeiotic DNA synthesis and meiosis in conjugated diploid zygotes under low nutrient conditions (Watanabe et al. 1988; Watanabe and Yamamoto 1994). Mei2 binds fission yeast Mip1/Raptor, marking it as a likely TORC1 substrate (Shinozaki-Yabana et al. 2000). Mei2 transcription is disrupted in Ste20/Rictor mutants, implicating it in TORC2 signaling (Hilti et al. 1999). Mei2 activity is regulated by phosphorylation (Watanabe et al. 1997).

There is a small family of Mei2-like proteins in plants (Anderson et al. 2004). All members share with Mei2 a pair of weakly conserved RRM-type N-terminal RNA recognition motifs and a single highly conserved C-terminal RRM. One member, AML1 (Arabidopsis Mei2-like), suppresses Mei2 signaling defects when overexpressed in fission yeast meiosis-deficient mutants (Hirayama et al. 1997).

4.3 Summary

Plant genomes encode TOR and the core TORC1 components, but no TORC2-specific components are identified. Additionally, some putative TOR effectors and a regulator are conserved. These results argue strongly for the presence of TOR signaling in plants through TORC1.

5 Rapamycin as a Tool

No plant species tested is sensitive to rapamycin (Menand et al. 2002). The soil dwelling nematode *C. elegans*, among others, is also rapamycin insensitive (Jia et al. 2004). Rapamycin sensitivity may be generally weak among organisms associated with soil.

In yeast three-hybrid assays and in vitro, AtTOR and rapamycin form a complex with human FKBP12 but not with AtFKBP12 (Menand et al. 2002; Robaglia et al. 2004; Mahfouz et al. 2006). This result indicates that Arabidopsis insensitivity to rapamycin is due to the failure of AtFKBP12 to form a TORC1-inhibiting complex with rapamycin rather than an insensitivity of AtTOR to rapamycin-FKBP12 complexes.

Unlike land plants, the unicellular green alga Chlamydomonas shows rapamycin sensitivity (Menand et al. 2002; Crespo et al. 2005). Growth is slowed but not arrested completely as in budding yeast. Disruption of the Chlamydomonas FKBP12 locus suppresses rapamycin sensitivity. Sophisticated analysis of Chlamydomonas FKBP12 revealed that the rapamycin-binding pocket in this protein makes fewer hydrogen bonds with rapamycin than human or yeast FKBP12. Complementing rapamycin-insensitive mutants with FKBP alleles engineered to increase the number of bonds with rapamycin yields lines that are more sensitive to rapamycin than are wild-type cells (Crespo et al. 2005).

From the work above one can conclude the following. Rapamycin requires FKBP to exert its effect on cell growth in the single-celled algal model Chlamydomonas as it does in yeast and mammals. Arabidopsis TOR binds rapamycin; its insensitivity to rapamycin treatment may be due to the failure of AtFKBP12 to form a TOR-inhibitory complex with rapamycin (the failure of plants to take up rapamycin has not been excluded). Thus, transgenic Arabidopsis lines expressing mutant FKBP alleles engineered to bind rapamycin, as was done in Chlamydomonas, may provide a means of hypersensitizing Arabidopsis to rapamycin, opening up an avenue to studying TOR signaling which has been so productive in other systems.

6 Biochemistry as a Tool: TOR and TORC Protein Interactions

Biochemical analysis of TOR binding partners is hindered by the fact that AtTOR accumulates in only a limited set of plant tissues. This set includes embryos, endosperm, shoot and root meristems, and cells recently emergent from the meristem (Menand et al. 2002). None of these cell types are conducive to biochemical analysis. *AtTOR* transcripts accumulate in a broader range of tissues (Robaglia et al. 2004). The discrepancy in TOR transcript accumulation and protein accumulation may be due to a small regulatory ORF in the TOR mRNA 5′ untranslated region (Menand et al. 2004; Robaglia et al. 2004). Alternately, a TOR-specific microRNA, mi34, conserved from Arabidopsis to rice (Bonnet et al. 2004) may regulate TOR translational regulation.

To overcome this obstacle, researchers (Mahfouz et al. 2006) have employed a tobacco transfection system (Voinnet et al. 2003) to test for interactions among TOR complex proteins. In this system, AtRaptor1B was shown to interact with the AtTOR N-terminal HEAT repeats (Mahfouz et al. 2006). This interaction is consistent with TOR-Raptor interactions in other systems, and is the only direct biochemical evidence to date for a plant TORC1 complex.

AtS6K1, one of two Arabidopsis ribosomal protein S6 kinases, interacts with AtRaptor1B in vivo in transfected tobacco (Mahfouz et al. 2006). In this system, AtS6K1 kinase activity was inhibited by osmotic stress. Cotransfection with AtRaptor1B restored AtS6K1 kinase activity under osmotic stress (Mahfouz et al. 2006).

The only other known binding partner of AtRaptor1B, identified thorough a targeted yeast two-hybrid assay, is AML1 (Anderson and Hanson 2005). As mentioned above, Mei2p interacts with fission yeast Mip1p/Raptor (Shinozaki-Yabana et al. 2000), suggesting that AML1 and Mei2p are conserved TOR substrates, and AML1 overexpression suppresses some fission yeast meiosis signaling defects, suggesting a conserved function (Hirayama et al. 1997).

In summary, work using a tobacco transfection system indicates that a Raptor-TOR interaction mediated by the TOR HEAT repeats is conserved in plants. This TOR-Raptor interaction is the hallmark of the nutrient- and rapamycin-sensitive complex TORC1. Two proteins, AtS6K1 and AML1, have been shown to interact with AtRaptor1B, marking them as conserved TOR effectors.

7 Genetic Analysis of TOR Signaling in Plants

7.1 Forward Genetics

Land plant insensitivity to rapamycin precludes its use in screens. Some progress has been made using the single-celled alga Chlamydomonas (Crespo et al. 2005), identifying FKBP12 as a critical locus for rapamycin sensitivity (see above). There is, however, the potential that a screen for rapamycin-insensitive Chlamydomonas mutants will uncover rapamycin-resistant alleles of CrTOR, some of which may shed light on TOR activity in this model.

7.2 Reverse Genetics

In the absence of a good forward genetic screen, researchers have relied on reverse genetics of AtTOR, its known binding partner AtRaptor1B and presumed binding partner AtRaptor1A, and on putative downstream effectors and an upstream regulator. This work was done using the insertionally mutagenized, insertion end-mapped Arabidopsis collections (Sussman et al. 2000; Alonso et al. 2003), and elsewhere, and using homologous recombination mediated gene targeting, a promising technique in Physcomitrella (Kamisugi et al. 2006).

7.2.1 Disruption of *AtTOR*

Insertions tagging the single *AtTOR* locus have been identified. Two of these insertions harboring alleles *tor1-1* and *tor1-2* were used to study the effects of *AtTOR* disruption on plant development (Menand et al. 2002). Both insertions disrupt the *AtTOR* coding region upstream of the kinase domain and likely represent null alleles. Consistent with this hypothesis, *tor1-1* and *tor1-2* are recessive to *AtTOR* in heterozygotes, and both mutant homozygotes showed the same phenotype.

AtTOR null mutant homozygotes are embryo lethal (Menand et al. 2002). *AtTOR-/-* zygotes divide to form a suspensor and embryo proper. The suspensor develops normally. Embryonic cell division occurs, but there is no pattern of cell division, daughter cells fail to gain any volume, and the embryo fails to establish an axis of polarity. Development arrests at this stage. *AtTOR-/-/-* endosperms (found in seeds of *AtTOR-/-* embryos) show severe developmental defects. Wild-type endosperm develops as a syncytium (a multinucleate cell) that contains up to 200 nuclei before undergoing cellularization. AtTOR mutant endosperm arrest with about 50 nuclei and fail to cellularize.

7.2.2 Disruption of *AtRaptor1A* and *AtRaptor1B*

Insertion alleles disrupting each *AtRaptor* locus have been described (Anderson et al. 2005; Deprost et al. 2005). Expression analysis of the two genes indicates that they show a similar pattern of expression but different relative levels. *AtRaptor1B* accounts for 80% of total *AtRaptor* transcript accumulation in most tissues. Given the high degree of similarity in their encoded proteins, single mutant homozygotes likely represent partial loss of Raptor function.

AtRaptor1A-/- insertion allele homozygotes show no phenotype (Anderson et al. 2005; Deprost et al. 2005). This phenotype is consistent with the similar pattern but low level of expression of this locus relative to *AtRaptor1B*.

AtRaptor1B-/- mutants described by Anderson et al. (2005) show a range of mild developmental phenotypes. Plants undergo leaf initiation more slowly than wild type and bolt later. *AtRaptor1B-/-* root development is mildly disrupted. Finally, *AtRaptor1B-/-* shoot architecture is altered. The primary shoot is shorter than wild type. Branches off of this primary shoot, as well as secondary shoots emerging from the rosette, are more abundant than wild type but do not differ from wild type in length. This phenotype points to a defect specifically in the maintenance of the primary shoot apex; the increased branching is consistent with a loss of repression of axial meristem activity upon the exhaustion of the primary apical meristem. There does not appear to be any defect in the maintenance of secondary, axial meristems in the *AtRaptor1B-/-* mutant.

AtRaptor1A-/- 1B-/- double mutants arrest development as seedlings with minimal post-embryonic growth on soil and on agar plates (Anderson et al. 2005). Seeds germinate slowly and yield seedlings that are smaller than wild type but otherwise fully formed. Roots show minimal growth on plates. Dark-germinated seedlings show a significant lengthening of the hyopcotyl, indicating that these plants are able to undergo vacuolar-expansion driven growth. The root apical meristem is easily recognized and fully formed in these mutants, although it is smaller than wild type and there is a reduction in the number of files of cells in the root elongation zone. Primordia for leaves one and two, formed embryonically, are present but do not grow. This indicates that the shoot apical meristem has formed but is unable to initiate the post-embryonic growth that characterizes vascular plants. Both the *AtRaptor1B-/-* and the *AtRaptor1A-/- 1B-/-* phenotypes described above have been confirmed by other researchers working with progeny or sibling progeny of the plants described above.

AtRaptor1B-/- single mutants as described by Deprost et al. (2005) show a much more severe phenotype than that described above. Single mutant *AtRaptor1B-/-* homozygotes (called *AtRaptor1*) show arrest at or immediately after fertilization. The authors did not detect even the first zygotic cell division that distinguishes the suspensor precursor cell from the precursor cell of the embryo proper. This phenotype, the authors note, is much more severe than that seen with *AtTOR-/-* disruption (Menand et al. 2002), and it is seen in a genetic background where the *AtRaptor1A* locus is intact and expressed. Significantly, this early embryonic arrest phenotype is also seen in a small fraction (7%) of wild-type embryos grown under similar conditions. The phenotype reported by Deprost et al. may indicate that *AtRaptor1B-/-* embryos are hypersensitive to a stress specific to their lab's growth conditions, consistent with a role for TOR in stress as well as nutrient signaling.

7.2.3 *AtLST8* Mutants have not been Reported

Insertion alleles are available for both *AtLST8* loci. No disruption mutants for either of these loci have been described.

7.2.4 Disruption of Putative TORC1 Regulators and Effectors

No Arabidopsis AMPK mutant has been reported. In the moss Physcomitrella, disruption of both AMPK genes via homologous recombination-mediated gene targeting yields plants which grow normally in continuous light but starve when grown on a light-dark cycle (Thelander et al. 2004). This phenotype is consistent with a failure to negatively regulate growth under low nutrient conditions.

There is biochemical evidence implicating AtS6K1 (Mahfouz et al. 2006) and AML1 (Anderson and Hanson 2005) as TOR effectors in Arabidopsis. Seeds of transgenic Arabidopsis plants expressing AtS6K1 driven by the viral 35S promoter germinate as wild-type under normal conditions but fail to germinate under osmotic stress. This result implicates TOR signaling via AtS6K1 in the response to osmotic stress. Combined with the biochemical work described above, the work on AtS6K suggests that TOR activation of AtS6K1 mediated by AtRaptor1B is sensitive to osmotic stress, and that an excess of inactive AtS6K1 inhibits germination under osmotic stress.

Genetic analysis of AML1 in Arabidopsis (Anderson and Hanson 2005) is complicated by the fact that it is one of five members of a gene family whose constituents show a high degree of similarity both in sequence and expression patterns (Anderson et al. 2004). Lines misexpressing AML1 have not been recovered. However, the accumulation of AML transcripts in the shoot and root meristem embryonic precursors speaks to their importance in plant development. Insertion alleles are available for all AML gene family members. Single and higher order insertion homozygotes grow at the same rate as wild-type but bolt earlier. More significantly, at a lower degree of penetrance, *AML* mutants show meiotic defects, sterility, defects in gametophyte development, and in some allele combinations, developmental arrest as seedlings (Kaur et al. 2006). The *AML* mutant meiotic defects argue strongly for a conserved role of Mei2-like proteins in meiosis regulation across a broad range of organisms. The *AML* seedling developmental arrest is strongly reminiscent of *AtRaptor1A-/- 1B-/-* double mutants, although unlike *AtRaptor* double mutants the *AML* seedling arrest is sometimes accompanied by an absence of root growth.

The larger family of Mei2-like genes in plants also includes the TELs, which share the conserved Mei2-like RRM and overall protein structure with the AMLs but are restricted to expression in the shoot and root meristems. Disruption of TE-1 in maize causes defects in leaf phylotaxy resulting in the terminal maize tassel being enclosed in an ear-like sheath of leaves (Veit et al. 1998).

8
Implications for TOR Signaling in Plants

From the work discussed above, one can make a few conclusions about TOR signaling in plants. All land plants tested are rapamycin insensitive (Menand et al. 2002); in Arabidopsis, this is likely due to the failure of FKBP12 to bind rapamycin and not a refection of a fundamental difference in plant TOR proteins (Menand et al. 2002; Mahfouz et al. 2006). Plants engineered to express rapamycin-binding FKBP proteins may be a valuable means of restoring the usefulness of rapamycin in dissecting plant TOR signaling. Such an approach has been used successfully in the green alga Chlamydomonas (Crespo et al. 2005) to increase the mild rapamycin sensitivity that this organism shows.

TOR is essential for organized cell growth early in embryonic development (Menand et al. 2002). In contrast, Raptor proteins are not essential for embryonic development in the absence of stress. *AtRaptor1A-/- 1B-/-* mutants show slow but otherwise near wild-type embryogenesis (Anderson et al. 2005). This phenotype contrasts with that of *AtTOR-/-* embryos, and indicates that TOR activity in embryogenesis is Raptor-independent.

The presence of Raptor-independent TOR activity in the embryo suggests that in plants as in animals, fungi, and slime molds TOR may function in a Raptor-independent complex. *AtTOR-/-* mutants do not show cell cycle arrest as one sees in rapamycin-treated yeast or in yeast *Tor1- Tor2-* double mutants, both of which lack TORC1 activity. Instead, *AtTOR-/-* embryos proceed through multiple rounds of unordered cell division with little or no cell growth (Menand et al. 2002), much like TORC2 disrupted yeast *Tor2-* mutants. There are no genes encoding TORC2-specific components Rictor/AVO3/Ste20 or hSin1/AVO1/Sin1 in any plant genome. These proteins may not be essential for plant Raptor-independent TOR activity; alternatively, plant homologues of these genes may have diverged so that they are no longer recognizable by sequence similarity searches.

Biochemical work indicates that a core TORC1 interaction between TOR and Raptor is conserved in plants (Mahfouz et al. 2006). The failure of *AtRaptor1A-/- 1B-/-* mutants to make the transition to meristem-driven growth after near normal embryonic development indicates that Raptor is not essential for all cell growth, but that it is essential for post-embryonic, meristem-driven growth. Notably, plant embryonic growth is largely determinate, while meristem-driven growth is remarkably plastic in response to environmental cues, nutrient cues, and stress. Collectively, these results suggest that a nutrient-sensitive TORC1 activity essential for all cell growth in yeast and mammals has been co-opted in land plants to drive cell growth specifically in the meristem, thus playing a major role in directing the plastic, nutrient-sensitive development of land plants rather than (or in addition to) cell-level responses to nutrient and environmental stimuli.

Biochemical work has identified two putative effectors of TORC1 signaling (Anderson and Hanson 2005; Mahfouz et al. 2006). The first of these, AtS6K1, interacts with AtRaptor1B physically, marking it as a putative TORC1 substrate. AtS6K1 is the homologue of a well-characterized TORC1 effector in mammals known to regulate ribosomal processivity. Work in plants indicates that a Raptor protein regulates AtS6K1 activity, and that AtS6K1 may play a role in growth in response to osmotic stress.

The second putative effector, AML1, interacts with AtRaptor1B in a yeast two-hybrid assay. Mei2-like proteins in plants have a conserved role in meiosis (Kaur et al. 2006) and in the transition to flowering (Anderson and Hanson 2005). A more divergent plant Mei2-like protein, TE1, is an important regulator of leaf initiation in maize (Veit et al. 1998). Mei2 in fission yeast acts by binding a noncoding, polyadenylated mRNA-like molecule that tethers it to a specific locus in the fission yeast nucleus (Watanabe and Yamamoto 1994; Yamashita et al. 1998; Shimada et al. 2003). Interestingly, a bioinformatics approach has identified a significant number of similar mRNA-like molecules in plants (MacIntosh et al. 2001), suggesting that much of this intriguing signaling pathway may be conserved from fission yeast to plants.

These interpretations present quite a few hypotheses that labs will no doubt address in the near future. The presence of distinct TORC complexes, suggested by the Raptor-independent TOR activity observed in the embryo, remains to be tested biochemically. The limited TOR expression pattern complicates biochemical analysis; however, work discussed above (Crespo et al. 2005; Mahfouz et al. 2006) suggests that tobacco transfection or algal models may prove useful tools in the biochemical dissection of TOR signaling.

Further resolution of the relative roles of Raptor-dependent vs. Raptor-independent TOR activity in plant development is also an important topic in TOR signaling. Rapamycin has thus far been of little use in the dissection of plant TOR signaling. Generation of a sensitized transgenic Arabidopsis line expressing rapamycin-binding FKBP could change this dramatically, allowing the observation of the effect of TORC1 disruption at specific stages of development. By comparing any rapamycin-induced effects with the inducible RNAi-mediated depletion of TOR, TOR binding partners or TOR effectors, much may be learned about how TOR has been adapted to mediate land plant growth.

References

AGI (2000) Analysis of the genome sequence of the flowering plant *Arabidopsis thaliana*. Nature 408:796–815

Alonso JM et al. (2003) Genome-wide insertional mutagenesis of *Arabidopsis thaliana*. Science 301:653–657

Altschul SF, Madden TL, Schaffer AA, Zhang J, Zhang Z, Miller W, Lipman DJ (1997) Gapped BLAST and PSI-BLAST: a new generation of protein database search programs. Nucleic Acids Res 25:3389–3402

Anderson GH, Hanson MR (2005) The Arabidopsis Mei2 homologue AML1 binds AtRaptor1B, the plant homologue of a major regulator of eukaryotic cell growth. BMC Plant Biol 5.2

Anderson GH, Veit B, Hanson MR (2005) The Arabidopsis AtRaptor genes are essential for post-embryonic plant growth. BMC Biol 3:12

Anderson GH, Alvarez ND, Gilman C, Jeffares DC, Trainor VC, Hanson MR, Veit B (2004) Diversification of genes encoding mei2-like RNA binding proteins in plants. Plant Mol Biol 54:653–670

Arsham AM, Neufeld TP (2006) Thinking globally and acting locally with TOR. Curr Opin Cell Biol 18:589–597

Bhalerao RP, Salchert K, Bako L, Okresz L, Szabados L, Muranaka T, Machida Y, Schell J, Koncz C (1999). Regulatory interaction of PRL1 WD protein with Arabidopsis SNF1-like protein kinases. Proc Natl Acad Sci USA 96:5322–5327

Bonnet E, Wuyts J, Rouze P, Van de Peer Y (2004) Detection of 91 potential conserved plant microRNAs in *Arabidopsis thaliana* and *Oryza sativa* identifies important target genes. Proc Natl Acad Sci USA 101:11511–11516

Bosotti R, Isacchi A, Sonnhammer EL (2000) FAT: a novel domain in PIK-related kinases. Trends Biochem Sci 25:225–227

Cafferkey R, Young PR, McLaughlin MM, Bergsma DJ, Koltin Y, Sathe GM, Faucette L, Eng WK, Johnson RK, Livi GP (1993) Dominant missense mutations in a novel yeast protein related to mammalian phosphatidylinositol 3-kinase and VPS34 abrogate rapamycin cytotoxicity. Mol Cell Biol 13:6012–6023

Chen J, Zheng XF, Brown EJ, Schreiber SL (1995) Identification of an 11-kDa FKBP12-rapamycin-binding domain within the 289-kDa FKBP12-rapamycin-associated protein and characterization of a critical serine residue. Proc Natl Acad Sci USA 92:4947–4951

Crespo JL, Hall MN (2002) Elucidating TOR signaling and rapamycin action: lessons from *Saccharomyces cerevisiae*. Microbiol Mol Biol Rev 66:579–591

Crespo JL, Diaz-Troya S, Florencio FJ (2005) Inhibition of target of rapamycin signaling by rapamycin in the unicellular green alga *Chlamydomonas reinhardtii*. Plant Physiol 139:1736–1749

Deprost D, Truong HN, Robaglia C, Meyer C (2005) An Arabidopsis homolog of RAPTOR/KOG1 is essential for early embryo development. Biochem Biophys Res Commun 326:844–850

Derelle E et al. (2006) Genome analysis of the smallest free-living eukaryote *Ostreococcus tauri* unveils many unique features. Proc Natl Acad Sci USA 103:11647–11652

Ginalski K, Zhang H, Grishin NV (2004) Raptor protein contains a caspase-like domain. Trends Biochem Sci 29:522–524

Goff SA et al. (2002) A draft sequence of the rice genome (*Oryza sativa* L. ssp. japonica). Science 296:92–100

Hara K, Yonezawa K, Weng QP, Kozlowski MT, Belham C, Avruch J (1998) Amino acid sufficiency and mTOR regulate p70 S6 kinase and eIF-4E BP1 through a common effector mechanism. J Biol Chem 273:14484–14494

Hara K, Maruki Y, Long X, Yoshino K, Oshiro N, Hidayat S, Tokunaga C, Avruch J, Yonezawa K (2002) Raptor, a binding partner of target of rapamycin (TOR), mediates TOR action. Cell 110:177–189

Hardie DG (2005) New roles for the LKB1->AMPK pathway. Curr Opin Cell Biol 17:167–173

Heitman J, Movva NR, Hall MN (1991) Targets for cell cycle arrest by the immunosuppressant rapamycin in yeast. Science 253:905–909

Helliwell SB, Howald I, Barbet N, Hall MN (1998) TOR2 is part of two related signaling pathways coordinating cell growth in *Saccharomyces cerevisiae*. Genetics 148:99–112

Helliwell SB, Wagner P, Kunz J, Deuter-Reinhard M, Henriquez R, Hall MN (1994) TOR1 and TOR2 are structurally and functionally similar but not identical phosphatidylinositol kinase homologues in yeast. Mol Biol Cell 5:105–118

Hilti N, Baumann D, Schweingruber AM, Bigler P, Schweingruber ME (1999) Gene ste20 controls amiloride sensitivity and fertility in *Schizosaccharomyces pombe*. Curr Genet 35:585–592

Hirayama T, Ishida C, Kuromori T, Obata S, Shimoda C, Yamamoto M, Shinozaki K, Ohto C (1997) Functional cloning of a cDNA encoding Mei2-like protein from *Arabidopsis thaliana* using a fission yeast pheromone receptor deficient mutant. FEBS Lett 413:16–20

Inoki K, Guan KL (2006) Complexity of the TOR signaling network. Trends Cell Biol 16:206–212

Jacinto E, Loewith R, Schmidt A, Lin S, Ruegg MA, Hall A, Hall MN (2004) Mammalian TOR complex 2 controls the actin cytoskeleton and is rapamycin insensitive. Nat Cell Biol 6:1122–1128

Jia K, Chen D, Riddle DL (2004) The TOR pathway interacts with the insulin signaling pathway to regulate *C. elegans* larval development, metabolism and life span. Development 131:3897–3906

Kamada Y, Fujioka Y, Suzuki NN, Inagaki F, Wullschleger S, Loewith R, Hall MN, Ohsumi Y (2005) Tor2 directly phosphorylates the AGC kinase Ypk2 to regulate actin polarization. Mol Cell Biol 25:7239–7248

Kamisugi Y, Schlink K, Rensing SA, Schween G, von Stackelberg M, Cuming AC, Reski R, Cove DJ (2006) The mechanism of gene targeting in *Physcomitrella patens*: homologous recombination, concatenation and multiple integration. Nucleic Acids Res 34:6205–6214

Kaur J, Sebastian J, Siddiqi I (2006) The Arabidopsis-mei2-like genes play a role in meiosis and vegetative growth in Arabidopsis. Plant Cell 18:545–559

Kim DH, Sarbassov DD, Ali SM, King JE, Latek RR, Erdjument-Bromage H, Tempst P, Sabatini DM (2002) mTOR interacts with raptor to form a nutrient-sensitive complex that signals to the cell growth machinery. Cell 110:163–175

Koltin Y, Faucette L, Bergsma DJ, Levy MA, Cafferkey R, Koser PL, Johnson RK, Livi GP (1991) Rapamycin sensitivity in *Saccharomyces cerevisiae* is mediated by a peptidyl-prolyl cis-trans isomerase related to human FK506-binding protein. Mol Cell Biol 11:1718–1723

Koser PL, Eng WK, Bossard MJ, McLaughlin MM, Cafferkey R, Sathe GM, Faucette L, Levy MA, Johnson RK, Bergsma DJ et al. (1993) The tyrosine89 residue of yeast FKBP12 is required for rapamycin binding. Gene 129:159–165

Kunz J, Schneider U, Howald I, Schmidt A, Hall MN (2000) HEAT repeats mediate plasma membrane localization of Tor2p in yeast. J Biol Chem 275:37011–37020

Kunz J, Henriquez R, Schneider U, Deuter-Reinhard M, Movva NR, Hall MN (1993) Target of rapamycin in yeast, TOR2, is an essential phosphatidylinositol kinase homolog required for G1 progression. Cell 73:585–596

Li H, Tsang CK, Watkins M, Bertram PG, Zheng XF (2006) Nutrient regulates Tor1 nuclear localization and association with rDNA promoter. Nature 442:1058–1061

Loewith R, Jacinto E, Wullschleger S, Lorberg A, Crespo JL, Bonenfant D, Oppliger W, Jenoe P, Hall MN (2002) Two TOR complexes, only one of which is rapamycin sensitive, have distinct roles in cell growth control. Mol Cell 10:457–468

MacIntosh GC, Wilkerson C, Green PJ (2001) Identification and analysis of Arabidopsis expressed sequence tags characteristic of non-coding RNAs. Plant Physiol 127:765–776

Mahfouz MM, Kim S, Delauney AJ, Verma DP (2006) Arabidopsis TARGET OF RAPAMYCIN interacts with RAPTOR, which regulates the activity of S6 kinase in response to osmotic stress signals. Plant Cell 18:477–490

Matsuzaki M et al. (2004) Genome sequence of the ultrasmall unicellular red alga *Cyanidioschyzon merolae* 10D. Nature 428:653–657

Menand B, Meyer C, Robaglia C (2004) Plant growth and the TOR pathway. Curr Top Microbiol Immunol 279:97–113

Menand B, Desnos T, Nussaume L, Berger F, Bouchez D, Meyer C, Robaglia C (2002) Expression and disruption of the Arabidopsis TOR (target of rapamycin) gene. Proc Natl Acad Sci USA 99:6422–6427

Morrison HG, Zamora G, Campbell RK, Sogin ML (2002) Inferring protein function from genomic sequence: *Giardia lamblia* expresses a phosphatidylinositol kinase-related kinase similar to yeast and mammalian TOR. Comp Biochem Physiol B Biochem Mol Biol 133:477–491

Nojima H, Tokunaga C, Eguchi S, Oshiro N, Hidayat S, Yoshino K, Hara K, Tanaka N, Avruch J, Yonezawa K (2003) The mammalian target of rapamycin (mTOR) partner, raptor, binds the mTOR substrates p70 S6 kinase and 4E-BP1 through their TOR signaling (TOS) motif. J Biol Chem 278:15461–15464

Oshiro N, Yoshino K, Hidayat S, Tokunaga C, Hara K, Eguchi S, Avruch J, Yonezawa K (2004) Dissociation of raptor from mTOR is a mechanism of rapamycin-induced inhibition of mTOR function. Genes Cells 9:359–366

Reinke A, Anderson S, McCaffery JM, Yates J 3rd, Aronova S, Chu S, Fairclough S, Iverson C, Wedaman KP, Powers T (2004) TOR complex 1 includes a novel component, Tco89p (YPL180w), and cooperates with Ssd1p to maintain cellular integrity in *Saccharomyces cerevisiae*. J Biol Chem 279:14752–14762

Robaglia C, Menand B, Lei Y, Sormani R, Nicolai M, Gery C, Teoule E, Deprost D, Meyer C (2004) Plant growth: the translational connection. Biochem Soc Trans 32:581–584

Sarbassov DD, Guertin DA, Ali SM, Sabatini DM (2005) Phosphorylation and regulation of Akt/PKB by the rictor-mTOR complex. Science 307:1098–1101

Sarbassov DD, Ali SM, Kim DH, Guertin DA, Latek RR, Erdjument-Bromage H, Tempst P, Sabatini DM (2004) Rictor, a novel binding partner of mTOR, defines a rapamycin-insensitive and raptor-independent pathway that regulates the cytoskeleton. Curr Biol 14:1296–1302

Schmelzle T, Hall MN (2000) TOR, a central controller of cell growth. Cell 103:253–262

Schmidt A, Kunz J, Hall MN (1996) TOR2 is required for organization of the actin cytoskeleton in yeast. Proc Natl Acad Sci USA 93:13780–13785

Shimada T, Yamashita A, Yamamoto M (2003) The fission yeast meiotic regulator Mei2p forms a dot structure in the horse-tail nucleus in association with the sme2 locus on chromosome II. Mol Biol Cell 14:2461–2469

Shinozaki-Yabana S, Watanabe Y, Yamamoto M (2000) Novel WD-repeat protein Mip1p facilitates function of the meiotic regulator Mei2p in fission yeast. Mol Cell Biol 20:1234–1242

Sugden C, Crawford RM, Halford NG, Hardie DG (1999) Regulation of spinach SNF1-related (SnRK1) kinases by protein kinases and phosphatases is associated with phosphorylation of the T loop and is regulated by 5′-AMP. Plant J 19:433–439

Sussman MR, Amasino RM, Young JC, Krysan PJ, Austin-Phillips S (2000) The Arabidopsis knockout facility at the University of Wisconsin-Madison. Plant Physiol 124:1465–1467

Thelander M, Olsson T, Ronne H (2004) Snf1-related protein kinase 1 is needed for growth in a normal day-night light cycle. Embo J 23:1900–1910

Thomas G, Sabatini D, Hall M (2004) TOR Target of Rapamycin. Springer, Berlin Heidelberg New York

Turck F, Kozma SC, Thomas G, Nagy F (1998) A heat-sensitive Arabidopsis thaliana kinase substitutes for human p70s6k function in vivo. Mol Cell Biol 18:2038–2044

Turck F, Zilbermann F, Kozma SC, Thomas G, Nagy F (2004) Phytohormones participate in an S6 kinase signal transduction pathway in Arabidopsis. Plant Physiol 134:1527–1535

Tuskan GA et al. (2006) The genome of black cottonwood, Populus trichocarpa (Torr. & Gray). Science 313:1596–1604

Veit B, Briggs SP, Schmidt RJ, Yanofsky MF, Hake S (1998) Regulation of leaf initiation by the terminal ear 1 gene of maize. Nature 393:166–168

Voinnet O, Rivas S, Mestre P, Baulcombe D (2003) An enhanced transient expression system in plants based on suppression of gene silencing by the p19 protein of tomato bushy stunt virus. Plant J 33:949–956

Watanabe Y, Yamamoto M (1994) *S. pombe* mei2+ encodes an RNA-binding protein essential for premeiotic DNA synthesis and meiosis I, which cooperates with a novel RNA species meiRNA. Cell 78:487–498

Watanabe Y, Lino Y, Furuhata K, Shimoda C, Yamamoto M (1988) The *S. pombe* mei2 gene encoding a crucial molecule for commitment to meiosis is under the regulation of cAMP. Embo J 7:761–767

Watanabe Y, Shinozaki-Yabana S, Chikashige Y, Hiraoka Y, Yamamoto M (1997) Phosphorylation of RNA-binding protein controls cell cycle switch from mitotic to meiotic in fission yeast. Nature 386:187–190

Wullschleger S, Loewith R, Hall MN (2006) TOR signaling in growth and metabolism. Cell 124:471–484

Wullschleger S, Loewith R, Oppliger W, Hall MN (2005) Molecular organization of target of rapamycin complex 2. J Biol Chem 280:30697–30704

Yamashita A, Watanabe Y, Nukina N, Yamamoto M (1998) RNA-assisted nuclear transport of the meiotic regulator Mei2p in fission yeast. Cell 95:115–123

Yang Q, Inoki K, Ikenoue T, Guan KL (2006) Identification of Sin1 as an essential TORC2 component required for complex formation and kinase activity. Genes Dev 20:2820–2832

Yu J et al. (2002) A draft sequence of the rice genome (*Oryza sativa* L. ssp. indica). Science 296:79–92

Plant Cell Monogr (10)
L. Bögre and G. Beemster: Plant Growth Signaling
DOI 10.1007/7089_2007_153/Published online: 26 January 2008

MAP Kinase Cascades Controlling Cell Division, Plant Growth and Development

Takamasa Suzuki · Yasunori Machida (✉)

Division of Biological Science, Graduate School of Science, Nagoya University, Furo-cho, Chikusa, 464-8602 Nagoya, Japan
yas@bio.nagoya-u.ac.jp

Abstract Mitogen-activated protein (MAP) kinase cascades are conserved among all eukaryotes. Plant genomes encode many components of the MAP kinase cascade, suggesting divergent roles in mediating various external and internal signals. In this chapter, we summarize the roles for MAP kinase cascades involved in plant cell division and development. Recently, a MAP kinase pathway, called the NACK–PQR pathway, has been shown to directly control the expansion of the phragmoplast by phosphorylating MAP65-1. In addition, there has been an advance in the understanding of the roles of MAP kinase pathways in embryogenesis, development of stomata and root hairs, and transmission of signals generated by ethylene and auxin.

1 Introduction

The mitogen-activated protein (MAP) kinase cascade is a signal transmission pathway conserved among eukaryotes. The cascade mediates external and internal signals for cellular responses. A typical cascade is composed of three classes of kinases, MAP kinase (MAPK), MAPK kinase (MAPKK) and MAPKK kinase (MAPKKK). MAPKKK phosphorylates MAPKK to activate the latter. MAPKK in turn phosphorylates MAPK to activate the latter. In the course of this signal transmission, the first, subtle signal is amplified sufficiently to induce various cellular responses including transcription, cytoskeleton dynamics and cell division. The MAP kinase cascade exists in higher plants and contributes to the mediation of a variety of signals, including those involving hormones, the environment, pathogenic attacks and internal signals generated during the cell cycle. While the MAP kinase pathways in plants have been studied intensively in research on cellular responses to environmental and pathogenic stimuli, understanding of the roles of MAP kinase pathways in cell division and plant growth and development is still very limited. In this section, the roles of MAP kinase pathways in these latter events are summarized.

2
The Plant Genome Encodes a Variety of MAPKs, MAPKKs and MAPKKKs

Members of each of the MAPK, MAPKK and MAPKKK families have been identified by genome projects using various model plants. In the *Arabidopsis* genome, 20 MAPKs, ten MAPKKs and 60 MAPKKKs were found (MAPK Group, 2002). In addition to the *Arabidopsis* genome, the rice (*Oryza sativa*) and poplar (*Populus trichocarpa*) genomes have been sequenced and shown to encode 15 and 21 MAPKs, and eight and 11 MAPKKs, respectively (Hamel et al. 2006). Thus, plants possess more MAPKs and MAPKKs than yeast (six for each) and the human (ten MAPKs and seven MAPKKs), indicating the complexity of the MAP kinase pathways in plants. Since the genomes of the three model plants described above encode fewer MAPKKs than MAPKs, a single MAPKK might activate multiple MAPKs. However, very few relationships between combinations of plant MAPKKs and MAPKs have been established.

Although more than 60 genes in the *Arabidopsis* genome are predicted to encode MAPKKKs, some of them may not function as MAPKKKs, such as MAP3Kε1, whose homolog, CDC7 in *Schizosaccharomyces pombe* and CDC15 in *Saccharomyces cerevisiae*, have been classified into a different family of protein kinases (Jouannic et al. 2001; Champion et al. 2004). Based on the conserved similarity of the kinase domain, more than 60 predicted proteins in *Arabidopsis* are classified as members of the MAPKKK family, but the complete primary structures of some members are far from the typical MAPKKK structure, which makes it difficult to classify all defined MAPKKK candidates in a single protein family (MAPK Group 2002). It seems necessary to reclassify these divergent members into suitable protein families. Although the precise number of MAPKKKs in *Arabidopsis* is still to be established, this plant apparently possesses a number of protein kinases that are structurally similar to MAPKKK.

3
The MAP Kinase Pathway as a Regulator of Plant Cytokinesis

3.1
Overview of Plant Cytokinesis

Cell division is a process that distributes the duplicated genome evenly to two prospective daughter cells and splits one mother cell into two daughter cells. Cell division is divided into two stages, namely karyokinesis and cytokinesis. Although the processes observed during karyokinesis (breakdown of the nuclear envelope, condensation of the chromosomes, alignment of the chromosomes on the equatorial plane, formation of the spindle and the movements of the chromosomes) are conserved in almost all eukaryotes, those of cytokinesis vary among different groups of organisms. Cytokinesis in animals and fungi is achieved by constriction of the cell membrane from the outside to the inside (Field et al. 1999); however, in plants it is achieved by the creation of cell walls from the inside to the outside during anaphase. The newly synthesized cell wall is called the cell plate.

The formation and development of the cell plates occurs in a plant-specific apparatus called the phragmoplast (Jürgens 2005). The phragmoplast is composed mainly of microtubules, whose plus ends are arranged head-to-head on the equatorial plane and the minus ends are directed towards each of the two daughter nuclei. The cell plate components are considered to be transported along the microtubules by vesicles derived from the Golgi body. However, it has recently been reported that endocytic delivery of cell surface material contributes significantly to the formation of the cell plate (Dhonukshe et al. 2006). Vesicles containing materials for the construction of new plasma membranes and cell walls accumulate at the cell plate and fuse to the pre-existing immature cell plate, resulting in expansion of the cell plate.

As the cell plate expands, vesicles are initially fused to each other to generate a fusion tube-generated network (FTN) at the edge of the cell plate (Samuels et al. 1995). The structure of the FTN gradually changes from that of a tubulovesicular network, through a tubular network and a fenestrated sheet, to the mature cell plate as the fusion of vesicles progresses. Finally, the cell plate reaches the parental plasma membrane and fuses with it. These changes in vesicular and tubular structures proceed from the edge to the center of the cell plate, coupled with the centrifugal expansion of microtubules from the center to the periphery.

The centrifugal expansion of the phragmoplast depends on the dynamic reconstitution of microtubules. In the early stage of phragmoplast development, microtubules form a cylinder-like structure between the two daughter nuclei. Upon expansion of the phragmoplast, this cylinder of microtubules becomes wider, eventually becoming a barrel-like structure. Such structural alteration of the phragmoplast is achieved by the depolymerization of microtubules and polymerization of tubulins in the inner and peripheral regions of the phragmoplast, respectively. The exponentially increasing turnover of microtubules has to support the transport of a vast number of vesicles to the mid-zone at the leading edge of the phragmoplast in order to provide sufficient material to supply the areas of the cell plate that are growing exponentially.

3.2 Regulation of Cytokinesis by a MAP Kinase Pathway

The involvement of a MAP kinase pathway in the cytokinesis of plant cells has been studied intensively. In this section, we describe the MAP kinase pathway that regulates cytokinesis in plant cells called the NACK–PQR pathway (Fig. 1).

NPK1 cDNA has been cloned from tobacco cells and has been shown to encode a member of the MAPKKK family (Banno et al. 1993). The promoter of the gene for NPK1 is active mainly in shoot and root apices in the tobacco plant, suggesting a role in cell proliferation (Nakashima et al. 1998). The NPK1 protein is localized to the nucleus during interphase; when the cells enter M-phase and the phragmoplast is formed from anaphase to cytokinesis, it is localized to the equatorial region of the phragmoplast (Nishihama et al. 2001; Ishikawa et al. 2002). The activity of the protein kinase is cell cycle-dependent and increases from anaphase to cytokinesis (Nishihama et al. 2001). The amount of protein and the activity of NPK1 decline markedly at the end of cytokinesis (Nishihama et al. 2001). Overexpression of the kinase-negative version of NPK1 in tobacco cells and plants results in multinucleate cells that contain stubs of cell walls but does not affect karyokinesis, suggesting that NPK1 plays an important role in the formation of cell walls (Nishihama et al. 2001).

Genetic studies of *Arabidopsis* homologs of NPK1 have provided direct evidence for a role of this MAPKKK in cytokinesis. *Arabidopsis* genome contains three genes for NPK1 homologs; *ANP1*, *ANP2* and *ANP3*, (Nishihama et al.

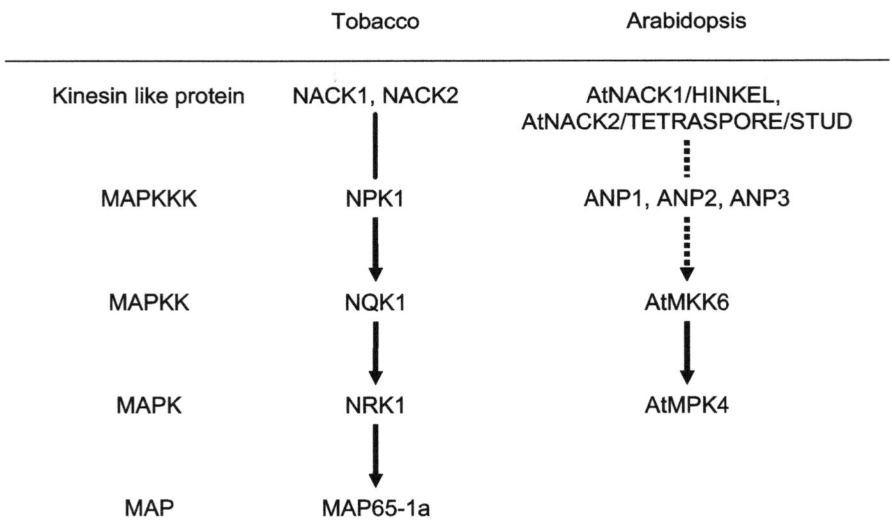

Fig. 1 MAP kinase pathways regulating cytokinesis in plants. All components of the MAP kinase pathways regulating phragmoplast expansion in tobacco and *Arabidopsis* are indicated. Each *line* indicates protein–protein binding between two components and each *arrow* indicates phosphorylation by the upper kinase. Pathways confirmed in biochemical experiments are shown as *solid lines* and *arrows*; unconfirmed pathways are shown as *dashed lines* and *arrows*

1997). *Arabidopsis* plants that have mutations in both *ANP2* and *ANP3* exhibit dwarfism of the plant body and weak defects in cytokinesis, and the presence of a mutation in all three genes, *ANP1*, *ANP2* and *ANP3*, is gametophytic lethal (Krysan et al. 2002). These observations support the idea proposed by the study with tobacco cells (Nishihama et al. 2001).

To identify a downstream factor(s) of NPK1, a tobacco cDNA library was screened using mutant yeast cells that expressed NACK1 and NPK1 and lacked the intrinsic MAPKK, PBS2, which is involved in the osmosensing signal transduction pathway (Soyano et al. 2003). Using functional screening, a cDNA for a MAPKK was obtained and this type of MAPKK was designated NQK1. NPK1 binds and phosphorylates NQK1. Overexpression of the kinase-negative NQK1 in tobacco cells results in cytokinesis defects similar to those produced by the kinase-negative NPK1, the cells become larger than wild-type cells and contain multinuclei and cell wall stubs.

AtMKK6, one of the *Arabidopsis* MAPKKs, is most similar to NQK1 of tobacco, and mutants of the *AtMKK6* gene also show multinucleation and stubs of cell walls resembling those found in *atnack1* mutants (described below) and double mutants for *ANP*s (Soyano et al. 2003). These results suggest a role of NQK1 in the formation of the cell plates. However, unlike the *anp1/anp2/anp3* triple mutant, the disruptions of *AtMKK6* (*AtMKK6-1* and *AtMKK6-2*) are not gametophytic lethal, suggesting the existence of another redundant gene (or genes) or that the currently studied disruptant is not a null mutant.

MAP kinase NRK1, which is located downstream of the NQK1 MAPKK, was identified using the yeast two-hybrid screening method (Soyano et al. 2003). NQK1 binds and phosphorylates NRK1 to activate the latter protein kinase in vitro. Both NQK1 and NRK1 are activated between anaphase and telophase, although the amounts of these proteins do not change throughout the cell cycle. NQK1 is very similar (only two amino acids are different) from the previously identified NtMEK1, which activates p43Ntf6 MAPK (Calderini et al. 2001). NQK1 and NRK1 are located in the equatorial region of the phragmoplast (Calderini et al. 1998; Takahashi et al., unpublished data; Sasabe et al., unpublished data), implying that NPK1, NQK1 and NRK1 are phosphorylated and activated in the equatorial plane of the phragmoplast. However, overexpression of kinase-negative NRK1 does not result in an abnormality of cytokinesis.

A MAP kinase that might function downstream of NQK1/AtMKK6 has been identified by the use of *Arabidopsis* mutants. Although the *Arabidopsis* genome encodes 20 homologs of MAP kinase, only those belonging to the group B MAP kinases exhibit great structural similarity to NRK1. Our data show that recombinant NQK1 and AtMKK6 proteins phosphorylate only AtMPK4 (Takahashi et al., unpublished data). In addition, the mutant *atmpk4-2* contains multinucleated cells with immature cell walls (Soyano et al., unpublished data). These results suggest that AtMPK4 is involved in cytokinesis in *Arabidopsis* cells. Other MAP kinases in group B might also be involved in this cell cycle process.

3.3 Factors Upstream of NPK1 MAPKKK

Activation of the MAP kinase pathway consisting of NPK1, NQK1 and NRK1 is observed only during the late M-phase of the cell division cycle, although the amounts of these protein kinases do not change during the M-phase. Two proteins, designated NACK1 and NACK2, have been identified as proteins activating NPK1 in yeast and have been shown to interact with NPK1 in vitro and vivo (Nishihama et al. 2002). Both proteins bind NPK1 to activate it. They are members of the kinesin-like protein family, which are motor proteins directed towards the plus end of microtubules. Overexpression of the truncated NACK1 protein, which lacks the motor domain, resulted in multinucleation and failure to complete cell plate development. In cells where this occurs, the localization of NPK1 to the phragmoplast equator is also disrupted, suggesting that NACK1 is required for the proper localization of NPK1 and that the activation of NPK1 in a limited region is important for phragmoplast-mediated cytokinesis.

The levels of mRNAs of *NACK1* and *NACK2* increase at the beginning of M-phase and decrease on exit from cytokinesis (Nishihama et al. 2002). The accumulation profile of NACK1 protein is similar to that of its transcripts. This cell cycle-dependent expression of *NACK1* and *NACK2* seems to be crucial in the regulation of the NPK1 activity that is required for phragmoplast expansion. Transcription of both *NACK1* and *NACK2* is regulated by a cis-element, designated the MSA element on its promoter, which mediates activation by a specific class of MYB proteins (Ito et al. 2001). These MYB proteins have three imperfect repeats in the DNA binding domain, as in animal c-MYB proteins, whereas the vast majority of plant MYB proteins have two repeats. At least one member of the MYB protein family, called NtMYBA2, is activated through phosphorylated by cyclin-dependent protein kinase (CDK; Araki et al. 2004). Since NPK1 and NACK1 (and NACK2) have potential sites of phosphorylation by CDKs (Nishihama et al. 1997, 2002), activation of NPK1 by protein–protein interaction with NACK1 might be controlled through phosphorylation by CDKs during anaphase.

Arabidopsis homologs of *NACK1* and *NACK2* have been identified by conventional molecular cloning and have been designated *AtNACK1* and *AtNACK2*, respectively (Nishihama et al. 2002). These genes have also been identified by formal genetic studies and named *HINKEL* and *TETRASPORE/STUD*, respectively (Hulskamp et al. 1997; Spielman et al. 1997; Strompen et al. 2002; Yang et al. 2003). The mutation in *AtNACK1/HINKEL*, transcripts of which can be detected in a number of somatic cells of *Arabidopsis* plants, causes defects in cell division with multinucleation and the formation of cell wall stubs at various developmental stages from embryogenesis to postembryonic development. The mutation in *AtNACK2/TETRASPORE/STUD*, transcripts of which are detected mainly in male gametes (Tanaka et al. 2004), however, causes only the generation of abnormal pollen. Double mutations in *AtNACK1* and *AtNACK2* cause both male and female gametophytic lethality, suggesting partial redundancy of *AtNACK1* and *AtNACK2* in cytokinesis during gamete formation (Tanaka et al. 2004). In contrast to the genomes of tobacco and *Arabidopsis*, the rice genome contains only one gene homologous to *NACK1*, designated *DBS1* (Sazuka et al. 2005). The *dbs1* mutant also shows defective cytokinesis, suggesting the conserved function of *NACK*-related genes among higher plants. The mutation observed in *dbs1* is a single-nucleotide substitution at the splicing donor site, decreasing the amount of mature spliced mRNA for DBS1, which produces a leaky phenotype.

3.4
Targets of the NACK–PQR Pathway

The MAP kinase pathway consisting of NPK1, NQK1 and NRK1, which is activated by NACK1, is designated the NACK–PQR pathway (Soyano et al. 2003). The NACK–PQR pathway identified in tobacco seems to be highly conserved in *Arabidopsis*. Since the MAP kinase pathway itself is a mediator that transmits and propagates the various signals to downstream factors, it is important to identify the targets whose activities are regulated by phosphorylation by the MAP kinase pathway. In the NACK–PQR pathway, one of the targets of MAPK has been identified as MAP65, a member of the microtubule-associated protein (MAP) family, a gene family that is conserved in eukaryotes, including humans, yeasts and plants in the bundling of microtubules (MTs; Sasabe and Machida 2006). In tobacco cells, NtMAP65-1 is phosphorylated during mitosis by CDK and NRK1. The phosphorylated NtMAP65-1 is located on the equatorial plane of the phragmoplast whereas unphosphorylated NtMAP65-1 is more broadly distributed (Sasabe et al. 2006). The overexpression of mutant MAP65-1 with a substitution of the amino acid residue that is phosphorylated by NRK1 results in a delay in cytokinesis. Bundling activity of NtMAP65-1 in vitro is inhibited by phosphorylation by NRK1, and the expression of a non-phosphorylatable form of NtMAP65-1 stabilizes MTs. These results suggest that one of the molecular processes controlled by NRK1 MAP kinase is the regulation of the stability of MTs, and that phosphorylation of NtMAP65-1 by NRK1 MAP kinase contributes to the destabilization of MTs, which might increase the dynamic instability of MTs (depolymerization of MTs and polymerization of tubulins).

The importance of phosphorylation of the MAP65 family protein is also becoming clear from studies with *Arabidopsis*. The *Arabidopsis* genome encodes nine members of the MAP65 protein family (Hussey et al. 2002).

AtMAP65-1 has been studied intensively and has been shown to be able to bundle MTs but not to polymerize tubulin (Smertenko et al. 2004, 2006). It is suggested that the bundling activity of AtMAP65-1 is also regulated by phosphorylation, although the responsible kinase has not been identified (Smertenko et al. 2006). This observation is consistent with the results of studies on NtMAP65-1 obtained by Sasabe et al. (2006).

Another member of the *Arabidopsis* MAP65 family, called AtMAP65-3/ PLEIADE, also has an important role in maintaining the structure of the phragmoplast (Müller et al. 2004). The *ple* mutant was originally identified as the mutant causing short roots and disorganized cell division (Müller et al. 2002). In the *ple* mutant root, multinucleated cells and stubs of cell walls were observed, phenotypes resembling those of other cytokinesis mutants. In contrast with other cytokinesis mutants including *atnack1* and *mkk6*, the *ple* mutant showed abnormal morphology in the roots but not the shoots, suggesting the existence of a homologous gene expressed in the shoots. The lack of biochemical analysis of AtMAP65-3 and mutant analysis of other members of this family prevents us from understanding the roles of this family in plant cell division and development.

4 YODA is a MAPKKK that is Part of a MAPK Signaling Pathway Regulating Cell Differentiation

The roles of MAPKKKs other than NPK1 in cell differentiation are becoming established (Table 1). The *yoda* mutants show distinctive changes in the pattern of cell division during embryogenesis, and the responsible gene encodes a MAPKKK. (Lukowitz et al. 2004). The cell division patterns of *Arabidopsis* embryos are well documented and stereotyped (Jürgens and Mayer 1994). After fertilization, a zygote elongates longitudinally and divides asymmetrically to produce apical and basal cells. The zygotes of *yoda* mutants are defective in this longitudinal elongation and divide into two cells of nearly the same sizes (Lukowitz et al. 2004). Furthermore, some basal cells of *yoda* mutants show longitudinal division planes, which are never observed in the wild-type *Arabidopsis* embryo. These defects in the cell division pattern are related to the lack in most *yoda* mutants of a suspensor, which is derived from basal cells and is not incorporated in the embryo, suggesting the importance of YODA in determining the fate of extra-embryonic cells.

The *yoda* mutation also affects the development of stomata. Since stomata are required for gas exchange and transpiration, which are essential for the survival of land plants, the number and distribution of stomata are regulated genetically. In the *yoda* mutant, the number of stomata increases and their distribution is affected (Bergmann et al. 2004). Although stomata are separated from each other by more than one pavement cell in wild-type plants,

Table 1 MAP kinase pathways in plant development

	MAPKKK	MAPKK	MAPK
Zygote development	YODA[1]		
Stomatal development	YODA[2]	MKK4/5[3]	MPK3/6[3]
Root hair elongation			SIMK[4]
Ethylene signaling	CTR1[5]	SIMKK[6]	SIMK, MKK3[6]
Auxin signaling	NPK1[7]		MKK7[8]

[1] Lukowitz et al. 2004
[2] Bergmann et al. 2004
[3] Wang et al. 2007
[4] Samaj et al. 2002
[5] Kieber et al. 1993
[6] Ouaked et al. 2003
[7] Kovtun et al. 1998
[8] Dai et al. 2006

clustered stomata are observed in *yoda* mutants. These phenotypes are explained by the loss of proper differentiation of the meristemoids into guard mother cells. Considering the abnormal cell divisions of zygotes of the *yoda* mutant, YODA may be an essential component of asymmetrical cell division in plant development.

In the context of stomatal development, two MAPKs and MAPKKs downstream of YODA have been identified recently. Mutations in both *Arabidopsis* MAPKs, namely MPK3 and MPK6, result in numerous stomata and no pavement cells in the epidermis of the cotyledons (Wang et al. 2007). Repression of both MKK4 and MKK5 MAPKKs also increases the number of stomata, as in *mpk3/mpk6* double mutants (Wang et al. 2007). These results demonstrate that the activation of the MAPKK upstream of MPK3 and MPK6 inhibits stomata development and represses the phenotypes caused by the *yoda* mutation. Thus, the MAP kinase pathway, consisting of YODA, MKK4/5 and MPK3/6, regulates the differentiation of stomata. Since clustered stomata are also found in cytokinetic mutants (Nishihama et al. 2001, 2002; Soyano et al. 2003), it might be intriguing to examine whether there is a direct relationship between the components responsible for cytokinesis and asymmetrical cell division.

5 MAP Kinase Signaling in Root Hair Development

A plant root has hairs that originate from a specific file of epidermal cells, called the trichoblast, by tip growth. In the growing tip of root hairs, actin filaments are abundant and are involved in polar growth (Hepler et al. 2001; Sieberer et al. 2005). The actin cytoskeleton is speculated to serve as a guide rail for the polar transport of vesicles. SIMK, a MAPK of alfalfa (lucerne), has been found to localize to the root tip and peripheral spots, suggesting a role in root hair outgrowth (Samaj et al. 2002). This localization is strengthened by the pharmacological stabilization of an actin mesh network by jasplakinolide and weakened by the inhibition of MAPKK by UO126. Expression of a gain-of-function version of SIMK overcomes the inhibition of tip growth by pharmacological inactivation of MAPKK. Regulation of the actin cytoskeleton by a MAP kinase pathway is also observed in animal cells. It is interesting that MAP kinase pathways regulate the cytoskeleton in both cytokinesis and tip growth, although the targets differ – microtubules for the former and actin filaments for the latter.

6 MAP Kinase Pathways in Phytohormone Signaling

The MAP kinase pathways are suggested to be involved in the intracellular transmission of the signals that are generated by various phytohormones, which play various roles in physiological and developmental processes in plants. Ethylene is a gas that regulates fruit ripening, leaf senescence and abscission, flowering and cell elongation (Chen et al. 2005). The *CTR1* gene encodes a protein kinase that is similar in amino acid sequence to an animal MAPKKK, Raf. The *ctr1* mutant shows a constitutive triple response, including short hypocotyls and roots, a thickened hypocotyl and an exaggerated apical hook; these are observed in wild-type plants only when grown in the dark in the presence of ethylene (Kieber et al. 1993). Thus, it was suggested that CTR1 is a negative regulator of the transduction pathway for ethylene signaling.

Although the direct downstream components of CTR1 have not been completely established, some MAPKKs and MAPKs have recently been reported to transduce the ethylene signal. In alfalfa, SIMK and MKK3 are MAPKs that were found to be activated upon ethylene treatment (Ouaked et al. 2003). The transgenic *Arabidopsis* plants expressing SIMKK, a MAPKK of SIMK and MKK3, could activate these downstream MAPKs and phenocopy ethylene treatment. Although this MAPK and MAPKK are genetically located downstream of CTR1 MAPKKK (Ouaked et al. 2003), CTR1 components have not been proven to phosphorylate SIMKK and to act as its negative regulators.

A role of MAP kinase pathways in the response to auxin has also been suggested. For example, it has been shown that tobacco NPK1 suppresses the activity of the auxin-responsive promoter in maize protoplasts (Kovtun et al. 1998). Some MAPKs from *Arabidopsis* are reported to be activated when auxin is applied to plants (Mizoguchi et al. 1994; Mockaitis and Howell 2000).

The AtMKK7, one of the MAPKKs of *Arabidopsis*, is suggested to be a negative regulator in the polar transport of auxin (Dai et al. 2005). However, these components have not been placed in the context of the respective signal cascades and the targets of the pathway have not been identified.

7 Multiple Roles of a MAP Kinase Pathway in Response to Various Signals

Several components described above have also been identified in other MAP kinase cascades, some of which are involved in responses against pathogen attacks and environmental stresses.

In alfalfa, SIMK appears in a variety of physiological processes. As mentioned above, SIMK is involved in the tip growth of root hairs and in response to ethylene. Since ethylene is known to stimulate the elongation of root hairs, the role of SIMK in the tip growth of root hairs might indicate its importance in the signal transduction pathway for ethylene. In addition, it is reported that SIMK is activated in response to osmotic stress (Munnik et al. 1999). It is interesting that two different signals – ethylene and hyper-osmotic stress – use the same MAP kinase cascade. It is an open question whether similar basic adaptations occur in response to different environmental changes or whether one signal represses or increases the response to another. It is also possible that the different responses depend on the types of cell expressing the genes that could be targets of a MAP kinase cascade.

NPK1 also appears in various biological processes. As described above, the NPK1 regulates phragmoplast expansion and is activated during the period from anaphase to telophase of mitosis. In addition, reduction in the expression of NPK1 by virus-induced gene silencing makes tobacco sensitive to a pathogen (Jin et al. 2002). Since the silenced plant exhibited reduced cell size and multinucleated cells, the authors report that the MAP kinase pathway regulating phragmoplast development was also impaired. However, it is uncertain whether the same MAPKKs and MAPKs are involved in both the defense response to pathogens and phragmoplast development.

A role of NPK1 in the stress response has also been suggested. Transgenic maize expressing the catalytic domain of NPK1 has superior freezing and drought tolerance, although in these studies it was not clear which MAP kinase cascade could be activated (Shou et al. 2004a,b). Since in these experiments, the regulatory domain was removed from NPK1 and its expression was ectopic, enhanced tolerance might have been induced by the activation of MAPKs and/or MAPKKs other than the intrinsic targets of NPK1.

ANP1 of *Arabidopsis*, which is orthologous to NPK1 of tobacco, is suggested to be involved in the response to oxidative stress (Kovtun et al. 2000). In mesophyll protoplast cells, ANP1 without the regulatory domain activates MPK3 and MPK6, which can be activated by H_2O_2, suggesting that ANP1 might invoke the response to H_2O_2. Since H_2O_2 is produced on pathogenic infection of plant cells, the signal transduction pathway that includes ANP1 is expected to have a similar effect to that of NPK1 in the defense system of plants against pathogen attack.

Although multiple roles have been suggested for NPK1 and its ortholog ANP1, whether NPK1 and its orthologs have roles in responses to stresses and pathogens other than cytokinesis is not clear. It is critical for our understanding of the functions of these proteins that we should know the sites of expression of the genes for NPK1 and its orthologs in plants. It is worth noting that NPK1 is expressed only in division-competent cells in plants, such as those in which CDK genes are expressed (Nakashima et al. 1998). However, responses to environmental stresses or pathogenic attacks may occur in developmentally mature tissues in plants. In addition, to understand the physiological roles of NPK1 MAPKKK it may also be crucial to identify molecules that are involved in the activation of NPK1 and those that are regulated downstream of NPK1. This may also be true for our understanding of physiological functions of the other MAP kinase cascades. The fact that there are many more MAPKKK and MAPKs than MAPKKs indicates that, rather than linear pathways, MAPK signaling presumably forms interacting networks. Although cross-talk mechanisms must therefore be operating among the several MAP kinase pathways in plants, they have yet to be demonstrated.

References

Araki S, Ito M, Soyano T, Nishihama R, Machida Y (2004) Mitotic cyclins stimulate the activity of c-Myb-like factors for transactivation of G2/M-phase-specific genes in tobacco. J Biol Chem 279:32979–32988

Banno H, Hirano K, Nakamura T, Irie K, Nomoto S, Matsumoto K, Machida Y (1993) NPK1, a tobacco gene that encodes a protein with a domain homologous to yeast BCK1, STE11, and Byr2 protein kinases. Mol Cell Biol 13:4745–4752

Bergmann DC, Lukowitz W, Somerville CR (2004) Stomatal development and pattern controlled by a MAPKK kinase. Science 304:1494–1497

Calderini O, Bögre L, Vicente O, Binarova P, Heberle-Bors E, Wilson C (1998) A cell cycle regulated MAP kinase with a possible role in cytokinesis in tobacco cells. J Cell Sci 111:3091–3100

Calderini O, Glab N, Bergounioux C, Heberle-Bors E, Wilson C (2001) A novel tobacco mitogen-activated protein (MAP) kinase kinase, NtMEK1, activates the cell cycle-regulated p43Ntf6 MAP kinase. J Biol Chem 276:18139–18145

Champion A, Picaud A, Henry Y (2004) Reassessing the MAP3K and MAP4K relationships. Trends Plant Sci 9:123–129

Chen YF, Etheridge N, Schaller GE (2005) Ethylene signal transduction. Ann Bot 95:901–915

Dai Y, Wang H, Li B, Huang J, Liu X, Zhou Y, Mou Z, Li J (2006) Increased expression of MAP KINASE KINASE7 causes deficiency in polar auxin transport and leads to plant architectural abnormality in *Arabidopsis*. Plant Cell 18:308–320

Dhonukshe P, Baluska F, Schlicht M, Hlavacka A, Samaj J, Friml J, Gadella TW (2006) Endocytosis of cell surface material mediates cell plate formation during plant cytokinesis. Dev Cell 10:137–150

Field C, Li R, Oegema K (1999) Cytokinesis in eukaryotes: a mechanistic comparison. Curr Opin Cell Biol 11:68–80

Hamel LP, Nicole MC, Sritubtim S, Morency MJ, Ellis M, Ehlting J, Beaudoin N, Barbazuk B, Klessig D, Lee J, Martin G, Mundy J, Ohashi Y, Scheel D, Sheen J, Xing T, Zhang S, Seguin A, Ellis BE (2006) Ancient signals: comparative genomics of plant MAPK and MAPKK gene families. Trends Plant Sci 11:192–198

Hepler PK, Vidali L, Cheung AY (2001) Polarized cell growth in higher plants. Annu Rev Cell Dev Biol 17:159–187

Hulskamp M, Parekh NS, Grini P, Schneitz K, Zimmermann I, Lolle SJ, Pruitt RE (1997) The STUD gene is required for male-specific cytokinesis after telophase II of meiosis in *Arabidopsis thaliana*. Dev Biol 187:114–124

Hussey PJ, Hawkins TJ, Igarashi H, Kaloriti D, Smertenko A (2002) The plant cytoskeleton: recent advances in the study of the plant microtubule-associated proteins MAP-65, MAP-190 and the Xenopus MAP215-like protein, MOR1. Plant Mol Biol 50:915–924

Ishikawa M, Soyano T, Nishihama R, Machida Y (2002) The NPK1 mitogen-activated protein kinase kinase kinase contains a functional nuclear localization signal at the binding site for the NACK1 kinesin-like protein. Plant J 32:789–798

Ito M, Araki S, Matsunaga S, Itoh T, Nishihama R, Machida Y, Doonan JH, Watanabe A (2001) G2/M-phase-specific transcription during the plant cell cycle is mediated by c-Myb-like transcription factors. Plant Cell 13:1891–1905

Jin H, Axtell MJ, Dahlbeck D, Ekwenna O, Zhang S, Staskawicz B, Baker B (2002) NPK1, an MEKK1-like mitogen-activated protein kinase kinase kinase, regulates innate immunity and development in plants. Dev Cell 3:291–297

Jouannic S, Champion A, Segui-Simarro JM, Salimova E, Picaud A, Tregear J, Testillano P, Risueño MC, Simanis V, Kreis M, Henry Y (2001) The protein kinases AtMAP3Kε1 and BnMAP3Kε1 are functional homologues of *S. pombe* cdc7p and may be involved in cell division. Plant J 26:637–649

Jürgens G (2005) Plant cytokinesis: fission by fusion. Trends Cell Biol 15:277–283

Jürgens G, Mayer U (1994) Arabidopsis. In: Bard J (ed) A colour atlas of develping embryos. Wolfe, London, pp 7–21

Kieber JJ, Rothenberg M, Roman G, Feldmann KA, Ecker JR (1993) CTR1, a negative regulator of the ethylene response pathway in *Arabidopsis*, encodes a member of the raf family of protein kinases. Cell 72:427–441

Kovtun Y, Chiu WL, Tena G, Sheen J (2000) Functional analysis of oxidative stress-activated mitogen-activated protein kinase cascade in plants. Proc Natl Acad Sci USA 97:2940–2945

Kovtun Y, Chiu WL, Zeng W, Sheen J (1998) Suppression of auxin signal transduction by a MAPK cascade in higher plants. Nature 395:716–720

Krysan PJ, Jester PJ, Gottwald JR, Sussman MR (2002) An Arabidopsis mitogen-activated protein kinase kinase kinase gene family encodes essential positive regulators of cytokinesis. Plant Cell 14:1109–1120

Lukowitz W, Roeder A, Parmenter D, Somerville C (2004) A MAPKK kinase gene regulates extra-embryonic cell fate in *Arabidopsis*. Cell 116:109–119

MAPK G (2002) Mitogen-activated protein kinase cascades in plants: a new nomenclature. Trends Plant Sci 7:301–308

Mizoguchi T, Gotoh Y, Nishida E, Yamaguchi-Shinozaki K, Hayashida N, Iwasaki T, Kamada H, Shinozaki K (1994) Characterization of two cDNAs that encode MAP kinase homologues in *Arabidopsis thaliana* and analysis of the possible role of auxin in activating such kinase activities in cultured cells. Plant J 5:111–122

Mockaitis K, Howell SH (2000) Auxin induces mitogenic activated protein kinase (MAPK) activation in roots of Arabidopsis seedlings. Plant J 24:785–796

Munnik T, Ligterink W, Meskiene I, Calderini O, Beyerly J, Musgrave A, Hirt H (1999) Distinct osmo-sensing protein kinase pathways are involved in signalling moderate and severe hyper-osmotic stress. Plant J 20:381–388

Müller S, Fuchs E, Ovecka M, Wysocka-Diller J, Benfey PN, Hauser MT (2002) Two new loci, PLEIADE and HYADE, implicate organ-specific regulation of cytokinesis in *Arabidopsis.* Plant Physiol 130:312–324

Müller S, Smertenko A, Wagner V, Heinrich M, Hussey PJ and Hauser MT (2004) The plant microtubule-associated protein AtMAP65-3/PLE is essential for cytokinetic phragmoplast function. Curr Biol 14:412–417

Nakashima M, Hirano K, Nakashima S, Banno H, Nishihama R, Machida Y (1998) The expression pattern of the gene for NPK1 protein kinase related to mitogen-activated protein kinase kinase kinase (MAPKKK) in a tobacco plant: correlation with cell proliferation. Plant Cell Physiol 39:690–700

Nishihama R, Banno H, Kawahara E, Irie K, Machida Y (1997) Possible involvement of differential splicing in regulation of the activity of *Arabidopsis* ANP1 that is related to mitogen-activated protein kinase kinase kinases (MAPKKKs). Plant J 12:39–48

Nishihama R, Ishikawa M, Araki S, Soyano T, Asada T, Machida Y (2001) The NPK1 mitogen-activated protein kinase kinase kinase is a regulator of cell-plate formation in plant cytokinesis. Genes Dev 15:352–363

Nishihama R, Soyano T, Ishikawa M, Araki S, Tanaka H, Asada T, Irie K, Ito M, Terada M, Banno H, Yamazaki Y, Machida Y (2002) Expansion of the cell plate in plant cytokinesis requires a kinesin-like protein/MAPKKK complex. Cell 109:87–99

Ouaked F, Rozhon W, Lecourieux D, Hirt H (2003) A MAPK pathway mediates ethylene signaling in plants. EMBO J 22:1282–1288

Samaj J, Ovecka M, Hlavacka A, Lecourieux F, Meskiene I, Lichtscheidl I, Lenart P, Salaj J, Volkmann D, Bögre L, Baluska F, Hirt H (2002) Involvement of the mitogen-activated protein kinase SIMK in regulation of root hair tip growth. EMBO J 21:3296–3306

Samuels AL, Giddings TH, Staehelin LA (1995) Cytokinesis in tobacco BY-2 and root tip cells: a new model of cell plate formation in higher plants. J Cell Biol 130:1345–1357

Sasabe M, Machida Y (2006) MAP65: a bridge linking a MAP kinase to microtubule turnover. Curr Opin Plant Biol 9:563–570

Sasabe M, Soyano T, Takahashi Y, Sonobe S, Igarashi H, Itoh TJ, Hidaka M, Machida Y (2006) Phosphorylation of NtMAP65-1 by a MAP kinase down-regulates its activity of microtubule bundling and stimulates progression of cytokinesis of tobacco cells. Genes Dev 20:1004–1014

Sazuka T, Aichi I, Kawai T, Matsuo N, Kitano H, Matsuoka M (2005) The rice mutant *dwarf bamboo shoot 1*: a leaky mutant of the NACK-type kinesin-like gene can initiate organ primordia but not organ development. Plant Cell Physiol 46:1934–1943

Shou H, Bordallo P, Fan JB, Yeakley JM, Bibikova M, Sheen J, Wang K (2004) Expression of an active tobacco mitogen-activated protein kinase kinase kinase enhances freezing tolerance in transgenic maize. Proc Natl Acad Sci USA 101:3298–3303

Shou H, Bordallo P, Wang K (2004) Expression of the *Nicotiana* protein kinase (NPK1) enhanced drought tolerance in transgenic maize. J Exp Bot 55:1013–1019

Sieberer BJ, Ketelaar T, Esseling JJ, Emons AM (2005) Microtubules guide root hair tip growth. New Phytol 167:711–719

Smertenko AP, Chang HY, Sonobe S, Fenyk SI, Weingartner M, Bögre L, Hussey PJ (2006) Control of the AtMAP65-1 interaction with microtubules through the cell cycle. J Cell Sci 119:3227–3237

Smertenko AP, Chang HY, Wagner V, Kaloriti D, Fenyk S, Sonobe S, Lloyd C, Hauser MT, Hussey PJ (2004) The Arabidopsis microtubule-associated protein AtMAP65-1: molecular analysis of its microtubule bundling activity. Plant Cell 16:2035–2047

Soyano T, Nishihama R, Morikiyo K, Ishikawa M, Machida Y (2003) NQK1/NtMEK1 is a MAPKK that acts in the NPK1 MAPKKK-mediated MAPK cascade and is required for plant cytokinesis. Genes Dev 17:1055–1067

Spielman M, Preuss D, Li FL, Browne WE, Scott RJ, Dickinson HG (1997) TETRASPORE is required for male meiotic cytokinesis in *Arabidopsis thaliana*. Development 124:2645–2657

Strompen G, El Kasmi F, Richter S, Lukowitz W, Assaad FF, Jürgens G, Mayer U (2002) The *Arabidopsis HINKEL* gene encodes a kinesin-related protein involved in cytokinesis and is expressed in a cell cycle-dependent manner. Curr Biol 12:153–158

Tanaka H, Ishikawa M, Kitamura S, Takahashi Y, Soyano T, Machida C, Machida Y (2004) The *AtNACK1/HINKEL* and *STUD/TETRASPORE/AtNACK2* genes, which encode functionally redundant kinesins, are essential for cytokinesis in *Arabidopsis*. Genes Cells 9:1199–1211

Wang H, Ngwenyama N, Liu Y, Walker JC, Zhang S (2007) Stomatal development and patterning are regulated by environmentally responsive mitogen-activated protein kinases in *Arabidopsis*. Plant Cell 19:63–73

Yang CY, Spielman M, Coles JP, Li Y, Ghelani S, Bourdon V, Brown RC, Lemmon BE, Scott RJ, Dickinson HG (2003) *TETRASPORE* encodes a kinesin required for male meiotic cytokinesis in *Arabidopsis*. Plant J 34:229–240

Plant Cell Monogr (10)
L. Bögre and G. Beemster: Plant Growth Signaling
DOI 10.1007/7089_2007_155/Published online: 29 January 2008

Protein Phosphatases in Plant Growth Signalling Pathways

Alois Schweighofer · Irute Meskiene (✉)

Max F. Perutz Laboratories, University of Vienna, Dr. Bohrgasse 9, 1030 Vienna, Austria
irute.meskiene@univie.ac.at

Abstract Intracellular signalling systems communicate the inputs perceived at the cell membranes to the nucleus to regulate cellular functions in developmental and stress responses. Pivotal to these transmissions are the reversible protein phosphorylations performed by opposing actions of protein kinases and protein phosphatases. Phosphorylation by protein kinases is an essential posttranslational modification mechanism for the majority of cellular proteins and can influence protein activity, localization and stability. The significance of protein phosphorylation by kinases is already established in Arabidopsis; but the importance of de-phosphorylation by phosphatases has not been studied equally intensively. Nevertheless, recent characterization of Arabidopsis protein phosphatase mutants and identification of interacting proteins/substrates highlights the important role of protein phosphatases in the pathways regulating stress, hormonal signalling, metabolism, cell cycle and plant growth. In this review we will focus principally on the involvement of plant protein phosphatases of PTP and PP2C-types in these processes.

1 Plant Protein Phosphatase Families

The importance of protein phosphorylation by protein kinases to regulate signalling responses is well recognized in plants; and it is becoming more and more evident that also the inverse, de-phosphorylation by protein phosphatases, is indispensable in these processes. The key phosphorylation targets in eukaryotic proteins, including plants, are serine, threonine, and tyrosine residues. Accordingly, protein phosphatases are organized into different groups dependent on the phosphor-amino acid residues that are targeted for dephosphorylation. Consequently, serine/threonine phosphatases, tyrosine phosphatases and dual specificity phosphatases (targeting both tyrosine and serine/threonine) are found in plants. Eukaryotic serine/threonine protein phosphatases are classified into the superfamily PPP (phospho protein phosphatases), the PPM (protein phosphatases magnesium- or manganese-dependent) family and the FCP family, according to the amino acid sequences of their catalytic subunits (Cohen 2004; Ingebritsen and Cohen 1983) (see Fig. 1). The PPP family of plant phosphatases includes the type 1 protein phosphatases (PP1), type 2A protein phosphatases (PP2A), protein phosphatases with Kelch-repeat domain (a conserved tertiary structure forming

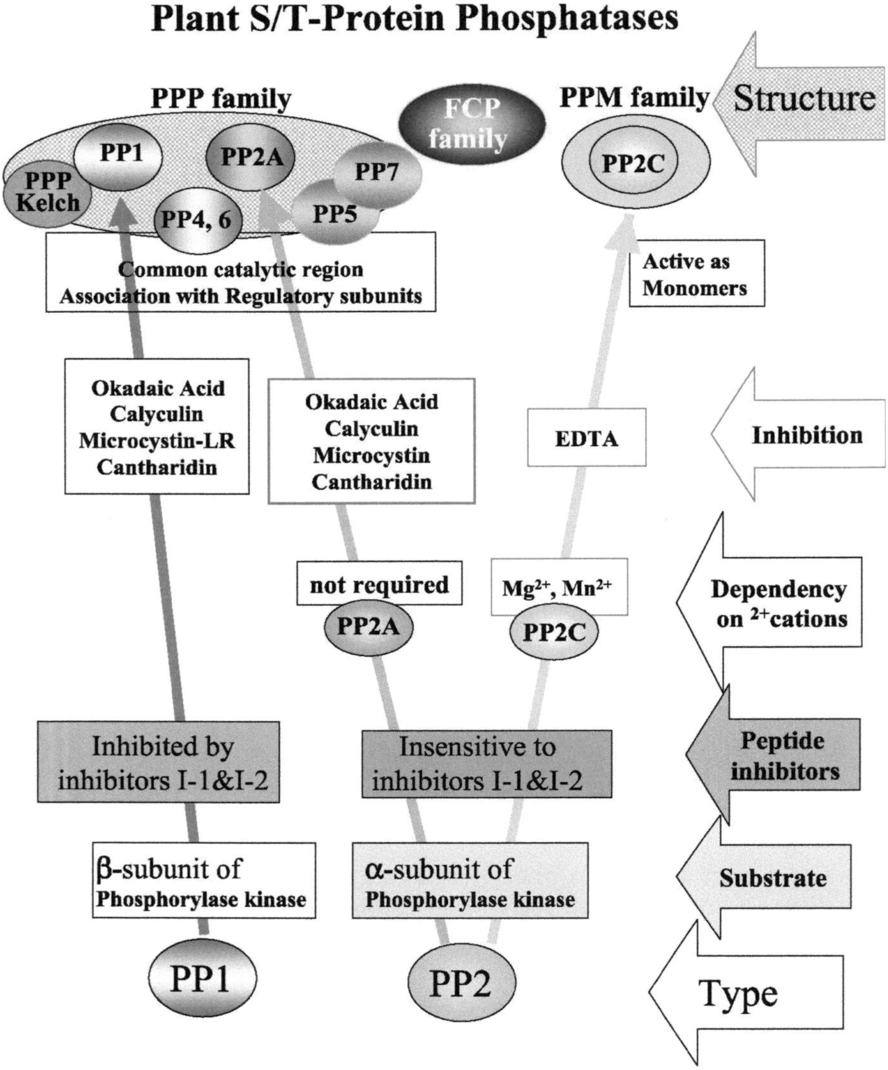

beta propellers) and protein phosphatases PP4, PP5, PP6, PP7. PP2B type phosphatases that are represented by calcineurin in other eukaryotes have not been detected in plants so far. The PPM superfamily consists of protein phosphatases 2C (PP2C). The new FCP family was recognized recently through CTD phosphatase-like (CPL) members with homology to the FCP1 (TFIIF-associating C-terminal domain of RNAP II phosphatase) protein serine phosphatase, which regulates transcription by dephosphorylating the carboxyl-terminal domain of the large subunit of RNA polymerase Pol II (Bang et al. 2006; Cohen 2004). Plant protein tyrosine phosphatases (PTP) are

◀ **Fig. 1** Classification and characterization of serine/threonine protein phosphatases. First classifications of S/T protein phosphatases were based on their different capabilities to dephosphorylate specific substrates and their responses to heat-stable inhibitor proteins (Cohen 1989; Ingebritsen and Cohen 1983). Thus, PP1 preferentially dephosphorylated the β-subunit of phosphorylase kinase, whereas type 2 phosphatases dephosphorylated the α-subunit of phosphorylase kinase. PP1 was inhibited by the heat-stable inhibitor-1 (I1) and inhibitor-2 (I2), whereas type 2 phosphatases (PP2A, PP2B and PP2C) were not. PP2Cs are Mg^{2+}/Mn^{2+}-dependent for their activity, while PP2As are active in the absence of bivalent cations. Although this classification remains true, the information about protein primary structure has led towards more comprehensive classification according to sequence similarity of the catalytic subunit. Thereby, PP2C belongs to a distinct PPM (protein phosphatases Mg^{2+}/Mn^{2+}-dependent) gene family, whereas the remaining type 1 and type 2 phosphatases belong to the same (PPP) family. In comparison with the PPP gene family, the PPM members are overrepresented in plants

structurally and catalytically distinct from the serine/threonine phosphatases. PTP superfamily consists of classical protein tyrosine phosphatases and dual-specificity protein phosphatases (DSPs) (see Fig. 2). The receptor-like PTPs that are abundant in animals have not been identified in plants so far.

Several reviews have been written on plant phosphatases (DeLong 2006; Farkas et al. 2007; Luan 1998; Smith and Walker 1996) and two reviews are available about Arabidopsis PP2C members (Rodriguez 1998; Schweighofer

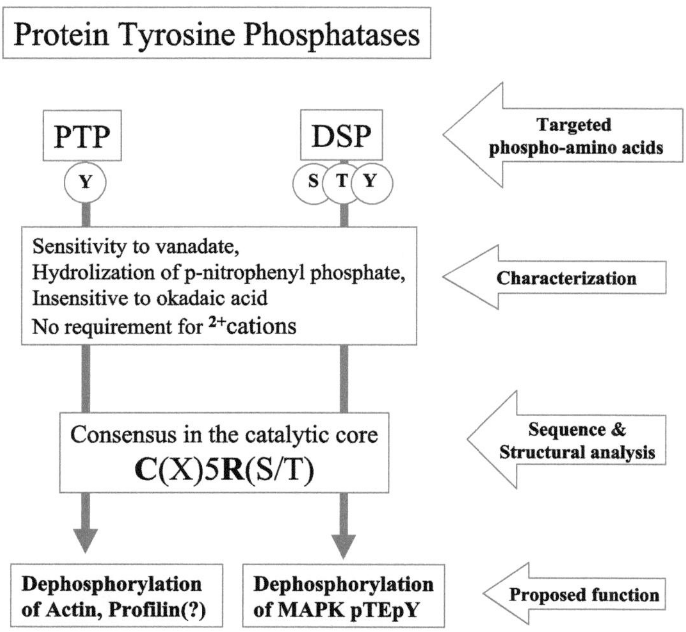

Fig. 2 Classification and characterization of tyrosine protein phosphatases

et al. 2004). Here we will update the information about plant PTP/DSP and PP2C members with the recent data, for the most part on Arabidopsis.

2 Protein Tyrosine Superfamily (PTP)

Currently 127 protein phosphatase candidates are predicted in the Arabidopsis genome (Kerk 2006; Kerk et al. 2002; Kerk, personal communication). Among them only one classical PTP, one low molecular weight LMW PTP and 23 DSP members were identified. The hallmark that defines the PTPs is a cysteine-containing signature motif HC(X)5R(S/T), which is the active site in the catalytic domain that hydrolyzes phosphor-tyrosine via a thiol phosphate intermediate (Farooq and Zhou 2004; Tonks 2005). A new group of dual-specificity tyrosine phosphatases that do not have the cysteine-containing signature but use an aspartate as a nucleophile in a metal-dependent reaction (Alonso et al. 2004), has also been recently described in plants by the Arabidopsis homologue of the Drosophila Eyes Absent (EYA) (Rayapureddi et al. 2005). Low numbers of plant PTPs compared to 107 PTPs encoded in the human genome (Alonso et al. 2004) may imply that tyrosine phosphorylation in higher plants is less common than in mammals; however, studies with anti-phosphotyrosine antibodies and tyrosyl-phosphorylation inhibitors suggested that plant proteins are phosphorylated on tyrosine residues at levels comparable to those seen in animals (Barizza et al. 1999). Tyrosine phosphorylation is necessary for activity of mitogen-activated protein kinases (MAPKs), as shown by phosphorylation of essential TEY motif and phosphor-amino acid analysis of phosphorylated MAPKs (Gupta and Luan 2003; Huang et al. 2000; Kiegerl et al. 2000). Both tyrosine and threonine must be phosphorylated to get the kinase into an active conformation (Canagarajah et al. 1997). Conversely, MAPKs are completely inactivated by dephosphorylation of either phosphor-amino acid residue. Thus, tyrosine-specific, dual-specificity and threonine-specific phosphatases are capable of inactivating the MAPKs (Camps et al. 2000; Cohen 2004; Farooq and Zhou 2004; Martin et al. 2005).

2.1 PTPs

In plants, salt activates MAPK signalling (Droillard et al. 2002; Munnik et al. 1999). Hence, it is not surprising that in Arabidopsis the only member of classical PTP, *AtPTP1* is a stress-responsive gene (Fordham-Skelton et al. 1999; Xu et al. 1998) up-regulated by salt, and down-regulated by cold treatment. Similar PTP genes are also found in other plants. Dephosphorylation of phosphor-tyrosine in Arabidopsis MPK4 by AtPTP1 in vitro suggests that

AtPTP1 phosphatase may function as a switch-off mechanism for salt-stress activated MAPKs (Huang et al. 2000); however, data in planta has to support this premise. Highly conserved cysteine and aspartate residues are essential for the AtPTP1 activity underlining the common catalytic mechanism between plant and other eukaryotic PTPs (Xu et al. 1998). A catalytic cysteine has to be in a reduced form for phosphatase activity. Thus, reactive oxygen species (ROS) produced during various stress conditions regulate mammalian PTPs by reversible oxidation of the active site cysteine and abrogation of its nucleophilic properties, thereby inhibiting PTP activity (Barford 2004; Tonks 2005). Similarly, the recombinant AtPTP1 protein can be reversibly inactivated by H_2O_2, suggesting that plant PTPs may also be regulated by the redox state of the cysteine, thereby serving as molecular targets during oxidative stress (Gupta and Luan 2003).

2.2 DSPs

In Arabidopsis, several dual-specificity phosphatases (DSPs) were described to control plant growth under stress conditions. A gene disruption of mitogen-activated protein kinase phosphatase MKP1 in Arabidopsis (*mkp1*) results in plant hypersensitivity to stress by UV-C and methyl methanesulfonate (Ulm et al. 2001). These factors activate repair mechanisms of DNA damage and cell recovery. UV-C activates the Arabidopsis MAPK, MPK6 but not MPK4. Though MPK6, MPK3 and MPK4 were all found to interact with MKP1 in yeast two hybrid assays, only the activity of MPK6 was effected in the *mkp1* mutant plants, suggesting that only MPK6 is a substrate for MKP1 in plants. The role of MKP1 in salt stress responses was demonstrated by expression profiling of wild-type vs. *mkp1* mutant lines and increased resistance to salinity of *mkp1* plants, both implicating functions for MKP1 in regulation of plant responses to environmental challenges (Ulm et al. 2002).

MKP1 contains the conserved catalytic part and a long carboxy-terminal extension with a domain similar to the actin-binding protein gelsolin, suggesting an association with the regulation of the cytoskeleton. Calmodulin (CaM) binding domain and interaction with CaM was identified in the orthologue of Arabidopsis MKP1 from tobacco NtMKP1 (Yamakawa et al. 2004). Its expression is regulated in response to cell death, pathogen infection and wounding. Wound-activation of the MAP kinase SIPK and cell death is compromised by NtMKP1 overexpression in tobacco. NtMKP1 binds SIPK through the N-terminal non-catalytic region. This binding strongly increases phosphatase activity and is partially dependent on the putative MAPK common docking domain of SIPK (Katou et al. 2005). Another Arabidopsis dual-specificity phosphatase DsPTP1 hydrolyzes both phosphor-threonine and phosphor-tyrosine and dephosphorylates/inactivates MPK4 in vitro. Phosphatase activity of DsPTP1 depends on the conserved catalytic cysteine (Gupta et al. 1998). DsPTP1 contains two Ca^{2+}-dependent CaM-binding domains, and binding to CaM inhibits its activity to dephosphorylate tyrosine on myelin basic protein (MBP) (Yoo et al. 2004).

Another DSP from Arabidopsis, PHS1 (propyzamide hypersensitive) controls microtubule organization and embryonic development (Naoi and Hashimoto 2004), as well as plant growth under stress conditions and regulation of ABA signalling (Quettier et al. 2006). The semidominant *phs1-1* mutation, which has a substitution of Arg to Cys amino acid in a putative MAPK interaction motif (KIM) in the N-terminal part of the protein, disrupts microtubule organization in roots. Similar KIM is present in MKPs from mammals and mediates binding to MAPKs (Farooq and Zhou 2004; Martin et al. 2005; Pulido et al. 1998). Whether putative KIM indeed mediates phosphatase binding to plant MAPKs and whether *phs1-1* mutation influences this binding has still to be tested.

T-DNA null plants *phs1-2* are recessive lethal embryonic mutants (Naoi and Hashimoto 2004). *phs1-3* mutant plants with T-DNA insertion in the promoter region, resulting in abridged gene expression, show inhibited seed germination in the presence of ABA and stronger inhibition of the light-induced opening of stomata by ABA, thereby suggesting PHS1 as a negative regulator of ABA signalling (Quettier et al. 2006). These examples demonstrate that plant dual-specificity phosphatases counteract the activity of MAPKs and control plant responses during stress and in development.

The DSP phosphatase IBR5 was identified as an Arabidopsis indole-3-butyric acid (IBA)-response mutant *ibr5* (Monroe-Augustus et al. 2003). The phenotype of this mutant is similar to other auxin-response mutants: a long root and a short hypocotyl when grown in the light, aberrant vascular patterning, increased leaf serration, and reduced accumulation of an auxin-inducible reporter. However, overexpression of IBR5 did not dramatically alter the auxin sensitivity of plants. No phosphatase activity in vitro was shown for IBR5, which could be due to the requirement of substrate binding to activate this enzyme (like for NtMKP1 or certain mammalian DSPs). The substrates of IBR5 remain to be identified.

AtPTEN1 encodes a dual-specificity phosphatase closely related to PTEN, a tumor suppressor in animals. The recombinant AtPTEN1 demonstrates phosphatase activity to dephosphorylate phosphotyrosine and phosphatidylinositol substrates, like its homologues in animals. *AtPTEN1* is expressed exclusively in pollen and is essential for pollen development, as shown by suppression of *AtPTEN1* expression by RNA interference that causes pollen cell death after mitosis (Gupta et al. 2002).

Another Arabidopsis DSP phosphatase is PTPKIS1/SEX4/DSP4, previously nominated PTPKIS1 (protein-tyrosine phosphatase kinase interaction sequence) due to the presence of a kinase interaction sequence (KIS), which mediates interaction with the plant SNF1-related kinase (SnRK), AKIN11 (Fordham-Skelton et al. 2002) and encodes a carbohydrate-binding domain, allowing it to bind starch granules (Kerk et al. 2006; Niittyla et al. 2006; Sokolov et al. 2006). In addition, SEX4/DSP4 protein has a plastid targeting sequence and is localized to chloroplasts, where it controls starch metabolism, primarily by regulation of starch breakdown. It associates with starch granules in a light-dependent manner and represents a major starch granule-bound phosphatase activity during the day. Phosphatase activity and the starch-binding capacity of SEX4/DSP4 are controlled by redox and pH. SEX4/DSP4 has close orthologues in other plant species and resembles the animal DSP laforin (Alonso et al. 2004) that regulates glycogen accumulation, indicating striking parallels in the regulation of starch metabolism in plants and glycogen metabolism in mammals. A future task will be to investigate whether the downstream targets of SEX4/DSP4 are glucans or proteins in plants (Kerk et al. 2006; Niittyla et al. 2006; Sokolov et al. 2006).

The animal Eyes Absent proteins (Eya) represent a novel family of dual-function enzymes with transcription factor and phosphatase activities (Jemc and Rebay 2007). Eya proteins are transcription factors with intrinsic phosphatase activity capable of modulating transcriptional complexes and are responsible for organ formation (Li et al. 2003). They also represent a mechanistically new class of tyrosine phosphatases (PTPs), which does not contain the cysteine-containing signature motif and includes an aspartate as a nucleophile in a metal-dependent reaction similar to the phosphoserine phosphatases of the haloacid dehalogenase (HAD) family (Lunn 2002). Eyes Absent homologues have been identified in rice, alfalfa and Arabidopsis (Takeda et al. 1999). Although the Arabidopsis homologue of animal Eya, AtEYA is a tyrosine-specific phosphatase, as demonstrated in vitro, the role of AtEYA-mediated dephosphorylation in plant biology remains to be elucidated (Rayapureddi et al. 2005).

2.3 Pseudo-phosphatase PTPL

PASTICCINO2 (PAS2) is a protein Tyr phosphatase-like member (PTPL) that is highly conserved in eukaryotes and is characterized by an inactive catalytic site of the phosphatase. PAS2 interacts directly with a cyclin-dependent kinase (CDK) in the CDK-Tyr-15 phosphorylation-dependent manner. Loss of the PAS2 function in Arabidopsis leads to dephosphorylation and enhancement of CDKA;1 activity, suggesting that PAS2, similar to other PTPL proteins, functions as an anti-phosphatase by binding the kinase and preventing its dephosphorylation by other activating phosphatases. PAS2 slows down cell division at the G2-to-M transition and holds cells in a differentiated state when overexpressed in plants. This suggests that the balance between cell division and differentiation is regulated through control of the CDKA;1 activity entailed by the PAS2 anti-phosphatase (Da Costa et al. 2006). No functional CDC25 phosphatases were found in plants so far to inactivate CDKs, and the association between PAS2 and Tyr-phosphorylated CDKs may be one of the mechanisms to regulate CDK temporal activation (Boudolf et al. 2006).

3 PP2C Family of Plant Protein Phosphatases

The PP2C family comprises a group of serine threonine phosphatases that are dependent on Mg^{2+} or Mn^{2+} for their activity, are active as monomers, insensitive to known phosphatase inhibitors and share no apparent amino acid sequence homology to the other types of phosphatases (Cohen 1989, 2004). However, similarities in protein architecture suggest a similar catalytic mechanism of these enzymes to other PPP proteins, whereby a metal-bound water molecule acts as a nucleophile to directly transfer phosphate from a specific amino acid of the substrate in an acid–base catalytic reaction (Barford et al. 1998). Arabidopsis PP2Cs represent the largest known protein phosphatase family in plants with 76 discrete members (Kerk 2006; Kerk et al. 2002; Schweighofer et al. 2004) (http://www.arabidopsis.org/browse/genefamily/PP2C.jsp). Bioinformatics aligned this family in clusters, where cluster A contains genes predominantly associated with ABA signal transduction, such as ABI1, ABI2, AtPP2CA, HAB1, HAB2 (Schweighofer et al. 2004) and *Fagus sylvatica* FsPP2C1 (Gonzalez-Garcia et al. 2003). Cluster B is characterized by the phosphatases containing MAPK interaction motif (KIM) in the N-terminal extensions. KIM $[(K/R)_{3-4}X_{1-6}(L/I)X(L/I)]$ is similar to the ones found in animal MAPK-interacting proteins, such as MAPK phosphatases, MAPK kinases (Farooq and Zhou 2004; Martin et al. 2005; Pulido et al. 1998) or transcription factors (Biondi and Nebreda 2003; Ho et al. 2003). Interestingly, plant PP2Cs have evolved similar MAPK interaction motifs, which otherwise are found in evolutionally unrelated animal or yeast PTP/DSP phosphatases. Arabidopsis AP2C1 and alfalfa MP2C regulate stress MAPK signalling. Cluster C includes the POLTERGEIST-type phosphatases (POL) that are involved in development. The most phylogenetically distant kinase-associated protein phosphatase (KAPP) represents a singleton in the PP2C family and it regulates receptor-like kinases (RLK).

One of the PP2C roles in eukaryotes is to reverse stress-induced protein kinase cascades. Below we will discuss the signalling pathways that plant PP2Cs participate in.

3.1 Regulation of MAPKs by PP2Cs

By interference of yeast pheromone-induced MAPK pathway (which leads to growth arrest in yeast) with an alfalfa cDNA library the *MP2C* gene was isolated (Meskiene et al. 1998). MP2C inactivates the MAPK SIMK through threonine dephosphorylation of the pTEpY motif. MP2C and its Arabidopsis orthologue AP2C1 dephosphorylate and inactivate stress-responsive MAPKs SIMK from alfalfa and MPK6 or MPK4 from Arabidopsis, respectively. Affinity towards MAPK is mediated through functional KIM domain at the N-terminal non-catalytic part in MP2C/AP2C1. Alfalfa SIMK and Arabidopsis MPK6 or MPK4 interact with MP2C and AP2C1, respectively, and form protein complexes (Meskiene et al. 2003; Schweighofer et al. 2007). The catalytic activity of the phosphatases or MAPK is not imperative for their interaction in yeast, but the N-terminal part and intact KIM are prerequisites. AP2C1 regulates ethylene and modulates plant innate immunity against a fungal necrotrophic pathogen. AP2C1 also negatively regulates wound jasmonates, which correlates with enhanced resistance to herbivores in its absence. *MP2C/AP2C1* is expressed in roots, flowers, young leaves and cultured suspension cells, but not in adult leaves. However, upon wounding gene expression is rapidly induced and correlates with the timing of MAPK inactivation, suggesting that MP2C/AP2C1 may be a part of a negative feedback responsible for resetting the MAPK signalling pathway (Bogre et al. 1997; Meskiene et al. 2003).

3.2 Regulation of ABA Signalling

The vast majority of research on PP2C functions is related to ABA signalling and is attributed to cluster A-type phosphatases. Several genes were described in this cluster, such as ABI1, ABI2, AtPP2CA, HAB1, HAB2 and FsPP2C1 from beech (*Fagus sylvatica*) that are up-regulated by ABA and involved in negative regulation of ABA signalling pathways.

3.2.1 ABI1 and ABI2

3.2.1.1 Studies of Mutations and Involved Pathways

The closely related related phosphatases ABI1 and ABI2 are negative regulators of ABA signalling according to genetic studies of several mutations and T-DNA insertion lines (Gosti et al. 1999; Merlot et al. 2001; Yoshida et al. 2006). (Semi)-dominant *abi1-1* and *abi2-1* mutations carry the same Gly to Asp amino acid substitution, which is located near to the Mg^{2+} binding site and reduces the phosphatase activity. These mutations cause an ABA-insensitivity, reduced seed dormancy, abnormal stomata regulation, defects in drought responses (Bertauche et al. 1996; Leung et al. 1994, 1997; Meyer et al. 1994; Rodriguez et al. 1998) and lead to salt tolerance (Ohta et al.

2003). Both ABI1 and abi1-1 proteins inhibit ABA signal transduction to ABA-responsive gene promoters HVA1 and RBCS in isolated protoplasts (Sheen 1998) as well as KIN2 and Rd29A promoters in cells microinjected with recombinant proteins (Wu et al. 2003). The N-terminal part of ABI1 contains a putative Ca^{2+}-binding domain (Leung et al. 1994; Meyer et al. 1994;), and ABI1 action was placed downstream of Ca^{2+} in the ABA pathway (Wu et al. 2003).

3.2.1.2 ABI1 and ABI2 Interacting Proteins and Compounds

Several protein kinases were found to interact with ABI1 and ABI2 in yeast two-hybrid screens. SOS2 (salt overly sensitive protein kinase, also known as Ca^{2+} sensor-interacting protein kinase, a member of the SNF1-related protein kinase SnRK3 group) and several SOS2-like protein kinases interact with ABI1 and ABI2 through a protein kinase interaction motif (PKI) identified within the catalytic part of the phosphatases. Interestingly, the *abi2-1* mutation located in the PKI domain disrupts this interaction and at the same time enhances plant tolerance to salt stress (Ohta et al. 2003). ABI2 and to a lesser extent ABI1 interact with PKS3 (SOS2-like protein kinase 3), a negative regulator of the ABA pathway. The *abi1-1* and *abi2-1* mutants are able to suppress the *pks3* ABA hypersensitive phenotype. Thereby, ABI2, ABI1 and PKS3 form a calcium-responsive negative regulatory loop together with the calcium binding protein CaBP5 in fine-tuning the plant sensitivity to ABA (Guo et al. 2002). Another ABA- and osmotic stress-activated protein kinase SnRK2E/OST1/SnRK2.6 (Mustilli et al. 2002; Yoshida et al. 2002) interacts with ABI1 possibly through PKI, as the *abi1-1* mutation reduces this binding (Yoshida et al. 2006). Distinct roles of ABI1 and ABI2 were demonstrated by suppression of the ABA-dependent activation of SRK2E/OST1 that was observed in *abi1-1* but not in *abi2-1* mutant. Similar wilty phenotypes of *srk2e/ost1* and *abi1-1* plants suggest that ABI1 positively regulates the activation of this kinase (Yoshida et al. 2002, 2006). A further ABI1-interacting protein is the ABA- and drought-inducible transcription factor ATHB6. This interaction depends on the catalytic activity of the phosphatase. ATHB6 negatively regulates ABA signalling. ABI1 acts upstream of *ATHB6*, as ABA induction of the ATHB6 promoter-reporter is abolished in *abi1-1* plants (Himmelbach et al. 2002). The interaction of ABI2 and, to a lesser extent, of ABI1 with AtGPX3, a glutathione peroxidase, provides a link between ABA and H_2O_2 signalling in stomatal closure. AtGPX3 is involved in ABA-mediated production of H_2O_2, which affects the redox states of ABI1, ABI2 and AtGPX3 (Miao et al. 2006). H_2O_2 reversible inactivation of ABI1 and ABI2 proteins and their susceptibility to phenylarsine oxide already suggested the oxidation of critical cysteine residue(s) in the phosphatase (Meinhard and Grill 2001; Meinhard et al. 2002). The ability of oxidized AtGPX3 to reduce the phosphatase activities of ABI1 and ABI2 and enhanced insensitivity of the *atgpx3 abi2-1* double mutants to ABA suggested that AtGPX3 may modulate the activities of ABI2 under oxidative stress in plants (Miao et al. 2006).

ABI1 interacts with phosphatidic acid (PA), which is produced in response to ABA application through phospholipase Dα1. PA binding decreases the PP2C activity and affects its membrane association. The Arg-73 is essential for the ABI1–PA interaction (Zhang et al. 2004). ABI1-R73A mutant plants are insensitive to ABA-induced stomatal closure, emphasizing the requirement for PA binding to ABI1 in this process, but not in the ABA-induced inhibition of stomatal opening, which is regulated by interaction of PA and PLDα1 with the Gα subunit of heterotrimeric G protein (Mishra et al. 2006). Involvement of other PP2Cs in the regulation of ABA pathway is indicated by the ABA-induced PP2C activity measurement, where ABI1 and ABI2 contributed by approx. 50% to total PP2C-related activity (Merlot et al. 2001) and is further supported by additional studies of other A-type PP2C members.

3.2.2 HAB1 and HAB2

HAB1 and HAB2 are closely related to ABI1 and ABI2 and also negatively regulate the ABA signalling pathway (Rodriguez et al. 1998; Saez et al. 2004). This was shown by studying the T-DNA insertion mutants *hab1-1* and *hab2ds* (Leonhardt et al. 2004; Saez et al. 2004; Yoshida et al. 2006). Double mutants *abi1-2/hab1-1* and *abi1-3/hab1-1* display enhanced responsiveness to ABA and sensitivity to NaCl or mannitol. The enhancement in ABA-mediated stomatal closure, leading to reduced water loss in these lines, indicates overlapping functions of ABI1 and HAB1 (Saez et al. 2006). Conversely, constitutive overexpression of HAB1 leads to ABA insensitivity, impaired stomatal closure, ABA-resistant root growth, and reduction of ABA-induced gene expression. These support a negative role in the ABA pathway. Seeds of HAB1 overexpressing plants germinate on inhibitory concentrations of ABA, mannitol and paclobutrazol, indicating that HAB1 may promote seed germination by negatively regulating ABA responses in seeds (Saez et al. 2004). Introduction of *abi1-1*-like mutation into HAB1 ($hab1^{G246D}$) led to reduced phosphatase activity and to a strong ABA insensitivity of seeds in $hab1^{G246D}$ overexpressing plants. The strong ABI phenotype of $hab1^{G246D}$, which is similar to $ahg3^{G145D}$ (Robert et al. 2006) and opposite to the ABA hypersensitivity of the HAB1 T-DNA insertion mutant, indicates a dominant mutation.

3.2.3 PP2CA/AHG3 and FsPP2C1

PP2CA/AHG3 phosphatase is able to rescue the *pde1 S. pombe* mutant, which is defective in cAMP phosphodiesterase that results in elevated cAMP amounts and activated cAMP-dependent protein kinase. Possibly, it counteracts the protein kinase A (PKA) in yeast (Kuromori and Yamamoto 1994). However, in plants PP2CA is a strong negative regulator of ABA signal transduction, controlling seed germination and stomatal aperture. This was revealed by studies of PP2CA overexpressing plants, *ahg3-1* point mutant (causing the loss of PP2C activity) and T-DNA insertion lines *pp2ca-1/ahg3-2* and *pp2ca-2* (Kuhn et al. 2006; Yoshida et al. 2006). PP2CA/AHG3 plays a major role among PP2Cs in the ABA response in seeds (Yoshida et al. 2006). PP2CA inhibits ABA signal transduction when transiently expressed in isolated protoplasts (Sheen 1998) and its overexpression confers ABA insensitivity in plants (Kuhn et al. 2006). Cold, drought, salt and ABA induce *PP2CA* expression. Cold- and drought-induced expressions are ABA-dependent as *PP2CA* expression is reduced in ABA-deficient *aba1-1* mutant. Drought-induced expression is ABI1-dependent. Down-regulation of PP2CA accelerates plant development and induces tolerance to cold, suggesting a negative control of the ABA pathway during cold acclimation (Tahtiharju and Palva 2001). PP2CA interacts with AKT2/AKT3 inward rectifying potassium channel protein (Cherel et al. 2002; Vranova et al. 2001) and the highest expression levels of both genes in the phloem vasculature suggests possible interaction in plants (Cherel et al. 2002). An orthologue of PP2CA/AHG3, FsPP2C1 from beech (*Fagus sylvatica*) (Lorenzo et al. 2001), indicates similarly negative regulation of ABA signalling, as its overexpression in Arabidopsis led to ABA insensitivity, reduced seed dormancy and resistance towards salt and osmotic stresses (Gonzalez-Garcia et al. 2003).

3.2.4 FsPP2C2

Another gene from beech, FsPP2C2, which is related to cluster D of Arabidopsis PP2Cs (Schweighofer et al. 2004) enhances sensitivity to ABA and abiotic stress according to overexpression studies (Reyes et al. 2006), suggesting an opposite trend of cluster D to cluster A phosphatases. The phenotypes could be reversed by gibberellic acid (GA), indicating that FsPP2C2 is a positive regulator of ABA signalling that may affect GA levels in transgenic plants and pointing towards a possible cross-talk between ABA signalling and GA biosynthesis.

It is established that ABI1, ABI2, PP2CA, HAB1, HAB2 from Arabidopsis and FsPP2C1 from beech negatively regulate ABA signal transduction and that the functions of these PP2Cs overlap, but their specific tissue- or developmental expressions confer distinct and indispensable physiological functions in the ABA response (Yoshida et al. 2006). Yet, challenging tasks for the future remain in order to uncover their biological substrates and to identify cross-talks with other signalling pathways.

3.3 Regulation of Receptor-Like Kinase Signalling

3.3.1 KAPP

Arabidopsis kinase-associated protein phosphatase, KAPP was identified through interaction with the receptor-like protein kinase HAESA/RLK5 (Stone et al. 1994). Beside a catalytic PP2C part, KAPP protein contains the N-terminal membrane anchor and a kinase-interacting forkhead-associated domain (KI-FHA). The FHA domain is a phosphorylation-dependent protein–protein interaction domain with specificity for phosphor-threonine and possibly phosphor-serine. Solution dynamics and phosphopeptide binding effects have illustrated the KAPP KI-FHA structure (Ding et al. 2005, 2007; Lee et al. 2003). Through this domain, KAPP interacts with several phosphorylated RLKs, such as RLK4 (Braun et al. 1997), embryogenesis-specific AtSERK1 (Rienties et al. 2005; Shah et al. 2001, 2002) and the cell wall kinase WAK1 (Park et al. 2001). The components of the brassinosteroid receptor (BR), brassinosteroid-insensitive kinase 1 (BRI1), and BRI1-associated kinase 1 (BAK1) also interact with KAPP (Ding et al. 2007). KAPP interacts with and negatively regulates transmembrane receptor kinases CLAVATA1 (CLV1), which controls meristem development (Li et al. 1999; Stone et al. 1998; Williams et al. 1997), and FLAGELLIN SENSITIVE2 (FLS2), which is essential for flagellin perception in pathogen-associated innate immune response signalling (Gomez-Gomez et al. 2001). Interactions with diverse receptors imply KAPP control of multiple signalling pathways; however, no growth and developmental phenotypes were observed in T-DNA *kapp-3* mutant (Ding et al. 2007). This is really surprising as KAPP has no close homologues, and at the same time suggests that there should be other phosphatases that act redundantly. A future task is to identify these proteins.

3.3.2 POLTERGEIST (POL) and Related PP2Cs

POLTERGEIST (POL) functions in CLAVATA1 (CLV1) and CLV1-related pathways regulating stem cell identity (Yu et al. 2000, 2003). *pol* mutations suppress meristem defects in *clv1* and *clv3* mutants, though all *pol* mutants are nearly indistinguishable from wild-type plants (Yu et al. 2000). POL in parallel with PLL1 regulates meristem and organ development. POL and PLL1 act downstream of the CLV signalling pathway through WUS to promote stem cell identity. This was demonstrated by *pol/pll1* tissue phenocopying *wus* (WUSCHEL) in grafted plants (grafting allowed to overcome the embryo lethality of *pol/pll1*). The ectopic expression of *WUS* in *pol/pll1* lines restored meristem activity, signifying that POL and PLL1 control maintenance of *WUS* expression. Analysis of POL/PLL1/CLV3 triple mutants revealed *pol/pll1* epistasis to *clv3* (Song et al. 2006). PLL4 and PLL5 are involved in regulation of leaf development (Song and Clark 2005).

3.4 Regulation of Transcription by PP2C

Tobacco DNA-binding protein phosphatase (DBP1) was isolated by interaction with the promoter region of tomato citrus exocortis viroid (CEVI1) gene, which is induced during the course of compatible plant–virus interactions. The C-terminal part of DBP1 shows homology to the PP2C catalytic domain, whereas the N-terminal region contains DNA binding sequences (Carrasco et al. 2003). Correspondingly, DBP1 shows Mg^{2+}-dependent phosphatase activity and contains a functional nuclear localization signal. DBP1 negatively controls transcription of *CEVI1* as demonstrated by constitutive up-regulation of the CEVI1 gene in DBP1-antisense transgenic tobacco plants. DBP1 and its distantly related Arabidopsis AtDBP1 possess in vitro DNA binding activity, mediated by the N-terminal region via the conserved DNC (DBP N-terminal core) motif (Carrasco et al. 2005). Interestingly, the 14-3-3 isoform G from tobacco and the 14-3-3 λ/GRF6 from Arabidopsis were identified by screenings in yeast using DBP1 and AtDBP1 as baits, respectively (Carrasco et al. 2006). The N-terminal region of DBP1 is necessary and sufficient for 14-3-3 G binding and confers nucleocytoplasmic shuttling in vivo. It is proposed that 14-3-3 isoform G positively regulates CEVI1 expression through interaction with DBP1 followed by nuclear export of the complex, thereby relieving repression of CEVI1 promoter (Carrasco et al. 2006).

Recently, important findings in the functional characterization of plant members of the PTP and PPM families have provided new perspectives on the regulation of signal transduction in stress and developmental pathways. It will be important to investigate the regulation of PTP and PP2C phosphatases by oxidation-related signalling in planta. Of particular significance would be to define the links between the disruption of PTP and/or PP2C function and the MAPK activities. Genetic analyses of knockout mutants, plant tilling lines and application of additional tools (such as the split ubiquitine system for protein interactions and protein complex analysis by proteomics) will certainly facilitate identification of new pathways and characterization of substrates that are targeted by these phosphatases in vivo. The united efforts of different groups in this respect will accelerate uncovering the roles of plant protein phosphatases in signalling.

Acknowledgements We thank Patricia T. W. Cohen, Alison DeLong and Pedro L. Rodriguez Egea for comments and David Kerk for providing the latest update of Arabidopsis phosphatase sequence prediction.

Our research is supported by grants from the Austrian Science Fund FWF.

References

Alonso A, Sasin J, Bottini N, Friedberg I, Osterman A, Godzik A, Hunter T, Dixon J, Mustelin T (2004) Protein tyrosine phosphatases in the human genome. Cell 117:699–711

Bang W, Kim S, Ueda A, Vikram M, Yun D, Bressan RA, Hasegawa PM, Bahk J, Koiwa H (2006) Arabidopsis carboxyl-terminal domain phosphatase-like isoforms share common catalytic and interaction domains but have distinct in planta functions. Plant Physiol 142:586–594

Barford D (2004) The role of cysteine residues as redox-sensitive regulatory switches. Curr Opin Struct Biol 14:679–686

Barford D, Das AK, Egloff MP (1998) The structure and mechanism of protein phosphatases: insights into catalysis and regulation. Annu Rev Biophys Biomol Struct 27:133–164

Barizza E, Lo Schiavo F, Terzi M, Filippini F (1999) Evidence suggesting protein tyrosine phosphorylation in plants depends on the developmental conditions. FEBS Lett 447:191–194

Bertauche N, Leung J, Giraudat J (1996) Protein phosphatase activity of abscisic acid insensitive 1 (ABI1) protein from *Arabidopsis thaliana*. Eur J Biochem 241:193–200

Biondi RM, Nebreda AR (2003) Signalling specificity of Ser/Thr protein kinases through docking-site-mediated interactions. Biochem J 372:1–13

Bogre L, Ligterink W, Meskiene I, Barker PJ, Heberle-Bors E, Huskisson NS, Hirt H (1997) Wounding induces the rapid and transient activation of a specific MAP kinase pathway. Plant Cell 9:75–83

Boudolf V, Inze D, De Veylder L (2006) What if higher plants lack a CDC25 phosphatase? Trends Plant Sci 11:474–479

Braun DM, Stone JM, Walker JC (1997) Interaction of the maize and Arabidopsis kinase interaction domains with a subset of receptor-like protein kinases: implications for transmembrane signaling in plants. Plant J 12:83–95

Camps M, Nichols A, Arkinstall S (2000) Dual specificity phosphatases: a gene family for control of MAP kinase function. FASEB J 14:6–16

Canagarajah BJ, Khokhlatchev A, Cobb MH, Goldsmith EJ (1997) Activation mechanism of the MAP kinase ERK2 by dual phosphorylation. Cell 90:859–869

Carrasco JL, Ancillo G, Castello MJ, Vera P (2005) A novel DNA-binding motif, hallmark of a new family of plant transcription factors. Plant Physiol 137:602–606

Carrasco JL, Ancillo G, Mayda E, Vera P (2003) A novel transcription factor involved in plant defense endowed with protein phosphatase activity. EMBO J 22:3376–3384

Carrasco JL, Castello MJ, Vera P (2006) 14-3-3 mediates transcriptional regulation by modulating nucleocytoplasmic shuttling of tobacco DNA-binding protein phosphatase-1. J Biol Chem 281:22875–22881

Cherel I, Michard E, Platet N, Mouline K, Alcon C, Sentenac H, Thibaud JB (2002) Physical and functional interaction of the Arabidopsis K(+) channel AKT2 and phosphatase AtPP2CA. Plant Cell 14:1133–1146

Cohen P (1989) The structure and regulation of protein phosphatases. Annu Rev Biochem 58:453–508

Cohen PT (2004) Overview of protein serine/threonine phosphatases. In: Arino JDRA (ed) Protein phosphatases. Springer, Berlin Heidelberg New York

Da Costa M, Bach L, Landrieu I, Bellec Y, Catrice O, Brown S, De Veylder L, Lippens G, Inze D, Faure JD (2006) Arabidopsis PASTICCINO2 is an antiphosphatase involved in regulation of cyclin-dependent kinase A. Plant Cell 18:1426–1437

DeLong A (2006) Switching the flip: protein phosphatase roles in signaling pathways. Curr Opin Plant Biol 9:470–477

Ding Z, Lee GI, Liang X, Gallazzi F, Arunima A, Van Doren SR (2005) PhosphoThr peptide binding globally rigidifies much of the FHA domain from Arabidopsis receptor kinase-associated protein phosphatase. Biochemistry 44:10119–10134

Ding Z, Wang H, Liang X, Morris ER, Gallazzi F, Pandit S, Skolnick J, Walker JC, Doren SR (2007) Phosphoprotein and phosphopeptide interactions with the FHA domain from Arabidopsis kinase-associated protein phosphatase. Biochemistry 46:2684–2696

Droillard M, Boudsocq M, Barbier-Brygoo H, Lauriere C (2002) Different protein kinase families are activated by osmotic stresses in *Arabidopsis thaliana* cell suspensions. Involvement of the MAP kinases AtMPK3 and AtMPK6. FEBS Lett 527:43–50

Farkas I, Dombradi V, Miskei M, Szabados L, Koncz C (2007) Arabidopsis PPP family of serine/threonine phosphatases. Trends Plant Sci 12:169–176

Farooq A, Zhou MM (2004) Structure and regulation of MAPK phosphatases. Cell Signal 16:769–779

Fordham-Skelton AP, Chilley P, Lumbreras V, Reignoux S, Fenton TR, Dahm CC, Pages M, Gatehouse JA (2002) A novel higher plant protein tyrosine phosphatase interacts with SNF1-related protein kinases via a KIS (kinase interaction sequence) domain. Plant J 29:705–715

Fordham-Skelton AP, Skipsey M, Eveans IM, Edwards R, Gatehouse JA (1999) Higher plant tyrosine-specific protein phosphatases (PTPs) contain novel amino-terminal domains: expression during embryogenesis. Plant Mol Biol 39:593–605

Gomez-Gomez L, Bauer Z, Boller T (2001) Both the extracellular leucine-rich repeat domain and the kinase activity of FSL2 are required for flagellin binding and signaling in Arabidopsis. Plant Cell 13:1155–1163

Gonzalez-Garcia MP, Rodriguez D, Nicolas C, Rodriguez PL, Nicolas G, Lorenzo O (2003) Negative regulation of abscisic acid signaling by the Fagus sylvatica FsPP2C1 plays a role in seed dormancy regulation and promotion of seed germination. Plant Physiol 133:135–144

Gosti F, Beaudoin N, Serizet C, Webb AAR, Vartanian N, Giraudat J (1999) ABI1 protein phosphatase 2C is a negative regulator of abscisic acid signaling. Plant Cell Oct 11:1897–1909

Guo Y, Xiong L, Song CP, Gong D, Halfter U, Zhu JK (2002) A calcium sensor and its interacting protein kinase are global regulators of abscisic acid signaling in Arabidopsis. Dev Cell 3:233–244

Gupta R, Huang Y, Kieber J, Luan S (1998) Identification of a dual-specificity protein phosphatase that inactivates a MAP kinase from Arabidopsis. Plant J 16:581–589

Gupta R, Luan S (2003) Redox control of protein tyrosine phosphatases and mitogen-activated protein kinases in plants. Plant Physiol 132:1149–1152

Gupta R, Ting JT, Sokolov LN, Johnson SA, Luan S (2002) A tumor suppressor homolog, AtPTEN1, is essential for pollen development in Arabidopsis. Plant Cell 14:2495–2507

Himmelbach A, Hoffmann T, Leube M, Hohener B, Grill E (2002) Homeodomain protein ATHB6 is a target of the protein phosphatase ABI1 and regulates hormone responses in Arabidopsis. EMBO J 21:3029–3038

Ho DT, Bardwell AJ, Abdollahi M, Bardwell L (2003) A docking site in MKK4 mediates high affinity binding to JNK MAPKs and competes with similar docking sites in JNK substrates. J Biol Chem 278:32662–32672

Huang Y, Li H, Gupta R, Morris PC, Luan S, Kieber JJ (2000) ATMPK4, an Arabidopsis homolog of mitogen-activated protein kinase, is activated in vitro by AtMEK1 through threonine phosphorylation. Plant Physiol 122:1301–1310

Ingebritsen TS, Cohen P (1983) Protein phosphatases: properties and role in cellular regulation. Science 221:331–338

Jemc J, Rebay I (2007) The eyes absent family of phosphotyrosine phosphatases: properties and roles in developmental regulation of transcription. Annu Rev Biochem 76:513–538

Katou S, Karita E, Yamakawa H, Seo S, Mitsuhara I, Kuchitsu K, Ohashi Y (2005) Catalytic activation of the plant MAPK phosphatase NtMKP1 by its physiological substrate salicylic acid-induced protein kinase but not by calmodulins. J Biol Chem 280:39569–39581

Kerk D (2006) Genome-scale discovery and characterization of class-specific protein sequences: an example using the protein phosphatases of *Arabidopsis thaliana*. Methods Mol Biol 365:347–370

Kerk D, Bulgrien J, Smith DW, Barsam B, Veretnik S, Gribskov M (2002) The complement of protein phosphatase catalytic subunits encoded in the genome of Arabidopsis. Plant Physiol 129:908–925

Kerk D, Conley TR, Rodriguez FA, Tran HT, Nimick M, Muench DG, Moorhead GB (2006) A chloroplast-localized dual-specificity protein phosphatase in Arabidopsis contains a phylogenetically dispersed and ancient carbohydrate-binding domain, which binds the polysaccharide starch. Plant J 46:400–413

Kiegerl S, Cardinale F, Siligan C, Gross A, Baudouin E, Liwosz A, Eklof S, Till S, Bogre L, Hirt H, Meskiene I (2000) SIMKK, a mitogen-activated protein kinase (MAPK) kinase, is a specific activator of the salt stress-induced MAPK, SIMK. Plant Cell 12:2247–2258

Kuhn JM, Boisson-Dernier A, Dizon MB, Maktabi MH, Schroeder JI (2006) The protein phosphatase AtPP2CA negatively regulates abscisic acid signal transduction in Arabidopsis, and effects of abh1 on AtPP2CA mRNA. Plant Physiol 140:127–139

Kuromori T, Yamamoto M (1994) Cloning of cDNAs from *Arabidopsis thaliana* that encode putative protein phosphatase 2C and a human Dr1-like protein by transformation of a fission yeast mutant. Nucleic Acids Res 22:5296–5301

Lee GI, Ding Z, Walker JC, Van Doren SR (2003) NMR structure of the forkhead-associated domain from the Arabidopsis receptor kinase-associated protein phosphatase. Proc Natl Acad Sci USA 100:11261–11266

Leonhardt N, Kwak JM, Robert N, Waner D, Leonhardt G, Schroeder JI (2004) Microarray expression analyses of Arabidopsis guard cells and isolation of a recessive abscisic acid hypersensitive protein phosphatase 2C mutant. Plant Cell 16:596–615

Leung J, Bouvier-Durand M, Morris PC, Guerrier D, Chefdor F, Giraudat J (1994) Arabidopsis ABA response gene ABI1: features of a calcium-modulated protein phosphatase. Science 264:1448–1452

Leung J, Merlot S, Giraudat J (1997) The Arabidopsis ABSCISIC ACID-INSENSITIVE2 (ABI2) and ABI1 genes encode homologous protein phosphatases 2C involved in abscisic acid signal transduction. Plant Cell 9:759–771

Li J, Smith GP, Walker JC (1999) Kinase interaction domain of kinase-associated protein phosphatase, a phosphoprotein-binding domain. Proc Natl Acad Sci USA 96:7821–7826

Li X, Oghi KA, Zhang J, Krones A, Bush KT, Glass CK, Nigam SK, Aggarwal AK, Maas R, Rose DW, Rosenfeld MG (2003) Eya protein phosphatase activity regulates Six1-Dach-Eya transcriptional effects in mammalian organogenesis. Nature 426:247–254

Lorenzo O, Rodriguez D, Nicolas G, Rodriguez PL, Nicolas C (2001) A new protein phosphatase 2C (FsPP2C1) induced by abscisic acid is specifically expressed in dormant beechnut seeds. Plant Physiol 125:1949–1956

Luan S (1998) Protein phosphatases and signaling cascades in higher plants. Trends Plant Sci 3:271–275

Lunn JE (2002) Evolution of sucrose synthesis. Plant Physiol 128:1490–1500

Martin H, Flandez M, Nombela C, Molina M (2005) Protein phosphatases in MAPK signalling: we keep learning from yeast. Mol Microbiol 58:6–16

Meinhard M, Grill E (2001) Hydrogen peroxide is a regulator of ABI1, a protein phosphatase 2C from Arabidopsis. FEBS Lett 508:443–446

Meinhard M, Rodriguez PL, Grill E (2002) The sensitivity of ABI2 to hydrogen peroxide links the abscisic acid-response regulator to redox signalling. Planta 214:775–782

Merlot S, Gosti F, Guerrier D, Vavasseur A, Giraudat J (2001) The ABI1 and ABI2 protein phosphatases 2C act in a negative feedback regulatory loop of the abscisic acid signalling pathway. Plant J 25:295–303

Meskiene I, Baudouin E, Schweighofer A, Liwosz A, Jonak C, Rodriguez PL, Jelinek H, Hirt H (2003) The stress-induced protein phosphatase 2C is a negative regulator of a mitogen-activated protein kinase. J Biol Chem 278:18945–18945

Meskiene I, Bogre L, Glaser W, Balog J, Brandstotter M, Zwerger K, Ammerer G, Hirt H (1998) MP2C, a plant protein phosphatase 2C, functions as a negative regulator of mitogen-activated protein kinase pathways in yeast and plants. Proc Natl Acad Sci USA 95:1938–1943

Meyer K, Leube MP, Grill E (1994) A protein phosphatase 2C involved in ABA signal transduction in *Arabidopsis thaliana*. Science 264:1452–1455

Miao Y, Lv D, Wang P, Wang XC, Chen J, Miao C, Song CP (2006) An Arabidopsis glutathione peroxidase functions as both a redox transducer and a scavenger in abscisic acid and drought stress responses. Plant Cell 18:2749–2766

Mishra G, Zhang W, Deng F, Zhao J, Wang X (2006) A bifurcating pathway directs abscisic acid effects on stomatal closure and opening in Arabidopsis. Science 312:264–266

Monroe-Augustus M, Zolman BK, Bartel B (2003) IBR5, a dual-specificity phosphatase-like protein modulating auxin and abscisic acid responsiveness in Arabidopsis. Plant Cell 15:2979–2991

Munnik T, Ligterink W, Meskiene I, Calderini O, Beyerly J, Musgrave A, Hirt H (1999) Distinct osmo-sensing protein kinase pathways are involved in signalling moderate and severe hyper-osmotic stress. Plant J 20:381–388

Mustilli AC, Merlot S, Vavasseur A, Fenzi F, Giraudat J (2002) Arabidopsis OST1 protein kinase mediates the regulation of stomatal aperture by abscisic acid and acts upstream of reactive oxygen species production. Plant Cell 14:3089–3099

Naoi K, Hashimoto T (2004) A semidominant mutation in an Arabidopsis mitogen-activated protein kinase phosphatase-like gene compromises cortical microtubule organization. Plant Cell 16:1841–1853

Niittyla T, Comparot-Moss S, Lue WL, Messerli G, Trevisan M, Seymour MD, Gatehouse JA, Villadsen D, Smith SM, Chen J, Zeeman SC, Smith AM (2006) Similar protein phosphatases control starch metabolism in plants and glycogen metabolism in mammals. J Biol Chem 281:11815–11818

Ohta M, Guo Y, Halfter U, Zhu JK (2003) A novel domain in the protein kinase SOS2 mediates interaction with the protein phosphatase 2C ABI2. Proc Natl Acad Sci USA 100:11771–11776

Park AR, Cho SK, Yun UJ, Jin MY, Lee SH, Sachetto-Martins G, Park OK (2001) Interaction of the Arabidopsis receptor protein kinase Wak1 with a glycine-rich protein, AtGRP-3. J Biol Chem 276:26688–26693

Pulido R, Zuniga A, Ullrich A (1998) PTP-SL and STEP protein tyrosine phosphatases regulate the activation of the extracellular signal-regulated kinases ERK1 and ERK2 by association through a kinase interaction motif. EMBO J 17:7337–7350

Quettier AL, Bertrand C, Habricot Y, Miginiac E, Agnes C, Jeannette E, Maldiney R (2006) The phs1-3 mutation in a putative dual-specificity protein tyrosine phosphatase gene provokes hypersensitive responses to abscisic acid in *Arabidopsis thaliana*. Plant J 47:711–719

Rayapureddi JP, Kattamuri C, Chan FH, Hegde RS (2005) Characterization of a plant, tyrosine-specific phosphatase of the aspartyl class. Biochemistry 44:751–758

Reyes D, Rodriguez D, Gonzalez-Garcia MP, Lorenzo O, Nicolas G, Garcia-Martinez JL, Nicolas C (2006) Overexpression of a protein phosphatase 2C from beech seeds in Arabidopsis shows phenotypes related to abscisic acid responses and gibberellin biosynthesis. Plant Physiol 141:1414–1424

Rienties IM, Vink J, Borst JW, Russinova E, de Vries SC (2005) The Arabidopsis SERK1 protein interacts with the AAA-ATPase AtCDC48, the 14-3-3 protein GF14lambda and the PP2C phosphatase KAPP. Planta 221:394–405

Robert N, Merlot S, N'Guyen V, Boisson-Dernier A, Schroeder JI (2006) A hypermorphic mutation in the protein phosphatase 2C HAB1 strongly affects ABA signaling in Arabidopsis. FEBS Lett 580:4691–4696

Rodriguez PL (1998) Protein phosphatase 2C (PP2C) function in higher plants. Plant Mol Biol 38:919–927

Rodriguez PL, Benning G, Grill E (1998) ABI2, a second protein phosphatase 2C involved in abscisic acid signal transduction in Arabidopsis. FEBS Lett 421:185–190

Rodriguez PL, Leube MP, Grill E (1998) Molecular cloning in *Arabidopsis thaliana* of a new protein phosphatase 2C (PP2C) with homology to ABI1 and ABI2. Plant Mol Biol 38:879–883

Saez A, Apostolova N, Gonzalez-Guzman M, Gonzalez-Garcia MP, Nicolas C, Lorenzo O, Rodriguez PL (2004) Gain-of-function and loss-of-function phenotypes of the protein phosphatase 2C HAB1 reveal its role as a negative regulator of abscisic acid signalling. Plant J 37:354–369

Saez A, Robert N, Maktabi MH, Schroeder JI, Serrano R, Rodriguez PL (2006) Enhancement of abscisic acid sensitivity and reduction of water consumption in Arabidopsis by combined inactivation of the protein phosphatases type 2C ABI1 and HAB1. Plant Physiol 141:1389–1399

Schweighofer A, Hirt H, Meskiene I (2004) Plant PP2C phosphatases: emerging functions in stress signaling. Trends Plant Sci 9:236–243

Schweighofer A, Kazanaviciute V, Scheikl E, Teige M, Doczi R, Hirt H, Schwanninger M, Kant M, Schuurink R, Mauch F, Buchala A, Cardinale F, Meskiene I (2007) The PP2C-type phosphatase AP2C1, which negatively regulates MPK4 and MPK6, modulates innate immunity, jasmonic acid, and ethylene levels in Arabidopsis. Plant Cell 19:2213–2224

Shah K, Russinova E, Gadella TW Jr, Willemse J, De Vries SC (2002) The Arabidopsis kinase-associated protein phosphatase controls internalization of the somatic embryogenesis receptor kinase 1. Genes Dev 16:1707–1720

Shah K, Vervoort J, de Vries SC (2001) Role of threonines in the *Arabidopsis thaliana* somatic embryogenesis receptor kinase 1 activation loop in phosphorylation. J Biol Chem 276:41263–41269

Sheen J (1998) Mutational analysis of protein phosphatase 2C involved in abscisic acid signal transduction in higher plants. Proc Natl Acad Sci USA 95:975–980

Smith RD, Walker JC (1996) Plant protein phosphatases. Annu Rev Plant Physiol Plant Mol Biol 47:101–125

Sokolov LN, Dominguez-Solis JR, Allary AL, Buchanan BB, Luan S (2006) A redox-regulated chloroplast protein phosphatase binds to starch diurnally and functions in its accumulation. Proc Natl Acad Sci USA 103:9732–9737

Song SK, Clark SE (2005) POL and related phosphatases are dosage-sensitive regulators of meristem and organ development in Arabidopsis. Dev Biol 285:272–284

Song SK, Lee MM, Clark SE (2006) POL and PLL1 phosphatases are CLAVATA1 signaling intermediates required for Arabidopsis shoot and floral stem cells. Development 133:4691–4698

Stone JM, Collinge MA, Smith RD, Horn MA, Walker JC (1994) Interaction of a protein phosphatase with an Arabidopsis serine- threonine receptor kinase. Science 266:793–795

Stone JM, Trotochaud AE, Walker JC, Clark SE (1998) Control of meristem development by CLAVATA1 receptor kinase and kinase-associated protein phosphatase interactions. Plant Physiol 117:1217–1225

Tahtiharju S, Palva T (2001) Antisense inhibition of protein phosphatase 2C accelerates cold acclimation in *Arabidopsis thaliana*. Plant J 26:461–470

Takeda Y, Hatano S, Sentoku N, Matsuoka M (1999) Homologs of animal eyes absent (eya) genes are found in higher plants. Mol Gen Genet 262:131–138

Tonks NK (2005) Redox redux: revisiting PTPs and the control of cell signaling. Cell 121:667–670

Ulm R, Ichimura K, Mizoguchi T, Peck SC, Zhu T, Wang X, Shinozaki K, Paszkowski J (2002) Distinct regulation of salinity and genotoxic stress responses by Arabidopsis MAP kinase phosphatase 1. EMBO J 21:6483–6493

Ulm R, Revenkova E, di Sansebastiano GP, Bechtold N, Paszkowski J (2001) Mitogen-activated protein kinase phosphatase is required for genotoxic stress relief in Arabidopsis. Genes Dev 15:699–709

Vranova E, Tahtiharju S, Sriprang R, Willekens H, Heino P, Palva ET, Inze D, Van Camp W (2001) The AKT3 potassium channel protein interacts with the AtPP2CA protein phosphatase 2C. J Exp Bot 52:181–182

Williams RW, Wilson JM, Meyerowitz EM (1997) A possible role for kinase-associated protein phosphatase in the Arabidopsis CLAVATA1 signaling pathway. Proc Natl Acad Sci USA 94:10467–10472

Wu Y, Sanchez JP, Lopez-Molina L, Himmelbach A, Grill E, Chua NH (2003) The abi1-1 mutation blocks ABA signaling downstream of cADPR action. Plant J 34:307–315

Xu Q, Fu HH, Gupta R, Luan S (1998) Molecular characterization of a tyrosine-specific protein phosphatase encoded by a stress-responsive gene in Arabidopsis. Plant Cell 10:849–857 [published erratum appears in Plant Cell 1998 Oct;10(10):1769]

Yamakawa H, Katou S, Seo S, Mitsuhara I, Kamada H, Ohashi Y (2004) Plant MAPK phosphatase interacts with calmodulins. J Biol Chem 279:928–936

Yoo JH, Cheong MS, Park CY, Moon BC, Kim MC, Kang YH, Park HC, Choi MS, Lee JH, Jung WY, Yoon HW, Chung WS, Lim CO, Lee SY, Cho MJ (2004) Regulation of the dual specificity protein phosphatase, DsPTP1, through interactions with calmodulin. J Biol Chem 279:848–858

Yoshida R, Hobo T, Ichimura K, Mizoguchi T, Takahashi F, Aronso J, Ecker JR, Shinozaki K (2002) ABA-activated SnRK2 protein kinase is required for dehydration stress signaling in Arabidopsis. Plant Cell Physiol 43:1473–1483

Yoshida R, Umezawa T, Mizoguchi T, Takahashi S, Takahashi F, Shinozaki K (2006) The regulatory domain of SRK2E/OST1/SnRK2.6 interacts with ABI1 and integrates abscisic acid (ABA) and osmotic stress signals controlling stomatal closure in Arabidopsis. J Biol Chem 281:5310–5318

Yoshida T, Nishimura N, Kitahata N, Kuromori T, Ito T, Asami T, Shinozaki K, Hirayama T (2006) ABA-hypersensitive germination3 encodes a protein phosphatase 2C (AtPP2CA) that strongly regulates abscisic acid signaling during germination among Arabidopsis protein phosphatase 2Cs. Plant Physiol 140:115–126

Yu LP, Miller AK, Clark SE (2003) POLTERGEIST encodes a protein phosphatase 2C that regulates CLAVATA pathways controlling stem cell identity at Arabidopsis shoot and flower meristems. Curr Biol 13:179–188

Yu LP, Simon EJ, Trotochaud AE, Clark SE (2000) POLTERGEIST functions to regulate meristem development downstream of the CLAVATA loci. Development 127:1661–1670

Zhang W, Qin C, Zhao J, Wang X (2004) Phospholipase Dα1-derived phosphatidic acid interacts with ABI1 phosphatase 2C and regulates abscisic acid signaling. Proc Natl Acad Sci USA 101:9508–9513

Plant Cell Monogr (10)
L. Bögre and G. Beemster: Plant Growth Signaling
DOI 10.1007/7089_2007_138/Published online: 28 July 2007

Armadillo Repeat Proteins: Versatile Regulators of Plant Development and Signalling

Juliet C. Coates

School of Biosciences, University of Birmingham, Edgbaston, Birmingham B15 2TT, UK
j.c.coates@bham.ac.uk

Abstract Proteins containing tandem copies of the Armadillo repeat motif are evolutionarily ancient and have diverse cellular roles. Genome sequencing has revealed a plethora of Armadillo repeat-containing proteins in plants. Studies in *Arabidopsis*, *Brassica*, tobacco, potato and rice confirm that plant Armadillo repeat proteins have a wide range of biological functions. A large number of plant Armadillo proteins appear to regulate targeted protein degradation, which is a fundamental part of many plant developmental processes. This chapter will review the current knowledge of plant Armadillo protein functions during growth and development.

1 Introduction

Hundreds of eukaryotic proteins possess tandem structural units called Armadillo (Arm) repeats. The name "Armadillo" reflects the mutant phenotype of the *Drosophila* segment polarity gene *armadillo*, the first member of the gene family to be characterised in detail (Riggleman et al. 1989). Armadillo and its mammalian homologue β-catenin are critical for multicellular development, regulating both gene expression and cell–cell adhesion (Conacci-Sorrell et al. 2002). Arm-repeat proteins share a conserved three-dimensional structure: tandem Arm-repeats form a right-handed superhelix of α-helices (Huber et al. 1997; Conti et al. 1998). In animals, protists and fungi, Arm repeats provide a versatile protein–protein interaction surface used to bind diverse target proteins during growth and development, recruiting molecules such as transcription factors and cytoskeletal regulators.

Arm repeats are structurally related to HEAT repeats, with which they have a common phylogenetic origin (Andrade et al. 2001). Both repeats are evolutionarily ancient, present in unicellular eukaryotes, animals, plants and prokaryotes (Sanger Institute 2007)). Arm family proteins fall into subfamilies that have characteristic sequences outside the Arm domain. Some subfamilies are found throughout eukaryotes, while others are specific to a particular taxonomic group (Coates 2003). Plants possess over a hundred Arm-repeat proteins (Mudgil et al. 2004; Samuel et al. 2006) and most have novel functions, which are reflected in their plant-specific domain architectures (Fig. 1). Unlike their animal counterparts, many plant Arm proteins

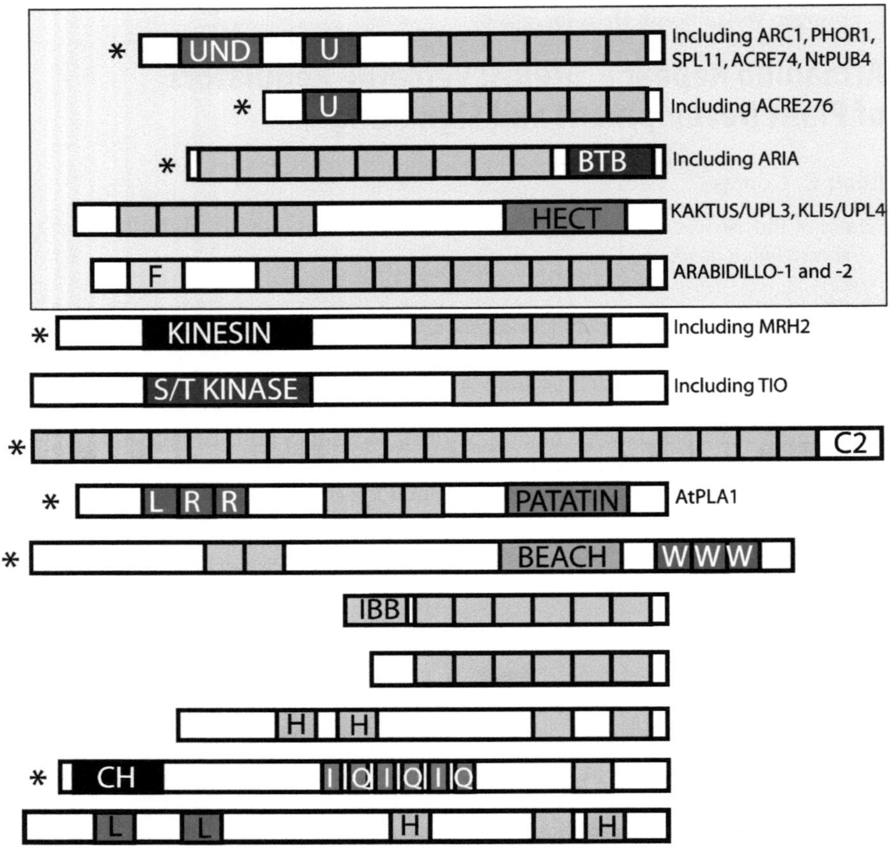

Fig. 1 Plant Armadillo repeat-containing protein families. Arm repeats are shown as *light grey boxes*. Other domains are labelled with their name/abbreviation and shown in *other shades of grey*. *U* U-box (Pfam04564); *UND* U-box N-terminal-associated (Mudgil et al. 2004); *BTB* BTB/POZ domain (PF00651); *HECT* homologous to E6AP C-terminus (PF00632); *F* F-box (PF00646); *S/T kinase*: serine/threonine kinase (PF00069); *C2* C2 domain (PF00168); *LRR* leucine-rich repeat (PF00560); *PATATIN* patatin domain (PF01734); *W* WD40 repeat (PF00400) *IBB* importin beta binding (PF01749); *H* HEAT repeat (PF02985); *CH* calponin homology domain (PF00307); *IQ* IQ calmodulin binding motif (PF00612); *L* Lis homology domain (PF08513). HEAT repeats and IBB domains are structurally related to Arm repeats. Protein families marked with an *asterisk* have plant-specific domain architecture. Proteins surrounded by the *pale grey box* are likely to be E3 ubiquitin ligases

appear to function as components of the ubiquitin–proteasome system, which targets proteins for regulated degradation.

Timely degradation of proteins is crucial for numerous physiological processes, including growth and development. Regulated protein degradation by the proteasome is a mechanism that has been conserved throughout eukaryotic evolution. Addition of polyubiquitin to a protein labels it for destruction by the constitutively active 26S proteasome. The process requires an E1 ubiquitin activating enzyme, an E2 ubiquitin conjugating enzyme, and an E3 ubiquitin–protein ligase, which transfers ubiquitin from the E2 to the target (reviewed in Vierstra 2003). A large subset ($\sim$ 40 proteins) of *Arabidopsis* Arm proteins contain a U-box, and are part of the PUB (plant U-Box) family (Mudgil et al. 2004; Samuel et al. 2006; and Fig. 1). Plants also possess Arm-HECT (Homology to E6-AP C-terminus) domain proteins, as do animals and fungi (El Refy et al. 2003). U-box and HECT proteins are single-subunit E3 ubiquitin ligases, contacting both the E2 ubiquitin conjugating enzyme and the target protein directly (Vierstra 2003).

In addition to these single-subunit E3 ligases, animals, plants and fungi also possess multiprotein E3 ubiquitin ligases (Petroski and Deshaies 2005). F-box proteins are subunits of SKP1-CULLIN1-F-box (SCF) E3 ligases. *Arabidopsis* possesses two F-box/Arm-repeat proteins, a domain combination also found in the protist *Dictyostelium*, but not in animals or fungi (Coates 2003). In addition, there is a class of two *Arabidopsis* proteins that contain a BTB/POZ (BR-C, tramtrack, bric-a-brac/Pox virus associated zinc finger) domain and Arm repeats (Fig. 1), which may act as specificity factors for CULLIN3-containing E3 ligases (Gingerich et al. 2005).

In this chapter, I will review current knowledge of the functions of plant Arm-repeat proteins in signal transduction and development, including their role in hormone signalling, morphogenesis, defence and cell death.

2 Arm-Repeat Proteins in Plant Growth, Development and Hormone Signalling

Arm-repeat proteins function during diverse aspects of plant development and morphogenesis, processes controlled by multiple intracellular signalling pathways. Plant Arm proteins function in gibberellin (GA) and abscisic acid (ABA) signalling, in addition to signalling through receptor kinases with as-yet unknown ligands. Interestingly, all plant Arm proteins so far implicated in developmental signal transduction are putative ubiquitin ligases, highlighting the key importance of regulated protein degradation during development. It is tempting to speculate that many hormones use plant Arm proteins for aspects of their signal transduction, to allow diverse cell type-specific responses.

2.1 NtPUB4 and Plant Development

Tobacco NtPUB4 is a U-box/Arm protein discovered by its binding to a receptor-like kinase, CHRK1 (Kim et al. 2003). Tobacco seedlings with reduced levels of CHRK1 have widespread developmental defects relating to cell proliferation, such as elevated cytokinin levels, callus formation, shoot overproduction, reduced apical dominance and abnormal flower development (Lee et al. 2003).

Like CHRK1, NtPUB4 localises to the plasma membrane. NtPUB4 is expressed in all tissues, but most highly in flowers (Kim et al. 2003). Loss-of-function of NtPUB4 has not been characterised, so its in vivo function is not clear. However, CHRK1 and NtPUB4 potentially represent players in a novel developmental signalling pathway, as plants with reduced CHRK1 function respond normally to exogenous cytokinins and auxin (Lee et al. 2003). Discovering ligands and interactors will be an exciting area of future research.

2.2 PHOR1 in Gibberellin and Light Signalling

The potato PHOTOPERIOD RESPONSIVE 1 (PHOR1) protein was the first plant Arm protein to be associated with hormone signalling and plant growth regulation, and is a U-box/Arm protein (Fig. 1). PHOR1 is a positive regulator of gibberellin (GA) signalling. GA is a key regulator of plant growth and development, including of cell elongation both in shoots and roots. Activation of an intracellular GA receptor causes the proteasomal degradation of DELLA proteins via an SCF–E3 ubiquitin ligase, and thus derepression of cellular GA responses (Hartweck and Olszewski 2006).

GA induces stem elongation in wild-type potato plants; PHOR1 antisense plants have shorter stems than wild-type and are less sensitive to exogenous GA. Conversely, PHOR1 overexpressing plants have elongated stems and show an overactive GA response (Amador et al. 2001). Interestingly, GA changes the subcellular distribution of a PHOR1–GFP (green fluorescent protein) fusion protein. PHOR1–GFP is present in both the nucleus and the cytosol, but moves transiently to the nucleus in response to GA. Nuclear localisation of PHOR1 requires the Arm repeats, whilst the U-box drives cytosolic localisation (Amador et al. 2001). Given the recent discovery of a soluble GA receptor (Hartweck and Olszewski 2006), it is intriguing to speculate whether PHOR1 (like NtPUB4 (Sect. 2.1) and ARC1 (Sect. 4.1)) binds to as-yet uncharacterised membrane-associated GA receptor(s), for which physiological evidence exists (Hartweck and Olszewski 2006). It is not yet known whether PHOR1 functions as an E3 ubiquitin ligase in vivo. Discovery of PHOR1 interaction partners will determine whether PHOR1 can degrade DELLA proteins or novel targets.

Potato tuberisation requires short day conditions, which also enhance *PHOR1* mRNA levels. Paradoxically, plants with reduced PHOR1 activity show earlier tuberisation and increased tuber yield in short days, suggesting that PHOR1 inhibits short day-induced tuberisation (Amador et al. 2001).

Thus, tuberisation is likely to be a complex process in which PHOR1 plays a regulatory role, integrating environmental and hormonal signals.

2.3
KAKTUS/UPL3 Regulates Trichome Development and Gibberellin Signalling

Trichomes are singled-celled hairs on the epidermis of leaves, stems and sepals that are thought to protect plants from damage by pathogens, light and drought. Wild-type *Arabidopsis* leaf trichomes have three branches, while stem trichomes are unbranched. Trichome branch number depends on GA signalling, nuclear DNA content (ploidy) and microtubules (Hulskamp et al. 1999). Plants with elevated GA levels or increased ploidy have overbranched trichomes, while plants with reduced GA signalling develop trichomes with fewer branches.

KAKTUS/UPL3 is an *Arabidopsis* HECT–Arm protein (Fig. 1; Downes et al. 2003; El Refy et al. 2003). The *kaktus/upl3* mutants produce overbranched trichomes with increased ploidy (Perazza et al. 1999). The *kaktus/upl3* mutants are also hypersensitive to GA-dependent hypocotyl cell elongation, and have many cells with elevated ploidy (Downes et al. 2003; El Refy et al. 2003). This suggests that, in contrast to PHOR1, KAKTUS/UPL3 is a negative regulator of GA signalling. KAKTUS/UPL3 also negatively regulates DNA content in a variety of cell types.

However, some GA responses are normal in *kaktus/upl3* mutants, suggesting that KAKTUS/UPL3 is involved only in specific aspects of GA signalling or that another protein can substitute for KAKTUS/UPL3 function in certain circumstances. *Arabidopsis* KLI5/UPL4 is 54% similar to KAKTUS/UPL3 (Downes et al. 2003; El Refy et al. 2003); however, the two genes are not entirely redundant given that *upl4* mutants have no trichome phenotype (Downes et al. 2003). As yet, KAKTUS/UPL3 has not been shown to function as a bona fide ubiquitin ligase, but the current hypothesis is that KAKTUS/UPL3 degrades protein(s) that are positive regulators of trichome branching, DNA replication and/or GA signalling.

2.4
ARIA and Abscisic Acid Signalling

Abscisic acid (ABA) signalling is involved in several abiotic stress responses, germination and seedling development. ABF2 is a transcription factor that binds to ABA-responsive promoter elements. ABF2 interacts with ARIA (Arm-repeat protein interacting with ABF2), which has a BTB/POZ domain (Kim et al. 2004a,b; Fig. 1).

Aria mutant seedlings show reduced sensitivity to ABA during germination and root growth. This suggests that ARIA functions as a positive regulator of ABA signalling, and accordingly *ARIA* mRNA is upregulated by ABA

Fig. 2 Functions and mechanisms of plant Arm-repeat protein signalling. *A* U-box/Arm proteins function as E3 ubiquitin ligases during a variety of physiological and developmental processes; interacting transmembrane receptors are known for NtPUB4 and ARC1 only. *B* UPL3/KAKTUS may respond to gibberellin to degrade inhibitors of trichome branching, DNA replication and cell expansion. *C* ARIA binds to ABF transcription factors that regulate stress- and glucose-responses in early development; they may also be part of CULLIN3-containing ubiquitin ligase complexes. *D* ARABIDILLO-1 and -2 are postulated to act as part of an SCF ubiquitin ligase complex to promote lateral root development. *E* MRH2 contains a kinesin domain and regulates microtubule-mediated root hair morphogenesis. TIO is required for cytokinesis and may interact with microtubules, possibly via phosphorylation ▶

(Kim et al. 2004a). However, *abf2* mutants are not ABA-insensitive (Kim et al. 2004b). Both *abf2* and *aria* mutants show reduced sensitivity to glucose (Kim et al. 2004a). Moreover, mutants in ABF2-related transcription factors, ABF3 and ABF4, are ABA-insensitive, suggesting that ARIA may regulate all three ABFs (Kim et al. 2004b; Fig. 2).

ARIA-overexpressing plants are hypersensitive to ABA, as would be predicted from the knockout phenotype. However, they also show hypersensitivity to osmotic stress during germination and salt-tolerance in later life (Kim et al. 2004a); both of these phenotypes could arise from ARIA interacting with ABF3 and ABF4 (Kim et al. 2004b).

Unlike ABF2, which is exclusively nuclear, an ARIA–GFP fusion is found both in the nucleus and at the cell periphery (Kim et al. 2004a), suggesting that ARIA could have non-nuclear functions. There is no evidence as yet that ARIA protein activity or localisation is regulated directly by ABA. In addition, ARIA affects only a subset of ABA-dependent processes (Kim et al. 2004a). *Arabidopsis* possesses one *ARIA*-related gene, and thus the relatively mild phenotypes seen in the *aria* mutant may be due to redundancy.

Although the BTB/POZ domain was originally characterised in transcription factors, it is also present in ubiquitin ligases (Gingerich et al. 2005). Thus, ARIA could degrade target proteins in addition to directly regulating ABF-mediated transcription (Fig. 2). In general, *aria* and *abf* mutants have similar phenotypes, suggesting ARIA does not degrade ABFs themselves. Interestingly, other Arm proteins may be involved in degradation responses to stress: in mangrove, *bg55* mRNA (encoding a U-box/Arm protein) is upregulated by salt stress (Banzai and Karube 2002).

2.5 ARABIDILLO Proteins Promote Root Branching

ARABIDILLO-1 and -2 are the *Arabidopsis* proteins most similar to animal and *Dictyostelium* β-catenins (Coates 2003). Loss-of-function *arabidillo-1/-2* mutants form fewer lateral roots than wild-type seedlings. Conversely, ARABIDILLO-1 overexpression promotes lateral root formation (Coates et al. 2006).

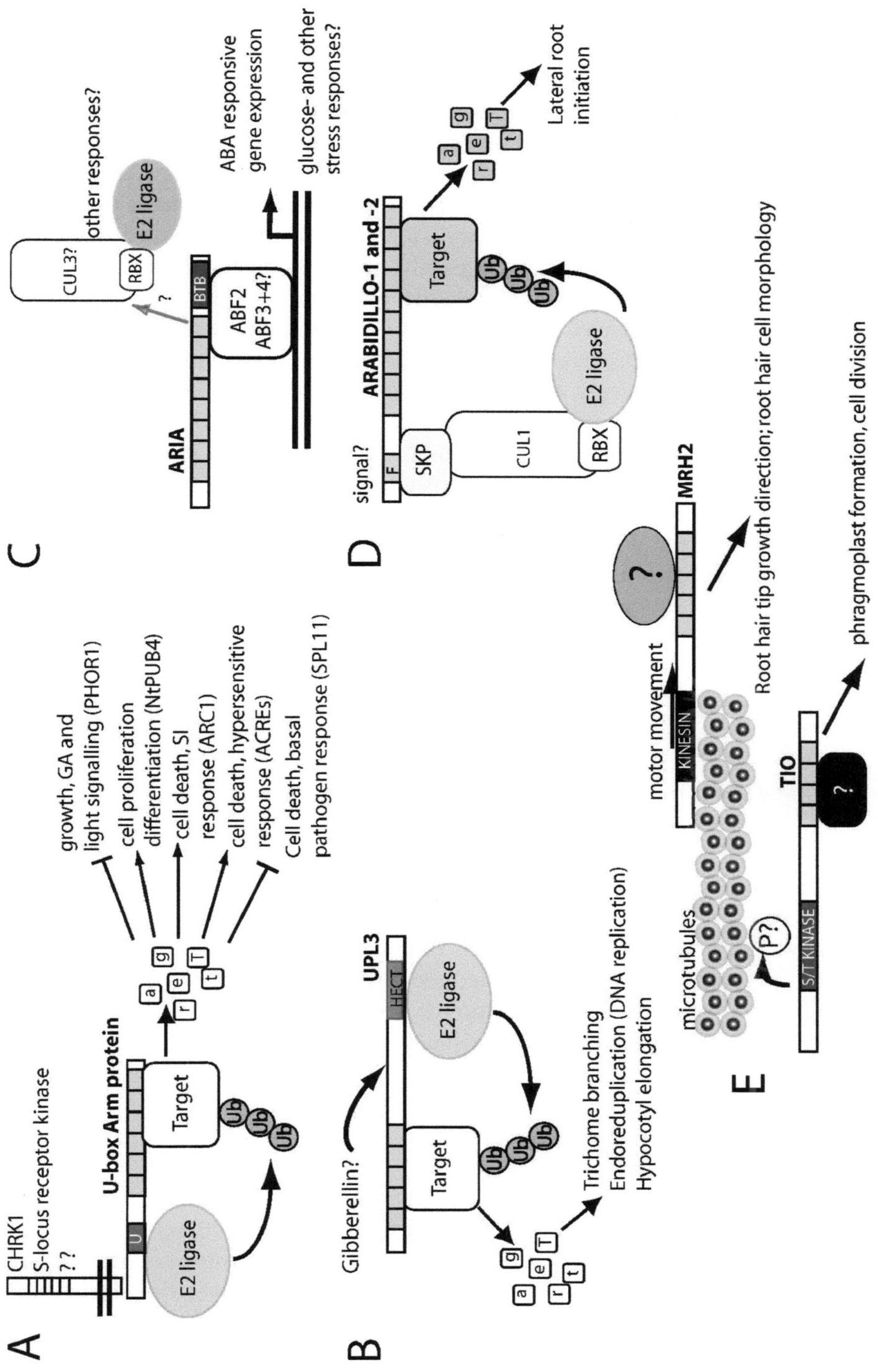
A
CHRK1
S-locus receptor kinase
? ?
U-box Arm protein
U
E2 ligase
Target
Ub
Ub
Ub
growth, GA and light signalling (PHOR1)
cell proliferation differentiation (NtPUB4)
cell death, SI response (ARC1)
cell death, hypersensitive response (ACREs)
Cell death, basal pathogen response (SPL11)
B
Gibberellin?
UPL3
HECT
E2 ligase
Target
Trichome branching
Endoreduplication (DNA replication)
Hypocotyl elongation
C
ARIA
BTB
ABF2
ABF3+4?
CUL3?
RBX
E2 ligase
other responses?
ABA responsive gene expression
glucose- and other stress responses?
D
signal?
F
SKP
CUL1
RBX
E2 ligase
ARABIDILLO-1 and -2
Target
Lateral root initiation
E
microtubules
P?
S/T KINASE
TIO
?
KINESIN
motor movement
MRH2
?
Root hair tip growth direction; root hair cell morphology
phragmoplast formation, cell division

Lateral roots initiate from coordinated cell divisions of selected pericycle cells in the centre of the primary root of the plant. This generates a lateral root primordium, and subsequently a new root meristem. This in turn generates a fully-formed root that emerges through the outer cell layers of the primary root (Casimiro et al. 2003). *Arabidillo-1/-2* mutants are defective in early stages of lateral root formation, when lateral roots first initiate (Coates et al. 2006). The hormone auxin is critical for lateral root development (Casimiro et al. 2003); however, *arabidillo-1* and *-2* appear to regulate lateral root formation in an auxin-independent manner (Coates et al. 2006), suggesting the existence of a novel lateral root regulatory pathway.

ARABIDILLO proteins localise to the nucleus, and this is dependent in part on the Arm repeats. Certain ARABIDILLO protein fragments, including the F-box region, are cytosolic (Coates et al. 2006). It is not yet known how ARABIDILLO subcellular localisation is regulated in vivo. ARABIDILLO-1 and -2 are F-box proteins but it is unclear whether ARABIDILLOs function as ubiquitin ligases. Preliminary data suggest that ARABIDILLO-1 can interact with *Arabidopsis* SKP1-related proteins, which are components of SCF ubiquitin ligases (DJ Gibbs and JC Coates, unpublished).

The current hypothesis is that ARABIDILLO proteins target an inhibitor of lateral root development (for example, an inhibitor of cell division) for degradation. Studies are currently underway to identify ARABIDILLO-interacting proteins.

3 Arm-Repeat Proteins Regulate Plant Cell Architecture

Many animal Arm-repeat proteins perform structural functions in the cell. For example, β-catenin was originally discovered in actin-containing cell–cell (adherens) junctions (Ozawa et al. 1989). β-Catenin also interacts indirectly with microtubules (Ligon et al. 2001). Other Arm-repeat proteins regulate actin, microtubule and intermediate filament organisation (e.g. Hatzfeld 2005). *Arabidopsis* Arm family proteins also regulate the cytoskeleton both during cell elongation and cell division.

3.1 MRH2 Regulates Root Hair Morphogenesis

Root hairs, like trichomes, are epidermal cell extensions, but root hairs are unbranched. Once initiated, root hairs grow to about 100 μm long by polarised tip growth, which requires vesicle trafficking of new cell membrane and cell wall components to the root hair tip, and involves both actin and microtubules (Bibikova et al. 1999; Baluska et al. 2000).

To date, about 40 genes required for root hair development have been identified by forward genetics. A complementary approach compared global gene expression in the root hair differentiation zone of wild-type roots with a root hair-defective mutant, *rhd2*, and defined the "root hair morphogenesis transcriptome" (Jones et al. 2006). Using these data, six new root hair mutants were isolated, including *mrh2*. Unlike wild-type root hairs, which are straight and unbranched, *mrh2* mutant root hairs are wavy and branched, and constantly reorient their direction of growth (Jones et al. 2006). Thus, *MRH2* is not required for growth itself, but is required to maintain a steady direction of tip growth and a single growing tip.

The *mrh2* root hairs resemble those with microtubule dynamics disrupted by chemical or genetic means (Bibikova et al. 1999; Bao et al. 2001; Whittington et al. 2001). MRH2 protein contains Arm repeats and a kinesin domain (Fig. 1). Kinesins are ATP-dependent microtubule motors that can generate force. The Arm repeats of MRH2 may bind to a specific cargo whose transport along microtubules is required for directed root hair tip growth (Fig. 2). MRH2 is one of three closely related kinesin–Arm proteins in *Arabidopsis* (Reddy and Day 2001). MRH2 and its relatives could have tissue-specific roles in plant development and morphogenesis, although functional redundancy is also possible.

Interestingly, a screen for mutants in another tip-growing cell type, pollen tubes, identified the *seth4* mutant. SETH4 has six Arm repeats and may affect pollen tube growth or guidance (Lalanne et al. 2004). However, the mechanism of SETH4 function is not yet clear.

3.2
An *Arabidopsis* Arm-Repeat Protein Regulates Cytokinesis

Another Arm-repeat protein identified in pollen, TIO, affects the cell division process. Formation of the male gametes (pollen) requires two consecutive mitotic divisions of haploid meiocyte cells. The first (asymmetric) division forms a small generative cell and a large vegetative cell. A second, symmetrical division of the generative cell makes two sperm cells that migrate inside the vegetative cell, forming a tricellular pollen grain (McCormick 2004). A screen for pollen cell division mutants identified the *two-in-one* (*tio*) mutant, which has aberrant cytokinesis (daughter cell separation) at the end of the first cell division. The cell plate (new cell membrane and cell wall) deposited between the vegetative and generative cell is incomplete, and the second mitotic division does not occur (Oh et al. 2005). *TIO* functions throughout the plant, as seedlings with reduced *TIO* function have a severe growth defect: the meristems cannot complete cell division and large, multinucleate cells form, indicating incomplete cytokinesis.

The TIO protein is a kinase related to human and *Drosophila* FUSED, which transduces developmental signals. *Arabidopsis* TIO contains four Arm repeats (Fig. 1), as do its orthologues in rice and the protozoan *Leishmania*. Human FUSED contains HEAT repeats, while *Drosophila* FUSED has no Arm/HEAT domain (Oh et al. 2005). TIO protein localises specifically to the midline of the phragmoplast, an actin- and microtubule-containing ring that determines the deposition of new cell plates during cytokinesis, allowing the formation of two independent daughter cells. These data combined with the incomplete cell plate phenotype of the *tio* mutant suggest that TIO is involved in the expansion of the phragmoplast as cytokinesis progresses. Interestingly, *Drosophila* FUSED associates with the microtubule cytoskeleton via a kinesin (Stegman et al. 2000), and the same may be true of TIO in *Arabidopsis*, despite its divergent function in a plant-specific cell division process (Fig. 2).

4 Arm-Repeat Proteins in Plant Cell Death

Cell death plays a crucial role in a number of responses during the life of a plant and it appears that in most cases Arm-repeat proteins play a key role in its activation or regulation.

4.1 ARC1 Signalling During Self-Incompatibility Promotes Pollen Cell Death

In many plant species, when pollen grains (containing the male gametes) land on the pistil (female reproductive organ) of flowers from the same plant or genetically identical plants, the self-incompatibility (SI) response is triggered and growth of "self" pollen is prevented, resulting in death of germinating pollen tubes before fertilisation can occur (Newbigin and Vierstra 2003). This mechanism maintains genetic diversity in plant populations by preventing inbreeding.

The first plant Arm protein to be discovered, *Brassica* Arm-repeat containing 1 (ARC1), a U-box/Arm protein, interacts with S-locus receptor kinases (SRKs) (Gu et al. 1998). Activation of transmembrane SRKs by pollen proteins triggers the stigmatic SI response in *Brassica* (Newbigin and Vierstra 2003) and ARC1 is expressed specifically in the stigma. Antisense inhibition of ARC1 reduces SI, allowing "self" pollen to grow, suggesting that ARC1 positively regulates SI (Stone et al. 1999). Interestingly, ARC1 is phosphorylated by SRKs; this phosphorylation is required for ARC1 binding to SRKs in vitro (Gu et al. 1998). ARC1 functions as an E3 ubiquitin ligase in vitro, and probably also in vivo, as ARC1 protein co-localises with the proteasome in a U-box-dependent manner when SRK is active (Stone et al. 2003). Interestingly, "self" pollen is able to germinate and grow on pistils treated with proteasome inhibitors. Together these data suggest that ARC1-mediated SI requires active proteasomal degradation in the stigma (Stone et al. 2003). Thus, during SI, active SRK activates ARC1, possibly by phosphorylation. ARC1 then probably interacts both with E2 ubiquitin ligases (via its U-box) and with target proteins (via the Arm repeats) that become ubiquitinated and degraded by the proteasome (Fig. 2). This allows the rejection and death of non-compatible "self" pollen.

Interestingly, a novel *Arabidopsis* U-box/Arm protein, AtPUB8, modifies the SI response in self-incompatible *Arabidopsis* species (Liu et al. 2007). AtPUB8 is not the closest *Arabidopsis* relative of ARC1 (see Sect. 4.2), suggesting that more than one U-box/Arm protein may regulate cell death during self-incompatibility in the Brassicaceae. Liu et al. (2007) also propose additional, non-SI functions for AtPUB8.

4.2 U-Box/Arm Proteins Mediate the Hypersensitive Response

Plants also activate cell death during pathogen infection. Cell death is induced during disease-causing pathogen infection (a "basal" pathogen response), and also during the hypersensitive response (HR), an immune response where localised cell death occurs at the infection site to prevent spread of the pathogen through the plant. Thus, plants are resistant to pathogens that they can mount the HR against. Regulated protein degradation is important for pathogen responses. Studies in several plants have identified U-box/Arm genes rapidly induced during pathogen infection, which are essential for the HR and disease resistance. These genes are the tobacco and tomato *ACRE276* (for Avr9/Cf-9 rapidly elicited), which is homologous to *Arabidopsis AtPUB17*; and also tobacco/tomato *ACRE74* (*Arabidopsis* counterparts *AtPUB20* and *AtPUB21*, parsley counterpart *CMPG1*) (Kirsch et al. 2001; Heise et al. 2002; Gonzalez-Lamothe et al. 2006; Yang et al. 2006). Reduction of *ACRE276* or *ACRE74* function abolishes tobacco and tomato HR, generating plants no longer resistant to leaf mould fungus. Correspondingly, *Arabidopsis atpub17* mutants have compromised resistance to bacterial infection. Thus, like ARC1, these Arm-repeat proteins are positive regulators of cell death (Gonzalez-Lamothe et al. 2006; Yang et al. 2006).

Both ACRE276/AtPUB17 and ACRE74 function as ubiquitin ligases in vitro (Gonzalez-Lamothe et al. 2006; Yang et al. 2006). In addition, U-box mutant forms of AtPUB17 cannot promote HR showing that E3 ligase activity is required in vivo (Yang et al. 2006). Interestingly, overexpression of ACRE74 increases the HR, and ACRE74 may be a limiting factor for HR-induced cell death. However, overexpression of U-box mutant ACRE74 acts in a dominant negative fashion, reducing the HR (Gonzalez-Lamothe et al. 2006). ACRE74 is required for resistance to *Pseudomonas* and *Phytophthora* elicitor proteins in tobacco and is thus likely to be involved in a widespread range of immunity responses (Gonzalez-Lamothe et al. 2006).

Interestingly, AtPUB17 is the closest *Arabidopsis* homologue of ARC1 (Azevedo et al. 2001), suggesting that the same U-box/Arm protein has been co-opted to function in more than one plant cell death pathway, as may be the case for AtPUB8 (see Sect. 4.1).

4.3 SPL11 as a Negative Regulator of Cell Death During Pathogen Attack

SPOTTED LEAF 11 (SPL11) is a rice U-box/Arm protein identified by its loss-of-function phenotype (Zeng et al. 2004). The *spl11* mutants display lesions resembling pathogen-induced cell death in the absence of pathogen attack. The *spl11* mutants show enhanced resistance to fungal and bacterial pathogens, suggesting that, in contrast to the ACRE proteins, SPL11 is a negative regulator of cell death in the defence against pathogens (Zeng et al. 2004). Importantly, SPL11 can act as an E3 ubiquitin ligase in vitro (Zeng et al. 2004).

The data described show that targeted protein degradation by plant U-box/Arm proteins is key to regulating cell death programmes in different contexts. In particular, U-box/Arm proteins provide a balance between disease resistance and activation of cell death pathways in the plant, and act at multiple levels within plant disease signalling networks. Discovering the target proteins of ACRE/PUB/SPL11 ubiquitin ligases is critical to understanding how they regulate cell death.

Other Arm-repeat and related proteins also function in pathogen signalling: *Arabidopsis* SNI1, a negative regulator of pathogen responses, is related in structure to Arm proteins (Mosher et al. 2006), while MOS6 is an importin-α required for disease resistance (Palma et al. 2005). In addition, *Arabidopsis* AtPLA1, an acyl hydrolase with Arm repeats, promotes resistance to a necrotrophic fungus (Yang et al. 2007). Discovering the mechanisms of action of these diverse Arm proteins will enhance our understanding of how plants respond to pathogens and biotic stress.

5 Conclusions

To date, only a small percentage of plant Arm family proteins have been functionally characterised. Those studied so far are functionally diverse, having signalling, transcriptional and structural roles as in animals. They are required for many aspects of plant development, including cell growth, cell division, cell polarity and cell death (Fig. 2).

Almost half of *Arabidopsis* Arm proteins may exert their function through regulation of proteolysis, a process key to plant development and environmental responses. Although the Arm–E3 ligases exist as protein families, the phenotypes and binding specificities discovered so far suggest that only limited functional redundancy occurs. The emerging picture is that these proteins provide fine-tuning and cell type-specificity to developmental signalling pathways. The challenge for the coming years is to find regulatory proteins and downstream targets to understand the mechanisms by which this intriguing family of proteins performs its roles in plant growth regulation.

Acknowledgements J.C. would like to thank Candida Nibau, Steve Thomas and Barend de Graaf for critical reading of the manuscript.

References

Amador V, Monte E, Garcia-Martinez JL, Prat S (2001) Gibberellins signal nuclear import of PHOR1, a photoperiod-responsive protein with homology to *Drosophila armadillo*. Cell 106:343–354

Andrade M, Petosa C, O'donoghue S, Muller C, Bork P (2001) Comparison of ARM and HEAT protein repeats. J Mol Biol 309:1–18

Azevedo C, Santos-Rosa MJ, Shirasu K (2001) The U-box protein family in plants. Trends Plant Sci 6:354–358

Baluska F, Salaj J, Mathur J, Braun M, Jasper F, Samaj J, Chua N, Barlow P, Volkmann D (2000) Root hair formation: F-actin-dependent tip growth is initiated by local assembly of profilin-supported F-actin meshworks accumulated within expansin-enriched bulges. Dev Biol 227:618–632

Banzai T, Hershkovits G, Katcoff DJ, Hanagata N, Dubinsky Z, Karube I (2002) Identification and characterization of mRNA transcripts differentially expressed in response to high salinity by means of differential display in the mangrove, *Bruguiera gymnorrhiza*. Plant Sci 162:499–505

Bao Y, Kost B, Chua N (2001) Reduced expression of alpha-tubulin genes in *Arabidopsis thaliana* specifically affects root growth and morphology, root hair development and root gravitropism. Plant J 28:145–157

Bibikova T, Blancaflor E, Gilroy S (1999) Microtubules regulate tip growth and orientation in root hairs of *Arabidopsis thaliana*. Plant J 17:657–665

Casimiro I, Beeckman T, Graham N, Bhalerao R, Zhang H, Casero P, Sandberg G, Bennett M (2003) Dissecting Arabidopsis lateral root development. Trends Plant Sci 8:165–171

Coates J (2003) Armadillo repeat proteins: beyond the animal kingdom. Trends Cell Biol 13:463–471

Coates J, Laplaze L, Haseloff J (2006) Armadillo-related proteins promote lateral root development in Arabidopsis. Proc Natl Acad Sci USA 103:1621–1626

Conacci-Sorrell M, Zhurinsky J, Ben-Ze'ev A (2002) The cadherin-catenin adhesion system in signaling and cancer. J Clin Invest 109:987–991

Conti E, Uy M, Leighton L, Blobel G, Kuriyan J (1998) Crystallographic analysis of the recognition of a nuclear localization signal by the nuclear import factor karyopherin alpha. Cell 94:193–204

Downes B, Stupar RM, Gingerich D, Vierstra R (2003) The HECT ubiquitin-protein ligase (UPL) family in Arabidopsis: UPL3 has a specific role in trichome development. Plant J 35:729–742

El Refy A, Perazza D, Zekraoui L, Valay J, Bechtold N, Brown S, Hulskamp M, Herzog M, Bonneville J (2003) The Arabidopsis KAKTUS gene encodes a HECT protein and controls the number of endoreduplication cycles. Mol Genet Genomics 270:403–414

Gingerich D, Gagne J, Salter DW, Hellmann H, Estelle M, Ma L, Vierstra R (2005) Cullins 3a and 3b assemble with members of the broad complex/tramtrack/bric-a-brac (BTB) protein family to form essential ubiquitin-protein ligases (E3s) in Arabidopsis. J Biol Chem 280:18810–18821

Gonzalez-Lamothe R, Tsitsigiannis D, Ludwig AA, Panicot M, Shirasu K, Jones J (2006) The U-box protein CMPG1 is required for efficient activation of defense mechanisms triggered by multiple resistance genes in tobacco and tomato. Plant Cell 18:1067–1083

Gu T, Mazzurco M, Sulaman W, Matias DD, Goring DR (1998) Binding of an arm repeat protein to the kinase domain of the S-locus receptor kinase. Proc Natl Acad Sci USA 95:382–387

Hartweck L, Olszewski N (2006) Rice GIBBERELLIN INSENSITIVE DWARF1 is a gibberellin receptor that illuminates and raises questions about GA signaling. Plant Cell 18:278–282

Hatzfeld M (2005) The p120 family of cell adhesion molecules. Eur J Cell Biol 84:205–214

Heise A, Lippok B, Kirsch C, Hahlbrock K (2002) Two immediate-early pathogen-responsive members of the AtCMPG gene family in *Arabidopsis thaliana* and the W-box-containing elicitor-response element of AtCMPG1. Proc Natl Acad Sci USA 99:9049–9054

Huber A, Nelson WJ, Weis W (1997) Three-dimensional structure of the armadillo repeat region of beta-catenin. Cell 90:871–882

Hulskamp M, Schnittger A, Folkers U (1999) Pattern formation and cell differentiation: trichomes in Arabidopsis as a genetic model system. Int Rev Cytol 186:147–178

Jones M, Raymond M, Smirnoff N (2006) Analysis of the root-hair morphogenesis transcriptome reveals the molecular identity of six genes with roles in root-hair development in Arabidopsis. Plant J 45:83–100

Kim M, Cho H, Kim D, Lee J, Pai H (2003) CHRK1, a chitinase-related receptor-like kinase, interacts with NtPUB4, an armadillo repeat protein, in tobacco. Biochim Biophys Acta 1651:50–59

Kim S, Choi H, Ryu H, Park J, Kim M, Kim S (2004a) ARIA, an Arabidopsis arm repeat protein interacting with a transcriptional regulator of abscisic acid-responsive gene expression, is a novel abscisic acid signaling component. Plant Physiol 136:3639–3648

Kim S, Kang J, Cho D, Park J, Kim S (2004b) ABF2, an ABRE-binding bZIP factor, is an essential component of glucose signaling and its overexpression affects multiple stress tolerance. Plant J 40:75–87

Kirsch C, Logemann E, Lippok B, Schmelzer E, Hahlbrock K (2001) A highly specific pathogen-responsive promoter element from the immediate-early activated CMPG1 gene in *Petroselinum crispum*. Plant J 26:217–227

Lalanne E, Michaelidis C, Moore J, Gagliano W, Johnson A, Patel R, Howden R, Vielle-Calzada JP, Grossniklaus U, Twell D (2004) Analysis of transposon insertion mutants highlights the diversity of mechanisms underlying male progamic development in Arabidopsis. Genetics 167:1975–1986

Lee J, Takei K, Sakakibara H, Sun Cho H, Kim Do M, Kim Y, Min S, Kim W, Sohn D, Lim Y, Pai H (2003) CHRK1, a chitinase-related receptor-like kinase, plays a role in plant development and cytokinin homeostasis in tobacco. Plant Mol Biol 53:877–890

Ligon L, Karki S, Tokito M, Holzbaur E (2001) Dynein binds to beta-catenin and may tether microtubules at adherens junctions. Nat Cell Biol 3:913–917

Liu P, Sherman-Broyles S, Nasrallah M, Nasrallah J (2007) A cryptic modifier causing transient self-incompatibility in *Arabidopsis thaliana*. Curr Biol 17:734–740

Mccormick S (2004) Control of male gametophyte development. Plant Cell 16(Suppl):S142–S153

Mosher R, Durrant W, Wang D, Song J, Dong X (2006) A comprehensive structure-function analysis of Arabidopsis SNI1 defines essential regions and transcriptional repressor activity. Plant Cell 18:1750–1765

Mudgil Y, Shiu S, Stone S, Salt J, Goring D (2004) A large complement of the predicted Arabidopsis ARM repeat proteins are members of the U-box E3 ubiquitin ligase family. Plant Physiol 134:59–66

Newbigin E, Vierstra R (2003) Plant reproduction: Sex and self-denial. Nature 423:229–230

Oh S, Johnson A, Smertenko A, Rahman D, Park S, Hussey P, Twell D (2005) A divergent cellular role for the FUSED kinase family in the plant-specific cytokinetic phragmoplast. Curr Biol 15:2107–2111

Ozawa M, Baribault H, Kemler R (1989) The cytoplasmic domain of the cell adhesion molecule uvomorulin associates with three independent proteins structurally related in different species. EMBO J 8:1711–1717

Palma K, Zhang Y, Li X (2005) An importin alpha homolog, MOS6, plays an important role in plant innate immunity. Curr Biol 15:1129–1135

Perazza D, Herzog M, Hulskamp M, Brown S, Dorne A, Bonneville J (1999) Trichome cell growth in *Arabidopsis thaliana* can be derepressed by mutations in at least five genes. Genetics 152:461–476

Petroski M, Deshaies R (2005) Function and regulation of cullin-RING ubiquitin ligases. Nat Rev Mol Cell Biol 6:9–20

Reddy A, Day I (2001) Kinesins in the Arabidopsis genome: a comparative analysis among eukaryotes. BMC Genomics 2:2

Riggleman B, Wieschaus E, Schedl P (1989) Molecular analysis of the armadillo locus: uniformly distributed transcripts and a protein with novel internal repeats are associated with a Drosophila segment polarity gene. Genes Dev 3:96–113

Samuel M, Salt J, Shiu S, Goring D (2006) Multifunctional arm repeat domains in plants. Int Rev Cytol 253:1–26

Sanger Institute (2007) Pfam database. Wellcome Trust, UK.
Available at http://www.sanger.ac.uk/Software/Pfam/, last visited: 20 July 2007

Stegman M, Vallance J, Elangovan G, Sosinski J, Cheng Y, Robbins D (2000) Identification of a tetrameric hedgehog signaling complex. J Biol Chem 275:21809–21812

Stone S, Anderson E, Mullen RT, Goring D (2003) ARC1 is an E3 ubiquitin ligase and promotes the ubiquitination of proteins during the rejection of self-incompatible Brassica pollen. Plant Cell 15:885–898

Stone S, Arnoldo M, Goring DR (1999) A breakdown of Brassica self-incompatibility in ARC1 antisense transgenic plants. Science 286:1729–1731

Vierstra R (2003) The ubiquitin/26S proteasome pathway, the complex last chapter in the life of many plant proteins. Trends Plant Sci 8:135–142

Whittington A, Vugrek O, Wei K, Hasenbein N, Sugimoto K, Rashbrooke M, Wasteneys G (2001) MOR1 is essential for organizing cortical microtubules in plants. Nature 411:610–613

Yang C, Gonzalez-Lamothe R, Ewan R, Rowland O, Yoshioka H, Shenton M, Ye H, O'donnell E, Jones J, Sadanandom A (2006) The E3 ubiquitin ligase activity of arabidopsis PLANT U-BOX17 and its functional tobacco homolog ACRE276 are required for cell death and defense. Plant Cell 18:1084–1098

Yang W, Devaiah S, Pan X, Isaac G, Welti R, Wang X (2007) AtPLAI is an acyl hydrolase involved in basal Jasmonic acid production and Arabidopsis resistance to botrytis cinerea. J Biol Chem 282:18116–18128

Zeng L, Qu S, Bordeos A, Yang C, Baraoidan M, Yan H, Xie Q, Nahm B, Leung H, Wang G (2004) Spotted leaf11, a negative regulator of plant cell death and defense, encodes a U-box/armadillo repeat protein endowed with E3 ubiquitin ligase activity. Plant Cell 16:2795–2808

Plant Cell Monogr (10)
L. Bögre and G. Beemster: Plant Growth Signaling
DOI 10.1007/7089_2007_154/Published online: 29 January 2008

Mass Spectrometry Based Proteomics as a Tool for the Analysis of Protein–Protein Interactions in Signaling Processes

Waltraud X. Schulze

Max Planck-Institut für Molekulare Pflanzenphysiologie, Am Mühlenberg 1, 14476 Golm, Germany
wschulze@mpimp-golm.mpg.de

Abstract This chapter focuses on the emerging technology of protein mass spectrometry and how it can be applied to the analysis of signaling pathways. Since proteins undergo posttranslational modifications, the analysis of the regulatory function of protein modification, especially phosphorylation, will be crucial for our view of plant signal transduction pathways. In addition, proteins are part of protein complexes and binary protein–protein interactions which need to be characterized in order to understand signal specificity and cross-talk. This chapter provides outlines of experiments involving (1) screening of interactions specific to protein modifications, (2) isolation of protein complexes, and (3) the analysis of kinases and their substrates. Finally, as signaling events are dynamic processes, the time component needs to be considered. Strategies implying quantitative proteomic methods are presented.

1 Introduction

Proteomics is the science of large-scale analysis of proteins. As proteins are the basis of structural, enzymatic, and many regulatory components of a cell, the direct study of proteins involved in signaling pathways is of great interest. Mass spectrometry has increasingly become the method of choice for the analysis of complex protein samples (Aebersold and Mann 2003; de Hoog and Mann 2004), especially in those cases in which the protein function is not yet fully understood. The success of protein mass spectrometry has been made possible by the development of soft protein ionization methods, such as electrospray ionization (ESI) or matrix-assisted laser desorption/ionization (MALDI), an achievement that has been acknowledged with the Nobel prize in chemistry 2002 to John Fenn and Koichi Tanaka. However, without the information derived from various full genome sequencing projects, and without efficient algorithms for peptide sequence determination from fragmentation spectra (Eng et al. 1994; Pevzner et al. 2001), proteomic experiments would be a great deal more difficult today.

The recent success of mass spectrometry based approaches in the elucidation of protein–protein interactions indicates that the technology has evolved to a crucial tool in signaling biology (Blagoev et al. 2003, 2004; Schulze et al. 2005). The field of mass spectrometry based proteomics is still under fast development; new and better instrumentation is being developed on almost a yearly basis. Thus, each technical breakthrough either allows new kinds of applications or improves the quality or throughput of traditional measurements. However, no method or instrument is capable of identifying and quantifying all the components in a complex protein extract in a single step. Therefore, careful experimental design involving the steps of protein separation, enrichment, and purification is essential for successful interpretation of proteomic datasets.

Among the variety of different technological aspects of mass spectrometry based proteomics, two major workflows have emerged, one involving separation of proteins on two-dimensional gels and subsequent identification of protein spots by mass spectrometry, and the other involving in-solution digestion of a complex protein mixture, subsequent separation of the peptides by one-dimensional or multidimensional liquid chromatography, and online analysis by mass spectrometry. Depending on the nature of the biological question, mixtures of the two major strategies are employed.

Protein mixtures today can routinely be characterized in terms of proteins present in a given sample. However, in order to allow biological interpretation, quantitative analyses are necessary. Peak integration, spectrum counts, or derived indices (Ishihama et al. 2005; Washburn et al. 2001) have been established as a basis for quantitative comparison of protein extracts derived from organisms under different conditions, or in defining a sub-proteome of interest (Andersen et al. 2003). Alternatively, stable isotope labeling strategies, such as metabolic labeling, $H_2{}^{18}O$ digest, iTRAQ or ICAT, are used. These techniques allow direct relative comparison of two or more samples in the same LC/MS/MS analysis (Dunkley et al. 2004; Engelsberger et al. 2006; Jones et al. 2006; Nelson et al. 2006). This chapter will give an overview of how quantitative proteomic strategies can be applied to the analysis of signaling pathways, with special respect to protein–protein interactions and protein phosphorylation.

2 Domains, Motifs, and Modification

2.1 Proteins Function with Partners

Most proteins do not function alone, but rather display their activity in protein complexes. Thus, each protein at some point or another undergoes protein–protein interactions with one or more other proteins. For example, membrane receptors undergo interactions with downstream signaling proteins. This has been especially well studied in the plant sensor histidine kinase receptors, for which the specificity and cross-talk between signaling pathways is defined through interaction specificity and generality. For example, the sensor histidine kinase CRE1 can interact with the phosphotransfer proteins AHP1 and AHP2, while the histidine kinase CKI1 specifically interacts only with AHP2, and the ethylene receptor ETR1 again can interact with AHP1 and AHP2 (Grefen and Harter 2004). Similarly, other receptor kinases, such as the LRR-receptor kinase BRI1, are starting to move into our focus as a model system for protein–protein interactions and phosphorylation in a plant signaling pathway (Ehsan et al. 2005; Wang et al. 2005).

2.2 The Role of Protein Modifications

While glycosylation is the most abundant protein modification, phosphorylation is the most important posttranslational modification with regulatory function. Over 50% of all proteins are thought to be able to undergo phosphorylation, but at a given time point, only 2% of all proteins are estimated to be phosphorylated (Reinders and Sickmann 2005). Knowledge about protein phosphorylation sites and the conditions under which these sites are being used is crucial for our understanding of signaling networks and their dynamics. In mammalian tissue, acid hydrolysis and autoradiography has revealed that about 2 to 4% of the acid-stable phosphate is present in the form of phosphotyrosine; the remaining majority is in serine and threonine (Galski et al. 1983). Several tools and techniques for robust identification of phosphorylation sites by mass spectrometry have been published recently, which also are readily applicable for proteins extracted from plant tissues (Beausoleil et al. 2004; Nühse et al. 2003; Olsen and Mann 2004; Peck 2006).

Sulfatation is a rather common modification of tyrosine residues that has not been well studied to date. In mammalian tissue it has been estimated that about 1% of all tyrosine residues are sulfated (Bäuerle and Hüttner 1985). The discovery of tyrosylprotein sulfotransferase activities in plant microsomal membrane fractions indicates that protein sulfatation also plays a role in a variety of plants (Hanai et al. 2000). However, the role of this modification for signaling processes needs to be further elucidated. Methylation and acetylation are known to have a regulatory function in the cell nucleus during growth, development, and under certain stress conditions (Butterbrodt et al. 2006; Tai et al. 2005; Tsuji et al. 2006; Xu et al. 2005). Other modifications, such as carboxylation and hydroxylation, are only now starting to move into our view. Their role in signaling processes is far less well understood, and is likely to yield plenty of novel and exciting discoveries in the near future.

2.3 Protein Domains

Regulation of cellular events requires temporal and spatial assembly and disassembly of protein–protein interactions. Control of these complex processes is often achieved through modification-dependent binding and activation. In particular, phosphorylation and sulfatation are considered to be involved. Proteins are not just linear amino acid sequences but rather consist of characteristic globular domains that also exert specific functions in protein–protein interaction. These modular domains are regions of proteins that are stable and foldable on their own, and they are characterized by a specific structure, amino acid sequence, or both. There are now up to 733 distinct domains described in the SMART database (smart.embl-heidelberg.de) for a variety of organisms ranging from prokaryotes to eukaryotes. Similarly, the Pfam database (www.sanger.ac.uk/Software/Pfam/) lists at least one characteristic domain associated with 75% of all proteins in *Arabidopsis*.

Before the so-called SH2 domain was discovered as the first phosphotyrosine binding domain, phosphorylation was thought to directly regulate protein activity through allosteric changes in protein structure and activity. Since then, the perception has changed to a view where important protein–protein interactions in cell signaling are frequently mediated by short, unstructured sequences, which specifically interact with peptide motif binding domains (Pawson and Scott 1997). Thus, phosphorylation also functions as a direct regulatory switch for protein–protein interactions.

Typical well-known examples from the mammalian field are the binding of tyrosyl-phosphorylated peptides to proteins containing Src homology domain 2 (SH2) or phosphotyrosyl binding (PTB) domain (Pawson and Gish 1992). Since the discovery of interactions of 14-3-3 proteins with characteristic peptide motifs around phosphoserine and phosphothreonine residues, the concept of phosphorylation-dependent protein–protein interactions is no longer only considered valid for phosphotyrosine, but also for phosphorylation in general. Regulatory binding of 14-3-3 proteins to characteristic phosphorylation sites in plant proteins has been well described for nitrate reductase (Atwal et al. 1998), sucrose phosphate synthase (Huber and Huber 1996), and the plasma membrane ATPases (Fuglsang et al. 1999; Toroser et al. 1998). Additional phosphoserine and phosphothreonine binding domains have since been discovered: WW domains, FHA domains, WD40 repeats, the Polo box domain, and BRCT repeats. These domains have also been characterized in plant proteins.

3 Strategies for Interaction Screening

Detection of protein–protein interactions is usually a balance between specificity (background reduction) and affinity (detection of weak interactions). Recently, the introduction of stable isotope labeling in mass spectrometric approaches allows one to distinguish specific from unspecific interaction partners in an immunoprecipitate or pull-down experiment. This principle has enabled detection of weak binders in the presence of background proteins, and it has led to a decrease in false positive detections (Blagoev et al. 2003; Gruhler et al. 2005).

However, there are still some problems associated with large-scale experimental interaction mapping: in most large-scale experimental datasets, false negative rates range up to 90% and false positive rates are still predicted to be higher than 50%, even in highly filtered datasets (von Mering et al. 2002). Bioinformatic methods will increasingly be necessary to contribute to the validation and prediction of protein interactions.

3.1 Interaction Partners to Short Modified Motifs

One of the main interests in signaling biology is to gain insight into how cells process information. In many signaling pathways, signaling molecules bind to receptor proteins and this information is passed on to ultimately induce transcriptional changes, often involving a series of protein phosphorylation events. A significant subset of these signal-dependent phosphorylation events creates new protein–protein interaction interfaces, which serve to recruit adaptor or effector molecules. Mechanistically, signal transduction pathways in all organisms are composed of modification-dependent protein–protein interactions between domains and extended peptide motifs. Consensus peptide sequences recognized by different protein domains have been studied in the past using oriented peptide libraries, peptide arrays, or phage display (Elia et al. 2003; Espejo et al. 2002; Kay et al. 2001). However, most of these methods lack specificity for modification-dependent interactions and few of them are unbiased or can be performed without already having candidate target proteins.

The use of synthetic modified and unmodified peptides as a bait in pull-down experiments to find peptide motif-based interaction partners appears to be a powerful method to dissect the role of certain peptide motifs in signaling events (Fig. 1). In this approach, synthetic peptide pairs comprising a motif of interest in modified (e.g., phosphorylated) and unmodified (e.g., unphosphorylated) form are used as bait in pull-down experiments. Eluted proteins are identified by mass spectrometry, and specific interaction partners are defined as being identified exclusively in pull-downs with the modified bait peptide and not the unmodified control bait peptide. This al-

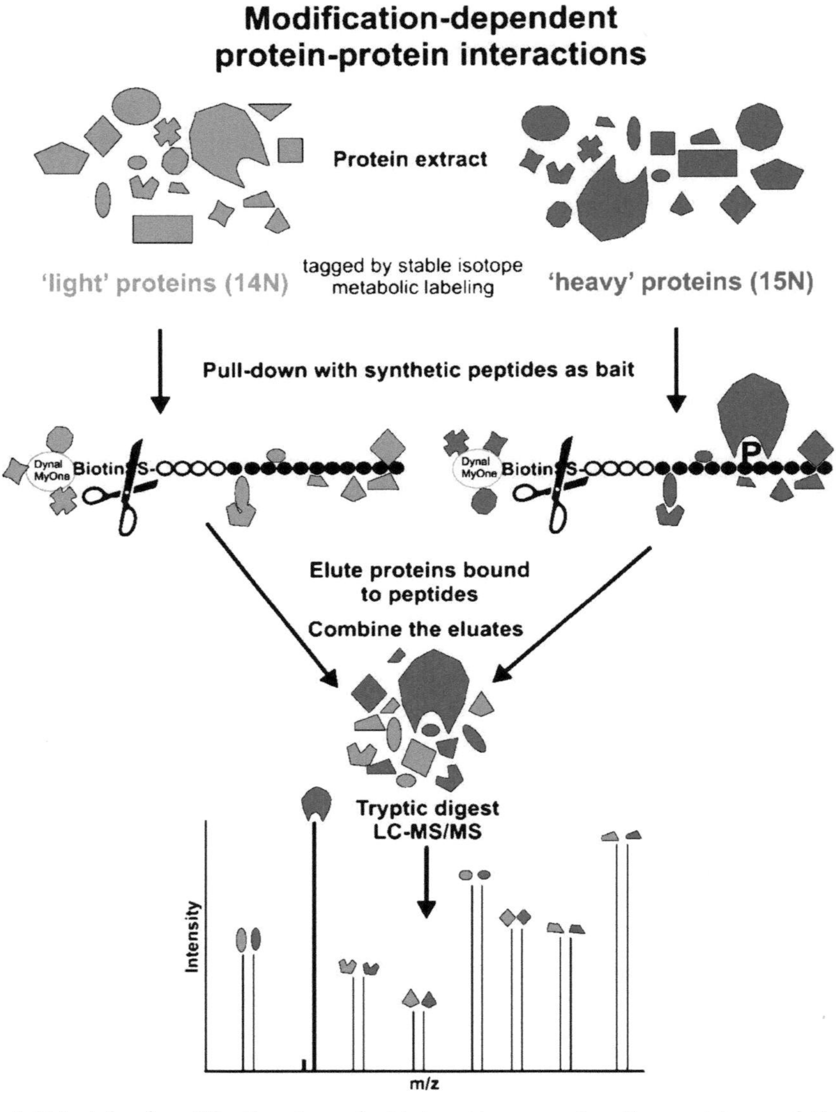

Fig. 1 Principle of modification-dependent interaction screening. Two protein populations are obtained by metabolic labeling (Engelsberger et al. 2006) and subjected to pull-down experiment using modified (indicated by P) and unmodified synthetic peptides as a bait. The bait peptides are chosen around a specific site of phosphorylation in a protein of interest. After incubation of protein extracts with the bait peptide of interest, eluted proteins are combined and digested with trypsin. Tryptic peptides are then detected by mass spectrometry. Those tryptic peptides from proteins specifically binding the phosphorylated bait will have a larger peak intensity of the ^{15}N-labeled form. Nonspecific binders will have a 1 : 1 ratio of both isotopic forms

lows identification of specific binding partners to the modified bait peptides, even in the presence of a large excess of background binders. Background proteins are characterized by also being present in the pull-down with the control peptide (Schulze and Mann 2004).

This peptide–protein interaction screen is specific and reproducible, as shown by a medium-scale analysis of the complete phosphotyrosine interactome of the ErbB-receptor tyrosine kinase family (Schulze et al. 2005), and in a proof-of-principle study using protein extracts from plant cell cultures, a 14-3-3 protein was identified as interaction partner to a bait peptide comprising the 14-3-3 binding site of nitrate reductase (Gruhler et al. 2005).

Compared with two-hybrid and array-based approaches, this peptide pull-down strategy has the advantages that fully processed and modified sequences can serve as baits, and that the interaction takes place in the native environment of the cell extract. However, it has to be kept in mind that mass spectrometry based affinity methods will only be able to detect a subset of the protein interactions that actually occurs in vivo.

3.2 Interaction Screening by Immunoprecipitation

Immunoprecipitation of tagged or untagged proteins and subsequent analysis of interacting proteins by mass spectrometry has led to the construction of organism-wide protein–protein interaction maps, for example in yeast (Gavin et al. 2002, 2006; Ho et al. 2002), and to the identification of novel players even in the well-studied EGF signaling pathway in mammals (Blagoev et al. 2003; Schulze et al. 2005).

The so-called tandem affinity purification tag (TAP-tag) has been developed to obtain highly purified protein complexes (Fig. 2). It consists of a protein A domain that can bind to IgG-agarose followed by a tobacco etch virus (TEV) protease recognition site, which allows the fusion protein to be cleaved from the IgG-agarose. The remaining protein complex is then purified again in a calcium-dependent manner over calmodulin-agarose, making use of the calmodulin binding protein domain (CBP domain). Proteins interacting with the bait protein can be identified by mass spectrometry after elution (Rigaut et al. 1999). For the efficient use of the TAP-tag principle in plants, a synthetic gene has been constructed with optimal codon usage and deletion of a cryptic nuclear localization signal (Rohila et al. 2004). This system was successfully applied to screen for interaction partners of 41 TAP-tagged protein kinases in rice (Rohila et al. 2006). However, full interpretation of this dataset suffers from poor annotation of the rice proteome.

Immunoprecipitation of epitope-tagged BRI1 and BAK1 (tagged either with FLAG tag or GFP) was used in a thorough analysis of receptor phosphorylation under different ligand concentrations (Wang et al. 2005). In these experiments the analysis of phosphorylation sites of the tagged bait proteins

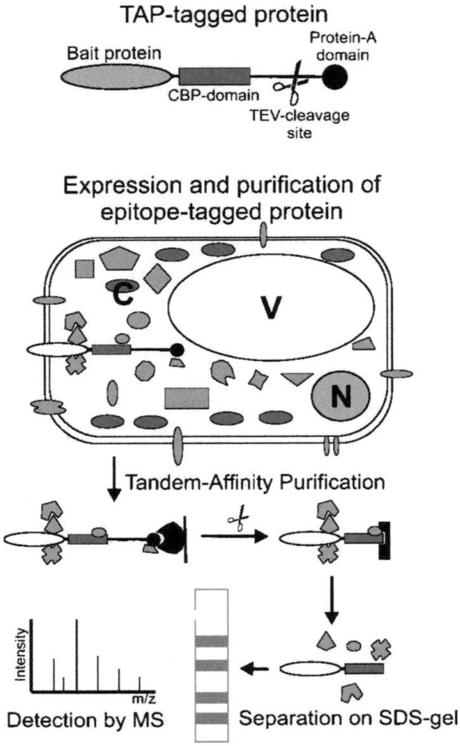

Fig. 2 Principle of TAP-tag purification. Overview of the experimental workflow using TAP-tagged proteins. The tagged protein is expressed in plants or cells and subsequently purified using IgG-agarose, TEV-protease cleavage, and calmodulin-agarose. Eluted proteins are separated on a one-dimensional SDS gel and individual bands are analyzed by mass spectrometry. C: chloroplast; N: nucleus; V: vacuole

themselves was the primary aim and not the identification of interaction partners; thus, tandem purification was not applied. This work on brassinosteroid signaling is one of the first clear examples in plants in which a receptor kinase was shown to be (auto)phosphorylated in a ligand-dependent manner.

3.3 Characterization of Macromolecular Complexes

Oligomerization of proteins in the plasma membrane has been postulated as a regulatory principle following the well-known example of oligomerization of mammalian receptor tyrosine kinases depending on tissue or ligand (Olayioye et al. 2000). In plants, oligomerization of transporters has been suggested to be a key process during regulation of sucrose transport (Barker et al. 2000; Reinders et al. 2002), and oligomerization of receptor kinases is discussed today as a general principle in plant signaling ranging from hormone signaling pathways to pathogen signaling. The oligomeric interaction between the receptor kinase BRI1 and the receptor kinase BAK1, as well as the endocytotic recycling of this signaling complex, is one of the first examples of a regulatory role of membrane protein oligomerization in plants (Russinova et al. 2004).

For the analysis of the oligomeric nature of protein complexes, blue native gel electrophoresis is very powerful (Eubel et al. 2005). Especially in the plant field, it has successfully been used to characterize the composition of the plant mitochondrial respiratory chain (Eubel et al. 2004) and, in combination with protein mass spectrometry, the composition of the plant photosystem was elucidated (Heinemeyer et al. 2004). In a systematic large-scale approach the oligomeric state of all stromal chloroplast proteins was characterized using blue native gel and subsequent protein identification by mass spectrometry (Peltier et al. 2006).

In future, and in combination with powerful stable isotope labeling techniques (Engelsberger et al. 2006), blue native gels and LC/MS/MS analysis will have high potential for the analysis of changes in oligomerization state upon external or internal stimulation. The analysis of signaling cascades may then become even more complex.

4 Search for Kinases

Kinases and their substrates are central components of signaling pathways. Most of the time, it is either known that a specific kinase is involved in a certain signaling pathway, or it is known that a certain protein of interest is being phosphorylated under a given condition. Thus, either the substrates of the kinase or the kinase itself remains to be identified. Candidates for kinase substrates have in the past been successfully identified by genetic screens and epistasis analyses or laborious biochemical purifications and in vitro assays (Manning and Cantley 2002).

Recently, an elegant method has been developed, which makes use of a series of synthetic peptides comprising the phosphorylation site of a protein of interest and the detection of changes in phosphorylation status of these standard peptides by mass spectrometry. In this approach, the standard peptides are incubated with fractions of cell extract and the phosphorylation status of the standard peptides is analyzed by multiple reaction monitoring in a triple-quadrupole mass spectrometer (Glinski et al. 2003; Glinski and Weckwerth 2005). This strategy was applied in a study to identify kinases involved in phosphorylation of trehalose-6-phosphate synthase (Glinski and Weckwerth 2005), and in addition it allowed testing for various parameters required for the kinase activity, such as calcium dependence. By carrying out the kinase

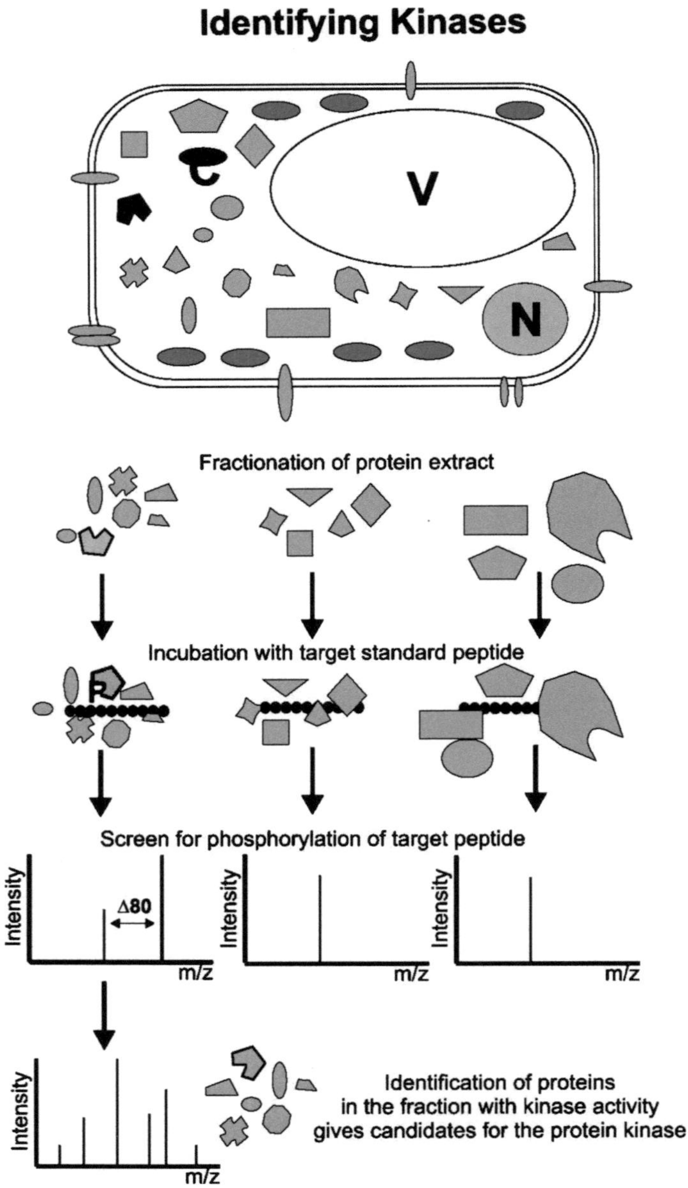

Fig. 3 Principle of kinase assay. Protein extract from plants is fractionated, e.g., by size exclusion chromatography. Each fraction is incubated with one or multiple target peptides of interest. The target peptides are chosen around potential phosphorylation sites and they are used in the assay in their unphosphorylated form. After incubation of protein extract fractions with the target peptide, phosphorylation of the target peptide is analyzed by mass spectrometry. The full analysis of proteins present in the fraction showing kinase activity will identify potential candidates for that kinase activity

assays on fractionated cell extract, the mass spectrometric analysis of the respective fractions exhibiting kinase activity can be used to narrow down the protein candidates for this kinase activity (Fig. 3).

The reverse approach, in which the kinase is known and the substrate is unknown, can be tackled by the use of peptide arrays. Synthetic peptides comprising a set of known and predicted phosphorylation sites from a variety of proteins are spotted on nylon membranes or glass slides. Upon incubation with the purified kinase of interest and the incorporation of radioactive or fluorescent phosphate analogs, the substrate specificity of the kinase can be determined. This approach is especially suitable for identifying the motif specificity of several kinases (Houseman et al. 2002). In plants, this method will gain more importance in the future, as the library of experimental phosphorylation sites is growing continuously, and may provide a basis for targeted array analysis.

An elegant method of finding specific kinase–substrate pairs in vivo involves genetic manipulation of the kinase pocket and the use of the respective synthetic ATP analogs (Shah et al. 1997). In that way, the genetically manipulated kinase can only use the synthetic ATP analog in its reactions, and the substrate proteins modified with the ATP analog can be affinity purified and identified by mass spectrometry. This strategy has been applied to a large-scale screen for specific substrates to the Pho85-Pcl1 cyclin-dependent kinase in yeast (Dephoure et al. 2005). Given the vast collections of knockout mutants, such a screen would certainly also be feasible in *Arabidopsis*.

5 Dynamics in Signaling Events

Signaling processes are dynamic. This inherent property has not yet been fully taken into account when linear signal transduction pathways are established and interaction networks are being interpreted. Taking dynamic aspects of protein–protein interactions and modifications into account is a big challenge to the biologist, as experiments need to be planned carefully and good controls need to be considered.

The first proteomic approach to address the temporal aspects of protein phosphorylation cascades came from the field of EGF signaling. Using metabolically labeled cell cultures, phosphorylated proteins were immunoprecipitated and quantified by mass spectrometry at different times after EGF stimulation (Blagoev et al. 2004; Olsen et al. 2006). The results indicate a clear hierarchy of phosphorylation events, starting with autophosphorylation of the EGF receptor and progressing to the MAP kinase signaling pathways within 10 min. In addition, at least in mammalian cells, early phosphorylation events are tyrosine phosphorylations, while later ones are in serine and threonine phosphorylation.

In similar experiments using liquid cultures of *Arabidopsis* seedlings, the phosphorylation status of membrane proteins was analyzed at different time points of sucrose resupply after a 2-day starvation period (Niittyla et al. 2007). The results reveal novel protein phosphorylation sites involved in sucrose response as well as changes in phosphorylation level of well-characterized proteins, such as the plasma membrane ATPase or phosphoenol-pyruvate carboxylase. These initial experiments will provide a broad basis for the identification of candidate proteins that may be involved in a particular signaling pathway.

6 Conclusion

Proteomics provides an ideal methodological supplement to existing genetic and physiological strategies for the analysis of signaling pathways in plants. The power of the proteomic tools lies in the potential to identify as yet unknown proteins to which no antibodies are available. In addition, the proteomic approaches have the potential for relative quantitation between two or more treatments. However, the success of proteomic experiments will to a great extent be dependent on good experimental design in combination with the analysis of specific plant mutants.

References

Aebersold R, Mann M (2003) Mass spectrometry-based proteomics. Nature 422:198–207

Andersen JS, Wilkinson CJ, Myoru T, Mortensen P, Nigg EA, Mann M (2003) Proteomic characterization of the human centrosome by protein correlation profiling. Nature 426:570–574

Atwal GS, Huber JL, Huber SC (1998) Phosphorylated nitrate reductase and 14-3-3 proteins. Site of interaction, effects of ions, and evidence for an amp-binding site on 14-3-3 proteins. Plant Physiol 118:1041–1048

Barker L et al (2000) SUT2, a putative sucrose sensor in sieve elements. Plant Cell 12:1153–1164

Bäuerle PA, Hüttner WB (1985) Tyrosine sulfation of yolk proteins 1, 2, and 3 in *Drosophila melanogaster*. J Biol Chem 260:6434–6439

Beausoleil SA et al (2004) Large-scale characterization of HeLa cell nuclear phosphoproteins. Proc Natl Acad Sci USA 101:12130–12135

Blagoev B, Kratchmarova I, Ong S-E, Nielsen M, Foster L, Mann M (2003) A proteomics strategy to elucidate functional protein–protein interactions applied to EGF signaling. Nat Biotechnol 21:315–318

Blagoev B, Ong S-E, Kratchmarova I, Mann M (2004) Temporal ordering of signaling networks by quantitative proteomics. Nat Biotechnol 22:1139–1145

Butterbrodt T, Thurow C, Gatz C (2006) Chromatin immunoprecipitation analysis of the tobacco PR-1a- and the truncated CaMV 35S promoter reveals differences in salicylic acid-dependent TGA factor binding and histone acetylation. Plant Mol Biol 61:665–674 de Hoog C, Mann M (2004) Proteomics. Annu Rev Genomics Hum Genet 5:267–293

Dephoure N, Howson RW, Blethrow JD, Shokat KM, O'Shea EK (2005) Combining chemical genetics and proteomics to identify protein kinase substrates. Proc Natl Acad Sci USA 102:17940–17945

Dunkley TPJ, Griffin JL, Dupree P, Lilley KS (2004) Localization of organelle proteins by isotope tagging (LOPIT). Mol Cell Proteomics 3:1128–1134

Ehsan H, Ray WK, Phinney BS, Wang X, Huber SC, Clouse SD (2005) Interaction of *Arabidopsis* BRASSINOSTEROID-INSENSITIVE 1 receptor kinase with a homolog of mammalian TGF-β receptor interacting protein. Plant J 43:251–261

Elia AEH, Cantley LC, Yaffe MB (2003) Proteomic screen finds pSer/pThr-binding domain localizing Plk1 to mitotic substrates. Science 299:1228–1231

Eng J, McCormack AL, Yates JRI (1994) An approach to correlate tandem mass spectral data of peptides with amino acid sequences in a protein database. J Am Soc Mass Spectrom 5:976–989

Engelsberger WR, Erban A, Kopka J, Schulze WX (2006) Metabolic labeling of plant cell cultures with $K^{15}NO_3$ as a tool for quantitative analysis of proteins and metabolites. Plant Methods 2:1–11

Espejo A, Cote J, Bednarek A, Richard S, Bedford MT (2002) A protein-domain microarray identifies novel protein–protein interactions. Biochem J 367:697–702

Eubel H, Heinemeyer J, Sunderhaus S, Braun HP (2004) Respiratory chain supercomplexes in plant mitochondria. Plant Physiol Biochem 42:937–942

Eubel H, Braun HP, Millar AH (2005) Blue-native PAGE in plants: a tool in analysis of protein–protein interactions. Plant Methods 1:1–13

Fuglsang AT et al (1999) Binding of 14-3-3 protein to the plasma membrane H(+)-ATPase AHA2 involves the three C-terminal residues Tyr(946)-Thr-Val and requires phosphorylation of Thr(947). J Biol Chem 274:36774–36780

Galski H, De Groot N, Ilan J, Hochberg AA (1983) Phosphorylation of tyrosine in cultured human placenta. Biochim Biophys Acta 761:284–290

Gavin AC et al (2002) Functional organization of the yeast proteome by systematic analysis of protein complexes. Nature 415:141–147

Gavin AC et al (2006) Proteome survey reveals modularity of the yeast cell machinery. Nature 440:631–636

Glinski M, Weckwerth W (2005) Differential multisite phosphorylation of the trehalose-6-phosphate synthase gene family in *Arabidopsis thaliana*: a mass spectrometry-based process for multiparallel peptide library phosphorylation analysis. Mol Cell Proteomics 4:1614–1625

Glinski M, Romeis T, Witte CP, Wienkoop S, Weckwerth W (2003) Stable isotope labeling of phosphopeptides for multiparallel kinase target analysis and identification of phosphorylation sites. Rapid Commun Mass Spectrom 17:1579–1584

Grefen C, Harter K (2004) Plant two-component systems: principles, functions, complexity and cross talk. Planta 219:733–742

Gruhler A, Schulze WX, Matthiesen R, Mann M, Jensen ON (2005) Stable isotope labeling of *Arabidopsis thaliana* cells and quantitative proteomics by mass spectrometry. Mol Cell Proteomics 4:1697–1702

Hanai H, Nakayama D, Yang H, Matsubayashi Y, Hirota Y, Sakagami Y (2000) Existence of a plant tyrosylprotein sulfotransferase: novel plant enzyme catalyzing tyrosine O-sulfation of preprophytosulfokine variants in vitro. FEBS Lett 470:97–101

Heinemeyer J, Eubel H, Wehmhoner D, Jansch L, Braun HP (2004) Proteomic approach to characterize the supramolecular organization of photosystems in higher plants. Phytochemistry 65:1683–1692

Ho Y et al (2002) Systematic identification of protein complexes in *Saccharomyces cerevisiae* by mass spectrometry. Nature 415:180–183

Houseman BT, Huh JH, Kron SJ, Mrksich M (2002) Peptide chips for the quantitative evaluation of protein kinase activity. Nat Biotechnol 20:270–274

Huber SC, Huber JL (1996) Role and regulation of sucrose-phosphate synthase in higher plants. Annu Rev Plant Physiol Plant Mol Biol 47:431–444

Ishihama Y et al (2005) Exponentially modified protein abundance index (emPAI) for estimation of absolute protein amount in proteomics by the number of sequenced peptides per protein. Mol Cell Proteomics 4:1265–1272

Jones AM, Bennett MH, Mansfield JW, Grant M (2006) Analysis of the defence phosphoproteome of *Arabidopsis thaliana* using differential mass tagging. Proteomics 6:4155–4165

Kay BK, Kasanov J, Yamabhai M (2001) Screening phage-displayed combinatorial peptide libraries. Methods 24:240–246

Manning F, Cantley LC (2002) Hitting the target: emerging technologies in the search for kinase substrates. Sci STKE 162:1–4

Nelson CJ, Hegeman AD, Harms AC, Sussman MR (2006) A quantitative analysis of *Arabidopsis* plasma membrane using trypsin-catalyzed ^{18}O labeling. Mol Cell Proteomics 5:1382–1395

Niittyla T, Frommer WB, Schulze WX (2007) Dynamics of phosphorylation in starved *Arabidopsis* seedlings upon sucrose resupply. Mol Cell Proteomics 6:1711–1726

Nühse TS, Stensballe A, Jensen ON, Peck J (2003) Large-scale analysis of in vivo phosphorylated membrane proteins by immobilized metal ion affinity chromatography and mass spectrometry. Mol Cell Proteomics 2:1234–1243

Olayioye M, Neve RM, Lane HA, Hynes NE (2000) The ErbB signaling network: receptor heterodimerization in development and cancer. EMBO J 19:3159–3169

Olsen JV, Mann M (2004) Improved peptide identification in proteomics by two consecutive stages of mass spectrometric fragmentation. Proc Natl Acad Sci USA 101:13417–13422

Olsen JV et al (2006) Global, in vivo, and site-specific phosphorylation dynamics in signaling networks. Cell 127:635–648

Pawson T, Gish GD (1992) SH2 and SH3 domains: from structure to function. Cell 71:359–362

Pawson T, Scott JD (1997) Signaling through scaffold, anchoring, and adaptor proteins. Science 278:2075–2080

Peck SC (2006) Analysis of protein phosphorylation: methods and strategies for studying kinases and substrates. Plant J 45:512–522

Peltier JB et al (2006) The oligomeric stromal proteome of *Arabidopsis thaliana* chloroplasts. Mol Cell Proteomics 5:114–133

Pevzner PA, Mulyukov Z, Dancik V, Tang CL (2001) Efficiency of database search for identification of mutated and modified proteins via mass spectrometry. Genome Res 11:290–299

Reinders J, Sickmann A (2005) State-of-the-art in phosphoproteomics. Proteomics 5:4052–4061

Reinders A, Schulze W, Thaminy S, Stagljar I, Frommer WB, Ward JM (2002) Intra- and intermolecular interactions in sucrose transporters at the plasma membrane detected by the split-ubiquitin system and functional assays. Structure 10:763–772

Rigaut G, Shevchenko A, Rutz B, Wilm M, Mann M, Seraphin B (1999) A generic protein purification method for protein complex characterization and proteome exploration. Nat Biotechnol 17:1030–1032

Rohila JS, Chen M, Cerny R, Fromm ME (2004) Improved tandem affinity purification tag and methods for isolation of protein heterocomplexes from plants. Plant J 38:172–181

Rohila JS et al (2006) Protein–protein interactions of tandem affinity purification-tagged protein kinases in rice. Plant J 46:1–13

Russinova E et al (2004) Heterodimerization and endocytosis of *Arabidopsis* brassinosteroid receptors BRI1 and AtSERK3 (BAK1). Plant Cell 16:3216–3229

Schulze W, Mann M (2004) A novel proteomic screen for peptide–protein interactions. J Biol Chem 279:10756–10764

Schulze W, Deng L, Mann M (2005) The phosphotyrosine interactome of the ErbB-receptor kinase family. Mol Syst Biol. doi:10.1038/msb4100012:E1–E13

Shah K, Liu Y, Deirmengian C, Shokat KM (1997) Engineering unnatural nucleotide specificity for Rous sarcoma virus tyrosine kinase to uniquely label its direct substrates. Proc Natl Acad Sci USA 94:3565–3570

Tai HH, Tai GC, Beardmore T (2005) Dynamic histone acetylation of late embryonic genes during seed germination. Plant Mol Biol 59:909–925

Toroser D, Athwal GS, Huber SC (1998) Site-specific regulatory interaction between spinach leaf sucrose-phosphate synthase and 14-3-3 proteins. FEBS Lett 435:110–114

Tsuji H, Saika H, Tsutsumi N, Hirai A, Nakazono M (2006) Dynamic and reversible changes in histone H3-Lys4 methylation and H3 acetylation occurring at submergence-inducible genes in rice. Plant Cell Physiol 47:995–1003

von Mering C et al (2002) Comparative assessment of large-scale data sets of protein–protein interactions. Nature 417:339–403

Wang X et al (2005) Identification and functional analysis of in vivo phosphorylation sites of the *Arabidopsis* brassinosteroid-insensitive 1 receptor kinase. Plant Cell 17:1685–1703

Washburn MP, Wolters D, Yates JRI (2001) Large-scale analysis of the yeast proteome by multidimensional protein identification technology. Nat Biotechnol 19:242–247

Xu CR et al (2005) Histone acetylation affects expression of cellular patterning genes in the *Arabidopsis* root epidermis. Proc Natl Acad Sci USA 102:14469–14474

Plant Cell Monogr (10)
L. Bögre and G. Beemster: Plant Growth Signaling
DOI 10.1007/7089_2007_152/Published online: 26 January 2008

Signal Transduction Networks During Stress Responses in *Arabidopsis*: High-Throughput Analysis and Modelling

Alessandra Devoto[1] (✉) · Alberto Paccanaro[2]

[1]School of Biological Sciences, Royal Holloway University of London, Egham Hill, Egham, London TW20 0EX, UK
alessandra.devoto@rhul.ac.uk

[2]Department of Computer Science, Royal Holloway University of London, Egham Hill, Egham, London TW20 0EX, UK

Abstract Plants ability to adapt to developmental phase transitions as well as changing environmental conditions is inherent to the existence of complex signalling networks that are activated by different types of exogenous and endogenous signals. Ultimately, the activation of signalling pathways leads to extensive transcriptional, proteomic and metabolic re-programming determining a specific plant response to the initial signalling input(s). This chapter presents the general basic concepts of signalling pathways and crosstalk as well as the high-throughput and bioinformatics methods which have been used for both inferring and modelling *Arabidopsis thaliana* signalling networks from gene expression data. Due to the large amount of data available, the discussion will focus on the latest published work in the field with particular emphasis to stress and defence related responses where excellent studies in this direction have been performed in recent times. The information provided will complement that from other chapters of this book where plant growth responses to environmental changes are considered (in particular see chapters from: Anderson GH, TOR signalling in plants; Doerner P, Signals and mechanisms in the control of plant growth; Durgardeyn J and Van Der Straeten D, Ethylene: inhibitor and stimulator of plant growth)

The authors will also discuss recent approaches that have been developed for inferring and modelling protein networks in other model organisms, with the aim of highlighting how integrating this information with gene expression data may lead to better inference of signalling networks. It is reasonable to think that, once large scale protein interaction data will be available, similar methods will be applicable to infer and model signalling networks in *Arabidopsis*.

1 Introduction – Signalling Networks and Signalling Pathways

Signalling networks allow cells to integrate external and/or internal signals perceived during changes in their environment and to respond to them by altering transcriptional activity, metabolism, or other regulatory mechanisms. The accurate functioning of these networks is vital for adaptation and continued existence under unstable conditions, as well as for differentiation and cell fate. The concept of signalling networks has evolved from that of pathways. A signalling pathway could be defined as a cascade of events connecting input elements (in this review: the environmental stimuli, biotic and abiotic) to output elements (in this review: the responses). Until recently, such cascade was viewed in isolation as a simple chain of consecutive steps. However, recent research has focused on the divergence that is observed at several steps in a pathway and on the crosstalk, feedback, pleiotropy and redundancy between signalling pathways (Schwartz and Baron 1999; Klipp and Liebermeister 2006). It has been shown that signalling pathways interact with each other, forming a network and a full understanding of cell signalling can only be achieved by considering such signalling networks rather than the isolated pathways. In order to understand the multifaceted behavior of signalling networks, researchers have implemented computational modelling approaches, ranging from abstract models that emphasize some key features of signalling pathways to detailed models that describe the dynamics of specific pathways in specific organisms.

2
The Use of High-Throughput Transcriptomics Data to Infer Signalling Networks Activated During Stress

To achieve understanding of complex biological systems it is necessary to integrate high-throughput biological studies. In plant research, systems biology is still in its dawn and very much at the stage of accumulating vast quantities of data, especially from high-throughput transcriptional profiling. An important problem is that of reconstruction of signalling networks from gene expression data.

Gene expression data convey information about pathways, as it is reasonable to presume that highly co-expressed genes work in the same pathway (Eisen et al. 1998; Marcotte et al. 1999). The standard way to select genes that belong to the same pathway is therefore to group them according to the correlation of their expression profiles over different conditions—this is sometimes referred to as "guilt by association" (Walker et al. 1999). Clearly, although helpful in identifying the components of a certain pathway, this methodology is still largely deficient in providing information on the hierarchical connection between these components.

Once some candidate genes have been identified using these in-silico methods, a necessary step is then to verify these predictions through reverse genetics and mutant analysis. In recent times, a number of excellent studies has been performed in the direction of constructing signalling networks activated during plant responses to stress and during defence. Plants respond to stimuli such as pathogen and pest attack, wounding, changes in light, temperature and availability of nutrients (see also chapters from Doerner P Signals and mechanisms in the control of plant growth; Durgardeyn J and Van Der Straeten D, Ethylene: inhibitor and stimulator of plant growth). A transition during development is also perceived by the plant as a change generating a signal. Activated responses generally lead to extensive transcriptional reprogramming of gene expression. In this review we will consider mainly responses at transcriptional level. Pathways historically involved in the perception and transduction of stress are those regulated by jasmonates (JAs), salicylic acid (SA), ethylene (ET) and abscisic acid (ABA). These molecules are involved in the local and/or systemic response of the plant (for relevant recent reviews see Fujita et al. (2006), Gfeller et al. (2006), Grant and Lamb (2006), Halim et al. (2006), Dreher and Callis (2007), Devoto and Turner (2005), Lorenzo and Solano (2005).

Particularly, in *Arabidopsis*, the availability of tools such as mutants and pathosystems as well as high-throughput technology such as transcription profiling by microarrays has greatly facilitated information gathering on the existence of components of signalling pathways as well as of crucial nodes facilitating communication within signalling networks. This approach has been used for example to identify regulatory nodes in the transcriptional network of systemic acquired resistance (SAR) in *Arabidopsis*. SAR is an inducible plant defence response involving a cascade of transcriptional events caused by SA through the transcription cofactor NPR1 (**n**on-expressor of **PR**; (Kinkema et al. 2000). To identify novel regulatory nodes in the SAR network (Wang et al. 2006), performed microarray analysis in a stepwise approach on *Arabidopsis* plants expressing the NPR1-GR (**G**lucocorticoid **R**eceptor) fusion protein. Since nuclear translocation of NPR1-GR requires dexamethasone (Wang et al. 2005), the authors were able to control NPR1-dependent transcription and to identify direct transcriptional targets of NPR1 acting as crucial regulatory nodes during SAR. Disrupting these regulatory nodes compromised various functions assigned to NPR1. Specifically it was found that NPR1 directly upregulates the expression of five WRKY transcription factor genes that had never been placed before in the SAR network. Among these WRKY factors, both positive (WRKY18 and 53) and negative regulators (WRKY58) of SAR were found. In addition, fine tuning of SAR occurs when SA levels are high: signalling through positive WRKY factors were found to overcome the negative effect of WRKY58 to activate downstream gene transcription and the action of WRKY70 and WRKY54 prevent excessive SA accumulation.

The nature of the mobile signal as well as of the remotely activated networks responsible for establishing SAR, remains unclear (Grant and Lamb 2006). In a recent study (Truman et al. 2007) have shown that in *Arabidopsis*, despite the absence of pathogen-associated molecular pattern (PAMPS, Nurnberger et al. 2004) contact, systemically responding leaves rapidly activate a SAR transcriptional signature with strong similarity to local basal defence responses to herbivory and wounding. The signature shares secondary metabolism components with late basal defence responses. The RPM1 (Resistance to *P. syringae* pv. *Maculicola 1*) pathosystem (Grant et al. 1995) has been used here to dissect both timing and nature of early transcriptional events in tissues associated with the establishment of systemic immunity after RPM1 recognition. A role as initiating signal for SAR has been attributed here to JAs which are suggested to act ahead of SA-dependent responses in systemic leaves. JAs, including the JA precursor oxophytodienoic acid (OPDA) and conjugated derivatives, such as methyl-JA (MeJA) or isoleucine-JA have been previously demonstrated to possess roles in defence signalling (Stintzi et al. 2001; Staswick and Tiryaki 2004; Sasaki-Sekimoto et al. 2005). These conclusions were reached by carrying out an extensive comparison of in-house microarray analysis with experiments representing host responses to biotic and abiotic stresses or hormone treatments from the ArrayExpress (www.ebi.ac.uk/arrayexpress/) and NASCArrays (affymetrix.arabidopsis.info) database repositories. The novel work from M. Grant's laboratory has shown that SAR can be mimicked by foliar JA application and was found to be abrogated in mutants impaired in JA synthesis or response. De novo JA biosynthesis was found to be associated with the induction of jasmonate-responsive genes in systemic tissues. Therefore, although JAs have been generally regarded as antagonizing SA-dependent responses, the plant can benefit from the advantages of both signalling pathways when their activation is separated during time or space.

The systematic comparison of the expression profiles between wild-type and mutants still represents one of the most exhaustive ways to elucidate critical steps in host responses, at least at the transcriptional level. A recent example of expression profiling and mutant analysis aimed at further dissecting *Arabidopsis* defence response to the necrotrophic pathogen *Botrytis cinerea* infection was provided by the laboratory of T. Mengiste (AbuQamar et al. 2006). In this study, mutants exhibiting enhanced susceptibility to Botrytis were used to correlate changes in mRNA profiles with impaired disease resistance responses of whole plants. *Arabidopsis* wild-type plants were compared to *coi1* (**co**ronatine **i**nsensitive 1) and *ein2* (**e**thylene **in**sensitive 1) mutants and to plants carrying the *nahG* (salicylate hydroxylase) gene. In wild-type plants, the expression of 621 genes representing approximately 0.48% of the *Arabidopsis* transcriptome was induced.

The expression of 181 Botrytis induced genes (BIGs) was dependent on a functional *COI1* gene, a well-known component of JA signalling (Feys et al. 1994; Xie et al. 1998), whereas the expression of 63 and 80 BIGs were dependent on ET signalling or SA accumulation, respectively. Thirty BIGs encode putative DNA-binding proteins previously found to be regulating ET responses such as zinc-finger, MYB, WRKY, and HD-ZIP family transcription-factor proteins. Importantly, T-DNA insertion mutants in two BIGs, encoding putative DNA-binding proteins ZFAR1 (At2g40140) and WRKY70 (At3g56400), showed increased susceptibility to Botrytis infection. ZFAR1 is also required for germination on ABA, and encodes a putative transcription-factor protein containing zinc-finger and ankyrin-repeat domains. The transcriptional activation of genes involved in plant hormone signalling and synthesis, removal of reactive oxygen species, and defence and abiotic-stress responses, coupled with the susceptibility of the *wrky70* and *zfar1* mutants, highlights the complex genetic network underlying defence responses to Botrytis in *Arabidopsis*. The above study by (AbuQamar et al. 2006) represents probably one of the latest examples about the previously demonstrated interconnection between signalling pathways such as JA, ET and SA (Devoto and Turner 2005; Devoto et al. 2005; Lorenzo and Solano 2005; Gfeller et al. 2006; Liechti et al. 2006).

Indeed, hormones rule every aspect of the biology of plants. Stress and development are regulated in comparable ways by multiple hormones and as also highlighted above by recent key studies, the existence of widespread crosstalk among different hormonal signalling pathways has been revealed (Finkelstein et al. 2002; Guo and Ecker 2004; Sun and Gubler 2004; Vert et al. 2005; Woodward and Bartel 2005). Crosstalk refers to the case that two inputs (in this review: stresses of biotic and abiotic origin) work through different signalling pathways but combine forces to regulate outputs and ultimately development. Intensive experimental work has revealed numerous potential paths for crosstalk. Despite the apparent integration of inputs from multiple hormones in regulating development, it has been recently shown that the level of convergence, defined by co-expression, on a common set of transcriptional targets is reduced only to a few genes (e.g., only seven genes were changed in the same direction by GA (gibberellic acid 3), IAA (auxin, indole acetic acid), and BL (brassinosteroids) treatments, none with known function) and that therefore there is not a core transcriptional growth-regulatory module in young *Arabidopsis* seedlings (Nemhauser et al. 2006). In this work, data produced by the AtGen- Express Consortium (http://web.uni-frankfurt.de/fb15/botanik/mcb/AFGN/atgenex.htm) in which the effects of seven plant hormones at three time points were surveyed with Affymetrix ATH1 GeneChips representing nearly all protein-coding transcripts of *Arabidopsis* were compared. The compounds assayed included ABA, GA, IAA, 1-amino-cyclopropane-1-carboxylic acid (ACC; ethylene precursor), zeatin (CK; cytokinin), BL and MeJA. These studies revealed that a major part of early hormone response in plants is specific and independent of the effects of other hormones. It has to be highlighted however that despite the low numbers of shared transcriptional targets, an ample evidence of one hormone-regulating genes involved in the metabolism of another hormone was observed. It is possible that this could be due to a knock-on effect from one hormone re-setting many systems within the plant. Caution is therefore needed in drawing conclusions regarding the existence of crosstalk from a limited number of genes that appear to be similarly regulated by different hormones. Despite that such a comparison has not been carried out yet in a similar manner for plants "under attack", it is possible that similar conclusions might be applicable. Interestingly, the work carried out by (Wang et al. 2006) highlights that even different members of the same gene family of transcription factors may have very specific functions within the same pathway.

3
The Next Phase of Microarray Analysis is to Add Structure to the Data: Novel Available Bioinformatics Tools for Database-Hopping Biologists

Given a set of transcriptomics data, different tools are available for inferring gene regulatory networks using the aforementioned correlative methodology. Several of these tools and databases have been developed for the analysis of *Arabidopsis* microarray data. A number of web-based services host gene expression data from *Arabidopsis* microarray experiments and provide information for individual genes or gene sets such as ArrayExpress (Parkinson et al. 2007), Botany Array Resource (Toufighi et al. 2005), Gene Expression Omnibus (GEO, Edgar et al. 2002), NASCArrays tools (Craigon et al. 2004), Stanford Microarray Database (Gollub et al. 2006; Demeter et al. 2007), TAIR (Rhee et al. 2003). In this review we will focus on those databases and tools that were most recently released or allowing a more comprehensive application of correlative methodology to retrieve gene-to-gene relationships. We regret that we are not able to report a series of surely useful minor databases due to space limitations.

The Comprehensive Systems Biology Database (CSB.DB; http://csbdb.mpimp-golm.mpg.de/csbdb/dbcor/ath/ath_tsgq.html; (Steinhauser et al. 2004)) presents the results of bio-statistical analyses on gene expression data in association with additional biochemical and physiological knowledge. It can be used to retrieve genes associated by co-response. Sets of co-response databases currently focus on the three key model organisms, *Escherichia coli*, *Saccharomyces cerevisiae* and *Arabidopsis thaliana*. By scanning for the best co-responses among changing transcript levels, CSB.DB allows to infer hypotheses on the functional interaction of genes. The database enables the search for pairs of genes and larger units of genes, which are under common transcriptional control making a regulon. Moreover, this database allows filtering according to the functional categories, which are reported together with the visualization tool MapMan (Thimm et al. 2004). The single gene query sGQ output includes the rank, the gene identifier of the co-responding gene, the correlation measure, the gene description, the number of pairs (n), the probability (P-value). The multiple gene query option (mGQ) allows predefinition of up to 15 genes of interest and returns the complete set of available correlations among these genes. This option may be used to discover interdependences of genes, which are known to contribute to a common function or pathway.

Genevestigator (https://www.genevestigator.ethz.ch/; Zimmermann et al. 2004) is a database and web-browser data mining interface for Affymetrix GeneChip data. Users can query the database to retrieve the expression patterns of individual genes throughout chosen environmental conditions, growth stages, or organs. In addition, mining tools allow users to identify genes specifically expressed during selected stresses, growth stages, or in particular organs. A useful feature is represented by Meta-Analyzer: this utility has been designed to study the gene expression profiles of several genes simultaneously in the context of environmental stresses, organs, and growth stages. It also represents a tool to verify correspondence between correlated and co-regulated genes identified through other tools. Genevestigator probably currently represents the most widely used tool of this kind used by the *Arabidopsis* community and has been subjected to continuous update since first published.

To facilitate the interpretation of the publicly available array data (Jen et al. 2006), have developed two new bioinformatic tools, the *Arabidopsis* Co-Expression Tool (ACT), which reports gene co-expression patterns across user-selected single or multiple arrays obtained from the Nottingham *Arabidopsis* Stock Centre and CLIQUE FINDER, a tool to determine the sets of genes most likely to be regulated in a similar manner, which provides a quantitative method for determination of correlation cut-offs to generate exclusive groups of genes which may share a common purpose.

In combination, these tools offer three levels of analysis: creation of correlation lists of co-expressed genes, refinement of these lists using two-dimensional scatter plots, and dissection into groups of co-regulated genes. Both tools are available at http://www.arabidopsis.leeds.ac.uk/ACT/.

To demonstrate the software, the authors analyzed genes encoding functionally related ribosomal proteins, followed by an analysis of heat-shock and cold-responsive genes to predict the involvement of uncharacterized genes in well-defined responses. Computational methods aimed at defining associations between gene expression levels and putative regulatory sequences in upstream regions of genes are increasingly used to consolidate genome-scale transcriptional regulatory networks (Jen et al. 2006) also demonstrates the integration of their analyses with promoter element detection software to identify conserved cis-regulatory elements potentially involved in the co-regulated expression of clustered genes. The analyses were applied to predict aspects of regulation and differential function within a group of genes involved in cell wall biosynthesis.

PathoPlant (http/www.pathoplant.de; Bulow et al. 2004, 2007) is the first platform for microarray expression data to analyze co-regulated genes involved in plant defence responses. The PathoPlant database was initially developed to display signal perception and signal transduction pathways on a molecular level during plant pathogenesis as well as the corresponding interactions between plants and pathogens on the organism level. Because only experimentally proven direct molecular interactions have been annotated, only a limited number of regulated genes were covered in PathoPlant. Therefore, the new version of PathoPlant has now been complemented by *A. thaliana* microarray gene expression data. The datasets chosen are plant pathogenesis related and represent not only endogenous plant signal molecules, such as SA and JA, but also include treatments with elicitors and infections with different pathogens to enable comparative studies. The PathoPlant gene expression function allows starting with combinations of up to three different stimuli to determine all overlapping genes being up-, down- or not regulated by these stimuli. An additional feature of PathoPlant is the integration of AthaMap (Steffens et al. 2005; Galuschka et al. 2007) for subsequent cis-regulatory element identification. AthaMap allows identification of putative functional cis-regulatory elements based on binding site specificities of transcription factors.

Another tool recently developed is the *Arabidopsis* trans-factor and cis-element prediction database (ATTED-II; http://www.atted.bio.titech.ac.jp; Obayashi et al. 2007), which can be used for retrieving gene-to-gene relationships similar to the other databases for co-expressed genes. ATTED-II contains in addition stored pre-calculated results for cis-element prediction linked to every gene At the time of publication ATTED-II contained co-expressed gene networks for 22 263 loci and for 1102 functional categories as well as predicted cis-elements represented by 304 heptamers.

At present discussion is still ongoing on the existence of the "most appropriate" statistical methods for co-expression analysis. It is clear however, that combinations of different tools offering different functions and providing visual representation of outputs are most needed to enable biologists to generate and test hypotheses.

4
The Modelling of Large Complex Networks

Novel important discoveries in the field of complex networks are providing the grounds for building new tools for the analysis of large networks, which could also prove to be very useful in the near future for the study of *Arabidopsis* responses during stress. Large complex networks are ubiquitous in many disciplines, such as biology, computer science, and social sciences, to name a few. In recent years, topological data about large complex networks has become increasingly available, and as a consequence we have witnessed great advances towards understanding the organizing principles of such networks. Strikingly, we are finding that many of the architectural features of large complex networks are shared by systems as different as the World Wide Web, social networks, and many types of biological networks.

Until recently, such networks were modelled by the Random Graph Theory, an elegant mathematical theory developed by Erdos and Renyi starting from the 1960s (Erdos and Renyi 1960). According to this theory, a complex network was identified as a random graph, i.e., a graph built starting from a set of nodes and then randomly adding edges with a certain probability. However, with the availability of large amounts of data describing the topology of real networks it has become clear that the theory of random graph does not explain many of their topological properties. Probably the most noticeable of them is constituted by the distribution of the degrees of the nodes (the degree of a node is the number of edges connected to it): while the distribution predicted by the random graph theory follows a Poisson distribution, the degree distribution found for real networks follows a power law distribution of the type $P(k) = a^{-k}$, where $P(k)$ indicates the probability of having a node of degree k, and a is a real value which for many complex networks has been found to be close to 2 (Barabasi and Albert 1999). An important consequence of this is that while for a random graph the probability of a node having high degree is virtually zero, the power law distribution indicates that there is a large probability to have a few nodes with a very high degree; and these few nodes will therefore dominate the connectivity in the network. Such highly connected nodes are commonly called "hubs", and networks with a power law degree distribution were named "scale-free" (Barabasi and Albert 1999).

Most networks within the cell have been found to be scale-free. For example, metabolic networks (in which the nodes are metabolites and the links are biochemical reactions), protein–protein interaction networks (in which the nodes correspond to proteins and the links represent a physical interaction between them), co-expression networks (where nodes are the genes and the links represent the amount of co-expression between them, as measured by correlations computed from microarray experiments), have all been shown to be scale-free (Barabasi and Oltvai 2004). Currently, there is no reason to doubt that these results hold true for transcriptomic networks in *Arabidopsis*.

An important consequence of the power law distribution of the degree is that scale-free networks are extremely robust to random noise. In other words, no significant decrease in performance can be seen even when a high number of nodes, chosen at random, is deleted from the network. This is due to the fact that there is a relatively low probability to knock-out simultaneously many of the network's hubs, which are its crucial functional elements. Interestingly, it has been shown that proteins that are hubs in protein–protein interaction networks in yeast (*S. cerevisiae*) have a tendency to be essential genes (Jeong et al. 2001). In *Arabidopsis* the robustness of the ABA network against perturbation was tested through simulating gene disruption and pharmacological intervention (Li et al. 2006).

5 Computational Frameworks to Model Signalling Networks in *Arabidopsis*

Another fundamental area of research is the development of mathematical models of signalling networks. These models are important as they allow simulating the transmission of signals in such networks, from the environmental stimuli to the cell responses.

Several approaches have been proposed for the inference of large regulatory networks from gene expression data. For example, several authors have used discrete models of Boolean networks (Rangel et al. 2004); Bayesian networks (Friedman 2003; Husmeier 2003); continuous models of neural networks (van Someren et al. 2002); differential equations (Kobayashi et al. 2002). An in-depth discussion of these methods goes beyond the scope of this review. We will instead focus on those approaches used to model gene networks in *Arabidopsis*. Strikingly, the models are still very limited, highlighting the need for further developments.

Several authors have started adapting Boolean language to represent and analyze interactions between pathways (Genoud and Metraux 1999; Genoud et al. 2001, 2002; Devoto et al. 2005). With this language the quantitative features of a molecular interaction may be described discontinuously by several qualitative steps. Using Boolean gates, signalling processes may be represented more accurately as network-like structures than with linear sequences of events in intuitive formalism. Interfering input signals reach a Boolean gate through switches with a molecular identity, generating an output signal that results from the combination of all inputs going through the gate (Arkin and Ross 1994). By using digital simulation programs, as described by Genoud et al. (2001), it is possible to predict the outputs of the logical gates by activating or inactivating input signals. Devoto and Turner (2005) represented using Boolean gates, the integration of the JA pathway with the SA, ET and light signalling pathways inferring the existence of multiple interferences and intersections from genetic and transcription profiling information. In addition, they have used simple Boolean language (one logical operator per molecule, or per complex of molecules) to identify groups of genes whose expression is differentially regulated by MeJA and/or wounding (Devoto et al. 2005). Further implementation on the digital simulation has been described by Trevino Santa Cruz et al. (2005) where it has been suggested in order to overcome the simplifications of digital networks by integrating noise and clock signals on a digital simulator in order account for the existence of signalling background and circadian rhythms in biological systems. In the first case a digital signal could be added to a source of random signal in order to simulate biological noise. The activation of an oscillating clock-like signal could be placed under the control of a simple ON/OFF switch through an AND operator.

One of the most studied abiotic stresses is water deprivation. During drought, the plant hormone ABA inhibits stomatal opening and promotes stomatal closure, thereby promoting water conservation. Abscisic acid signal transduction in guard cells is therefore one of the best characterized signalling systems in plants. Li et al. (2006) have formalized the large amount of information that has been gathered on ABA induction of stomatal closure from individual experiments and used this information to reconstruct the ABA signalling network. An advantage of the method used here over other methods such as those used in Science's Signal Transduction Knowledge Environment (STKE) connection maps (Assmann 2004) is the inclusion of intermediate nodes when direct physical interactions between two components have not been demonstrated. In this model, the dynamics of state changes are governed by Boolean rules providing the state transition of each node given the state of its regulators (upstream nodes). The model obtained sums up the regulation of more than 40 identified network components, and it is in agreement with previous experimental results. By simulating gene disruptions and pharmacological interventions, the robustness of the network against perturbations was also assessed. Simulations of stomatal response as derived from the proposed model provide an efficient tool for the identification of candidate manipulations that have the best chance of conferring increased drought stress tolerance and for the prioritization of future experimental analyses.

Recently Wang et al. (2006) have inferred a network in *Arabidopsis* based on 35 links generated from stress response datasets in shoots. GNR (Gene Network Reconstruction tool; http://zhangorup.aporc.org/bioinfo/grninfer/, http://digbio.missouri.edu/grninfer/ and http://intelligent.eic.osaka-sandai.ac.jp) is based on linear programming and differential equations aimed to reconstruct gene network using multiple datasets from different sources without normalization among the datasets. One of the main limitations of gene expression datasets consists of relatively few time points with respect to a large number of genes (generally in thousands). In addition to the dimensionality problem of the data, another problem is that the derived gene networks often have heavily connected gene regulatory associations among nodes. This method provides a general scaffold to analyze microarray data by fully exploiting all available microarray data for a given species, so as to improve the problem of dimensionality or data scarceness. An assumption for the proposed method is that the structure of the regulatory network is stationary, and does not "rewire" under the environmental conditions for those different datasets. Nine whole-genome Affymetrix chips microarray datasets related to the stress responses, each with six or more time points and each for root and shoot experiments (ATGenExpress database, TAIR, http://www.arabidopsis.org/) were used to test this method.

Despite that the main focus of this review is to examine latest progresses in modelling stress regulatory networks, it is worth highlighting the impact that spatial structures have on gene expression dynamics and the advancements made in growth modelling. Artificial life simulations provide a basis for evaluating methods to reconstruct regulatory networks based on gene expression measurements. The effects of spatial growth on gene expression have to be expected to be significant for network reconstruction. Jan Kim (2005) proposed to use "transsys" simulations (Kim 2001) to explore the impact of morphogenesis and of other parameters on network reconstruction using the algorithm by Rung et al. (2002). This approach simulates reconstruction of a target network that does not organize morphogenesis, but may be informed by it. In this approach, it was chosen to enable attribution of differences to individual morphological structures, rather than to collections of mutant structures with complex and unfavorable statistical properties. This algorithm assumes that significant changes in expression levels resulting from a gene knockout indicate a direct target gene. Generation of knockout mutants and collection of gene expression measurements was implemented in Python (http://www.python.org/), based on the transsys framework (Kim 2001). The "R" language (Ihaka and Gentleman 1996) was used for programming data analysis and visualization. The code underlying the results presented is available on the transsys website (http://www2.cmp.uea.ac.uk/~jtk/transsys/).

Another approach that has been used to model gene regulation and interaction has been provided by fuzzy logic (Du et al. 2005). This work models interactions (also referred to as edges or links) in the network as fuzzy functions depending on the detail known about the network.

6
Protein Interaction Networks

Inferring signalling networks solely from transcriptomics data has several limitations. For example, the discrete nature of the data could limit the complexity of the networks that can be inferred. Moreover, transcriptomics data can provide only a limited picture of the actual physiological changes underlying a living organism.

This has been clearly shown very recently, through proteomic analysis of *Arabidopsis* suffering biotic stress. Jones et al. (2006) have analyzed the alterations in the proteome of *Arabidopsis* leaves during responses to challenge by *Pseudomonas syringae* pv tomato DC3000 using two-dimensional gel electrophoresis. The abundance of each protein identified was compared with that of selected transcripts obtained from comparable GeneChip experiments (Truman et al. 2006). Changes were reported in total soluble protein, chloroplast-enriched, and mitochondria-enriched over four time points (1.5–6 h after inoculation). In total, 73 differential spots representing 52 unique proteins were successfully identified. Significantly, many of the changes in protein spot density occurred before transcriptional reprogramming. The high proportion of proteins represented by more than one spot indicated that many of the changes to the proteome can be attributed to post-transcriptional modifications. One further strength of this proteomic analysis was the ability to separate components of basal defence (by inclusion of the hrpA mutant; de Torres et al. 2003) from disease and resistance responses, DC3000, and DC3000 (avrRpm1) inoculations.

In recent years, large-scale protein–protein interaction data have become available for some model organisms, and such data have proven extremely useful for inferring gene regulatory networks. The effective integration of data from different sources appears to be one of the most important approaches for unravelling the cell dynamics. Unfortunately, protein–protein interaction data are still very limited for *Arabidopsis*.

A promising approach for expanding a given dataset of protein–protein interaction is that of the "in silico" prediction of interactions from a set of genomic features using machine learning techniques.

For example, Bayesian Networks (Jensen 1997) have been used to predict genome-wide protein–protein interactions in yeast by integrating information from different genomic features, ranging from co-expression relationships to similar phylogenetic profiles (Jansen et al. 2003; Lu et al. 2005). These results were particularly important because it was possible to show that at a certain level of sensitivity the predictions were more accurate than the existing high-throughput experimental dataset.

On the other hand, when experimental data for a given organism are available, it is often necessary to combine experimental results in order to create an interaction network. In fact, when different techniques are used to identify protein interactions, the process of creating a unique protein–protein interaction network involves combining the results of separate experiments. Moreover, the problem can be complicated by the fact that the data may not to be directly comparable and is likely to have different amounts of noise.

A technique that has been successfully applied to solve this problem involves using a machine learning algorithm to learn the parameters of a model that combines the different experimental results. In general, using a small set of well-known protein–protein interactions (a.k.a. gold standard), the system is trained to output a probability of a protein–protein interaction given the different experimental data. Recently, this method has been used for integrating the results of two (possibly repeated) purifications (matrix-assisted laser desorption/ionization-time of flight mass spectrometry (MALDI) and liquid chromatography tandem mass spectrometry (LCMS)) of 4,562 different tagged proteins of the yeast *S. cerevisiae* (Krogan et al. 2006). Using the hand-curated protein complexes in the MIPS (Munich Information Center for Protein Sequences) reference database (Mewes et al. 2006), a machine learning system was trained to assign a probability that each pairwise interaction is true based on experimental reproducibility and mass spectrometry scores from the relevant purifications. In this way, from the two "incomplete" graphs obtained using the LC-MS and MALDI technique it was possible to generate a single combined protein–protein interaction network for *S. cerevisiae*. Notice that the edges of this network are labelled with a number that is the probability of interaction between the two proteins they connect. In other words, the network is an undirected weighted graph in which individual proteins are nodes and the weight of the edge connecting two nodes is the probability that the interaction is correct.

Interaction data are noisy, and therefore the protein–protein interaction networks obtained from them will contain many errors in the form of links which can be either missing or incorrect (von Mering et al. 2002). A very interesting question is whether it is possible to use the network topology to reduce the amount of noise in the experimental data that is, to "correct" some of the experimental errors.

A positive answer to this question for PPI networks has been given recently by Paccanaro et al. (Paccanaro et al. 2005; Yu et al. 2006). The basic idea of the method derives from the way in which large-scale PPI experiments are carried out and particularly from the matrix model interpretation of their results (Bader and Hogue 2002). In these experiments, one protein (the bait), is used to pull out the set of proteins interacting with it (the preys) in the form of a list. When such lists differ only in a few elements, it is reasonable to assume that this is because of experimental errors, and the missing elements should therefore be added. Each list can be represented as a fully connected graph in which proteins occupy the nodes. Then the problem of identifying lists that differ in only a few elements is equivalent to finding a clique (a completely connected subgraph) in a graph with a few missing edges, which was named a "defective clique". Therefore the algorithm searches the network for defective cliques (i.e., nearly complete complexes of pairwise interacting proteins) and predicts the interactions that complete them. This method was shown to have a very good predictive performance, thus allowing the correction of many errors present in large-scale experiments.

Once a network has been obtained, it can be used as a model to answer important biological questions. For example, it is well known that proteins carry out their function by interacting with other proteins and that they tend to act in complexes. Identifying these complexes is therefore a crucial step in understanding the cell dynamics and can give important clues to protein function.

One way to identify such complexes is by identifying tight clusters in PPI networks. This approach has been recently used in (Krogan et al. 2006) to identify protein complexes in *S. cerevisiae*. Particularly, the Markov cluster algorithm (van Dongen 2000) (which simulates random walks within graphs) was used to identify highly connected modules within the global protein–protein interaction network. The algorithm identified 547 protein complexes, about half of which were previously unknown.

Finally, we would point out to a recent work which builds a slightly different type of network that has been used for function prediction. Some biological problems or data do not have a natural representation as networks. However, sometimes they can be remapped onto a network formalism and this representation can offer an efficient solution.

An interesting case is represented by the problem of clustering protein sequences. Clustering protein sequences based on their evolutionary relationship is important for sequence annotation as structural and functional relationships can be potentially inferred. This problem can be easily mapped into that of clustering the nodes of a weighted undirected graph in which each node corresponds to a protein sequence and the weights on the edges correspond to a measure of distance between two sequences. The goal is to partition such a graph into a set of discrete clusters whose members are homologs.

Recently, a method has been introduced for solving this problem that is based on spectral graph theory. Such method partitions the graph into clusters by considering the random walk formulation on the graph, and analyzing the perturbations to the stationary distribution of a Markov relaxation process. This is done by looking at the eigenvectors of the Markov transition matrix. A detailed explanation of the technique is beyond the scope of this review, and we refer the interested reader to the work of Paccanaro et al. (2003, 2006). When this algorithm was tested on difficult sets of proteins whose relationships were known from the SCOP database (Structural Classification of Proteins, http://scop.mrc-lmb.cam.ac.uk/scop/) the method correctly identified many of the family/superfamily relationships. Results obtained using this approach were much better than those obtained using other methods on the same datasets. On average, when quantifying the quality of the clusters using a measure that combines sensitivity and specificity, this approach showed improvements of 84% over hierarchical clustering (Everitt 1993), 34% over Connected Component Analysis (CCA) (similar to GeneRAGE; Enright and Ouzounis 2000) and 72% over another global method, TribeMCL (Enright et al. 2002).

7 Conclusions

The increased complexity of a biological system is achieved through multiple regulatory input points. Plant responses to stress are orchestrated through a network that integrates signalling pathways characterized primarily by the production of JA, SA and ET. The identified regulatory steps in the network highlight the intricacy of the signalling networks involving various levels of control ranging from transcriptional to post-translational.

In this chapter, we have looked mainly at stress related signalling pathways in *Arabidopsis thaliana* and at the way in which they are combined into large signalling networks. Particularly, we have shown examples of how bioinformatics techniques combining microarray data offer a novel way to identify genes in such networks. Experimental techniques, using reverse genetic and mutant analysis have been used to verify "in silico" predictions. An important goal of current research has also been that of defining mathematical models that can be used for simulating the transmission of signals in such networks, from the environmental stimuli to the cell responses.

However, it is becoming evident that inference of signalling networks solely from transcriptomics data has several limitations. It has been shown in some model organisms that the integration of transcriptomics data together with protein–protein interaction data is extremely useful for inferring signalling networks. Unfortunately this type of data is still very limited for *Arabidopsis*. We believe that when this data will become available it will lead to a better mechanistic explanation and identification of crucial nodes in signalling pathways.

Acknowledgements The authors are grateful to Dr. Virginia Balbi for critical reading of the manuscript. Alessandra Devoto and Alberto Paccanaro research is supported by the Biotechnological and Biological Sciences Research Council (BBSRC) and Engineering and Physical Sciences Research Council (EPSRC).

References

AbuQamar S, Chen X, Dhawan R, Bluhm B, Salmeron J, Lam S, Dietrich RA, Mengiste T (2006) Expression profiling and mutant analysis reveals complex regulatory networks involved in *Arabidopsis* response to Botrytis infection. Plant J 48:28–44

Arkin A, Ross J (1994) Computational functions in biochemical reaction networks. Biophys J 67:560–578

Assmann SM (2004) Plant G proteins, phytohormones, and plasticity: three questions and a speculation. Sci STKE 2004: re20

Bader GD, Hogue CW (2002) Analyzing yeast protein–protein interaction data obtained from different sources. Nat Biotechnol 20:991–997

Barabasi AL, Albert R (1999) Emergence of scaling in random networks. Science 286:509–512

Barabasi AL, Oltvai ZN (2004) Network biology: understanding the cell's functional organization. Nat Rev Genet 5:101–113

Bulow L, Schindler M, Choi C, Hehl R (2004) PathoPlant: a database on plant-pathogen interactions. In Silico Biol 4:529–536

Bulow L, Schindler M, Hehl R (2007) PathoPlant: a platform for microarray expression data to analyze co-regulated genes involved in plant defense responses. Nucleic Acids Res 35:D841–D845

Craigon DJ, James N, Okyere J, Higgins J, Jotham J, May S (2004) NASCArrays: a repository for microarray data generated by NASC's transcriptomics service. Nucleic Acids Res 32:D575–D577

de Torres M, Sanchez P, Fernandez-Delmond I, Grant M (2003) Expression profiling of the host response to bacterial infection: the transition from basal to induced defence responses in RPM1-mediated resistance. Plant J 33:665–676

Demeter J, Beauheim C, Gollub J, Hernandez-Boussard T, Jin H, Maier D, Matese JC, Nitzberg M, Wymore F, Zachariah ZK, Brown PO, Sherlock G, Ball CA (2007) The Stanford Microarray Database: implementation of new analysis tools and open source release of software. Nucleic Acids Res 35:D766–D770

Devoto A, and Turner JG (2005) Jasmonate – regulated *Arabidopsis* stress signalling network. Physiologia Plantarum 123:161–172

Devoto A, Ellis C, Magusin A, Chang HS, Chilcott C, Zhu T, Turner JG (2005) Expression profiling reveals COI1 to be a key regulator of genes involved in wound- and methyl jasmonate-induced secondary metabolism, defence, and hormone interactions. Plant Mol Biol 58:497–513

Dreher K, Callis J (2007) Ubiquitin, hormones and biotic stress in plants. Ann Bot (Lond)

Du P, Gong J, Syrkin Wurtele E, Dickerson JA (2005) Modeling gene expression networks using fuzzy logic. IEEE Trans Syst Man Cybern B Cybern 35:1351–1359

Edgar R, Domrachev M, Lash AE (2002) Gene Expression Omnibus: NCBI gene expression and hybridization array data repository. Nucleic Acids Res 30:207–210

Eisen MB, Spellman PT, Brown PO, Botstein D (1998) Cluster analysis and display of genome-wide expression patterns. Proc Natl Acad Sci USA 95:14863–14868

Enright AJ, Ouzounis CA (2000) GeneRAGE: a robust algorithm for sequence clustering and domain detection. Bioinformatics 16:451–457

Enright AJ, Van Dongen S, Ouzounis CA (2002) An efficient algorithm for large-scale detection of protein families. Nucleic Acids Res 30:1575–1584

Erdos P, Renyi A (1960) On the evolution of Random Graphs. Publ Math Inst Hung Acad Sci 5:17–61

Everitt BS (1993) Cluster Analysis, 3rd edn. Edward Arnold, London

Feys B, Benedetti CE, Penfold CN, Turner JG (1994) *Arabidopsis* mutants selected for resistance to the phytotoxin coronatine are male sterile, insensitive to methyl jasmonate, and resistant to a bacterial pathogen. Plant Cell 6:751–759

Finkelstein RR, Gampala SS, Rock CD (2002) Abscisic acid signalling in seeds and seedlings. Plant Cell 14 Suppl: S15–45

Friedman N (2003) Probabilistic models for identifying regulation networks. Bioinformatics 19(2):II57

Fujita M, Fujita Y, Noutoshi Y, Takahashi F, Narusaka Y, Yamaguchi-Shinozaki K, Shinozaki K (2006) Crosstalk between abiotic and biotic stress responses: a current view from the points of convergence in the stress signalling networks. Curr Opin Plant Biol 9:436–442

Galuschka C, Schindler M, Bulow L, Hehl R (2007) AthaMap web tools for the analysis and identification of co-regulated genes. Nucleic Acids Res 35:D857–D862

Genoud T, Buchala AJ, Chua NH, Metraux JP (2002) Phytochrome signalling modulates the SA-perceptive pathway in *Arabidopsis*. Plant J 31:87–95

Genoud T, Metraux JP (1999) Crosstalk in plant cell signalling: structure and function of the genetic network. Trends Plant Sci 4:503–507

Genoud T, Trevino Santa Cruz MB, Metraux JP (2001) Numeric simulation of plant signalling networks. Plant Physiol 126:1430–1437

Gfeller A, Liechti R, Farmer EE (2006) *Arabidopsis* jasmonate signalling pathway. Sci STKE 2006: cm1

Gollub J, Ball CA, Sherlock G (2006) The Stanford Microarray Database: a user's guide. Methods Mol Biol 338:191–208

Grant M, Lamb C (2006) Systemic immunity. Curr Opin Plant Biol 9:414–420

Grant MR, Godiard L, Straube E, Ashfield T, Lewald J, Sattler A, Innes RW, Dangl JL (1995) Structure of the *Arabidopsis* RPM1 gene enabling dual specificity disease resistance. Science 269:843–846

Guo H, Ecker JR (2004) The ethylene signalling pathway: new insights. Curr Opin Plant Biol 7:40–49

Halim VA, Vess A, Scheel D, Rosahl S (2006) The role of salicylic acid and jasmonic acid in pathogen defence. Plant Biol (Stuttg) 8:307–313

Husmeier D (2003) Sensitivity and specificity of inferring genetic regulatory interactions from microarray experiments with dynamic Bayesian networks. Bioinformatics 19:2271–2282

Jansen R, Yu H, Greenbaum D, Kluger Y, Krogan NJ, Chung S, Emili A, Snyder M, Greenblatt JF, Gerstein M (2003) A Bayesian networks approach for predicting protein-protein interactions from genomic data. Science 302:449–453

Jen CH, Manfield IW, Michalopoulos I, Pinney JW, Willats WG, Gilmartin PM, Westhead DR (2006) The *Arabidopsis* co-expression tool (ACT): a WWW-based tool and database for microarray-based gene expression analysis. Plant J 46:336–348

Jensen FV (1997) Introduction to Bayesian Networks. Springer,

Jeong H, Mason SP, Barabasi AL, Oltvai ZN (2001) Lethality and centrality in protein networks. Nature 411:41–42

Jones AM, Thomas V, Bennett MH, Mansfield J, Grant M (2006) Modifications to the *Arabidopsis* defense proteome occur prior to significant transcriptional change in response to inoculation with Pseudomonas syringae. Plant Physiol 142:1603–1620

Kim JT (2001) transsys: A generic formalism for modelling regulatory networks in morphogenesis. In: Kelemen J, Sosík P (eds) Advances in Artificial Life (Proceedings of the 6th European Conference on Artificial Life), vol 2159 of Lecture Notes in Artificial Intelligence. Springer, Berlin Heidelberg New York, pp 242–251

Kim JT (2005) Effects of spatial growth on gene expression dynamics and on regulatory network reconstruction. ECAL 2005, LNAI 3630:825–834

Kinkema M, Fan W, Dong X (2000) Nuclear localization of NPR1 is required for activation of PR gene expression. Plant Cell 12:2339–2350

Klipp E, Liebermeister W (2006) Mathematical modeling of intracellular signalling pathways. BMC Neurosci 7(1):S10

Kobayashi T, Chen L, Aihara K (2002) Design of genetic switches with only positive feedback loops. Proc IEEE Comput Soc Bioinform Conf 1:151–162

Krogan NJ, Cagney G, Yu H, Zhong G, Guo X, Ignatchenko A, Li J, Pu S, Datta N, Tikuisis AP, Punna T, Peregrin-Alvarez JM, Shales M, Zhang X, Davey M, Robinson MD, Paccanaro A, Bray JE, Sheung A, Beattie B, Richards DP, Canadien V, Lalev A, Mena F, Wong P, Starostine A, Canete MM, Vlasblom J, Wu S, Orsi C, Collins SR, Chandran S, Haw R, Rilstone JJ, Gandi K, Thompson NJ, Musso G, St Onge P, Ghanny S, Lam MH, Butland G, Altaf-Ul AM, Kanaya S, Shilatifard A, O'Shea E, Weissman JS, Ingles CJ, Hughes TR, Parkinson J, Gerstein M, Wodak SJ, Emili A, Greenblatt JF (2006) Global landscape of protein complexes in the yeast *Saccharomyces cerevisiae*. Nature 440:637–643

Ihaka R, Gentleman R (1996) R: A language for data analysis and graphics. J Comp Graph Statistics 5:299–314

Li S, Assmann SM, Albert R (2006) Predicting essential components of signal transduction networks: a dynamic model of guard cell abscisic acid signalling. PLoS Biol 4:e312

Liechti R, Gfeller A, Farmer EE (2006) Jasmonate signalling pathway. Sci STKE 2006: cm2

Lorenzo O, Solano R (2005) Molecular players regulating the jasmonate signalling network. Curr Opin Plant Biol 8:532–540

Lu LJ, Xia Y, Paccanaro A, Yu H, Gerstein M (2005) Assessing the limits of genomic data integration for predicting protein networks. Genome Res 15:945–953

Marcotte EM, Pellegrini M, Thompson MJ, Yeates TO, Eisenberg D (1999) A combined algorithm for genome-wide prediction of protein function. Nature 402:83–86

Mewes HW, Frishman D, Mayer KF, Munsterkotter M, Noubibou O, Pagel P, Rattei T, Oesterheld M, Ruepp A, Stumpflen V (2006) MIPS: analysis and annotation of proteins from whole genomes in 2005. Nucleic Acids Res 34:D169–D172

Nemhauser JL, Hong F, Chory J (2006) Different plant hormones regulate similar processes through largely nonoverlapping transcriptional responses. Cell 126:467–475

Nurnberger T, Brunner F, Kemmerling B, Piater L (2004) Innate immunity in plants and animals: striking similarities and obvious differences. Immunol Rev 198:249–266

Obayashi T, Kinoshita K, Nakai K, Shibaoka M, Hayashi S, Saeki M, Shibata D, Saito K, Ohta H (2007) ATTED-II: a database of co-expressed genes and cis elements for identifying co-regulated gene groups in *Arabidopsis*. Nucleic Acids Res 35:D863–D869

Paccanaro A, Casbon JA, Saqi MA (2006) Spectral clustering of protein sequences. Nucleic Acids Res 34:1571–1580

Paccanaro A, Chennubhotla C (2003) Spectral clustering of protein sequences. In: International Joint Conference on Neural Networks – IJCNN. Portland, Oregon, USA

Paccanaro A, Trifonov V, Yu H (2005) Inferring protein–protein interactions using interaction network topologies. In: International Joint Conference on Neural Networks – IJCNN 2005. Montreal, Canada

Parkinson H, Kapushesky M, Shojatalab M, Abeygunawardena N, Coulson R, Farne A, Holloway E, Kolesnykov N, Lilja P, Lukk M, Mani R, Rayner T, Sharma A, William E, Sarkans U, Brazma A (2007) ArrayExpress—a public database of microarray experiments and gene expression profiles. Nucleic Acids Res 35:D747–D750

Rangel C, Angus J, Ghahramani Z, Lioumi M, Sotheran E, Gaiba A, Wild DL, Falciani F (2004) Modeling T-cell activation using gene expression profiling and state-space models. Bioinformatics 20:1361–1372

Rhee SY, Beavis W, Berardini TZ, Chen G, Dixon D, Doyle A, Garcia-Hernandez M, Huala E, Lander G, Montoya M, Miller N, Mueller LA, Mundodi S, Reiser L, Tacklind J, Weems DC, Wu Y, Xu I, Yoo D, Yoon J, Zhang P (2003) The *Arabidopsis* Information Resource (TAIR): a model organism database providing a centralized, curated gateway to *Arabidopsis* biology, research materials and community. Nucleic Acids Res 31:224–228

Rung J, Schlitt T, Brazma A, Freivalds K, Vilo J (2002) Building and analysing genome-wide gene disruption networks. Bioinformatics 18(2):S202–S210

Sasaki-Sekimoto Y, Taki N, Obayashi T, Aono M, Matsumoto F, Sakurai N, Suzuki H, Hirai MY, Noji M, Saito K, Masuda T, Takamiya K, Shibata D, Ohta H (2005) Coordinated activation of metabolic pathways for antioxidants and defence compounds by jasmonates and their roles in stress tolerance in *Arabidopsis*. Plant J 44:653–668

Schwartz MA, Baron V (1999) Interactions between mitogenic stimuli, or, a thousand and one connections. Curr Opin Cell Biol 11:197–202

Staswick PE, Tiryaki I (2004) The oxylipin signal jasmonic acid is activated by an enzyme that conjugates it to isoleucine in *Arabidopsis*. Plant Cell 16:2117–2127

Steffens NO, Galuschka C, Schindler M, Bulow L, Hehl R (2005) AthaMap web tools for database-assisted identification of combinatorial cis-regulatory elements and the display of highly conserved transcription factor binding sites in *Arabidopsis thaliana*. Nucleic Acids Res 33:W397–W402

Steinhauser D, Usadel B, Luedemann A, Thimm O, Kopka J (2004) CSB.DB: a comprehensive systems-biology database. Bioinformatics 20:3647–3651

Stintzi A, Weber H, Reymond P, Browse J, Farmer EE (2001) Plant defense in the absence of jasmonic acid: the role of cyclopentenones. Proc Natl Acad Sci USA 98:12837–12842

Sun TP, Gubler F (2004) Molecular mechanism of gibberellin signalling in plants. Annu Rev Plant Biol 55:197–223

Thimm O, Blasing O, Gibon Y, Nagel A, Meyer S, Kruger P, Selbig J, Muller LA, Rhee SY, Stitt M (2004) MAPMAN: a user-driven tool to display genomics data sets onto diagrams of metabolic pathways and other biological processes. Plant J 37:914–939

Toufighi K, Brady SM, Austin R, Ly E, Provart NJ (2005) The Botany Array Resource: e-Northerns, Expression Angling, and promoter analyses. Plant J 43:153–163

Trevino Santa Cruz MB, Genoud D, Metraux JP, Genoud T (2005) Update in bioinformatics. Toward a digital database of plant cell signalling networks: advantages, limitations and predictive aspects of the digital model. Phytochemistry 66:267–276

Truman W, Bennett MH, Kubigsteltig I, Turnbull C, Grant M (2007) *Arabidopsis* systemic immunity uses conserved defense signalling pathways and is mediated by jasmonates. Proc Natl Acad Sci USA 104:1075–1080

Truman W, de Zabala MT, Grant M (2006) Type III effectors orchestrate a complex interplay between transcriptional networks to modify basal defence responses during pathogenesis and resistance. Plant J 46:14–33

van Dongen S (2000) Graph clustering by flow simulation. PhD Thesis, University of Utrecht

van Someren EP, Wessels LF, Backer E, Reinders MJ (2002) Genetic network modeling. Pharmacogenomics 3:507–525

Vert G, Nemhauser JL, Geldner N, Hong F, Chory J (2005) Molecular mechanisms of steroid hormone signalling in plants. Annu Rev Cell Dev Biol 21:177–201

von Mering C, Krause R, Snel B, Cornell M, Oliver SG, Fields S, Bork P (2002) Comparative assessment of large-scale data sets of protein–protein interactions. Nature 417:399–403

Walker MG, Volkmuth W, Sprinzak E, Hodgson D, Klingler T (1999) Prediction of gene function by genome-scale expression analysis: prostate cancer-associated genes. Genome Res 9:1198–1203

Wang Y et al. (2006) Inferring gene regulatory networks from multiple microarray datasets. Bioinformatics 22:2413–2420

Wang D, Amornsiripanitch N, Dong X (2006) A genomic approach to identify regulatory nodes in the transcriptional network of systemic acquired resistance in plants. PLoS Pathog 2:e123

Wang D, Weaver ND, Kesarwani M, Dong X (2005) Induction of protein secretory pathway is required for systemic acquired resistance. Science 308:1036–1040

Woodward AW, Bartel B (2005) Auxin: regulation, action, and interaction. Ann Bot (Lond) 95:707–735

Xie DX, Feys BF, James S, Nieto-Rostro M, Turner JG (1998) COI1: an *Arabidopsis* gene required for jasmonate-regulated defense and fertility. Science 280:1091–1094

Yu H, Paccanaro A, Trifonov V, Gerstein M (2006) Predicting interactions in protein networks by completing defective cliques. Bioinformatics 22:823–829

Zimmermann P, Hirsch-Hoffmann M, Hennig L, Gruissem W (2004) GENEVESTIGATOR. *Arabidopsis* microarray database and analysis toolbox. Plant Physiol 136:2621–2632

Plant Cell Monogr (10)
L. Bögre and G. Beemster: Plant Growth Signaling
DOI 10.1007/7089_2007_159/Published online: 26 January 2008

Cell Growth Control in an Algal Model

Kateřina Bišová

Institute of Microbiology, Academy of Sciences of the Czech Republic,
Opatovický mlýn, 379 81 Třeboň, Czech Republic
bisova@alga.cz

Abstract Algae provide a useful set of model organisms. The most versatile of them is *Chlamydomonas reinhardtii*, a powerful genetic model with a genome sequence available. This chapter discusses the regulation of cell growth and the cell cycle in the green algae *Chlamydomonas reinhardtii*, *Scenedesmus quadricauda*, and *Ostreococcus tauri*, and the red alga *Cyanidioschyzon merolae*.

1 Introduction

Algae form a diverse polyphyletic group of uni- and multicellular organisms. Traditional taxonomy divides algae into red, brown, and green lineages based on their plastid pigment composition. More recently algae have been divided into two main groups based on the origin of their plastids: primary and secondary algae. The primary algae, or archaeplastida, are comprised of three monophyletic groups: green algae (including land plants), red algae, and glaucophytes, all of which arose from a single endosymbiotic event between a cyanobacterium and a flagellate protist. Secondary algae result from the engulfment of a primary alga by another protist; they include seven groups of single-celled organisms, two having arisen from symbiosis with a green archaeplastid alga (Euglenozoans, Chlorarachniophytes) and the other five having arisen from one or more endosymbiotic events between a red alga and another protist (Cryptophytes, Haptophytes, Dinoflagellates, Heterokonts, Apicomplexans) (for reviews see Bhattacharya et al. 2003; Palmer 2003). While primary algal plastids are surrounded by only two membranes (the outer membrane of the cyanobacterium and a host endosomal membrane), secondary algal plastids are surrounded by three or four membranes and sometimes even contain nucleomorph, a relict of the primary alga nucleus (Chlorarachniophytes, Cryptophytes). Each endosymbiosis is followed by massive gene transfer from the endosymbiont plastid/nucleomorph genome into the host nuclear genome and an introduction of signal peptides (or composite signal peptides) allowing trafficking of plastid proteins from cytoplasm back to the plastids.

Some algal species have been model organisms for decades (*Chlamydomonas reinhardtii*, *Euglena gracilis*, *Scenedesmus quadricauda*, *Volvox car-*

teri). Others have more recently drawn scientists' attention (*Cyanidioschyzon merolae, Ostreococcus tauri, Thalassiosira pseudonana*). Here, I will discuss the merits of a handful of algal species as models for the study of the cell cycle. I will focus on *Chlamydomonas reinhardtii*, the most popular algal model organism used mainly for studies of photosynthesis and flagellar movement but also cell growth and cell cycle regulation. I will also point out the merits of other algal models for the study of the cell cycle.

Scenedesmus quadricauda and *Ostreococcus tauri*, like *Chlamydomonas reinhardtii*, belong to the green algae and *Cyanidioschyzon merolae* is a primary red alga; all therefore represent archaeplastid algae. *Chlamydomonas reinhardtii* and *Scenedesmus quadricauda* are closely related to one another and are also the closest relatives to land plants among the foursome; *Ostreococcus tauri* is evolutionarily placed at the base of the green algae (see below). *Cyanidioschyzon merolae* is a primary red alga that therefore diverged from the other three models earlier in archaeplastidian evolution. While the older models *Chlamydomonas reinhardtii* and *Scenedesmus quadricauda* were chosen based on their abundance in nature, the two recent models *Ostreococcus tauri* and *Cyanidioschyzon merolae* were chosen due to their evolutionary position and their small genomes that could be easily sequenced.

2 The Basic Cell Cycle

The eukaryotic cell cycle has been divided into two "active" phases, DNA replication (S phase) and mitosis (M phase), and two gap phases, G1 and G2 (Howard and Pelc 1953). The G1 phase delimits the previous mitosis from entry into the next S phase, whereas the G2 phase separates the S phase from the subsequent M phase. The progression into each of the active phases is permitted only upon completion of the previous active phase; the major regulatory points in the cell cycle operate at the G1/S and G2/M boundaries. These regulatory points are governed by a class of serine–threonine protein kinases, which require binding of regulatory protein, cyclin, and are therefore designated cyclin-dependent kinases (CDKs).

CDKs are homologs of proteins encoded in fission yeast by the cdc2 locus and in budding yeast by CDC28. In both yeasts only one CDK, possessing a canonical PSTAIRE motif in its cyclin-binding domain, is sufficient to drive the cell cycle (Mendenhall and Hodge 1998; Moser and Russell 2000). However, cell cycle regulation in more complex models requires activity of several CDKs. In humans, there are three PSTAIRE CDKs (CDK1/cdc2, CDK2, and CDK3) and a variant CDK4/6 subfamily with a P(I/L)ST(V/I)RE motif, all of which function in cell cycle regulation (Meyerson et al. 1992; Pines 1995; Reed 1997; Lee and Yang 2003). Higher plants encode two subclasses of CDK involved in the cell cycle regulation, CDKA and CDKB. CDKA, the better characterized of the two, possesses a PSTAIRE motif. CDKA is ubiquitously expressed in dividing tissues (Fobert et al. 1996; Segers et al. 1996) and in suspension cells (Martinez et al. 1992; Hemerly et al. 1993; Magyar et al. 1997; Richard et al. 2001; Sorrell et al. 2001; Menges et al. 2002, 2003; Menges and Murray 2002) and can functionally substitute for its ortholog cdc2/CDC28 in both budding and fission yeast (Ferreira et al. 1991; Hirt et al. 1991). B-type CDKs are unique to plants and algae (Mironov et al. 1999); they are expressed exclusively during G2 and M phase (Fobert et al. 1996; Segers et al. 1996; Magyar et al. 1997; Porceddu et al. 2001; Sorrell et al. 2001; Menges and Murray 2002; Lee et al. 2003; Boudolf et al. 2004). This expression pattern is unique to plant CDKBs and has not been observed for any type of CDKs in other eukaryotes (Dewitte and Murray 2003).

CDKs are activated by binding to the cyclins (Sherr 1994). As the name implies, the cyclins' accumulation levels oscillate during the cell cycle; cyclin accumulation levels are regulated both transcriptionally and posttranslationally. There are three main classes of cyclins: G1, S, and M phase-specific cyclins. G1-specific D-type cyclins are not conserved between animals and plants while S-phase-specific A-type and M-phase-specific B-type cyclins are orthologous in animals and plants (for review see Renaudin et al. 1996; Mironov et al. 1999; Murray 2004).

The activity of the CDK–cyclin complexes is further regulated by binding of the complex by inhibitor proteins and phosphorylation of CDKs themselves (Morgan 1995). The phosphorylation of CDK is mediated by two groups of proteins: cyclin-dependent activating kinases (CAKs) and Wee1 kinases. CAKs are responsible for the activating phosphorylation within the T-loop of the CDKs (Ducommun et al. 1991; Gould et al. 1991). On the contrary, Wee1 kinase executes the inhibitory phosphorylation within the ATP-binding site of CDK (Gould and Nurse 1989; Jin et al. 1996). Phosphorylation by Wee1 enables the inactivation of CDKs until the G2/M transition, when they are abruptly dephosphorylated by Cdc25 phosphatase leading to the activation of CDK–cyclin complexes and triggering of mitosis (Russell and Nurse 1986, 1987; Kumagai and Dunphy 1991).

3 Multiple Fission: A Variation on the Theme

The cells of *Chlamydomonas reinhardtii* and *Scenedesmus quadricauda* divide by a variant of the common cell cycle, multiple fission. *C. reinhardtii*, *S. quadricauda*, and their relatives grown in light undergo a prolonged G1 phase during which they may grow to many (2^n) times their original size. The n is determined by a combination of the growth rate and the species limitations (in most algae n can range from 3 to 5; in some species it can reach 10). With each doubling of size cells attain a size-governed control point, called commitment. Commitment is the formal equivalent of start in budding yeast and the restriction point in mammalian cells (Donnan and John 1983, 1984), which will lead to one round of DNA replication, nuclear and cellular division (Fig. 1).

The multiple fission cell cycle is shared with other algae (Pickett-Heaps 1975) and exists in two different patterns: clustered (*C. reinhardtii* and most of the order *Volvocales*) and consecutive (*S. quadricauda* and most of the genus *Hydrodictyon*) (Šetlík and Zachleder 1984; Zachleder et al. 2002) (Fig. 2).

Green algae are excellent model organisms for cell cycle studies. They are usually unicellular, grow fast, and can be easily synchronized by alternating light/dark periods. The synchrony reached this way is very high, especially for the species dividing by multiple fission. While synchronized plant cell suspension cultures reach only about 60% synchrony, that is, 60% of cells in mitosis occur over an interval of 5–6 h (Nagata et al. 1992; Samuels et al. 1998;

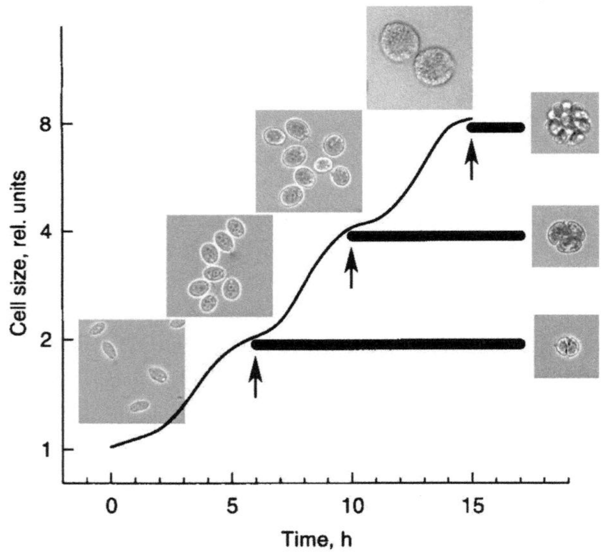

Fig. 1 Schematic illustration of attaining/determination of commitment points to cellular division in synchronous populations of *C. reinhardtii*. The idealized curve represents the growth of cells in continuous light during the cell cycle; at times marked by *arrows*, the subcultures were put into the dark (indicated by *black stripes*). The microphotographs above the curve show typical cells from the synchronized culture at the time of transfer of subcultures into the dark; the vertical lane of microphotographs illustrates the microcolonies of daughter cells released from the mother cell during the corresponding dark interval on agar plates. The moments of transfer into the dark correspond to the attainment of the first (5 h of light), second (10 h of light), and third (15 h of light) commitment points; two, four, and eight cells were released during the dark period, respectively. Reprinted from Vítová and Zachleder (2005) with the permission of the authors and publisher, modified

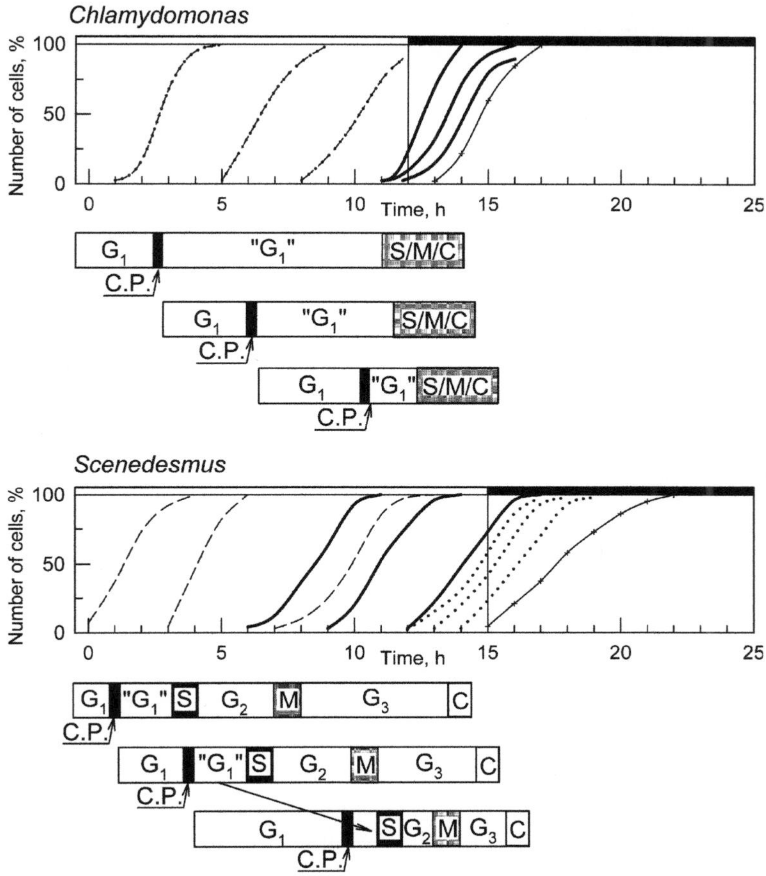

Fig. 2 Schematic illustration of the *Chlamydomonas* (clustered) and *Scenedesmus* (consecutive) patterns of the multiple fission cell cycle. The *graphs* show the fraction of cells that passed commitment point (*dashed lines*), finished nuclear division (*thick solid black lines*) or protoplast division (*dotted lines*), and released daughter cells/coenobia (*solid lines with crosses*). In the case of *Chlamydomonas* the curves representing the protoplast division were omitted because the nuclear and protoplast divisions overlap. Notice accumulation of nuclear divisions in the *Chlamydomonas* clustered cell cycle as opposed to the spreading of the nuclear divisions during the *Scenedesmus* consecutive cell cycle. The *stripes* under the graphs represent individual sequences of the common cell cycle running simultaneously within one multiple fission cell cycle. C.P.: commitment point; G1, S, G2, M: phases of the common cell cycle; "G1": G1-like phase after the cells passed commitment; G3: gap phase separating nuclear and protoplast divisions in *Scenedesmus* cell cycle (Zachleder et al. 1997); C: cytokinesis

Menges and Murray 2002), *C. reinhardtii* cell culture can be synchronized so that > 95% of cells proceed through three rounds of mitosis within 4–5 h. The inhibitors routinely used for synchronization of plant cells provide an additional means to manipulate the outcome of the cell cycle (see below). Of the species described below, *C. reinhardtii* is the best established model system for genetics with an improving molecular toolkit (Harris 2001). Work thus far indicates that regulators of the cell cycle seem to be conserved between algae and land plants but much simpler. Algae can serve not only as a model for higher plant cell cycle regulation, they can also be useful in other fields like the study of the relationship between cell and organellar cycles/division or the study of organellar division.

4 *Chlamydomonas*

4.1 *Chlamydomonas* Basics

C. reinhardtii is a unicellular freshwater green alga with two flagella, one haploid nucleus, and one cup-shaped chloroplast containing one or more pyrenoids. The cell is enclosed within a cell wall consisting mainly of hydroxyproline-rich glycoproteins (Harris 2001). The *C. reinhardtii* genome is available (Merchant et al. 2007). At an estimated 125 Mb it is comparable in size to the genome of *Arabidopsis thaliana*, but possesses significantly fewer gene families, facilitating forward genetic analysis. Both the chloroplast (Maul et al. 2002) and mitochondrial (GenBank accession U03843) genomes have been sequenced.

Wild-type *C. reinhardtii* cells grow phototrophically in minimal media without any carbon supplement. Additionally, they are able to take up acetate as a carbon source for heterotrophic or mixotrophic growth. This has allowed isolation of mutants blocked in photosynthesis, making *C. reinhardtii* an excellent model for the genetic analysis of this fundamental process. *C. reinhardtii* cells are of two mating types, mt+ or mt−. When the cells are starved of nitrogen and exposed to blue light they differentiate into morphologically identical gametes and enter the sexual cycle. Gametes of opposite mating types recognize each other by sex-specific agglutinin proteins on their flagella. Mating pairs adhere to one another's flagella followed by morphological change in the flagellar tips and dissolution of the cell wall by a gamete-specific lytic enzyme. The mating partners then fuse to form a diploid zygote (Harris 2001). In favorable conditions, the mature zygote goes through meiosis and germinates, yielding four haploid progeny. Separation and analysis of the meiotic progeny (tetrad analysis) is the basis of traditional *C. reinhardtii* genetics. Indeed, tetrad analysis was first used in *C. reinhardtii*. The nucleus can be easily transformed using glass beads (Kindle 1990) or electroporation (Shimogawara et al. 1998); chloroplasts (Boynton et al. 1988) and mitochondria (Randolph-Anderson et al. 1993; Remacle et al. 2006) can be specifically transformed by biolistic bombardment. Tetrad analysis, together with the ease and speed of forward genetic screens in haploids and the availability of stable DNA transformation of all three genomes, forms the foundation of *C. reinhardtii* as a genetic model organism.

4.2 The *Chlamydomonas reinhardtii* Cell Cycle

Vegetative *C. reinhardtii* cells grow in a long G1 phase followed by n alternating rounds of S and M phases terminated by cell division into 2^n daughter cells. The alternating rounds of S/M phases occur very rapidly, without G1 or G2 phases, and are punctuated by cell divisions preventing any multinuclear division intermediates (Lien and Knutsen 1976, 1979; Coleman 1982; Craigie and Cavalier-Smith 1982; Donnan and John 1983). Under physiological conditions of alternating light and dark periods the cell cycle becomes synchronized so that growth occurs during the light phase and cell division occurs during the dark phase (Lien and Knutsen 1976, 1979).

It was proposed that the *C. reinhardtii* cell cycle is governed by a combination of a "sizer" and "timer" (Donnan and John 1983, 1984), which was recently confirmed by observation of a single cell (Matsumura et al. 2003). The sizer measures attainment of a commitment size (Spudich and Sager 1980; Donnan and John 1983, 1984; Umen and Goodenough 2001) (Fig. 2), which is followed by a postcommitment period of more or less fixed duration (measured by a timer) preceding the beginning of the S/M cycles (Donnan and John 1983, 1984; Umen and Goodenough 2001; Matsumura et al. 2003). The sizer determines not only when the cells commit but also how many times they will divide. Cells that have passed commitment will complete one round of S phase and mitosis even in the absence of nutrients and exogenous energy (Donnan and John 1983, 1984). On the contrary, cells that have not passed commitment when growth is stopped by withdrawal of light or nutrients do not divide and stay in G1 phase until resupplied. If the light supply continues after the attaining of the first commitment point, cells can attain additional commitment points, each of them allowing one round of DNA replication and mitosis (Fig. 2).

After passing commitment, cells remain in a G1-like phase, the postcommitment period, which lasts for another $\sim$5–8 h and precedes the initiation of the first S/M phase. There is no obvious difference between the pre- and postcommitment period and both of them are therefore usually designated as G1. However, committed cells require neither growth nor energy to divide so the two periods strictly differ physiologically. The postcommitment period is an equivalent of late G1 phase; late G1 phase is the period when the origins of replication become licensed for replication and S phase CDKs become activated (Nasmyth 1996). It has been proposed to call this phase pre-S (pre-synthetic) in algal cell cycles (Zachleder et al. 1997).

Attaining commitment is regulated by growth conditions: higher amounts of light or a longer light period allow more commitment points to be attained (Zachleder et al. 1997). Interestingly, attainment of commitment can also be modulated by light properties. Blue light delays the attainment of commitment leading to a larger cell size at commitment followed by a delay in the cell division and division into more cells (Munzner and Voigt 1992; Oldenhof et al. 2004a,b, 2006), while under red light cells attain commitment faster and divide sooner into fewer cells (Oldenhof et al. 2004a,b, 2006). The attainment of commitment point/s and cell division probably involves active CDKs because CDK activity is correlated with the increase in the amount of p34 protein and its phosphorylation (John et al. 1989) and also with the increase in CDK-like histone H1 kinase activity (Zachleder et al. 1997).

4.3 Cell Cycle Genes

The *C. reinhardtii* genome project has allowed the comprehensive analysis and expression profiling of its core cell cycle genes (Bisova et al. 2005). This work identified clear homologs of cyclins A, B, and D, CDKA, and the plant-specific CDKB. There are also a few highly divergent CDKs and cyclins that are specific to *C. reinhardtii.* In contrast to *Ostreococcus tauri* (see below), no clear homolog of Cdc25 phosphatase was identified. The expression profiling unraveled two different profiles among cell cycle genes: a constitutive profile that was seen for CDKA and CYCD; and a regulated profile with peaks of mRNA amounts at commitment point and cell division, typical for CDKB, CYCA, and CYCB. The overall organization of *C. reinhardtii* cell cycle genes was more plant-like and metazoan-like than are yeasts' cell cycle genes. Conveniently, most cell cycle genes are present in only one copy.

4.4 Cell Cycle Mutants

C. reinhardtii is a powerful genetic model and during the past few decades a number of cell cycle mutants have been recovered (for review see Harper 1999). Only a small portion of these mutations have been mapped; this number will undoubtedly increase with the completion of the *C. reinhardtii* genome sequence.

Recently, two groups of mutants in cell cycle related genes have been mapped. Both show an alteration in daughter cell size. The founding member of the first group is the *mat3–4* mutant described by Umen and Goodenough (2001); mat3–4 cells are tiny compared to wild-type cells. This size phenotype is due to defects in cell size perception and/or regulation; mat3–4 cells attain commitment at a smaller cell size, and for a given mother cell size, mat3–4 cells undergo more rounds of cell division than would wild-type cells (that is, their division number n is larger). The mat3–4 phenotype is caused by a deletion in the *C. reinhardtii* homolog of the retinoblastoma protein. Mutations in E2F and DP protein, downstream of the retinoblastoma protein in other systems, were isolated as suppressors of the mat3–4 phenotype. Both DP and E2F mutants showed cell size alterations; DP1 null mutants show bigger commitment cell size than wild type (Fang et al. 2006).

The second group of size mutants arose from two members of the NIMA kinase family, FA2 and CNK2 (Mahjoub et al. 2002; Bradley and Quarmby 2005; Quarmby and Parker 2005). The *fa2* mutant was isolated in a genetic screen for mutants defective in deflagellation; fa2 cells are bigger than wild-type cells. The commitment size is the same as that of wild type; the cells divide into more cells (the division number is higher) but that is in line with bigger mother cells (Mahjoub et al. 2002). Therefore, it seems that the length of the postcommitment period, rather than cell size perception, leads to cell division later at bigger cell size. CNK2, another member of the NIMA kinase family, is probably involved in the regulation of both the cell size and flagellar length. An increase in the amount of Cnk2p results in small cells and short flagella. Cells with less Cnk2p are larger and have longer flagella than wild type (Bradley and Quarmby 2005).

C. reinhardtii is powerful genetic model. It is also a remarkable model for the study of cell size because it allows one to discriminate between cell size change due to growth and/or alteration of cell cycle progression and changes due to alterations of the sizing mechanism (Umen 2005). Its set of cell cycle genes is complete yet without unnecessary duplications providing for a simple cell cycle model. However, its cell cycle organization differs from the classical G1–S–G2–M pattern. It would therefore be useful to have an organism with cell cycle organization in between the common cell cycle pattern and that of *C. reinhardtii*. Also, *C. reinhardtii* is a very close relative of land plants and it would be interesting to see which of the cell cycle regulators have been conserved in more distant algal species. The other model organisms described below therefore complement *C. reinhardtii* as a model organism.

5 Other Models

5.1 *Scenedesmus quadricauda*

Scenedesmus is a genus of common nonmotile chlorophyte green freshwater alga that has been used for decades as a model organism for the regulation of the cell cycle, photosynthesis, and a variety of toxicity studies. The daughter cells coming from a division of a single mother cell stay connected by a common cell wall in structured clusters called coenobia; in *S. quadricauda* coenobia could be either four- or eight-celled with the terminal cells having two spines. Individual cells contain a nucleus and a single chloroplast. The chloroplast and mitochondrial genomes of another member of the genus *Scenedesmus*, *S. obliquus*, have been sequenced (Kuck et al. 2000; Nedelcu et al. 2000; de Cambiaire et al. 2006).

5.1.1 Cell Cycle

Similarly to *C. reinhardtii*, *S. quadricauda* divides by a multiple fission. However, the mechanism is slightly different (Fig. 2). While *C. reinhardtii* divides by so-called clustered multiple fission, *S. quadricauda* divides by a consecutive multiple fission pattern (see above, Fig. 2) (Šetlík and Zachleder 1984; Zachleder et al. 2002). The *S. quadricauda* cells grow during a G1 phase and attain consecutive commitment points, each of which allows one round of DNA replication and nuclear division to occur. Individual commitment points are with only a short delay followed by DNA replication and nuclear division but not by cell division (Fig. 2). *S. quadricauda* cells are routinely multinuclear during the cell cycle (Fig. 3) with the cell division occurring in several rounds only after all the nuclear divisions have finished.

The *S. quadricauda* cell cycle has been characterized in detail in the work of Zachleder and colleagues (Šetlík and Zachleder 1984; Zachleder et al. 2002; Zachleder and Šetlík 1990). When put in light to allow photosynthetic growth the cells attain a commitment (for a description of commitment see above) at a critical cell size implying the involvement of a sizer. At commitment the processes start which will eventually lead to DNA replication, mitosis, and cell division. The period immediately following commitment could therefore

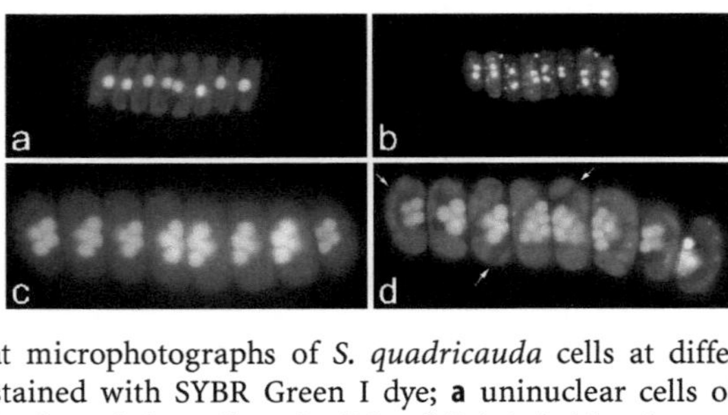

Fig. 3 Fluorescent microphotographs of *S. quadricauda* cells at different stages of the cell cycle. DNA stained with SYBR Green I dye; **a** uninuclear cells of daughter coenobium at the beginning of the cell cycle (0 h of light); **b** binuclear cells of the coenobium after 9 h of light; **c** quadrinuclear cells of the coenobium after 11 h of light; **d** quadri- and octonuclear cells of the coenobium after 13 h of light. The terminal cells are usually smaller and their cell cycle progression is delayed compared to the central cells, here, still with only four nuclei. *Arrows* show where the chloroplasts have already divided; the chloroplast fission will be shortly followed by protoplast division. Photographs are courtesy of M. Vítová

be called the pre-S (pre-synthetic) phase (Zachleder et al. 1997). Depending on the growth rate (e.g., amount of light and length of the light period) the cells can attain additional consecutive commitment point/s before DNA replication and mitosis take place; under optimal conditions the *S. quadricauda* cells can attain four commitment points which will all lead to DNA replication and mitosis. It is an intrinsic property of overlapping reproductive events that the time period from each following commitment point to its corresponding DNA replication and mitosis (S/G2 phases) is shorter than that of the previous one, probably due to the intervening of growth with the division processes in the earlier reproductive events.

A more detailed analysis of commitment point in *S. quadricauda* has shown that it is possible to separate commitment point for DNA replication (S phase) from a commitment point for nuclear division (M phase and S phase), with the former being related to a threshold in the accumulation of total RNA and the latter to a similar threshold in the accumulation of total protein. The two processes can be separated by timely withdrawal of light during which the cells will either only replicate their DNA or replicate DNA and divide their nuclei/cells (Zachleder and Šetlík 1988). Similar commitment/restriction points for G2 phase progression were also described in *Euglena gracilis* (Hagiwara et al. 2001) and in mouse embryonic fibroblasts (Foijer and te Riele 2006).

Alternating light/dark regimes are a way to synchronize algal cultures. However, different light/dark regimes could also be used to manipulate the outcome of the cell cycle in order to get cell cycle patterns differing in the number of attained commitment points and/or length of postcommitment period. Combined with the use of inhibitors of protein synthesis and/or DNA replication, synchronization is a tool providing for a wide range of cell cycle patterns (Bišová et al. 2000; Zachleder et al. 2002; Zachleder and Šetlík 1990). This has been used to modify the number of attained commitment points and mitosis/es and also to alter the timing of nuclear division in relation to the attainment of commitment point. Analysis of histone H1 kinase activity in *S. quadricauda* under these conditions clearly uncovered the presence of at least two different histone H1 (CDK-like) kinase complexes. The activity of one of them correlates with growth and reaches its maximum just before attainment of the commitment point; the activity of the other one is related to mitosis/es (Bišová et al. 2000).

5.1.2 Coordination of Cell and Chloroplast Division Cycles

S. quadricauda has also been extensively used to study the relationship between cell and chloroplast division. The division of a cell and its organelles is coordinated; in plants there are two types of organelles, the division of which has to be coordinated: chloroplasts and mitochondria. Algae with a single chloroplast provide a nice model to uncover the basis of this coordination. Very early in the study of green algae it was observed that there is a very tight correlation between the division of chloroplasts and of cells; the chloroplast divides first, and is followed by cell division. In *S. quadricauda* the numbers of chloroplast and nuclear replications/divisions are also usually equal. However, a detailed analysis has shown that chloroplast DNA replication and the number of nucleoids (chloroplast nuclei) depends on growth rate. Depending on the growth rate there could be both increase or decrease in the number of nucleoids per daughter cell compared to the mother cell (Zachleder and Cepák 1987a,b; Zachleder et al. 1995). More importantly, the processes of chloroplast DNA replication and nucleoid division can be uncoupled from nuclear DNA replication and division by application of 5-fluorodeoxyuridine (Zachleder et al. 1996), an inhibitor of thymidylate synthase that in *S. quadricauda* specifically inhibits nuclear DNA replication and does not affect the chloroplast DNA replication (Zachleder 1994). In the presence of 5-fluorodeoxyuridine replication of nuclear DNA is stopped immediately, while chloroplast DNA replicates, the nucleoids divide, and ultimately the chloroplasts are also able to divide in a giant cell with uninuclear genomic content of DNA. This implies that there is no checkpoint control that interconnects chloroplast and nuclear DNA replication/division. The same phenomenon was also observed in *C. reinhardtii* (Harper and John 1986) and *C. merolae* (Itoh et al. 1996). Therefore, it seems that despite a strikingly tight correlation between chloroplast and nuclear division cycles there is no regulatory checkpoint interconnecting both processes.

I propose the following hypothetical model to explain the correlation between chloroplast and nuclear cycles. In phototrophically growing green algae photosynthesis leads primarily to the growth of chloroplast which may trigger chloroplast DNA replication/division. Since chloroplast occupies the majority of the cell volume, its growth directly affects the cell size which in turn leads to attaining of the commitment point and to nuclear DNA replication and division.

S. quadricauda has proven to be a useful model for unraveling the relationship between different events of the multiple fission cell cycle and also between cell and chloroplast division cycles. Unfortunately, no efforts have been made so far to establish genetic tools and/or stable DNA transformation techniques or to sequence its nuclear genome. Since *C. reinhardtii* and *S. quadricauda* are close relatives, one could assume that methods used in *C. reinhardtii* should work for *S. quadricauda*. *S. quadricauda* is definitely a potent biochemical model for cell cycle studies. Since its cell cycle organization is slightly different from that of *C. reinhardtii*, it would be interesting to see how the cell cycle is regulated genetically.

5.2
Ostreococcus tauri

The marine green alga *Ostreococcus tauri* was isolated only recently (in 1994); with a diameter of about 1 μm it is the smallest free-living eukaryote known (Courties et al. 1994). It is a member of an ancient group of green algae, Prasinophyceae, that dominated the Paleozoic oceans but are now only minor contributors to marine productivity. Phylogenetically, the Prasinophyceae are positioned at the base of the primary green algal clade. *O. tauri* cells lack cell walls, are nonflagellated, and contain only a single mitochondrion and chloroplast (Chrétiennot-Dinet et al. 1995). These properties together with its evolutionary position attracted an interest which yielded the completion of sequencing of *O. tauri*'s strikingly small (approx. 13 Mb) genome (Derelle et al. 2006).

Since *O. tauri* is a promising new algal model, efforts have been made to establish/improve both the cultivation conditions and conditions for synchronizing the cultures (Courties et al. 1998; Farinas et al. 2006). *O. tauri* divides by binary fission (in contrast to the multiple fission of *C. reinhardtii* and *S. quadricauda*) with chloroplast dividing first during S phase, followed by division of the mitochondrion and nucleus (Farinas et al. 2006). Like *C. reinhardtii* and *S. quadricauda*, *O. tauri* can be synchronized by alternating light/dark regimes. Due to its binary fission cell cycle, synchronization is far more difficult (maximum of 35% of cells in S phase). However, an increase in the cell synchrony can be achieved by application of hydroxyurea, aphidicolin, or propyzamide (Farinas et al. 2006).

5.2.1
Cell Cycle Genes

The sequencing of the *O. tauri* genome has allowed the analysis of core cell cycle genes. This analysis revealed that, like *C. reinhardtii*, the *O. tauri* genome encodes one homolog of each group of CDKs and cyclins including a plant-specific CDK of B-type (Robbens et al. 2005). The expression analysis of some cell cycle genes showed that CDKA, CYCA, and CYCD are transcribed ubiquitously during the cell cycle while mRNA for CDKB and CYCB accumulate only during S and M phases (Corellou et al. 2005; Farinas et al. 2006). The expression pattern therefore mimics that of the corresponding homologs in *Arabidopsis thaliana* and *C. reinhardtii* (Segers et al. 1996; Mironov et al. 1999; Bisova et al. 2005). *O. tauri* CDKB has been characterized in more detail (Corellou et al. 2005). This study has shown that CDKB protein is present only during S and M phases and CDKB (and not CDKA) is also regulated by tyrosine phosphorylation. In line with the nature of tyrosine phosphorylation in other systems, this modification has an inhibitory effect and is removed prior to mitosis leading to the activation of CDKB/cyclin complex. CDKB histone H1 kinase activity is about tenfold higher than that of CDKA; also CDKB is responsible for the main peak of mitotic activity, while CDKA has a more steady-state activity that probably accounts for the regulation of S phase. Therefore it seems that in *O. tauri*, CDKB is the main mitotic kinase while in higher plants this type of kinase has evolved to fill in more specialized functions outside the cell cycle (Boudolf et al. 2004). It would be interesting to see what is the function of CDKB in *C. reinhardtii* to discriminate which of these two scenarios represents the ancestral state.

O. tauri is interesting in another aspect of the cell cycle regulation because it is the first member of the green lineage to encode a functional Cdc25 phosphatase (Khadaroo et al. 2004). While CDKs, cyclins, and Wee1 kinases are present in all eukaryotes, there is no distinguishable homolog of Cdc25 phosphatase in higher plants, a fact that has been puzzling researchers in the field for years. The *O. tauri* genome encodes a protein whose sequence shows clear similarity to Cdc25 phosphatases from yeast and animals. Moreover, this protein complements a cdc25 mutation in *Schizosaccharomyces pombe* and activates CDK1/cyclin B complexed from starfish oocytes both in vitro and in vivo (Khadaroo et al. 2004). The sequence of *O. tauri* Cdc25 phosphatase shares Cdc25 features in its C terminus but diverges from yeast and animal homologs in its N terminus. Using *O. tauri*'s Cdc25 a putative Cdc25-like protein can be found in *C. reinhardtii* (Khadaroo et al. 2004; Bisova et al. 2005) and also in *Arabidopsis* (Landrieu et al. 2004a,b); however, it will require a functional verification to prove that any of these putative candidates have Cdc25 phosphatase activity.

As mentioned above, *O. tauri* is a promising new model for cell cycle studies. The main advantage of this system is its evolutionary position at the base of green lineage. *O. tauri* has already proven its usefulness as an intermediate model between plants and fungi/animals. Although no stable DNA transformation has been reported, one can assume that, as in the case of *S. quadricauda*, techniques used for *C. reinhardtii* or *C. merolae* (see below) may be successful.

5.3 *Cyanidioschyzon merolae*

Another new rising model among algal species is the primary red alga *Cyanidioschyzon merolae. C. merolae* is another small unicellular organism (diameter 1.5 μm). It lives in sulfate-rich hot springs (pH 1.5, 45 °C) (De Luca et al. 1978, as cited in Misumi et al. 2005). Similarly to *O. tauri*, *C. merolae* is wall-less and contains single mitochondrion, chloroplast, and nucleus. At 17 Mb, its nuclear genome is slightly bigger than that of *O. tauri* but still very small; a completed genomic sequence is available (Matsuzaki et al. 2004), and the chloroplast and mitochondrial genomes are also completed (Ohta et al. 1998, 2003).

C. merolae divides by binary fission with chloroplast dividing first, followed by mitochondrion and nucleus. The cell cycle can be synchronized by light/dark cycles to a degree similar to *O. tauri* cells (approx. 40% dividing cell at one time point). Synchrony can be increased by the use of aphidicolin which blocks nuclear and cellular division (but not chloroplast division) (Itoh et al. 1996). On the contrary, propyzamide and nocodazole have no effect on the cell cycle progression (Terui et al. 1995). Recently, a protocol for nuclear DNA transformation by homologous recombination has been reported (Minoda et al. 2004). It is the first case of a DNA transformation by homologous recombination of the nuclear genome being reported in a unicellular alga. This technique will undoubtedly speed up the reverse genetics of this model (Minoda et al. 2004).

5.3.1 Cell Cycle Genes

The *C. merolae* genome encodes homologs of both A- and B-type CDKs, G1 and G2/M phase cyclins, Wee1 kinase, and retinoblastoma protein (http://merolae.biol.s.u-tokyo.ac.jp/). The set of cell cycle regulators therefore resembles that of *C. reinhardtii* and *O. tauri*. The presence of a B-type CDK implies that this plant-specific family of CDKs was present in the common ancestor of the primary algae.

5.3.2 Organellar Division

C. merolae has been largely used as a model for the study of division of both mitochondrion and chloroplast (Kuroiwa et al. 1995, 1998; Miyagishima et al. 1999, 2001a,b,c, 2004; Takahara et al. 2000; Nishida et al. 2003, 2004; Yoshida et al. 2006). Mitochondria in higher plants and animals divide by the combined action of mitochondrion division and dynamin rings. On the contrary, chloroplasts in higher plants divide by the combined action of protoplast division and FtsZ rings. It has therefore been proposed that during evolution dynamin replaced FtsZ in the mitochondrial division (Erickson 2000; Margolin 2000). However, the *C. merolae* mitochondrion divides by a combined action of mitochondrion division, dynamin, and FtsZ rings (Beech et al. 2000; Takahara et al. 2000; Nishida et al. 2003). Similarly, the *C. merolae* chloroplast divides by a combined action of protoplast division, FtsZ, and dynamin rings (Miyagishima et al. 2003, 2004; Misumi et al. 2005) implying that the organization is ancestral and *C. merolae* is the only model organism that retains the ancestral set of organellar division proteins.

With a completed, compact genome sequence and nuclear transformation by homologous recombination available, *C. merolae* is a promising model organism. It has already proven its usefulness for the study of organellar division. It will be interesting to see whether it can also become a model for the study of the cell cycle, which would no doubt be attractive due to its unique evolutionary position right after the split of primary red and green lineages (including land plants). As discussed above, algae offer quite a few model organisms for the study of different aspects of cell growth and cell cycle regulation. In the three models with sufficient genomic information, the sets of cell cycle genes are very similar. The genomes encode for a complete set of cell cycle genes including plant-specific B-type CDK. Conveniently, most of these genes are present in single copy. This implies that (1) B-type CDKs were present in the ancestor of primary algae very early after the acquisition of chloroplast or even before it, (2) the cell cycle regulation of these unicellular organisms is more complicated than that of yeasts, and (3) the complex cell cycle regulation by gene families seen in land plants is not necessary for coordination of cell and chloroplast divisions.

All of the discussed models would be valuable in the characterization of plant cell cycle regulation and more specifically in unraveling why plant-specific B-type CDK evolved. We can also expect some more insight into the coordination of the cell cycle with the division cycles of mitochondria and chloroplast.

Acknowledgements This work was supported by the Grant Agency of the Czech Republic (grant 204/06/0102), the Grant Agency of the Academy of Sciences of the Czech Republic (grant IAA500200614) and the Institutional Research Concept (no. AV0Z50200510).

References

Beech PL, Nheu T, Schultz T, Herbert S, Lithgow T, Gilson PR, McFadden GI, Lutkenhaus J (2000) Mitochondrial FtsZ in a chromophyte alga. Science 287:1276–1279

Bhattacharya D, Yoon HS, Hackett JD (2003) Photosynthetic eukaryotes unite: endosymbiosis connects the dots. Bioessays 26:50–60

Bišová K, Vítová M, Zachleder V (2000) The activity of total histone H1 kinases is related to growth and commitment points while the p13(suc1)-bound kinase activity relates to mitoses in the alga *Scenedesmus quadricauda*. Plant Physiol Biochem 38:755–764

Bišová K, Krylov DM, Umen JG (2005) Genome-wide annotation and expression profiling of cell cycle regulatory genes in *Chlamydomonas reinhardtii*. Plant Physiol 137:475–491

Boudolf V, Vlieghe K, Beemster GTS, Magyar Z, Acosta JAT, Maes S, Van Der Schueren E, Inzé D, de Veylder L (2004) The plant-specific cyclin-dependent kinase CDKB1;1 and transcription factor E2Fa-DPa control the balance of mitotically dividing and endoreduplicating cells in *Arabidopsis*. Plant Cell 16:2683–2692

Boynton J, Gillham N, Harris E, Hosler J, Johnson A, Jones A, Randolph-Anderson B, Robertson D, Klein T, Shark K (1988) Chloroplast transformation in *Chlamydomonas* with high velocity microprojectiles. Science 240:1534–1538

Bradley BA, Quarmby LM (2005) A NIMA-related kinase, Cnk2p, regulates both flagellar length and cell size in *Chlamydomonas*. J Cell Sci 118:3317–3326

Chrétiennot-Dinet MJ, Courties C, Vaquer A, Neveux J, Claustre H, Lautier J, Machado MC (1995) A new marine picoeucaryote—*Ostreococcus tauri* gen. and sp. nov. (Chlorophyta, Prasinophyceae). Phycologia 34:285–292

Coleman AW (1982) The nuclear-cell cycle in *Chlamydomonas* (Chlorophyceae). J Phycol 18:192–195

Corellou F, Camasses A, Ligat L, Peaucellier G, Bouget FY (2005) Atypical regulation of a green lineage-specific B-type cyclin-dependent kinase. Plant Physiol 138:1627–1636

Courties C, Vaquer A, Troussellier M, Lautier J, Chretiennotdinet MJ, Neveux J, Machado C, Claustre H (1994) Smallest eukaryotic organism. Nature 370:255

Courties C, Perasso R, Chretiennot-Dinet MJ, Gouy M, Guillou L, Troussellier M (1998) Phylogenetic analysis and genome size of *Ostreococcus tauri* (Chlorophyta, Prasinophyceae). J Phycol 34:844–849

Craigie RA, Cavalier-Smith T (1982) Cell volume and the control of the *Chlamydomonas* cell cycle. J Cell Sci 54:173–191

de Cambiaire J-C, Otis C, Lemieux C, Turmel M (2006) The complete chloroplast genome sequence of the chlorophycean green alga *Scenedesmus obliquus* reveals a compact gene organization and a biased distribution of genes on the two DNA strands. BMC Evol Biol 6:37

De Luca P, Taddei R, Varano L (1978) *Cyanidioschyzon merolae*: a new alga of thermal acidic environments. Webbia 33:37–44

Derelle E, Ferraz C, Rombauts S, Rouze P, Worden AZ, Robbens S, Partensky F, Degroeve S, Echeynie S, Cooke R, Saeys Y, Wuyts J, Jabbari K, Bowler C, Panaud O, Piegu B, Ball SG, Ral J-P, Bouget F-Y, Piganeau G, De Baets B, Picard A, Delseny M, Demaille J, Van de Peer Y, Moreau H (2006) Genome analysis of the smallest free-living eukaryote *Ostreococcus tauri* unveils many unique features. Proc Natl Acad Sci USA 103:11647–11652

Dewitte W, Murray JA (2003) The plant cell cycle. Annu Rev Plant Biol 54:235–264

Donnan L, John PCL (1983) Cell cycle control by timer and sizer in *Chlamydomonas.* Nature 304:630–633

Donnan L, John PCL (1984) Timer and sizer controls in the cell cycles of *Chlamydomonas* and *Chlorella.* In: Streiblová E (ed) Microbial cell cycle. CRC, Boca Raton, pp 231–251

Ducommun B, Brambilla P, Felix MA, Franza BR, Karsenti E, Draetta G (1991) Cdc2 phosphorylation is required for its interaction with cyclin. EMBO J 10:3311–3319

Erickson HP (2000) Dynamin and FtsZ: missing links in mitochondrial and bacterial division. J Cell Biol 148:1103–1105

Fang SC, de la Reyes C, Umen JG (2006) Cell size checkpoint control by the retinoblastoma tumor suppressor pathway. PLos Genet 2:e167

Farinas B, Mary C, de O Manes C-L, Bhaud Y, Peaucellier G, Moreau H (2006) Natural synchronisation for the study of cell division in the green unicellular alga *Ostreococcus tauri.* Plant Mol Biol 60:277–292

Ferreira P, Hemerly AS, Villarroel R, Van Montagu M, Inzé D (1991) The *Arabidopsis* functional homolog of the p34cdc2 protein kinase. Plant Cell 3:531–540

Fobert PR, Gaudin V, Lunness P, Coen ES, Doonan JH (1996) Distinct classes of cdc2-related genes are differently expressed during the cell division cycle in plants. Cell 8:1465–1476

Foijer F, te Riele H (2006) Restriction beyond the restriction point: mitogen requirement for G2 passage. Cell Div 1:8

Gould KL, Nurse P (1989) Tyrosine phosphorylation of the fission yeast Cdc2+ protein kinase regulates entry into mitosis. Nature 342:39–45

Gould KL, Moreno S, Owen DJ, Sazer S, Nurse P (1991) Phosphorylation at Thr167 is required for *Schizosaccharomyces pombe* p34cdc2 function. EMBO J 10:3297–3309

Hagiwara S, Takahashi M, Yamagishi A, Zhang YL, Goto K (2001) Novel findings regarding photoinduced commitments of G1-, S- and G2-phase cells to cell-cycle transitions in darkness and dark-induced G1-, S- and G2-phase arrests in *Euglena*. Photochem Photobiol 74:726–733

Harper J (1999) *Chlamydomonas* cell cycle mutants. Int Rev Cytol 189:131

Harper JDI, John PCL (1986) Coordination of division events in the *Chlamydomonas* cell cycle. Protoplasma 131:118–130

Harris EH (2001) *Chlamydomonas* as a model organism. Annu Rev Plant Physiol Plant Mol Biol 52:363–406

Hemerly AS, Ferreira P, de Almeida Engler J, Van Montagu M, Engler G, Inzé D (1993) cdc2a expression in *Arabidopsis* is linked with competence for cell division. Plant Cell 5:1711–1723

Hirt H, Pay A, Györgyey J, Bakó L, Neméth K, Bögre L, Schweyen RJ, Heberle-Bors E, Dudits D (1991) Complementation of a yeast cell cycle mutant by an alfalfa cDNA encoding a protein kinase homologous to p34cdc2. Proc Natl Acad Sci USA 88:1636–1640

Howard A, Pelc SR (1953) Synthesis of deoxyribonucleic acid in normal and irradiated cells and its relation to chromosome breakage. Heredity Suppl 6:261–273

Itoh R, Takahashi H, Toda K, Kuroiwa H, Kuroiwa T (1996) Aphidicolin uncouples the chloroplast division cycle from the mitotic cycle in the unicellular red alga *Cyanidioschyzon merolae*. Eur J Cell Biol 71:303–310

Jin P, Gu Y, Morgan DO (1996) Role of inhibitory CDC2 phosphorylation in radiation-induced G2 arrest in human cells. J Cell Biol 134:963–970

John PC, Sek FJ, Lee MG (1989) A homolog of the cell cycle control protein p34cdc2 participates in the division cycle of *Chlamydomonas*, and a similar protein is detectable in higher plants and remote taxa. Plant Cell 1:1185–1193

Kathir P, LaVoie M, Brazelton WJ, Haas NA, Lefebvre PA, Silflow CD (2003) Molecular map of the *Chlamydomonas reinhardtii* nuclear genome. Eukaryot Cell 2:362–379

Khadaroo B, Robbens S, Ferraz C, Derelle E, Eychenie S, Cooke R, Peaucellier G, Delseny M, Demaille J, Van de Peer Y, Picard A, Moreau H (2004) The first green lineage cdc25 dual-specificity phosphatase. Cell Cycle 3:513–518

Kindle KL (1990) High-frequency nuclear transformation of *Chlamydomonas reinhardtii*. Proc Natl Acad Sci USA 87:1228–1232

Kuck U, Jekosch K, Holzamer P (2000) DNA sequence analysis of the complete mitochondrial genome of the green alga *Scenedesmus obliquus*: evidence for UAG being a leucine and UCA being a non-sense codon. Gene 253:13–18

Kumagai A, Dunphy WG (1991) The Cdc25 protein controls tyrosine dephosphorylation of the Cdc2 protein in a cell-free system. Cell 64:903–914

Kuroiwa T, Suzuki K, Itou R, Toda K, O'Keefe TC, Kuroiwa H (1995) Mitochondria-dividing ring: ultrastructural basis for the mechanism of mitochondrial division in *Cyanidioschyzon merolae*. Protoplasma 186:12–23

Kuroiwa T, Kuroiwa H, Sakai A, Takahashi H, Toda K, Itoh R (1998) The division apparatus of plastids and mitochondria. Int Rev Cytol 181:1–41

Landrieu I, da Costa M, de Veylder L, Dewitte F, Vandepoele K, Hassan S, Wieruszeski J-M, Faure J-D, Van Montagu M, Inzé D, Lippens G (2004a) A small CDC25 dual-specificity tyrosine-phosphatase isoform in *Arabidopsis thaliana*. Proc Natl Acad Sci USA 101:13380–13385

Landrieu I, Hassan S, Sauty M, Dewitte F, Wieruszeski JM, Inzé D, de Veylder L, Lippens G (2004b) Characterization of the *Arabidopsis thaliana* Arath;CDC25 dual-specificity tyrosine phosphatase. Biochem Biophys Res Commun 322:734–739

Lee J, Das A, Yamaguchi M, Hashimoto J, Tsutsumi N, Uchimiya H, Umeda M (2003) Cell cycle function of a rice B2-type cyclin interacting with a B-type cyclin-dependent kinase. Plant J 34:417–425

Lee MH, Yang HY (2003) Regulators of G1 cyclin-dependent kinases and cancers. Cancer Metastasis Rev 22:435–449

Lien T, Knutsen G (1976) Synchronized cultures of a cell wall-less mutant of *Chlamydomonas reinhardtii.* Arch Microbiol 108:189–194

Lien T, Knutsen G (1979) Synchronous growth of *Chlamydomonas reinhardtii* (Chlorophyceae): a review of optimal conditions. J Phycol 15:191–200

Magyar Z, Meszaros T, Miskolczi P, Deak M, Feher A, Brown S, Kondorosi E, Athanasiadis A, Pongor S, Bilgin M, Bako L, Koncz C, Dudits D (1997) Cell cycle phase specificity of putative cyclin-dependent kinase variants in synchronized alfalfa cells. Plant Cell 9:223–235

Mahjoub MR, Montpetit B, Zhao L, Finst RJ, Goh B, Kim AC, Quarmby LM (2002) The FA2 gene of *Chlamydomonas* encodes a NIMA family kinase with roles in cell cycle progression and microtubule severing during deflagellation. J Cell Sci 115:1759–1768

Margolin W (2000) Organelle division: self-assembling GTPase caught in the middle. Curr Biol 10:R328–R330

Martinez MC, Jorgensen JE, Lawton MA, Lamb CJ, Doerner PW (1992) Spatial pattern of cdc2 expression in relation to meristem activity and cell proliferation during plant development. Proc Natl Acad Sci USA 89:7360–7364

Matsumura K, Yagi T, Yasuda K (2003) Role of timer and sizer in regulation of *Chlamydomonas* cell cycle. Biochem Biophys Res Commun 306:1042–1049

Matsuzaki M, Misumi O, Shin-i T, Maruyama S, Takahara M, Miyagishima S, Mori T, Nishida K, Yagisawa F, Nishida K, Yoshida Y, Nishimura Y, Nakao S, Kobayashi T, Momoyama Y, Higashiyama T, Minoda A, Sano M, Nomoto H, Oishi K, Hayashi H, Ohta F, Nishizaka S, Haga S, Miura S, Morishita T, Kabeya Y, Terasawa K, Suzuki Y, Ishii Y, Asakawa S, Takano H, Ohta N, Kuroiwa H, Tanaka K, Shimizu N, Sugano S, Sato N, Nozaki H, Ogasawara N, Kohara Y, Kuroiwa T (2004) Genome sequence of the ultrasmall unicellular red alga *Cyanidioschyzon merolae* 10D. Nature 428:653–657

Maul JE, Lilly JW, Cui L, de Pamphilis CW, Miller W, Harris EH, Stern DB (2002) The *Chlamydomonas reinhardtii* plastid chromosome: islands of genes in a sea of repeats. Plant Cell 14:2659–2679

Mendenhall MD, Hodge AE (1998) Regulation of Cdc28 cyclin-dependent protein kinase activity during the cell cycle of the yeast *Saccharomyces cerevisiae.* Microbiol Mol Biol Rev 62:1191–1243

Menges M, Murray JAH (2002) Synchronous *Arabidopsis* suspension cultures for analysis of cell-cycle gene activity. Plant J 30:203–212

Menges M, Hennig L, Gruissem W, Murray JAH (2002) Cell cycle-regulated gene expression in *Arabidopsis.* J Biol Chem 277:41987–42002

Menges M, Hennig L, Gruissem W, Murray JA (2003) Genome-wide gene expression in an *Arabidopsis* cell suspension. Plant Mol Biol 53:423–442

Merchant SS, Prochnik SE, Vallon O, Harris EH, Karpowicz SJ, Witman GB, Terry A, Salamov A, Fritz-Laylin LK, Marechal-Drouard L, Marshall WF, Qu L-H, Nelson DR, Sanderfoot AA, Spalding MH, Kapitonov VV, Ren Q, Ferris P, Lindquist E, Shapiro H, Lucas SM, Grimwood J, Schmutz J, Cardol P, Cerutti H, Chanfreau G, Chen C-L, Cognat V, Croft MT, Dent R, Dutcher S, Fernandez E, Fukuzawa H, Gonzalez-Ballester D, Gonzalez-

Halphen D, Hallmann A, Hanikenne M, Hippler M, Inwood W, Jabbari K, Kalanon M, Kuras R, Lefebvre PA, Lemaire SD, Lobanov AV, Lohr M, Manuell A, Meier I, Mets L, Mittag M, Mittelmeier T, Moroney JV, Moseley J, Napoli C, Nedelcu AM, Niyogi K, Novoselov SV, Paulsen IT, Pazour G, Purton S, Ral J-P, Riano-Pachon DM, Riekhof W, Rymarquis L, Schroda M, Stern D, Umen J, Willows R, Wilson N, Zimmer SL, Allmer J, Balk J, Bisova K, Chen C-J, Elias M, Gendler K, Hauser C, Lamb MR, Ledford H, Long JC, Minagawa J, Page MD, Pan J, Pootakham W, Roje S, Rose A, Stahlberg E, Terauchi AM, Yang P, Ball S, Bowler C, Dieckmann CL, Gladyshev VN, Green P, Jorgensen R, Mayfield S, Mueller-Roeber B, Rajamani S, Sayre RT, Brokstein P, Dubchak I, Goodstein D, Hornick L, Huang YW, Jhaveri J, Luo Y, Martinez D, Ngau WCA, Otillar B, Poliakov A, Porter A, Szajkowski L, Werner G, Zhou K, Grigoriev IV, Rokhsar DS, Grossman AR (2007) The *Chlamydomonas* genome reveals the evolution of key animal and plant functions. Science 318:245–250

Meyerson M, Enders GH, Wu CL, Su LK, Gorka C, Nelson C, Harlow E, Tsai LH (1992) A family of human cdc2-related protein kinases. EMBO J 11:2909–2917

Minoda A, Sakagami R, Yagisawa F, Kuroiwa T, Tanaka K (2004) Improvement of culture conditions and evidence for nuclear transformation by homologous recombination in a red alga, *Cyanidioschyzon merolae* 10D. Plant Cell Physiol 45:667–671

Mironov V, de Veylder L, Van Montagu M, Inzé D (1999) Cyclin-dependent kinases and cell division in plants—the nexus. Plant Cell 11:509–529

Misumi O, Matsuzaki M, Nozaki H, Miyagishima S, Mori T, Nishida K, Yagisawa F, Yoshida Y, Kuroiwa H, Kuroiwa T (2005) *Cyanidioschyzon merolae* genome. A tool for facilitating comparable studies on organelle biogenesis in photosynthetic eukaryote. Plant Physiol 137:567–585

Miyagishima S, Itoh R, Aita S, Kuroiwa H, Kuroiwa T (1999) Isolation of dividing chloroplasts with intact plastid-dividing rings from a synchronous culture of the unicellular red alga *Cyanidioschyzon merolae*. Planta 209:371–375

Miyagishima S, Kuroiwa H, Kuroiwa T (2001a) The timing and manner of disassembly of the apparatuses for chloroplast and mitochondrial division in the red alga *Cyanidioschyzon merolae*. Planta 212:517–528

Miyagishima S, Takahara M, Kuroiwa T (2001b) Novel filaments 5 nm in diameter constitute the cytosolic ring of the plastid division apparatus. Plant Cell 13:707–721

Miyagishima S, Takahara M, Mori T, Kuroiwa H, Higashiyama T, Kuroiwa T (2001c) Plastid division is driven by a complex mechanism that involves differential transition of the bacterial and eukaryotic division rings. Plant Cell 13:2257–2268

Miyagishima S, Nishida K, Mori T, Matsuzaki M, Higashiyama T, Kuroiwa H, Kuroiwa T (2003) A plant-specific dynamin-related protein forms a ring at the chloroplast division site. Plant Cell 16:655–665

Miyagishima S, Nozaki H, Nishida K, Matsuzaki M, Kuroiwa T (2004) Two types of FtsZ proteins in mitochondria and red-lineage chloroplasts: The duplication of FtsZ is implicated in endosymbiosis. J Mol Evol 58:291–303

Morgan DO (1995) Principles of CDK regulation. Nature 374:131–134

Moser BA, Russell P (2000) Cell cycle regulation in *Schizosaccharomyces pombe*. Curr Opin Microbiol 3:631–636

Munzner P, Voigt J (1992) Blue light regulation of cell division in *Chlamydomonas reinhardtii*. Plant Physiol 99:1370–1375

Murray A (2004) Recycling the cell cycle: cyclins revisited. Cell 116:221–234

Nagata T, Nemoto Y, Hasezawa S (1992) Tobacco BY-2 cell line as the HeLa cell in the cell biology of higher plants. Int Rev Cytol 132:1–30

Nasmyth K (1996) Viewpoint: putting the cell cycle in order. Science 274:1643–1645

Nedelcu AM, Lee RW, Lemieux C, Gray MW, Burger G (2000) The complete mitochondrial DNA sequence of *Scenedesmus obliquus* reflects an intermediate stage in the evolution of the green algal mitochondrial genome. Genome Res 10:819–831

Nishida K, Takahara M, Miyagishima S, Kuroiwa H, Matsuzaki M, Kuroiwa T (2003) Dynamic recruitment of dynamin for final mitochondrial severance in a primitive red alga. Proc Natl Acad Sci USA 100:2146–2151

Nishida K, Misumi O, Yagisawa F, Kuroiwa H, Nagata T, Kuroiwa T (2004) Triple immunofluorescent labeling of FtsZ, dynamin, and EF-Tu reveals a loose association between the inner and outer membrane mitochondrial division machinery in the red alga *Cyanidioschyzon merolae.* J Histochem Cytochem 52:843–849

Ohta N, Sato N, Kuroiwa T (1998) Structure and organization of the mitochondrial genome of the unicellular red alga *Cyanidioschyzon merolae* deduced from the complete nucleotide sequence. Nucleic Acids Res 26:5190–5198

Ohta N, Matsuzaki M, Misumi O, Miyagishima S, Nozaki H, Tanaka K, Shin IT, Kohara Y, Kuroiwa T (2003) Complete sequence and analysis of the plastid genome of the unicellular red alga *Cyanidioschyzon merolae.* DNA Res 10:67–77

Oldenhof H, Bišová K, van den Ende H, Zachleder V (2004a) Effect of red and blue light on the timing of cyclin-dependent kinase activity and the timing of cell division in *Chlamydomonas reinhardtii.* Plant Physiol Biochem 42:341–348

Oldenhof H, Zachleder V, van den Ende H (2004b) Blue light delays commitment to cell division in *Chlamydomonas reinhardtii.* Plant Biol 6:689–695

Oldenhof H, Zachleder V, van den Ende H (2006) Blue- and red-light regulation of the cell cycle in *Chlamydomonas reinhardtii* (Chlorophyta). Eur J Phycol 41:313–320

Palmer JD (2003) The symbiotic birth and spread of plastids: how many times and whodunit? J Phycol 39:4–12

Pickett-Heaps JD (1975) Green algae: structure, reproduction and evolution in selected genera. Sinauer, Sunderland

Pines J (1995) Cyclins and cyclin-dependent kinases: a biochemical view. Biochem J 308:697–711

Porceddu A, Stals H, Reichheldt JP, Segers G, de Veylder L, Barroco RD, Casteels P, Van Montagu M, Inzé D, Mironov V (2001) A plant-specific cyclin-dependent kinase is involved in the control of G(2)/M progression in plants. J Biol Chem 276:36354–36360

Quarmby LM, Parker JDK (2005) Cilia and the cell cycle? J Cell Biol 169:707–710

Randolph-Anderson B, Boynton J, Gillham N, Harris E, Johnson A, Dorthu M, Matagne R (1993) Further characterization of the respiratory deficient dum-1 mutation of *Chlamydomonas reinhardtii* and its use as a recipient for mitochondrial transformation. Mol Gen Genet 236:235–244

Reed SI (1997) Control of the G1/S transition. Cancer Surv 29:7–23

Remacle C, Cardol P, Coosemans N, Gaisne M, Bonnefoy N (2006) High-efficiency biolistic transformation of *Chlamydomonas* mitochondria can be used to insert mutations in complex I genes. Proc Natl Acad Sci USA 103:4771–4776

Renaudin J, Doonan J, Freeman D, Hashimoto J, Hirt H, Inze D, Jacobs T, Kouchi H, Rouze P, Sauter M, Savoure A, Sorrell D, Sundaresan V, Murray J (1996) Plant cyclins: a unified nomenclature for plant A-, B- and D-type cyclins based on sequence organization. Plant Mol Biol 32:1003–1018

Richard C, Granier C, Inze D, de Veylder L (2001) Analysis of cell division parameters and cell cycle gene expression during the cultivation of *Arabidopsis thaliana* cell suspensions. J Exp Bot 52:1625–1633

Robbens S, Khadaroo B, Camasses A, Derelle E, Ferraz C, Inzé D, Van de Peer Y, Moreau H (2005) Genome-wide analysis of core cell cycle genes in the unicellular green alga *Ostreococcus tauri.* Mol Biol Evol 22:589–597

Russell P, Nurse P (1986) cdc25+ functions as an inducer in the mitotic control of fission yeast. Cell 45:145–153

Russell P, Nurse P (1987) Negative regulation of mitosis by wee1+, a gene encoding a protein kinase homolog. Cell 49:559–567

Samuels AL, Meehl J, Lipe M, Staehelin LA (1998) Optimizing conditions for tobacco BY-2 cell cycle synchronization. Protoplasma 202:232–236

Segers G, Gadisseur I, Bergounioux C, de Almeida Engler J, Jacqmard A, Van Montagu M, Inzé D (1996) The *Arabidopsis* cyclin-dependent kinase gene cdc2bAt is preferentially expressed during S and G2 phases of the cell cycle. Plant J 10:601–612

Šetlík I, Zachleder V (1984) The multiple fission cell reproductive patterns in algae. In: Streiblová E (ed) The microbial cell cycle. CRC, Boca Raton, pp 253–279

Sherr CJ (1994) G1 phase progression: cycling on cue. Cell 79:551–555

Shimogawara K, Fujiwara S, Grossman A, Usuda H (1998) High-efficiency transformation of *Chlamydomonas reinhardtii* by electroporation. Genetics 148:1821–1828

Sorrell DA, Menges M, Healy JMS, Deveaux Y, Amano C, Su Y, Nakagami H, Shinmyo A, Doonan JH, Sekine M, Murray JAH (2001) Cell cycle regulation of cyclin-dependent kinases in tobacco cultivar bright yellow-2 cells. Plant Physiol 126:1214–1223

Spudich JL, Sager R (1980) Regulation of the *Chlamydomonas* cell-cycle by light and dark. J Cell Biol 85:136–146

Takahara M, Takahashi H, Matsunaga S, Miyagishima S, Takano H, Sakai A, Kawano S, Kuroiwa T (2000) A putative mitochondrial ftsZ gene is present in the unicellular primitive red alga *Cyanidioschyzon merolae.* Mol Gen Genet 264:452–460

Terui S, Suzuki K, Takahashi H, Itoh R, Kuroiwa T (1995) Synchronization of chloroplast division in the ultramicroalga *Cyanidioschyzon merolae* (Rhodophyta) by treatment with light and aphidicolin. J Phycol 31:958–961

Umen JG (2005) The elusive sizer. Curr Opin Cell Biol 17:435–441

Umen JG, Goodenough UW (2001) Control of cell division by a retinoblastoma protein homolog in *Chlamydomonas.* Genes Dev 15:1652–1661

Vítová M, Zachleder V (2005) Points of commitment to reproductive events as a tool for analysis of the cell cycle in synchronous cultures of algae. Folia Microbiol (Praha) 50:141–149

Yoshida Y, Kuroiwa H, Misumi O, Nishida K, Yagisawa F, Fujiwara T, Nanamiya H, Kawamura F, Kuroiwa T (2006) Isolated chloroplast division machinery can actively constrict after stretching. Science 313:1435–1438

Zachleder V (1994) The effect of hydroxyurea and fluorodeoxyuridine on cell cycle events in the chlorococcal alga *Scenedesmus quadricauda* (Chlorophyta). J Phycol 30:274–279

Zachleder V, Cepák V (1987a) The effect of light on the number of chloroplast nucleoids in daughter cells of the alga *Scenedesmus quadricauda.* Protoplasma 138:37–44

Zachleder V, Cepák V (1987b) Variations in chloroplast nucleoid number during the cell cycle in the alga *Scenedesmus quadricauda* grown under different light conditions. Protoplasma 141:74–82

Zachleder V, Šetlík I (1988) Distinct controls of DNA replication and of nuclear division in the cell cycles of the chlorococcal alga *Scenedesmus quadricauda.* J Cell Sci 91:531–539

Zachleder V, Šetlík I (1990) Timing of events in overlapping cell reproductive sequences and their mutual interactions in the alga *Scenedesmus quadricauda.* J Cell Sci 97:631–638

Zachleder V, Kawano S, Kuroiwa T (1995) The course of chloroplast DNA replication and its relationship to other reproductive processes in the chloroplast and nucleocytoplasmic compartment during the cell cycle of the alga *Scenedesmus quadricauda*. Protoplasma 188:245–251

Zachleder V, Kawano S, Kuroiwa T (1996) Uncoupling of chloroplast reproductive events from cell cycle division processes by 5-fluorodeoxyuridine in the alga *Scenedesmus quadricauda*. Protoplasma 192:228–234

Zachleder V, Schäfli O, Boschetti A (1997) Growth-controlled oscillation in activity of histone H1 kinase during the cell cycle of *Chlamydomonas reinhardtii* (Chlorophyta). J Phycol 33:673–681

Zachleder V, Bišová K, Vítová M, Kubín Š, Hendrychová J (2002) Variety of cell cycle patterns in the alga *Scenedesmus quadricauda* (Chlorophyta) as revealed by application of illumination regimes and inhibitors. Eur J Phycol 37:361–371

Subject Index

MIX
Papier aus verantwortungsvollen Quellen
Paper from responsible sources
FSC® C105338

If you have any concerns about our products,
you can contact us on
ProductSafety@springernature.com

In case Publisher is established outside the EU,
the EU authorized representative is:
Springer Nature Customer Service Center GmbH
Europaplatz 3, 69115 Heidelberg, Germany

Printed by Libri Plureos GmbH
in Hamburg, Germany